NORWICH CITY COLLEGE LIBRARY
Stock No. 186243
Class 540 RAM
Cat. Proc. OS

Chemistry

EILEEN RAMSDEN BSc PhD DPhil
Formerly of Wolfreton School, Hull

Eileen Ramsden © 2001
Original Illustrations © Nelson Thornes Ltd 1997, 2001

The right of Eileen Ramsden to be identified as author of this work has been asserted by her in accordance with the Copyright, Designs and Patents Act 1988.

All rights reserved. No part of this publication may be reproduced or transmitted in any form or by any means, electronic or mechanical, including photocopy, recording, or any information storage and retrieval system, without permission in writing from the publisher or under licence from the Copyright Licensing Agency Limited. Further details of such licenses (for reprographic reproduction) may be obtained from the Copyright Licensing Agency Limited, of 90 Tottenham Court Road, London W1P 4LP.

First published in 1994
Stanley Thornes (Publishers) Ltd
Second edition 1997
This edition published in 2001 by:
Nelson Thornes Ltd,
Delta Place,
27 Bath Road,
CHELTENHAM,
GL50 7TH
United Kingdom

A catalogue record for this book is available from the British Library.

ISBN 0 7487 6242 6

01 02 03 04 05 / 10 9 8 7 6 5 4 3 2 1

Related titles
Key Science: Biology (0–7487–6241–8) Third edition
Key Science: Physics (0–7487–6243–4) Third edition
Key Science: Chemistry Extension File (0–7487–6254–X)

Artwork by Barking Dog Art, Peters & Zabransky, Harry Venning, Cauldron Design Studio, Tech Set Ltd
Typeset by Tech Set Ltd, Gateshead, Tyne & Wear
Printed and bound in Spain by Graficas Estella

Acknowledgements

I would like to express my thanks to Mr Gavin Cameron, Mr Steve Hewitt and Mrs Sheila Rogers for their constructive comments on the Earth Science topic. I thank Professor Jerry Wellington for providing an extensive list of CD-Roms and websites to amplify the content of each topic. These references are to be found on the website: www.keyscience.co.uk. An icon in the margin of the text indicates that there is relevant material on the website.

I would like to thank the following organisations and people who have supplied photographs.
Action Plus 1.1A, p55; Ariel Plastics: 1.6D; Austin Rover: 13.4B; British Rail: 20.5A; British Steel: 4.4A; Casio Electronics Co. Ltd.: 21.2A; Environmental Picture Library: 16.10A (David Townend), 17.1A (Alan Greig); Fisons plc: 10.4A; Gavin Cameron: 19.2C; Geological Museum: 19.5B, 20.8A; Geoscience Features Picture Library: 20.8F; ICI Chemicals and Polymers Ltd: 1.2A, 1.5D, 3.2A, 3.8B, 7.6C, 23.7B; Image Bank: x (top – Bill Varie) (bottom – Jaime Villaseca); Impact Photos: 10.4C (Tom Webster) International Centre for Conservation Education: 16.4A, 17.4A (Andy Purcell), 17.6D, 28.1A; Mark Boulton: 29A, 22.1E; Martyn Chillmaid: 15.13A, 22.1G, 25.8A; Mary Evans Picture Library: p53, p54; Philip Harris 9.6B; Metropolitan Police: 30.1D; Mountain Camera: 1.5G; NASA: 13.4A, 22.3B, p124; National Portrait Gallery: portrait of Michael Faraday p103 reproduced by courtesy; Northern Irish Tourist Board: 19.1E; Oxford Scientific Films: 3.6A (G.I. Bernard); Panasonic (UK) Ltd: 21.2B, 21.2C; Panos Pictures: 17.3A (J Hartley); Peter Newark's Western Americana: 4.2A; Robert Harding's Picture Library: 13.4B, 13.5B; Pilkington Glass plc: 22.2B; Popperfoto: 2.1C, 17.6C; Rex Features: 13.8A; Science Photo Library: 2.1B (Prof. Stewart Lowther), 2.4B (Ben Johnson), 2.4C (Dr Mitsuo Ohtsuki), 2.4D (Dr M B Hursthouse), 4.1A (Ferranti Electronics/A Sternberg), 10.4D (CNRI), 13.1A (Susan Leavines), 17.6B (Simon Fraser), 19.2A, p208, 22.1D, p316 (Martin Bond), 19.3D (Sinclair Stammers), 20.1A (John Ross), 3.3B, 17.11B, 30.6B, 30.6C, p270; Sotherbys: 4.2B; Still Pictures: 20.11A (Martin Edwards); Sue Boulton: 29B, 30.1A; Sygma: 18.4A (Les Stone); Tony Stone Images: 4.3A (Rohan), 17.6D (Alan Levenson), 17.6E (Ben Osborne), 17.6A (David Woodfall), 24.18; Topham Picturepoint: 29.1G (John Maier); Woodmansterne Ltd: 19.1D, 19.2F; www.JohnBirdsall.co.uk: 4.1C, 30.1A.
Picture research by johnbailey@axonimages.com

Every effort has been made to contact copyright holders to clear permission for reproduction of copyright material. The publishers should like to apologise if any such material has not been fully acknowledged and shall endeavour to rectify the situation at the earliest opportunity.

I thank the following awarding bodies for permission to reproduce questions from examination papers:
London Examinations, a division of Edexcel, Assessment and Qualifications Alliance, Oxford, Cambridge and RSA Examinations, Northern Ireland Council for the Curriculum Examination and Assessment, Welsh Joint Education Committee.

The Awarding Bodies bear no responsibility for the answers to questions taken from their question papers contained in this publication.

The production of a science text book from manuscript involves considerable effort, energy and expertise from many people, and I wish to acknowledge the work of all the members of the publishing team, especially Sarah Ryan and Debra Goring.

Finally I thank my family for the tolerance which they have shown and the encouragement which they have given during the preparation of this book.

Eileen Ramsden.

Theme D Planet Earth

Theme E Using Earth's resources

KEY Science: Chemistry is a comprehensive and up-to-date textbook designed to meet the requirements of the five awarding bodies for all their GCSE science syllabuses. The textbook can be used for the chemistry component of the Single or Double Award and for all GCSE Science: Chemistry specifications.

Topics are differentiated into core material for Single or Double Science and extension material for Science: Chemistry (blue/grey margin). The awarding bodies have chosen different extension topics in addition to the Programme of Study for Key Stage 4; therefore teachers and students need to be familiar with the specific requirements of their own specification. *Key Science: Chemistry* contains the extension topics for all awarding bodies.

Key Science: Chemistry is organised in 7 Themes covering 30 major topics. Each topic is sub-divided into numbered sections covering all aspects of the topic. Each section includes special features, usually located in the margin, such as:

- **First Thoughts,** setting the scene and reminding students of the background knowledge to the section,
- **It's a Fact,** sharpening interest and enhancing text,
- **Key Scientist,** providing a historical perspective on how scientific ideas have developed,
- **Icons,** A key icon in the margin indicates an opportunity for the exercise of one or more of the key skills: application of number (AoN), communication (Comm) and information and communication technology (ICT). A star icon suggests that the topic can be used for the development of scientific ideas and the weighing of evidence. A mouse icon refers the reader to the book's website.
- **Commentary and Summaries,** within each section of a topic and at the end of each section, helping students to review their work and highlighting the key points they should understand and learn,
- **Checkpoints,** testing knowledge and understanding through sets of short questions at regular intervals at the end of each section,
- **Theme Questions,** in the style of examination questions.

Answers to all numerical questions and Checkpoints and a comprehensive **Index** are at the end of the book.

Support for the text book
Key Science: Chemistry is supported by an *Extension File*, which contains a *Teacher's Guide*, a bank of photocopiable material and key diagrams from the text in colour on CD-Rom. There are 80 photocopiable Activities for Sc1 and Assignments for Sc3 and Science: Chemistry topics. There are 60 Exam File questions for homework and revision, with mark schemes provided. The *Extension File* in the third edition gives a list of activities which students can use to generate evidence for their key skills portfolio.

The *Teacher's Guide* offers advice on how students can use the activities provided in *Key Science: Chemistry* and in the *Extension File* to gain evidence for their key skills portfolio. It also gives advice on topics that can be used for developing the theme of ideas and evidence in science. Guidance on the use of computers in science is provided. The website http: www.keyscience.co.uk has a selection of CD-Roms and educational websites which provide support for each topic.

The separate student booklet *Key Skills through Key Science* gives a selection of activities which will enable the student to demonstrate and record his or her performance in the three main skill areas (AoN, Comm, ICT). This booklet for GCSE higher–level students

- explains the nature of key skills and the portfolio requirements
- offers advice on gathering evidence for the student's portfolio
- provides a range of suitable science activities that cover level 2 key skills, with advice to the student on how to use these activities
- provides a record of coverage and commentaries by teachers as assessors.

The booklet covers topics in biology, chemistry and physics.

Student's Note

I hope that this book will provide all the information you need and enough practice to enable you to do well in the tests and examinations in your GCSE course. However, although examinations are an important part of your course, they are not the only part. I hope also that you will develop a real interest in science, which will continue after your course has finished. Finally, I hope that you enjoy using the book. If you do, all the effort will have been worth while.

Eileen Ramsden

What is chemistry?

Chemistry is the study of **matter** and the ways in which matter changes. Matter is anything that takes up space and has mass. Chemistry therefore takes in all that you see around you: the chair you are sitting on, the book you are reading, the clothes you are wearing and the flowers on the windowsill. Matter appears in many different forms, including metals, plastics, ceramics, living plants and animals, clouds, oceans and stars. To understand the world around you, you need to have a knowledge of chemistry.

The scientific method

Chemistry is a branch of science. Discoveries are often made in chemistry through the application of the **scientific method**. You will find many examples in your course of the application of the scientific method. The first step is to make measurements and collect results on the topic under study. If a pattern can be seen in the results, it is described in the form of a scientific **law**. This law simply summarises the observations that have been made. The next step is to put forward a hypothesis (a scientific suggestion) which explains the observations. The hypothesis is then tested. Predictions are made on the basis of the hypothesis and experiments are done to test it. If the results of the experiments agree with the predictions, this is support for the hypothesis, and a theory is formulated. This theory is an explanation of the law. Sometimes theories need to be revised if new experiments yield results that do not fit in with the theory.

Enough to eat and drink – thanks to chemistry

We have a safe supply of drinking water, thanks to chemistry. Water treatment makes use of some of the separation techniques which you will meet in Topic 3. It also involves chemical reactions: neutralisation of excess acidity, oxidation of impurities, precipitation of magnesium and calcium ions, coagulation of colloids and chlorination to kill bacteria.

Without help from chemistry, the Earth could not grow enough food to feed the population. Thanks to fertilisers made by the chemical industry, farmers in most parts of the world can grow enough crops to feed the people. Chemical pesticides protect these crops from competition from weeds and attack by insects so that they survive to be harvested. Chemical preservatives ensure that the foods we eat are safe and appetising.

The clothes you wear are made from chemicals

The supply of natural fibres, such as cotton and wool, is insufficient to provide for our clothing needs. Chemists have invented many synthetic fibres from which much of our clothing is made.

Nylon is a polymer (that is, it consists of large molecules made when many small molecules combine). Since nylon is water-repellant, it is ideal for swimwear and waterproof jackets. In the texturised nylon called

Tactel® there are air spaces between the fibres. They make the nylon bulkier, softer and able to absorb moisture, such as perspiration. Tactel® is widely used for sportswear such as skiwear, jogging suits and tennis outfits.

We wear sweaters, coats and trousers made of **acrylics** such as Orlon®, Acrilan® and Courtelle®. These polymers are hard-wearing and soft with a wool-like feel.

Bullet-proof vests are not an everyday item of clothing, but some people never go to work without one. They are made from a polymer called **Kevlar®**. Kevlar® is also a favourite with mountaineers because it is used to make ropes which are twenty times as strong as steel of the same weight.

A healthy life – thanks to chemistry

Chemotherapy (the use of chemicals in the treatment of diseases) allows us to live longer, healthier lives, more free from pain than in any previous century. Most of us have occasion to use pain-killers such as aspirin and antiseptics such as TCP. For more serious complaints we have antibiotics such as penicillin. Surgery is made safe by anaesthetics such as fluothane.

It is unlikely that GCSE students will be in need of hip replacements. However, in older people hip joints wear out, causing pain and lameness. The remedy is replacement of the hip joint by an artificial joint. The hip joint is a ball and socket joint. Surgeons cut out the worn-out socket and replace it with a socket of the polymer poly(ethene). They bond this socket to the hip bone by a chemical cement. The ball of the joint is a steel ball attached to a shaft which is inserted into the thigh bone and bonded by the chemical cement. The operation brings mobility and relief from pain to forty thousand patients a year in the UK.

Our visits to the dentist also benefit from chemistry. When tooth enamel decays dentists have the job of taking out decayed material and filling the cavities that are left. Gold was the first material used for dental fillings, but only rich people could afford it; others had to lose their teeth. The development of an amalgam of mercury with silver, tin and other metals made dental fillings available to the population as a whole. Dental amalgam has a low melting temperature so it can be moulded at body temperature (much better than having molten metal poured into the cavity!). An amalgam filling does not look at all like a natural tooth. For front teeth, chemists have developed natural-looking materials by combining a glass powder with a polymer. People can still smile with confidence, even after having a front tooth filled. Sometimes people lose a tooth completely. It is now possible to insert an implant into the socket left by an extracted tooth. Porous ceramics are favoured as implants because bone can grow into the implant.

Your leisure activities depend on chemistry

Sports equipment was traditionally made from natural materials, such as metals, wood and ox-gut. The supply of these materials was sufficient when only a small fraction of the population played sports. Chemists have developed new, cheaper synthetic materials which bring the price of sports equipment down and open up sport to a vast number of people. In addition, new materials enhance the performance of sportspeople and athletes and make it possible for new records to be set.

Polymers called **polyesters** can be combined with glass fibres to form a material which is tough and strong and which can be moulded during its manufacture into a required shape. These **glass-fibre-reinforced-polyesters** (GFRP) are used to make small boats, skis and some car bodies.

Racing cyclists benefit from modern materials invented by chemists. The first bicycles had wood frames. A century ago, frames made of steel tubes took over. New materials are now available to make lighter bicycles for racing cyclists. They may choose a very light cycle made from a magnesium alloy, which has only 1/3 the density of aluminium. Another choice for a light cycle is titanium alloy. It is used in top-of-the-range mountain bikes because it is low in density, strong, tough and resistant to corrosion. Some bicycles are made entirely of plastics. They have a frame made of a **glass-fibre-reinforced polymer**, which is moulded in one piece, and wheels of a different GFRP. Another material is polyester resin strengthened with fibres of carbon, Kevlar or ceramic. Such frames can be cast in one piece. Chris Boardman rode a cycle with a one-piece frame of carbon-fibre-reinforced epoxy resin when he won the Olympic Gold Medal in 1992. In 2000 Jan Ullrich won the Olympics on a titanium alloy cycle with a rear disc wheel. Substituting a disc of aluminium alloy or carbon-fibre-reinforced polymer for a wheel with spokes makes for less wind resistance. All these new materials are more costly than steel. Racing cyclists are prepared to pay a high price for a lighter bicycle and for a model with less wind resistance.

The original materials from which tennis racquets were made were wood for the frame and ox gut for the strings. There is no longer enough ox gut to supply the manufacturers of tennis racquets. Nylon strings are widely used. Wood is still used to make racquets, but top-class players use racquets made from synthetic materials. A tennis racquet needs a frame which is strong, stiff, light and tough. The material must be capable of being moulded to the required shape. It must also damp the vibrations which occur when the player hits a ball, preventing vibrations passing through the frame to damage the player's elbow. Materials composed of polymer resins strengthened with fibres of carbon, glass, ceramics and Kevlar are used. They are stiff and can hit the ball strongly. The vibration damping is less good than wood and is improved by filling the frame with the polymer, polyurethane foam.

Space travel depends on chemistry

The crew of a space shuttle produce carbon dioxide through respiration. It is removed from the air by means of canisters of lithium hydroxide which, being basic, reacts with the acidic gas carbon dioxide to form lithium carbonate and water. In a space station, the crew may be in for a long stay and unable to take enough oxygen to last the mission. They must regenerate oxygen from carbon dioxide. They do this by means of the oxidising agent potassium superoxide, KO_2. This oxidises carbon dioxide to oxygen. And what do crew members wear? Spacesuits are made from strong tough polymers.

A space shuttle is lifted from the ground by booster rockets. The fuel is aluminium powder which is oxidised by ammonium chlorate(VII). The shuttle is sent into orbit by the reaction between hydrogen and oxygen. They are stored as liquids, then vaporised and ignited as they mix. When a space shuttle re-enters Earth's atmosphere from space, friction arises between the shuttle and air, and this friction is converted into heat. The temperature of the surface can reach 1500 ^{o}C. Steel melts at this temperature. To insulate the shuttle, it is covered with a layer of ceramic tiles made of silicon oxide fibres. These ceramic tiles are probably the most heat-resistant materials ever made.

Whatever career you choose to follow, you will almost certainly need to use your GCSE science skills and knowledge. Science is a feature of everyday life, at work or at home. Industry, hospitals, transport and agriculture are examples of major sectors of the economy that depend heavily on science. At home, almost all the things you do make use of science in some way.

Your GCSE science course is designed to give you a good scientific background so that you can, if you wish, carry on with science studies after GCSE.

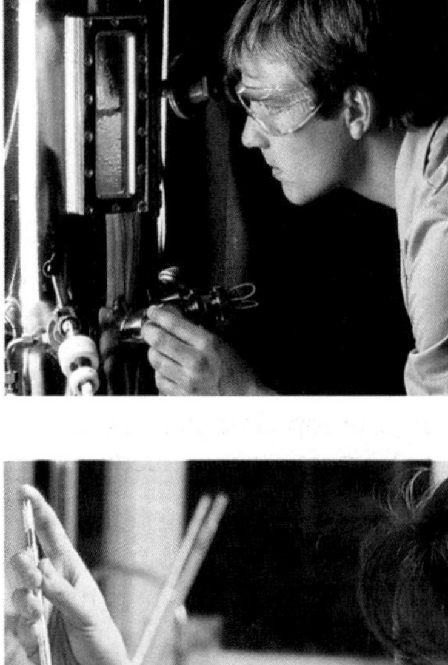

To work in science requires specific personal qualities in addition to academic qualifications. Scientists are very creative and imaginative people and the work of an individual scientist can bring huge benefits to everyone. For example, Alexander Fleming's discovery of penicillin has saved countless lives. But do not be misled into thinking that the life of a scientist is one of continual discoveries. Scientists have to be very patient and methodical to discover anything; they have to be good at working together and at communicating their ideas to each other and to other people. The qualifications needed to become a scientist are outlined on p. xii; the qualities needed to become a scientist are just the same as you need in your GCSE course – enthusiasm, hard work, imagination, awareness and concern.

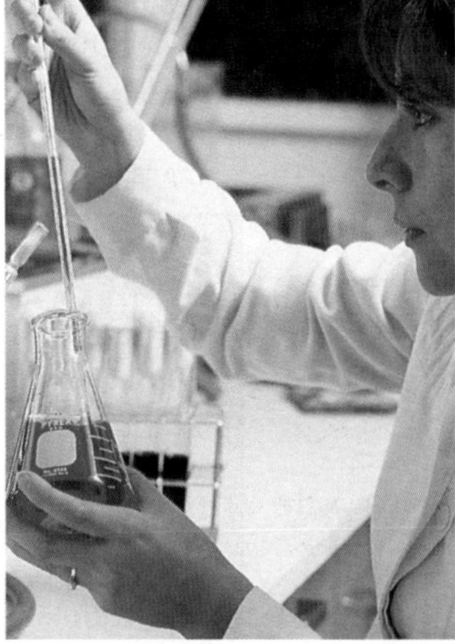

What jobs are done by scientists? In industry, scientists design, develop and test new products. For example, scientists in the glass industry are developing amazingly clear glass for use as **optical fibres** in communication links. In medicine, scientists are continually finding applications for scientific discoveries; for example, medical scientists have developed a high-power ultrasonic transmitter for destroying kidney stones, thus avoiding a surgical operation. These are just two examples of the work of scientists. You will find scientists at work in research laboratories, industrial laboratories, forensic laboratories, hospitals, schools, on field trips, expeditions, radio, TV and lots of other places. Scientists have to be very versatile as science is a very wide and varied field.

The skills and knowledge you gain through studying science will enable you to gain the benefits of new technologies. Ask your parents what aspects of life have **not** changed since they were children – they may be stuck for an answer! New technologies force the pace of change and if you do not learn to use them, you will not share their benefits. Studying science encourages you to develop an open mind and to seek new approaches. That is why a wide range of careers involve further studies in science.

For many careers further studies in science are essential, for example, medicine, dentistry, pharmacy, engineering and computing. For other careers such as law, business studies, administration, the armed forces and retail management, studying science after GCSE will be helpful. Thus by continuing your science studies after GCSE, you keep many career options open, which is important when you are considering your choice of career.

The next steps after GCSE

Read this section carefully, bearing in mind that your working life will probably be about forty years. If you want your life ahead to be interesting, if you want to make your own decisions about what you do, if you want to make the most of your talents, then you should continue your studies. After GCSE, you can continue in full-time education at school or at college, or you can train in a job through part-time study. If you take a job without training, you will soon find that your friends who stayed on have much better prospects.

Most students aiming for a career in science or technology continue full-time study for two years, taking either GCE A- and AS-levels or a GNVQ course in Science. Successful completion of a suitable combination of these courses can then lead to a degree course at a university or a college of higher education or, alternatively, straight into employment.

The **A-level route** to higher education requires successful completion of a two-year full-time course. AS-level (subsidiary) examinations (four or five) are taken at the end of the first year and A-level examinations (three or four) at the end of the second year. Students taking A-level Chemistry frequently choose their other subjects from Biology, Physics, Mathematics, Geography, Geology and Home Economics. Students with an interest in taking further courses in chemistry, biochemistry or chemical engineering are well advised to choose Mathematics. Students who plan further studies in chemistry or chemical engineering should choose Physics.

The **GNVQ science route** is also a two-year full-time course at advanced level, leading to a qualification equivalent to double award A-level science. All students study some biology, chemistry and physics and can then specialise. The course is assessed continuously through tests and laboratory work. Students who do not have grade CC in science to enter advanced level GNVQ can take the one-year GNVQ science intermediate course as a preparation for the advanced course.

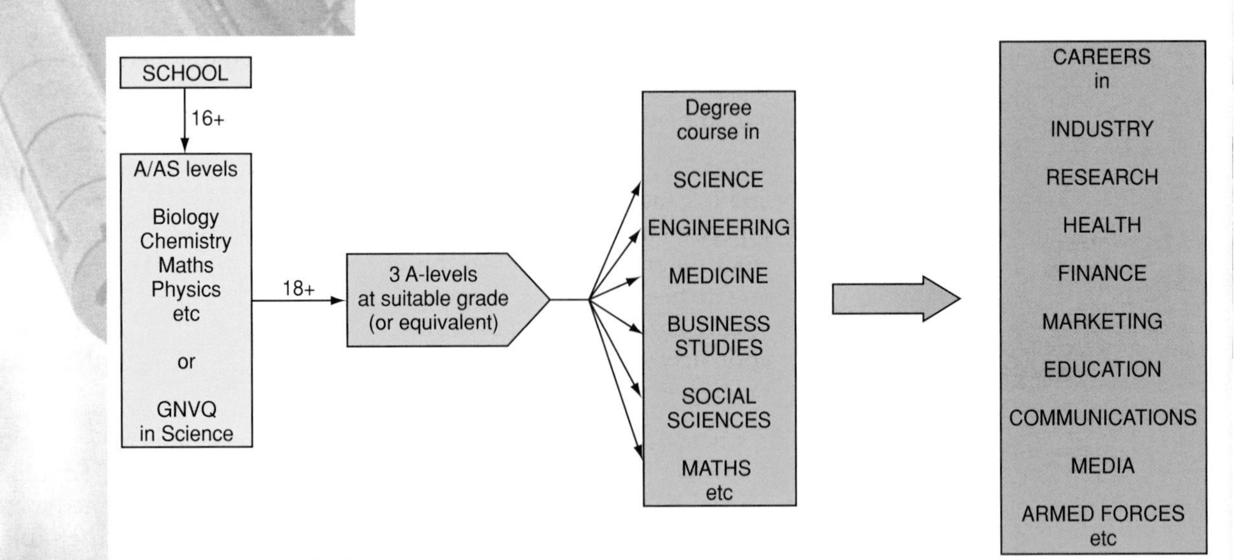

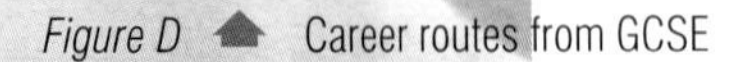
Figure D Career routes from GCSE

Matter

Earth and all the other planets in the universe are made of matter. How many different kinds of matter are there? They are countless: far too many for one book to mention, but in Theme A we start to answer the question. What does matter do: how does it behave? The study of how matter behaves is called chemistry. Can one kind of matter change into a different kind of matter? The answer to the question is 'yes, it can': provided that it receives energy.

1.1 The states of matter

FIRST THOUGHTS

What is a state of matter? How do states of matter differ? These are questions for you to explore in this topic.

It's lucky these skaters know how to change one state of matter into another! How do they do it?

www.keyscience.co.uk

Figure 1.1A ▲ Ice skaters

Everything you see around you is made of matter. The skaters, the skates, the ice and all the other things in Figure 1.1A are different kinds of matter.

The skaters are overcoming friction as they glide across the frozen pond. *How do they manage this?* They change one kind of matter, ice, into another kind of matter, water. A thin layer of water forms between a skate and the ice, and this reduces friction and enables the skater to glide across the pond. The water beneath the skate refreezes as the skater moves on. *How are skaters able to melt ice? Why does it refreeze behind them?* Read on to find out.

The different kinds of matter are **solid** and **liquid** and **gaseous** matter. These are called the **states of matter**. Table 1.1 summarises the differences between the three chief states of matter. The symbols (s) for solid, (l) for liquid and (g) for gas are called **state symbols**. Later you will use another state symbol, (aq), which means 'in aqueous (water) solution'.

Table 1.1 ▼ States of Matter

State	Description
Solid (s)	Has a fixed volume and a definite shape. The shape is usually difficult to change.
Liquid (l)	Has a fixed volume. Flows easily; changes its shape to fit the shape of its container.
Gas (g)	Has neither a fixed volume nor a fixed shape; changes its volume and shape to fit the size and shape of its container. Flows easily; liquids and gases are called **fluids**. Gases are much less dense than solids and liquids.

SUMMARY

The three chief states of matter are:

- solid (s): fixed volume and shape,
- liquid (l): fixed volume; shape changes,
- gas (g): neither volume nor shape is fixed.

Liquids and gases are fluids.

1.2 Pure substances

Most of the solids, liquids and gases which you see around you are mixtures of substances.

- Rock salt, the impure salt which is spread on roads in winter, is a mixture of salt and sand and other substances.
- Crude oil is a mixture of petrol, paraffin, diesel fuel, lubricating oil and other liquids.
- Air is a mixture of gases.

Note the differences between a pure substance and a mixture.

Some substances, however, consist of one substance only. Such substances are **pure substances**. For example, from the mixture of substances in rock salt chemists can obtain pure salt which is 100% salt.

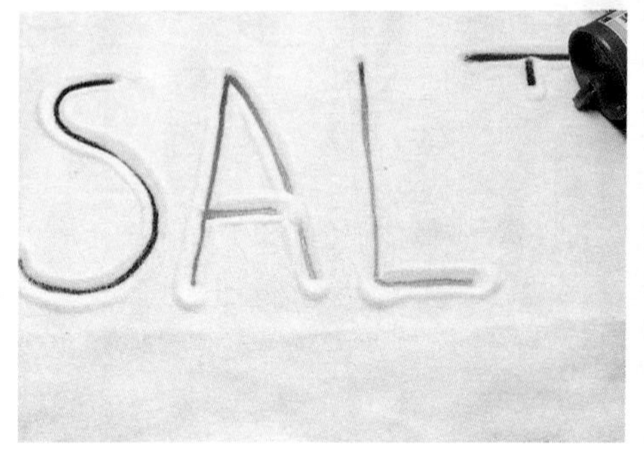

Figure 1.2A ▲ Rock salt and pure salt

SUMMARY

A pure substance is a single substance.

1.3 Density

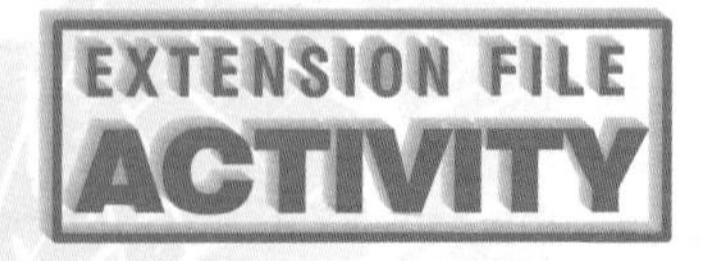

Table 1.1 tells you that gases are much less **dense** than solids and liquids. What does **dense** mean? What is **density**? The two lengths of car bumper shown in Figure 1.3A have the same volume. You can see that they do not have the same mass. The steel bumper is heavier than the plastic bumper. This is because steel is a more **dense** material than the plastic; steel has a higher **density** than the plastic has.

$$\text{Density} = \frac{\text{Mass}}{\text{Volume}}$$

The unit of density is kg/m^3 or g/cm^3. The density values of some common substances are shown in Table 1.2.

Table 1.2 ▼ Density

Substance	Density (g/cm^3)
Air	1.2×10^{-3}
Aluminium	2.7
Copper	8.92
Ethanol	0.789
Gold	19.3
Hydrogen	8.33×10^{-5}
Iron	7.86
Lead	11.3
Methane	6.67×10^{-4}
Oxygen	1.33×10^{-3}
Silver	10.5
Water	1.00

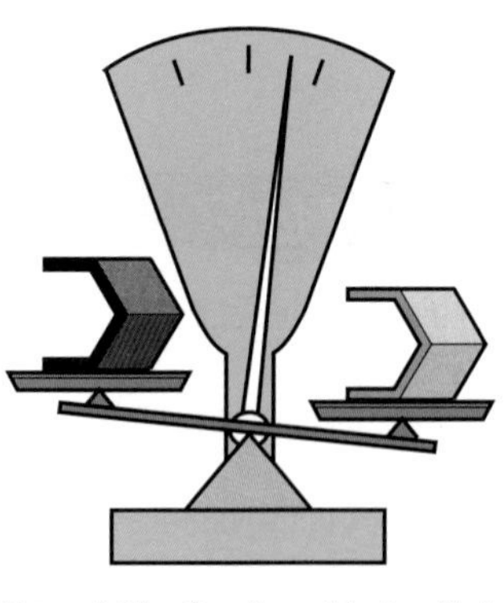

Figure 1.3A ▲ Two objects with the same volume

Archie says that aluminium is a 'light' material. Can you explain to him why this is incorrect? An aeroplane made of aluminium is not a light object! What term should Archie use in describing aluminium?

You can see that the gases are much less dense than any of the solid or liquid substances.

SUMMARY

$$\text{Density} = \frac{\text{Mass}}{\text{Volume}}$$

$$\text{Mass} = \text{Volume} \times \text{Density}$$

$$\text{Volume} = \frac{\text{Mass}}{\text{Density}}$$

The density triangle:

$$\frac{M}{V \times D}$$

To find the quantity you want, cover up that letter.
The other letters in the triangle show you the formula.

CHECKPOINT

1. A worker in an aluminium plant taps off 300 cm^3 of the molten metal. It weighs 810 g. What is the density of aluminium?
2. An object has a volume of 2500 cm^3 and a density of 3.00 g/cm^3. What is its mass?
3. Mercury is a liquid metal with a density of 13.6 g/cm^3. What is the mass of 200 cm^3 of mercury?
4. 50.0 cm^3 of metal A weigh 43.0 g

   52.0 cm^3 of metal B weigh 225 g

   Calculate the density of each metal. Say whether they will float or sink in water.

1.4 Change of state

Matter can change from one state into another. Some changes of state are summarised in Figure 1.4A.

EXTENSION FILE ACTIVITY

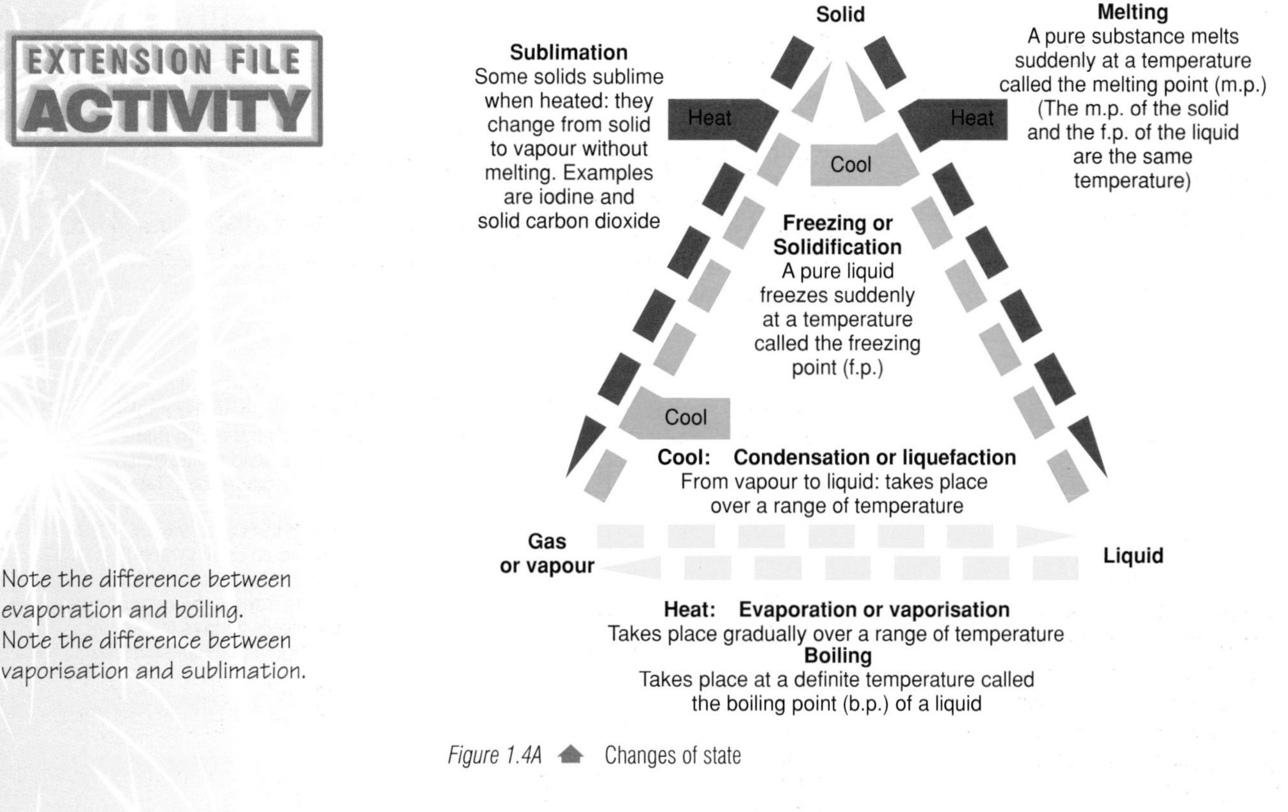

Figure 1.4A ▲ Changes of state

Note the difference between evaporation and boiling.
Note the difference between vaporisation and sublimation.

Sometimes gases are described as **vapours**. A liquid evaporates to form a vapour. A gas is called a vapour when it is cool enough to be liquefied.

$$\text{Vapour} \xrightarrow{\text{Either cool or compress without cooling}} \text{Liquid}$$

SUMMARY

Matter can change from one state into another.
The changes of state are:
- melting
- freezing or solidification
- evaporation or vaporisation
- condensation or liquefaction
- sublimation

The hotter a liquid is, the faster it evaporates. At a certain temperature, it becomes hot enough for vapour to form in the body of the liquid and not just at the surface. Bubbles of vapour appear inside the liquid. When this happens, the liquid is boiling, and the temperature is the boiling point of the liquid.

CHECKPOINT

1 Copy the diagram below. Fill in the names of the changes of state. (Some of them have two names.)

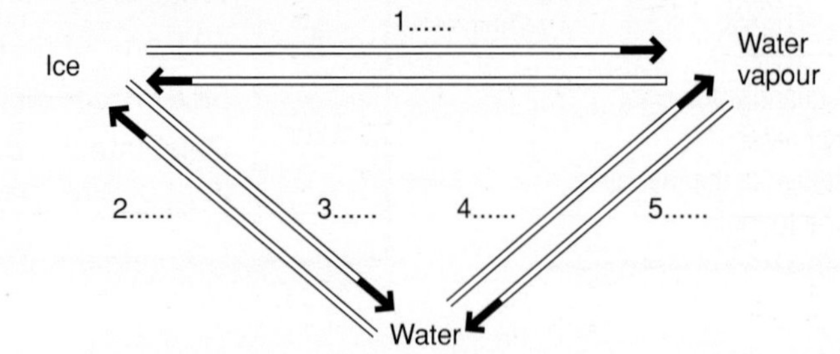

2 Give the scientific name for each of these changes.
(a) A puddle of water gradually disappears.
(b) A mist appears on your glasses.
(c) A mothball gradually disappears from your wardrobe.
(d) The change that happens when margarine is heated.

1.5 Finding melting points and boiling points

FIRST THOUGHTS

You can identify substances by finding their melting points and boiling points. Accuracy is essential, as you will see in this section

Finding the melting point of a solid

Figure 1.5A shows an apparatus which can be used to find the melting point of a solid. To get an accurate result:

- first note the temperature at which the solid melts,
- allow the liquid which has been formed to cool and note the temperature at which it freezes.

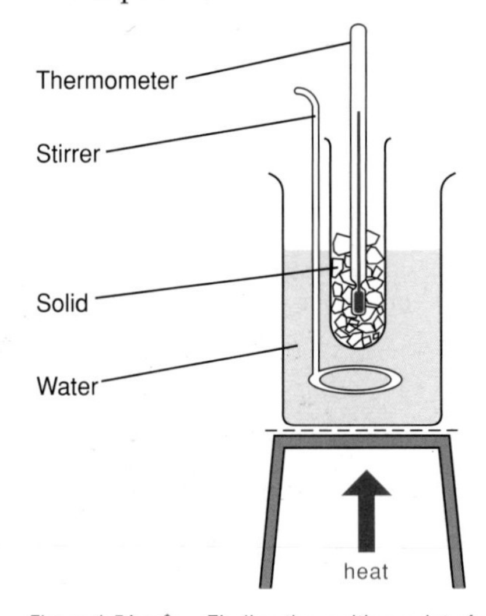

First, find the m.p. of the solid. Heat the water in the beaker. Stir. Watch the thermometer. When the solid melts, note the temperature. Stop heating

Now find the f.p. of the liquid. Let the liquid cool. Watch the thermometer. When the liquid begins to freeze, the temperature stops falling. It stays the same until all the liquid has solidified

Figure 1.5A Finding the melting point of a solid (for solids which melt above 100 °C, a liquid other than water must be used)

The apparatus shown in Figure 1.5A will work between 20 °C and 100 °C. For solids with melting points above 100 °C, a liquid with a higher boiling point than water must be used. For liquids which freeze below room temperature, a liquid with a lower freezing point than water must be used. A mixture of ice and salt can be used down to −18 °C (see Figure 1.5B).

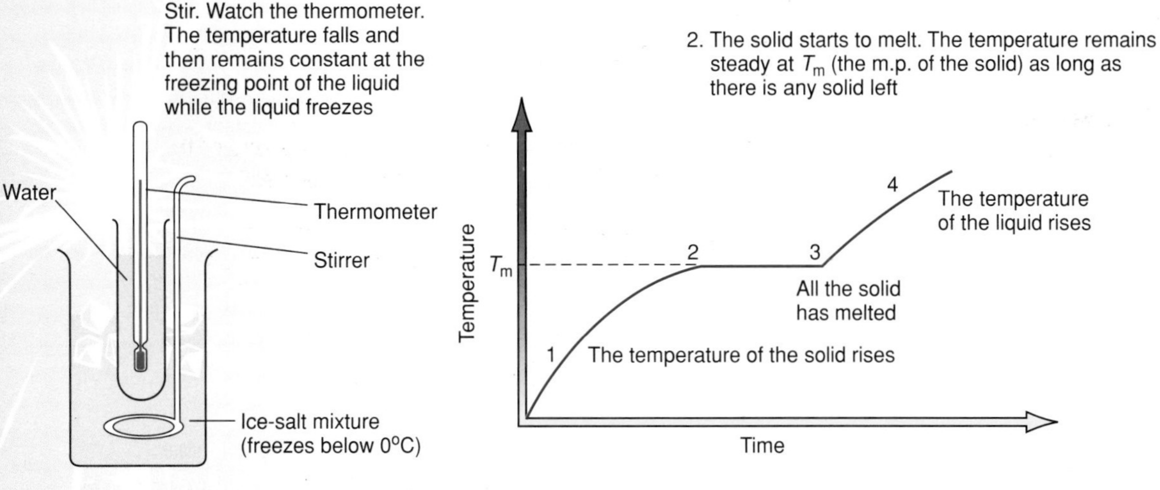

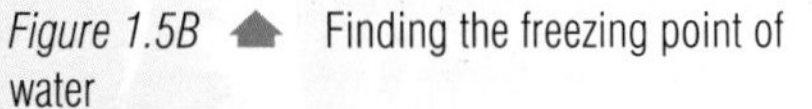

Figure 1.5B ▲ Finding the freezing point of water

Figure 1.5C ▲ A graph of temperature against time when a pure solid melts

SUMMARY

The apparatus shown in Figure 1.5A or 1.5B can be used to find the melting point of a solid and the freezing point of a liquid.

- The temperature of a pure solid stays constant while it is melting.
- The temperature of a pure liquid stays constant while it is freezing.

If the solid is a pure substance, a graph of temperature against time as the solid melts will look like Figure 1.5C.

If you have a pure solid and you do not know what it is, you can use its melting point to find out. Chemists have drawn up lists of pure substances with their melting points. You find the melting point of the unknown solid and compare it with the listed melting points. *Which of the solids could be X?*

Solid	Melting point (°C)
Unknown solid X	116
Benzamide	132
Butanamide	116
Ethanamide	82

Figure 1.5D ▲ A truck spreading salt on an icy road in winter

If a solid is not pure, the melting point will be low, and the impure solid will melt gradually over a range of temperature. Look at Figure 1.5D and explain why the ice on the road melts.

SUMMARY

The melting point can be used to identify an unknown pure solid. The presence of an impurity lowers the melting point

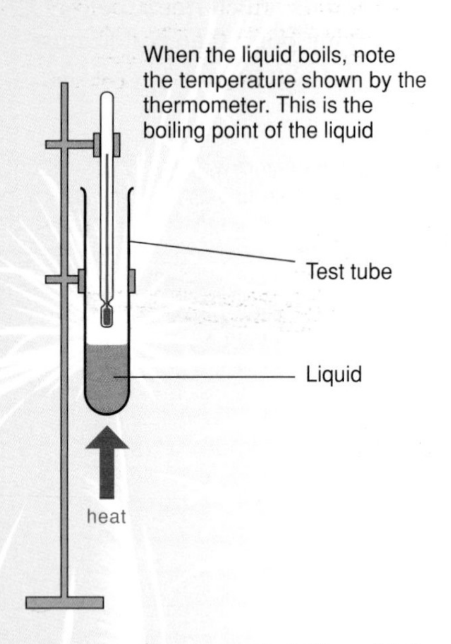

Figure 1.5E Finding the boiling point of a non-flammable liquid

SUMMARY

The apparatus in Figure 1.5E can be used to find the boiling point of a liquid. The temperature of a pure liquid stays constant while it is boiling. Dissolving a solid in a liquid raises the boiling point.

Why is the mountaineer's water boiling even before it reaches 100 °C?

Finding the boiling point of a liquid

Figure 1.5E shows an apparatus which can be used to find the boiling point of a non-flammable liquid. For a flammable liquid, a distillation apparatus, e.g. Figure 3.6A must be used.

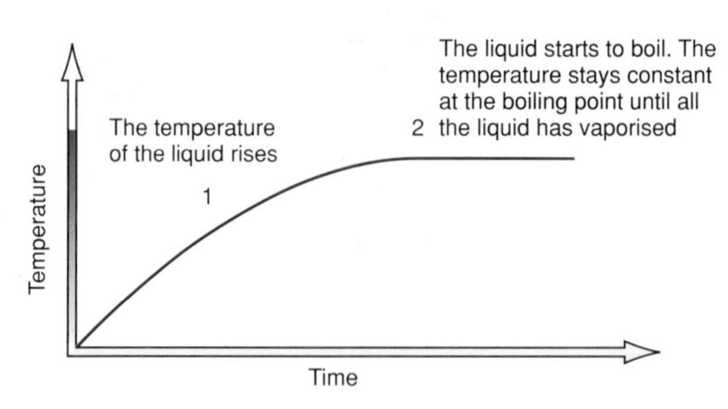

Figure 1.5F A graph of temperature against time when a pure liquid is heated

While a pure liquid is boiling, the temperature remains steady at its boiling point. All the heat going into the liquid is used to vaporise the liquid and not to raise its temperature. Figure 1.5F shows a graph of temperature against time when a pure liquid is heated.

A mixture of liquids, such as crude oil, boils over a range of temperature. If a solid is dissolved in a pure liquid, it raises the boiling point.

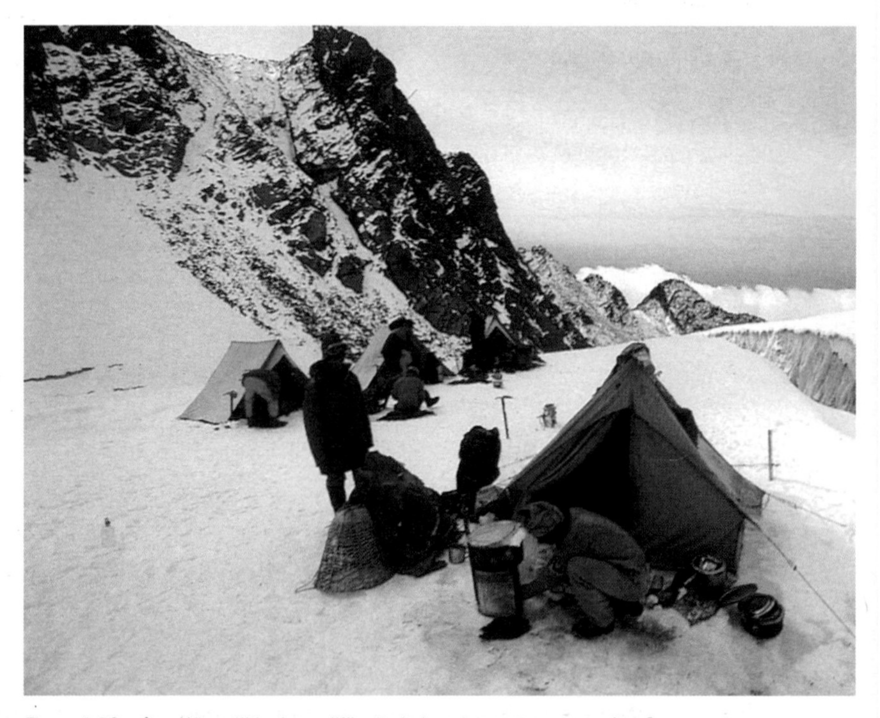

Figure 1.5G Why will he have difficulty in brewing a strong cup of tea?

The boiling point of a liquid depends on the surrounding pressure. If the surrounding pressure falls, the boiling point falls. The boiling point of water on a high mountain is lower than 100 °C. An increase in the surrounding pressure raises the boiling point (see Figure 1.5H).

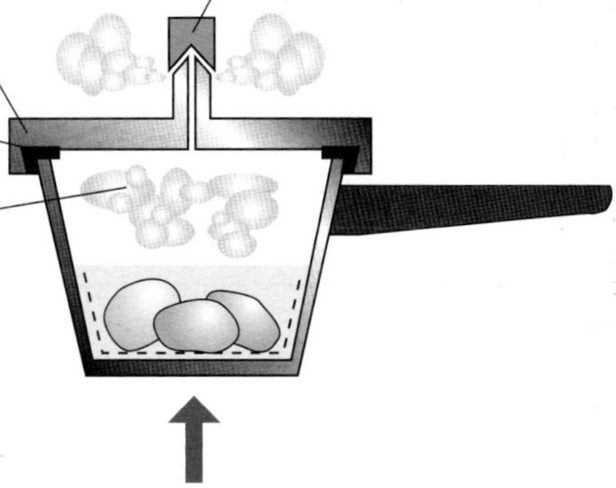

Figure 1.5H ▲ How a pressure cooker works

Why does the pressure cooker cut down on cooking times?

Boiling points are stated at standard pressure (atmospheric pressure at sea level). Boiling points can be used to identify pure liquids. You take the boiling point of the liquid you want to identify; then you look through a list of boiling points of known liquids and find one which matches. *Which of the liquids could be the unknown substance X?*

Solid	*Melting point (°C)*
Substance X	111
Benzene	80
Methylbenzene	111
Naphthalene	218

No two substances have both the same boiling point and the same melting point.

SUMMARY

The boiling point of a liquid is stated at standard pressure.

- At lower pressures, the boiling point is lower.
- At higher pressure, the boiling point is higher.

CHECKPOINT

▶ **1** A student heated a beaker full of ice (and a little cold water) with a Bunsen burner. She recorded the temperature of the ice at intervals until the contents of the beaker had turned into boiling water. The table shows the results which the student recorded.

Time (minutes)	0	2	4	6	8	10	12	14	16
Temperature (°C)	0	0	0	26	51	76	100	100	100

(a) On graph paper, plot the temperature (on the vertical axis) against time (on the horizontal axis).
(b) On your graph, mark the m.p. of ice and the b.p of water.
(c) What happens to the temperature while the ice is melting?
(d) What happens to the temperature while the water is boiling?
(e) The Bunsen burner gives out heat at a steady rate. Explain what happens to the heat energy (i) when the beaker contains a mixture of ice and water at 0 °C, (ii) when the beaker contains water at 100 °C and (iii) when the beaker contains water at 50 °C.

▶ **2** Bacteria are killed by a temperature of 120 °C. One way of sterilising medical instruments is to heat them in an autoclave (a sort of pressure cooker: see the figure overleaf). The table shows the effect of pressure on the boiling point of water.

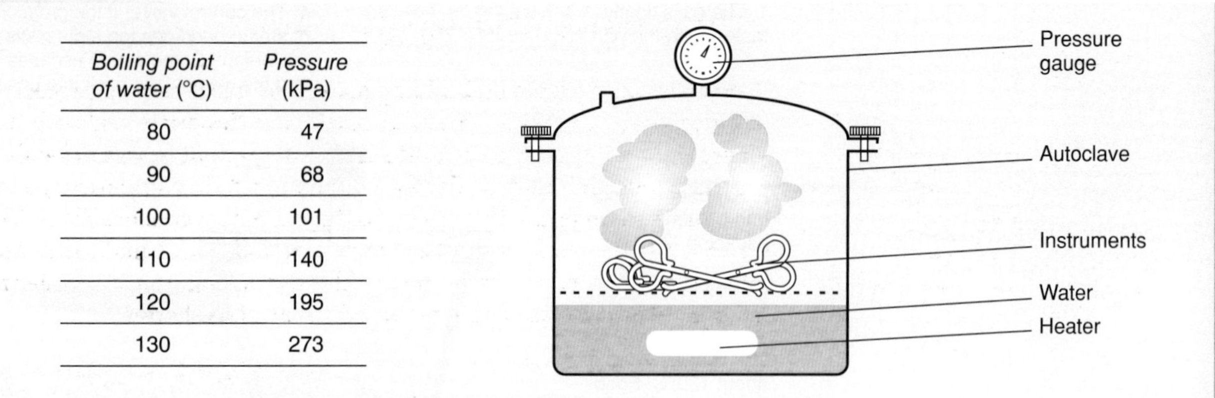

Boiling point of water (°C)	*Pressure* (kPa)
80	47
90	68
100	101
110	140
120	195
130	273

(a) Why are the instruments not simply boiled in a covered pan?
(b) What pressure must the autoclave reach to sterilise the instruments?
(c) What is the value of standard pressure in kPa (kilopascals)?

1.6 Properties and uses of materials

FIRST THOUGHTS

Properties of materials
- Hardness
- Toughness
- Strength
- Flexibility
- Elasticity
- Solubility
- Density (Topic 1.3)
- Melting point (Topic 1.5)
- Boiling point (Topic 1.5)
- Conduction of heat (KS: Physics, Topics 5.5 and 5.6)
- Conduction of electricity (Topic 7.1)

Types of matter from which things are made are called **materials**. Different materials are used for different jobs. The reason is that their different **properties** (characteristics) make them useful for different purposes. You will see a list of properties in the margin. Some of them have been mentioned earlier in this topic. Conduction of heat and electricity will be covered in later topics. Now let us look at the rest.

Hardness

It is difficult to change the shape of a hard material. A hard material will dent or scratch a softer material. A hard material will withstand impact without changing. Table 1.3 shows the relative hardness of some materials on a 1 – 10 scale.

Table 1.3 Relative hardness of materials

Material	Relative hardness	Uses
Diamond	10.0	Jewellery, cutting tools
Silicon carbide	9.7	Abrasives
Tungsten carbide	8.5	Drills
Steel	7–5	Machinery, vehicles, buildings
Sand	7.0	Abrasives, e.g. sandpaper
Glass	5.5	Cut glass can be made by cutting glass with harder materials.
Nickel	5.5	Used in coins; hard-wearing
Concrete	5–4	Building material
Wood	3–1	Construction, furniture
Tin	1.5	Plating steel food cans

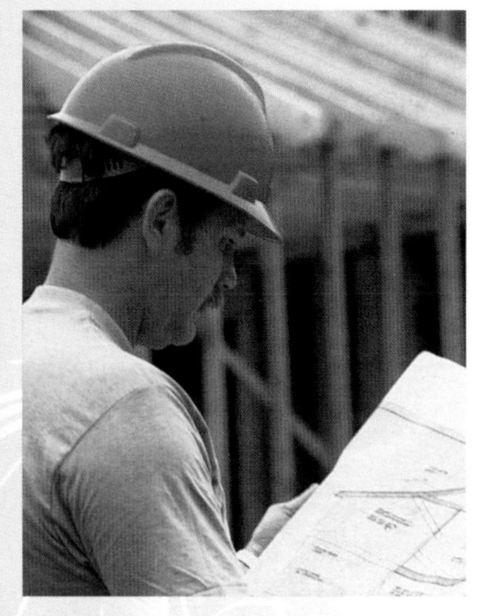

Figure 1.6A ▲ Hardness

Toughness and brittleness

Construction workers on building sites wear 'hard hats' to protect themselves from falling objects. A hard hat is designed to absorb the energy of an impact. The hat material is **tough**, that is, it is difficult to break, although it may be dented by the impact. In comparison, a brick is difficult to dent and will shatter if dropped onto a concrete floor. The brick is **brittle**. Glass is another brittle material. These materials cannot absorb the energy of a large force without cracking. If a still larger force is applied the cracks get bigger and the materials shatter.

Composite materials are widely used because they combine the good points of their components.

Composite materials

To make brittle materials tougher you have to try to stop them cracking. Mixing a brittle material with a material made of fibres, e.g. glass fibre or paper, will often do this. The fibres are able to absorb the energy of a force and the brittle material does not crack. Plaster is a brittle material. Plasterboard is much tougher. It is made by coating a sheet of plaster with paper fibres. It is a **composite material**. Glass-reinforced plastic (GRP) is a mixture of glass fibre and a plastic resin.

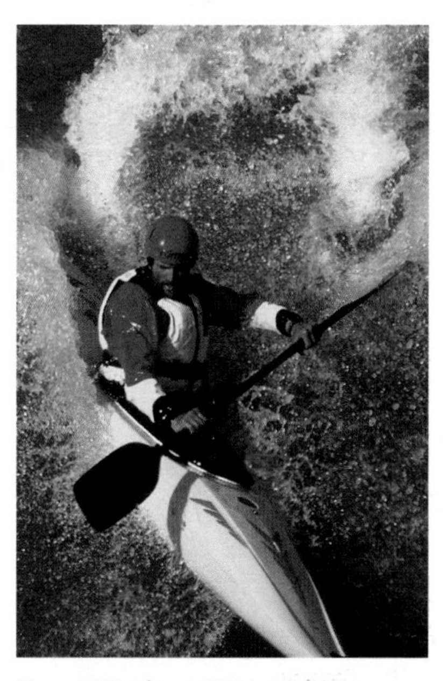

Figure 1.6C ▲ A GRP canoe

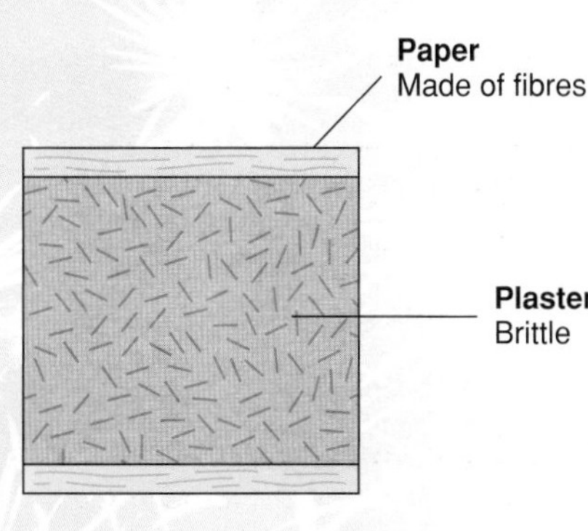

Figure 1.6B ▲ Plasterboard

Figure 1.6D ▲ Corrugated plastic roof

Concrete has great compressive strength, but it can crack if it is stretched. For construction purposes, **reinforced concrete** is used. Running through this composite material are steel rods which act like the glass fibres in GRP.

The shape of a piece of material alters its strength. Corrugated cardboard is used for packaging. Corrugated iron sheet and corrugated plastic sheet are used for roofing.

SUMMARY

Materials may be:
- hard – resistant to impact, difficult to scratch,
- tough – difficult to break, will 'give' before breaking,
- brittle – will break without 'giving',
- composite – made of more than one substance.

What do you mean exactly when you say a material is 'strong'? Can you stretch it? Can you squeeze it? Is it difficult to crush? Is it easy to bend without breaking? Note the difference between a plastic material and an elastic material.

Strength

A strong material is difficult to break by applying force. The force may be a stretching force (e.g. a pull on a rope), or a squeeze (e.g. a vice tightening round a piece of wood), or a blow (e.g. a hammer blow on a lump of stone). A material which is hard to break by stretching has good **tensile strength**; a material which is hard to break by crushing has good **compressive strength**. The tensile strength of a material depends on its cross-sectional area.

Flexibility

While a material is pulled it is being stretched: it is under **tension**. While a material is squashed it is being compressed: it is under **compression**. When a material is bent, one side of the material is being stretched while the opposite side is being compressed. A material which is easy to bend without breaking has both tensile strength and compressive strength. It is **flexible**.

Figure 1.6E ▲ It's flexible

Elasticity

You can change the shape of a material by applying enough force. When you stop applying the force, some materials retain their new shapes; these are **plastic** materials. Other materials return to their old shape when you stop applying the force; these are **elastic materials**.

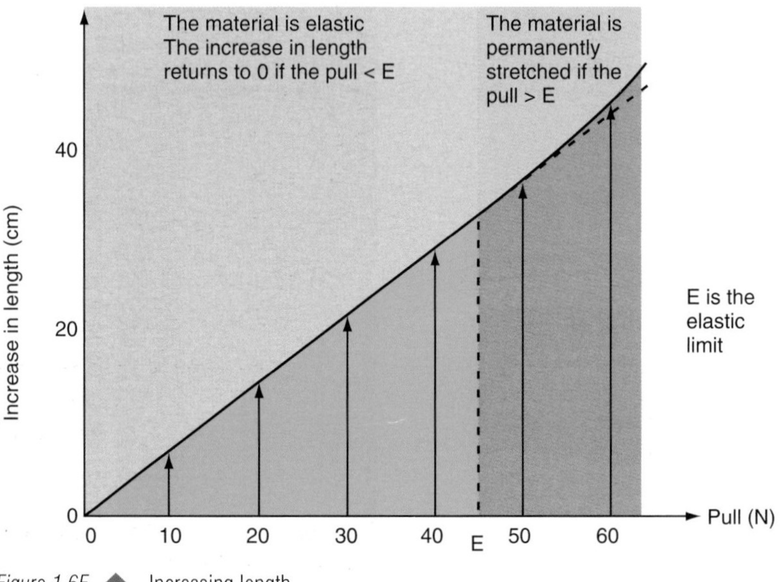

Figure 1.6F ▲ Increasing length

When you pull an elastic material, it stretches – increases in length. At first, when you double the pull, you double the increase in length. As the pull increases, however, you reach a point where the material no longer returns to its original shape. This pull is the **elastic limit** of the material. Increasing the pull still more eventually makes the material break (see Figure 1.6F).

Science at work

Materials are tested to destruction to find their tensile strength, brittleness, etc. The results are fed into a computer to form a data base which can be used in the computer-aided design of articles from nuts and bolts to battleships.

SUMMARY

Materials may have tensile strength (resistance to stretching), compressive strength (resistance to pressure), flexibility (both tensile and compressive strength) or elasticity (the ability to return to their original shape after being stretched).

Solubility

A solution consists of a **solvent** and a **solute**. The solute, which may be a solid or a liquid or a gas, dissolves in the solvent. Water is the most common solvent, but there are many others such as ethanol (alcohol) and trichloroethane (trichlor). A **concentrated** solution contains a high proportion of solute; a **dilute** solution contains a small proportion of solute. The **concentration** of a solution is the mass of solute dissolved in a certain volume, say one litre, of the solution (see Figure 1.6G).

A solution that contains as much solute as it is possible to dissolve at that temperature is a **saturated** solution (Figure 1.6H).

Solute 10.00g has been weighed out

Graduated flask Note the mark at 1.000 l

Solvent Distilled water

The solute has dissolved. Distilled water has been added to make the volume of solution up tp the 1.000 l mark. The *concentration* of the solution is 10.00 g/l

Figure 1.6G Concentration

Undissolved crystals

The solution must be *saturated* because no more solute will dissolve

Figure 1.6H A saturated solution

A concentrated solution contains more solute per unit volume than a dilute solution.

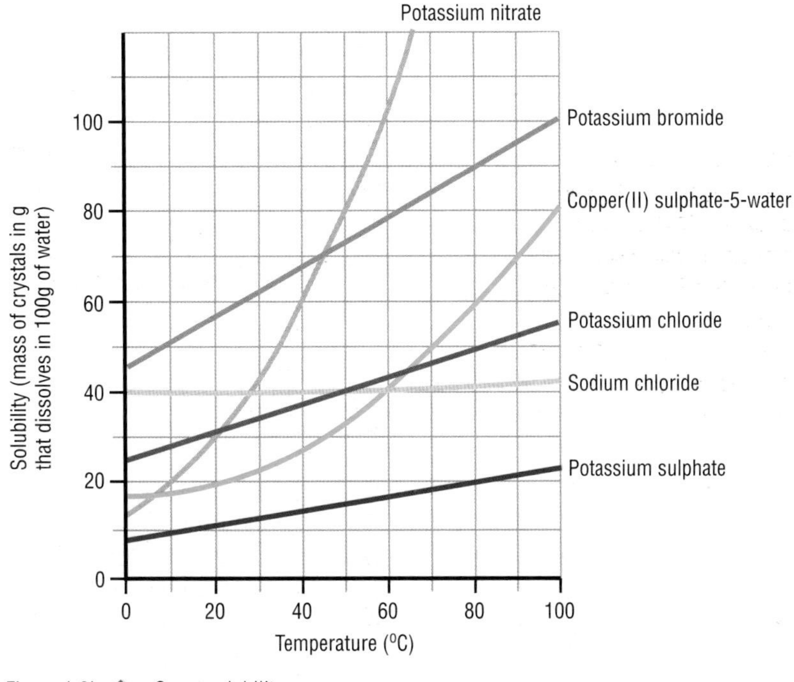

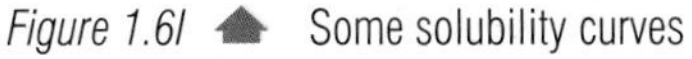
Figure 1.6I Some solubility curves

The concentration of solute in a saturated solution is the **solubility** of the solute at that temperature. Solubility is stated as the mass in grams of the solute that will saturate 100 grams of solvent at a certain temperature. A graph of solubility against temperature is called a **solubility curve**.

As you can see from Figure 1.6I the solubilities of most solutes increase with temperature. When a saturated solution is cooled, it can hold less solute at the lower temperature and some solute comes out of solution: it crystallises. Gases, on the other hand, are less soluble at higher temperatures.

SUMMARY

Solubility = the mass of solute that will saturate 100 g of solvent at a stated temperature.

CHECKPOINT

1 (a) Name two materials than can be used to drill through steel.
(b) Why can a steel blade slice through tin?
(c) Name a material that is used to make wood smooth.
(d) Why are diamond-tipped saws used to slice through concrete?
(e) Name two materials which can be used to cut glass.
(f) Why is nickel a better coinage material than tin?

2 (a) Name two tough materials. Say what they are used for.
(b) Name two brittle materials. Say what can be done to make them tougher.

3 Use the information in the table below to answer this question.

Cross-sectional area (mm^2)	1	2	4	6	8	10
Breaking force (N)	0.5	1	2	3	4	5

(a) On a piece of graph paper, plot the breaking force (on the vertical axis) against the cross-sectional area (on the horizontal axis).
(b) From the shape of the graph, say how the breaking force alters when the cross-sectional area (i) doubles, (ii) increases by a factor of 10.
(c) Write a sentence saying how the tensile strength of the material depends on the cross-sectional area.

4 Name a material to fit each of the following descriptions:
hard, soft, strong, flexible, tough, brittle, composite.

5 Do the following materials need good tensile strength or good compressive strength?
a tent rope, a tow bar, a stone wall, an anchor chain, a concrete paving stone, a building brick, a rope ladder.

6 Classify the following materials as either plastic or elastic:
plasticine, pottery clay, a rubber band, a balloon, 'potty putty'.

7 Taran needs to make a solution of some large crystals. Which of the following suggestions would help her?
(a) Crush the crystals before adding them to water.
(b) Add the crystals to water one at a time.
(c) Warm the water.
(d) Stir the mixture of water and crystals.
(e) Add the water dropwise to the crystals.

8 Refer to Figure 1.6I.
(a) What mass of potassium bromide will dissolve in 100 g of water at (i) 20 °C and (ii) 100 °C?
(b) What will happen when the solution in (a) is cooled from 100 °C to 20 °C?
(c) What mass of water is needed to dissolve 40 g of potassium sulphate at 80 °C?
(d) When 1.00 kg of water saturated with copper(II) sulphate-5-water is cooled from 70 °C to 20 °C, what mass of solid crystallises?
(e) A 100 g mass of water is saturated with sodium chloride and potassium chloride at 100 °C. When the solution is cooled to 20 °C, what mass of (i) sodium chloride and (ii) potassium chloride crystallises?

9 On graph paper, plot a solubility curve for potassium chlorate(V) from the data given below.

Solubility (g/100 g water)	8	11	14	18	24	31	39	50
Temperature (°C)	20	30	40	50	60	70	80	90

One kilogram of water is saturated with potassium chlorate(V) at 90 °C and then cooled to 40 °C. What mass of crystals will separate from solution?

2.1 The atomic theory

FIRST THOUGHTS

What exactly is an atom, and why did the atomic theory take nearly 2000 years to catch on? Combine your imagination and your experimental skills in this section to find out why.

The idea that matter consists of tiny particles is very, very old. It was first put forward by the Greek thinker Democritus in 500 BC. For centuries the theory met with little success. People were not prepared to believe in particles which they could not see. The theory was revived by a British chemist called John Dalton in 1808. Dalton called the particles **atoms** from the Greek word for 'cannot be split'. According to Dalton's **atomic theory**, all forms of matter consist of atoms.

The atomic theory explained many observations which had puzzled scientists. Why are some substances solid, some liquid and others gaseous? When you heat them, why do solids melt and liquids change into gases? Why are gases so easy to compress? How can gases diffuse so easily? In this topic, you will see how the atomic theory provides answers to these questions and many others.

2.2 Elements and compounds

All the models in the photograph show molecules of elements. Each contains only one type of atom.

There are two kinds of pure substances: **elements** and **compounds**. An element is a simple substance which cannot be split up into simpler substances. Iron is an element. Whatever you do with iron, you cannot split it up and obtain simpler substances from it. All you can do is to build up more complex substances from it. You can make it combine with other elements. You can make iron combine with the element sulphur to form iron sulphide. Iron sulphide is made of two elements chemically combined: it is a compound.

The smallest particle of an element is an **atom**. In some elements, atoms do not exist on their own: they join up to form groups of atoms called **molecules**. Figure 2.2A shows models of the molecules of some elements.

Figure 2.2A ▲ Models of molecules (helium, He; oxygen, O_2; phosphorus, P_4; sulphur, S_8)

SUMMARY

Matter is made up of particles. Pure substances can be classified as elements and compounds. Elements are substances which cannot be split up into simpler substances. The smallest particle of an element is an atom. In many elements, groups of atoms join to form molecules.

A compound is a pure substance that contains two or more elements. The elements are not just mixed together: they are chemically combined. Many compounds consist of molecules, groups of atoms which are joined together by **chemical bonds**. All gases, whether they are elements or compounds, consist of molecules. Figure 2.2B shows models of molecules of some compounds.

The models in the photograph show molecules of compounds. You can see that each molecule contains more than one type of atom.

Figure 2.2B ▲ Models of molecules of some compounds (carbon dioxide, CO_2; methane, CH_4; ammonia, NH_3; water, H_2O; hydrogen chloride, HCl)

SUMMARY

Compounds are pure substances that contain two or more elements chemically combined. Many compounds are made up of molecules; others are made up of ions. All gases consist of molecules.

Some compounds do not consist of molecules; they are made up of electrically charged particles called **ions**. The word particle can be used for an atom, a molecule and an ion. Gaseous compounds and compounds which are liquid at room temperature consist of molecules.

2.3 How big are atoms?

Hydrogen atoms are the smallest. One million hydrogen atoms in a row would stretch across one grain of sand. Five million million hydrogen atoms would fit on a pinhead. A hydrogen atom weighs 1.7×10^{-24} g; the heaviest atoms weigh 5×10^{-22} g.

How big is a molecule?

You can get an idea of the size of a molecule by trying the experiment shown in Figure 2.3A. This experiment uses olive oil, but a drop of detergent will also work. You can try this experiment at home.

Please try this experiment. You will be fascinated to find that you yourself can measure a molecule.

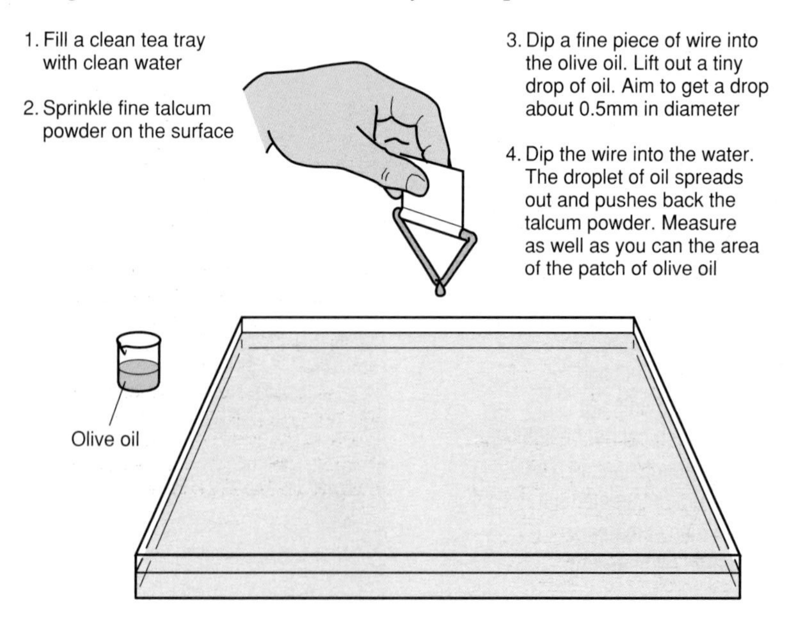

Figure 2.3A ▲ Estimating the size of a molecule

Sample results

Diameter of drop = 0.5 mm
Volume of drop = $(0.5\,\text{mm})^3 = 0.125\,\text{mm}^3$
Area of patch = $(25\,\text{cm})^2 = (250\,\text{mm})^2 = 6.25 \times 10^4\,\text{mm}^2$
Volume of patch = area × depth (d)
$0.125\,\text{mm}^3 = 6.25 \times 10^4\,\text{mm}^2 \times d$

$$d = \frac{0.125\,\text{mm}^3}{6.25 \times 10^4\,\text{mm}^2}$$

$$d = 2 \times 10^{-6}\,\text{mm}$$

The layer is only 2×10^{-6} mm deep (two millionths of a millimetre). We assume that it is one molecule thick.

SUMMARY

Atoms are tiny! A pinhead would hold 5×10^{12} hydrogen atoms. Olive oil molecules are 2×10^{-6} mm in diameter.

2.4 ▶ The kinetic theory of matter

FIRST THOUGHTS

Particles in motion: what does this idea explain? The difference between solids, liquids and gases for a start, and the beauty of crystalline solids.

The **kinetic theory of matter** states that matter is made up of small particles which are constantly in motion. (Kinetic comes from the Greek word for 'moving'.) The higher the temperature, the faster they move. In a solid, the particles are close together and attract one another strongly. In a liquid the particles are further apart and the forces of attraction are weaker than in a solid. Most of a gas is space, and the particles shoot through the space at high speed. There are almost no forces of attraction between the particles in a gas.

Scientists have been able to explain many things with the aid of the kinetic theory.

Solid, liquid and gaseous states

The differences between the solid, liquid and gaseous states of matter can be explained on the basis of the kinetic theory.

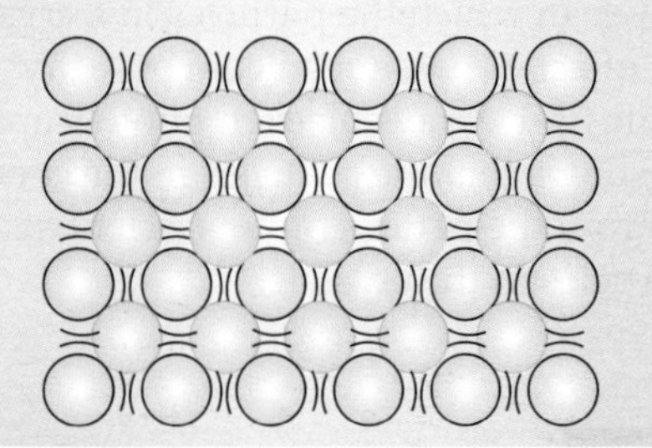

A solid is made up of particles arranged in a regular 3-dimensional structure. There are strong forces of attraction between the particles. Although the particles can vibrate, they cannot move out of their positions in the structure.

When a solid is heated, the particles gain energy and vibrate more and more vigorously. Eventually they may break away from the solid structure and become free to move around. When this happens, the solid has turned into liquid: it has melted.

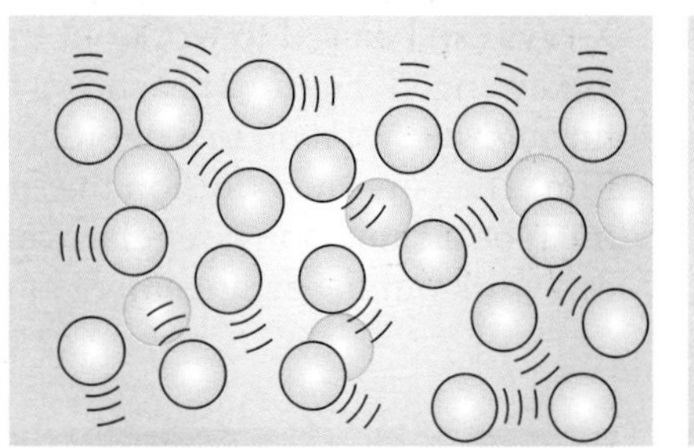

In a liquid the particles are free to move around. A liquid therefore flows easily and has no fixed shape. There are still forces of attraction between the particles.

When a liquid is heated, some of the particles gain enough energy to break away from the other particles. The particles which escape from the body of the liquid become a gas.

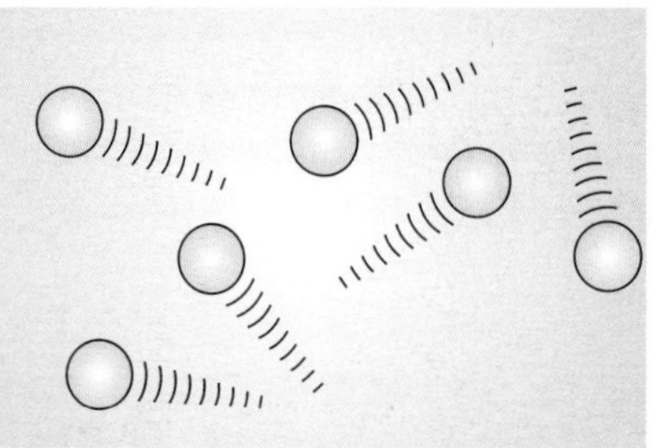

In a gas, the particles are far apart. There are almost no forces of attraction between them. The particles move about at high speed. Because the particles are so far apart, a gas occupies a very much larger volume than the same mass of liquid.

The molecules collide with the container. These collisions are responsible for the pressure which a gas exerts on its container.

Figure 2.4A ▲ The arrangement of particles in a solid, a liquid and a gas

Crystals

Crystals are a very beautiful form of solid matter. A crystal is a piece of solid which has a regular shape and smooth faces (surfaces) which reflect light (see Figure 2.4B). Different salts have differently shaped crystals. *Why are many solids crystalline?* Viewing a crystal with an electron microscope, scientists can actually see individual particles arranged in a regular pattern (see Figure 2.4C). It is this regular pattern of particles which gives the crystal a regular shape.

You will enjoy watching crystals form under the microscope. The Extension File gives details.

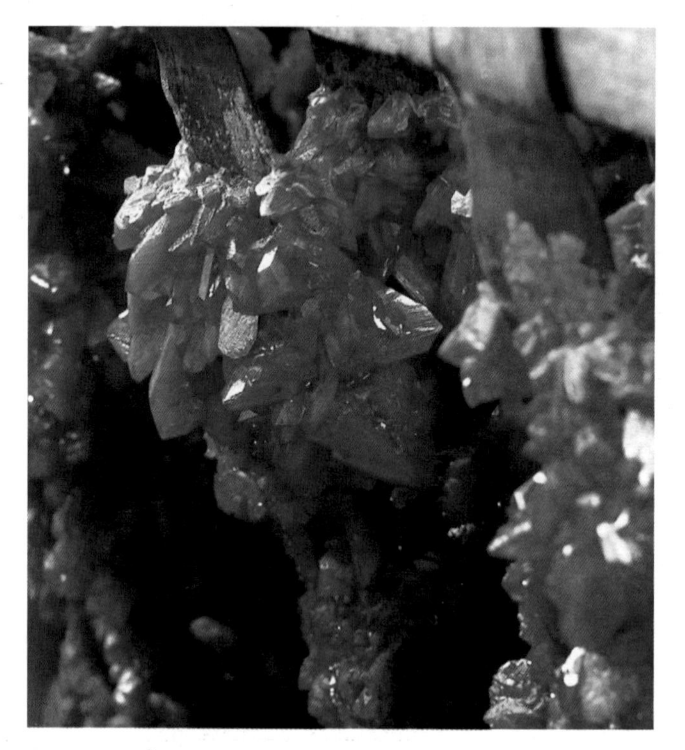

Figure 2.4B ▲ Crystals of copper(II) sulphate

X-ray crystallographers have their own ways of studying crystals

X-rays can be used to work out the way in which the particles in a crystal are arranged. Figure 2.4D shows the effect of passing a beam of X-rays through a crystal on to a photographic film. X-rays blacken photographic film. The pattern of dots on the film shows that the particles in the crystal must be arranged in a regular way. From the pattern of dots, scientists can work out the arrangement of particles in the crystals.

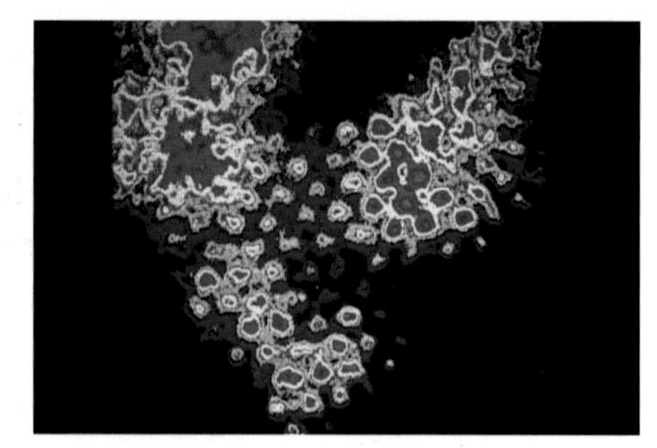

Figure 2.4C ▲ An electron microscope picture of uranyl ethanoate: each spot represents a single uranium atom

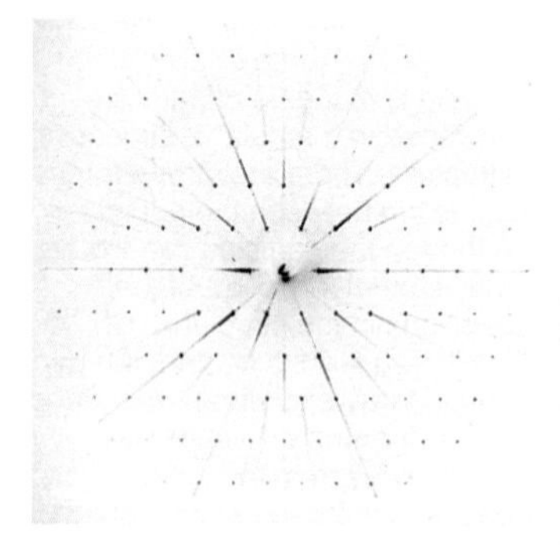

Figure 2.4D ▲ X-ray pattern from crystals of the metal palladium

SUMMARY

The kinetic theory of matter can explain the differences between the solid, liquid and gaseous states, and also how matter can change state. X-ray photographs show that crystals consist of a regular arrangement of particles.

FIRST THOUGHTS

How do solids dissolve, how do liquids vaporise and how do gases diffuse? Imagine particles in motion, and you will be able to explain all these changes.

Dissolving

Crystals of many substances dissolve in water. You can explain how this happens if you imagine particles splitting off from the crystal and spreading out through the water. See Figure 2.4E.

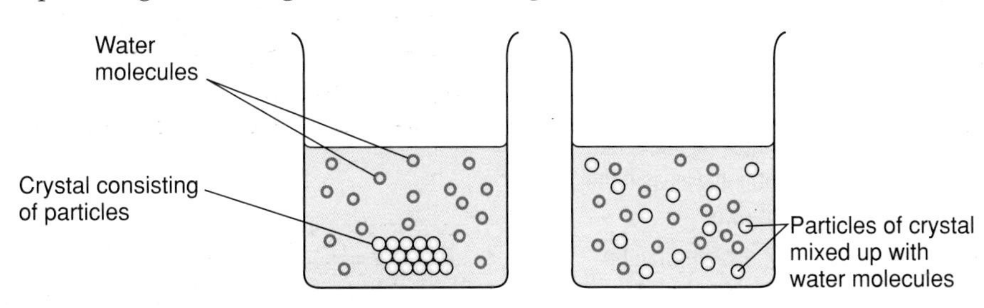

Figure 2.4E ▲ A coloured crystal dissolving

Diffusion

Chlorine

TOXIC

SUMMARY

Gases diffuse: they spread out to occupy all the space available to them. The kinetic theory of matter explains gaseous diffusion.

What evidence have we that the particles of a gas are moving? The diffusion of gases can be explained. **Diffusion** is the way gases spread out to occupy all the space available to them. Figure 2.4F shows what happens when a jar of the dense green gas, chlorine, is put underneath a jar of air.

On the theory that gases consist of fast-moving particles, it is easy to explain how diffusion happens. Moving molecules of air and chlorine spread themselves between the two gas jars.

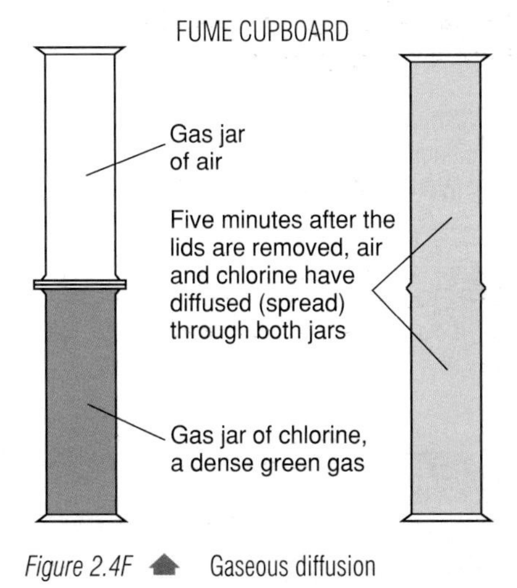

Figure 2.4F ▲ Gaseous diffusion

FIRST THOUGHTS

This section describes Brownian motion. You will see how neatly the kinetic theory can explain it.

Brownian motion

Figure 2.4G shows a smoke cell and the erratic path followed by a particle of smoke.

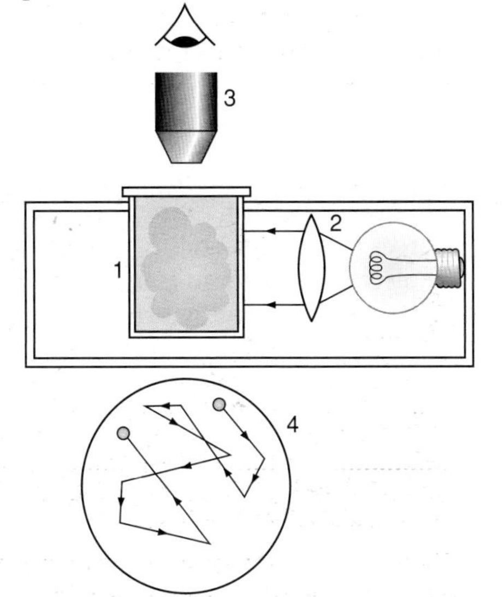

1 A small glass cell is filled with smoke

2 Light is shone through the cell

3 The smoke is viewed through a microscope

4 You see the smoke particles constantly moving and changing direction. The path taken by one smoke particle will look something like this

Figure 2.4G ▲ A smoke cell

Brownian motion in a liquid

In 1785, Robert Brown was using a microscope to observe pollen grains floating on water. He was amazed to see that the pollen grains were constantly moving about and changing direction. It was as if they had a life of their own.

Brown could not explain what he saw. You have the kinetic theory of matter to help you. Can you explain, with the aid of a diagram, what was making the pollen grains move?

SUMMARY

Brownian motion puzzled scientists until the kinetic theory of matter offered an explanation.

FIRST THOUGHTS

Why can you cool a hot cup of tea by blowing on it? Read this section to see if your idea is correct.

We call this kind of motion **Brownian motion** after the botanist, Robert Brown, who first observed it. Figure 2.4H shows the explanation of Brownian motion.

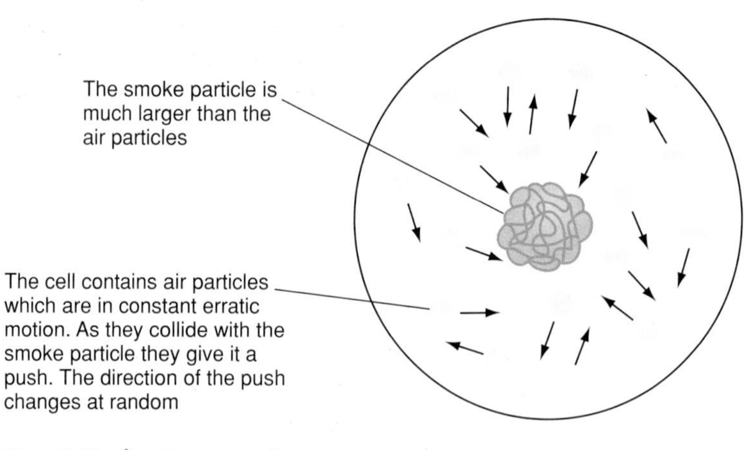

Figure 2.4H ▲ Brownian motion

Evaporation

When a liquid evaporates, it becomes cooler (see Figure 2.4I).

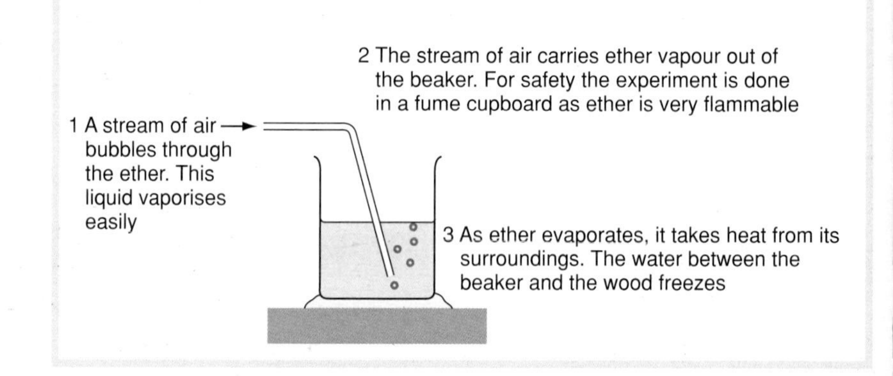

Figure 2.4I ▲ The cooling effect produced when a liquid evaporates

The kinetic theory can explain this cooling effect. Attractive forces exist between the molecules in a liquid (see Figure 2.4J). Molecules with more energy than average can break away from the attraction of other molecules and escape from the liquid. After the high energy molecules have escaped, the average energy of the molecules which remain is lower than before: the liquid has become cooler.

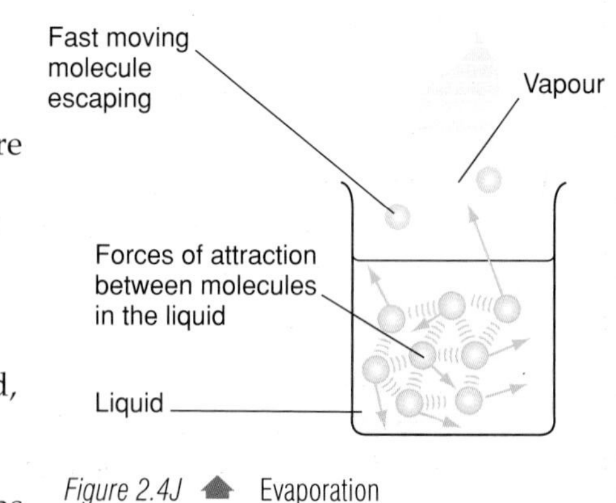

Figure 2.4J ▲ Evaporation

SUMMARY

When a liquid evaporates, it takes heat from its surroundings. You can speed up evaporation by heating the liquid and by blowing air over it.

What happens when you raise the temperature? More molecules have enough energy to break away from the other molecules in the liquid. The rate of evaporation increases.

What happens if you pass a stream of dry air across the surface of the liquid? The dry air carries vapour away. The particles in the vapour are prevented from re-entering the liquid, that is, condensing. The liquid therefore evaporates more quickly.

CHECKPOINT

1 Explain the following statements in terms of the kinetic theory.
   (a) Water freezes when it is cooled sufficiently.
   (b) Heating a liquid makes it evaporate more quickly.
   (c) Heating a solid makes it melt.

2 The solid X does not melt until it is heated to a very high temperature. What can you deduce about the forces which exist between particles of X?

3 Of the five substances listed in the table below which is/are (a) solid (b) liquid (c) gaseous (d) unlikely to exist?

Substance	*Distance between particles*	*Arrangement of particles*	*Movement of particles*
A	Close together	Regular	Move in straight lines
B	Far apart	Regular	Random
C	Close together	Random	Random
D	Far apart	Random	Move in straight lines
E	Close together	Regular	Vibrate a little

4 Supply words to fill in the blanks in this passage.
A solid has a fixed ____ and a fixed ____. A liquid has a fixed ____, but a liquid can change its ____ to fit its container. A gas has neither a fixed ____ nor a fixed ____. Liquids and gases flow easily; they are called ____. There are forces of attraction between particles. In a solid, these forces are ____, in a liquid they are ____ and in a gas they are ____.

5 Imagine that you are one of the millions of particles in a crystal. Describe from your point of view as a particle what happens when your crystal is heated until it melts.

6 Which of the two beakers in the figure opposite represents (a) evaporation, (b) boiling? Explain your answers.

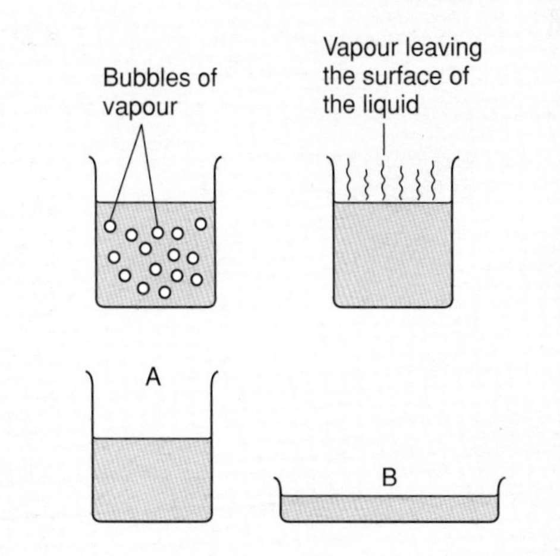

7 When a stink bomb is let off in one corner of a room, it can soon be smelt everywhere. Why?

8 Beaker A and dish B contain the same volume of the same liquid. Will the liquid evaporate faster in A or B? Explain your answer.

9 Why can you cool a cup of hot tea by blowing on it?

10 A doctor dabs some ethanol (alcohol) on your arm before giving you an injection. The ethanol makes your arm feel cold. How does it do this?

2.5 Gases

Gases have much lower densities than solids and liquids. The reason is that most of a gas is space. The particles are far apart, and move through the space at high speed. From time to time they collide with the walls of the container and with other particles. Gases are much more compressible than solids and liquids:

- Increase the pressure → the volume decreases.
- Decrease the pressure → the volume increases.

The kinetic theory explains this by saying that there is so much space between the particles that it is easy for the particles to move closer together when the gas is compressed.

Gases exert pressure because their particles are colliding with the walls of the container. If the volume of gas is decreased, the particles will hit the walls more often and the pressure will increase. The pressure of a given mass of gas changes with temperature:

- Increase the temperature, while keeping the volume constant → the pressure increases.
- Increase the temperature while keeping the pressure constant → the gas expands.

The kinetic theory explains these observations because, as the temperature rises, the particles have more energy, move faster and collide more frequently with the walls of the container.

A gas is mostly space! The particles of gas are far apart, with a lot of space between them. The particles move through the space at high speed. As a result, gases can be compressed.

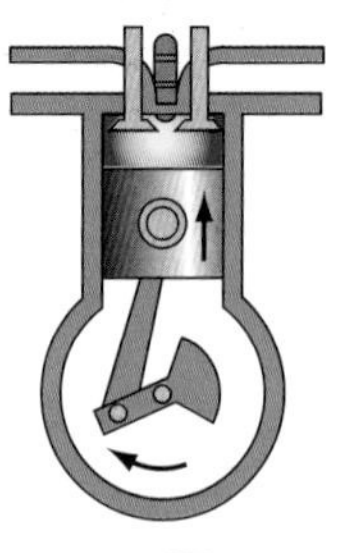

One cylinder of a four-stroke car engine. During the compression stroke, the piston moves up the cylinder and decreases the volume of the mixture of petrol vapor and air; therefore the pressure increases

(a) The volume decreases therefore the pressure increases

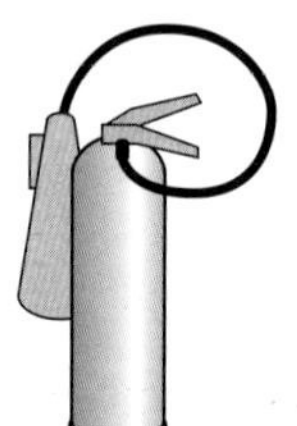

A cylinder of gas under pressure. When the valve is opened, the pressure decreases. The gas expands out of the cylinder

(b) The gas expands as the pressure decreases

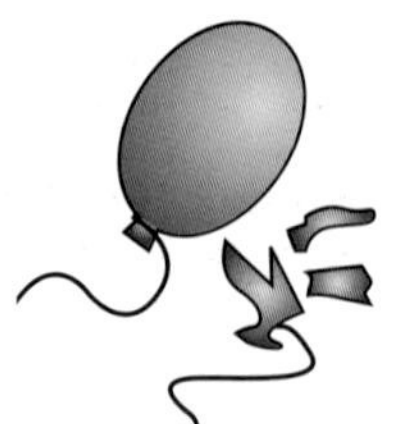

When a balloon full of gas is heated, the pressure of gas increases until the balloon bursts

(c) The pressure increases as the temperature rises

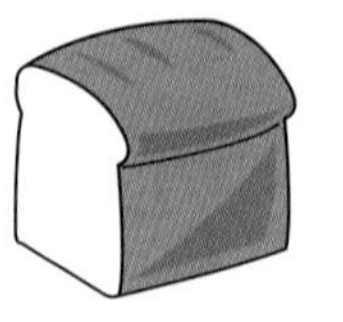

A lump of bread dough contains air and carbon dioxide. When the dough is heated in an oven, the volume of gas increases and the dough 'rises'

(d) The gas expands as the temperature rises

Figure 2.5A Factors which determine the volume of a gas

An increase in pressure makes the particles move closer together, and the volume of the gas decreases.

When the pressure is reduced, the particles move further apart; the volume increases.

An increase in temperature makes the particles move more rapidly and collide more often with the walls of the container. If the volume is held constant the pressure increases.

The relationship between the volume, pressure and temperature of a gas is expressed by the equation:

$$\frac{\text{Pressure} \times \text{Volume}}{\text{Absolute temperature}} = \text{constant}$$

$$\frac{pV}{T} = \text{constant}$$

(for a fixed mass of gas).

The pressure, volume and temperature of a gas are related by the equation pV/T = constant

Stating gas volume

It would not be very informative to state the volume of a gas without stating the temperature and pressure at which the volume is measured. It is the custom to quote gas volumes either at standard temperature and pressure (s.t.p., 0 °C and 1 atm) or at room temperature and pressure (r.t.p., 20 °C and 1 atm).

Figure 2.5B ▲ Hot air balloon

Diffusion of gases

A gas which has large molecules has a higher density than a gas which has small molecules. For example, chlorine molecules are larger than hydrogen molecules, and chlorine is denser than hydrogen. Under the same conditions of temperature and pressure, chlorine is 35.5 times as dense as hydrogen. Large, heavy molecules move more slowly than small, light molecules, and dense gases therefore diffuse more slowly than gases of low density. Under the same conditions, the rate of diffusion of chlorine is one-sixth that of hydrogen.

The sport of hot air ballooning depends on the fact that gases expand and become less dense when they are heated. The air inside the balloon is heated by a propane torch. This makes the air inside the balloon less dense than the air outside, and the balloon rises.

CHECKPOINT

1. Explain why people are advised to let some of the air out of their car tyres when driving from the UK to a very hot country.
2. A weather balloon is released into the atmosphere, where it will rise to a height of 10 km. It is only partially inflated at ground level. Why should it not be filled completely?
3. A gas syringe holds 100 cm^3 of nitrogen at a pressure of 1 atm. The gas is allowed to expand into a 1 l flask. What can you say about the pressure of gas in the 1 l flask?

3.1 Raw materials from the Earth's crust

FIRST THOUGHTS

The Earth provides all the raw materials we use. The problem is to separate the substances we want from the mixture of substances which makes up the Earth's crust.

The Earth's crust and atmosphere provide us with all the raw materials that we use: metals, oil, salt, sand, limestone, coal and many other resources. We use these substances as the raw materials for the manufacture of the houses, clothing, tools, machines, means of transport, medicines and all the other goods which we need. Few useful raw materials are found in a pure form in the Earth's crust. It is usual to find raw materials that we want mixed up with other materials. Chemists have worked out methods of separating substances from mixtures. Table 3.1 summarises some of them.

The table summarises methods of separating the components of different types of mixture.

Table 3.1 Methods of separating substances from mixtures

Mixture	Type	Method
Solid + Solid		Make use of a difference in properties, e.g. solubility or magnetic properties
Solid + Liquid	Mixture	Filter
Solid + Liquid	Solution	Crystallise to obtain the solid Distil to obtain the liquid
Liquid + Liquid	Miscible (form one layer)	Use fractional distillation
Liquid + Liquid	Immiscible (form two layers)	Use a separating funnel
Solid + Solid	In solution	Use chromatography

www.keyscience.co.uk

3.2 Pure substances from a mixture of solids

EXTENSION FILE ACTIVITY

Dissolving one of the substances

There are vast deposits of **rocksalt** in Cheshire. A salt mine is shown in Figure 3.2A. One method of mining salt is to insert charges of explosives in the rock face and then detonate them.

Rocksalt is crushed and used for spreading on the roads in winter to melt the ice. For many uses, pure salt (sodium chloride) is needed. It can be obtained by using water to dissolve the salt in rocksalt, leaving the rock and other impurities behind.

Figure 3.2A Winsford salt mine

SUMMARY

A soluble substance can be separated from a mixture by dissolving it to leave insoluble substances behind.

3.3 ▶ Solute from solution by crystallisation

EXTENSION FILE ACTIVITY

While the solvent is evaporating, dip a cold glass rod into the solution from time to time. When small crystals form on the rod, take the solution off the water bath and leave it to cool.

Solution in evaporating basin

Water bath (steam bath)

Heating element

Figure 3.3A ▲ Evaporating a solution to obtain crystals of solute

SUMMARY

A solute crystallises out of a saturated solution. If a solution is unsaturated, some of the solvent must be evaporated before crystals form.

A laboratory method of evaporating a solution until it crystallises is shown in Figure 3.3A. The salt industry uses large scale evaporators which run non-stop. Figure 3.3B shows another method.

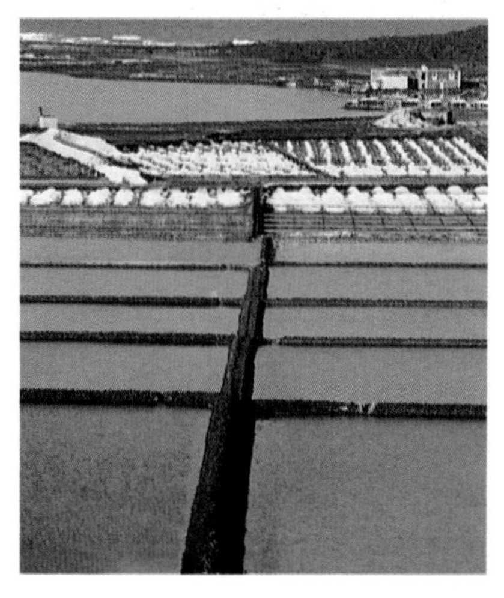

Figure 3.3B ▶ Salt pans in the Canary Islands. Sea water flows into the 'pans' and much of the water evaporates in the hot sun. When the brine has became a saturated solution, salt crystallises. The sea water is pumped through sluice gates where the salt crystals are filtered out.

3.4 ▶ Filtration: separating a solid from a liquid

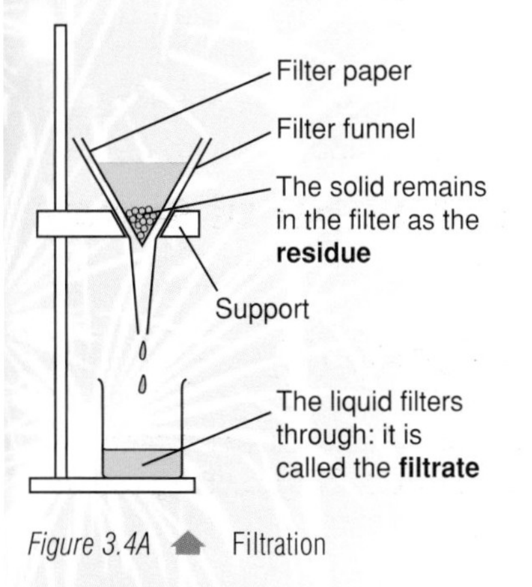

Figure 3.4A ▲ Filtration

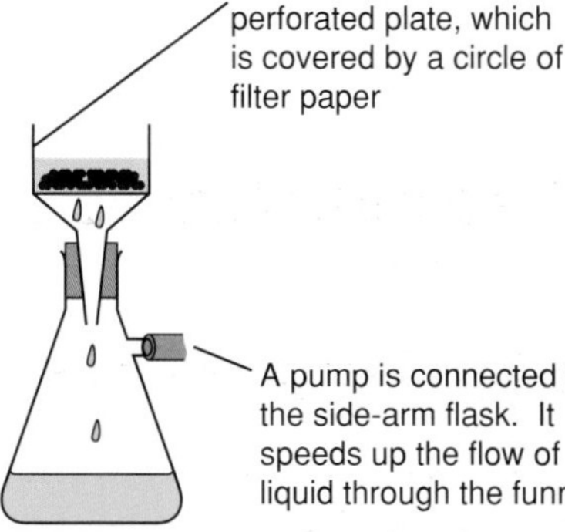

Figure 3.4B ▲ Filtration under reduced pressure

Filtration can be used to separate a solid and a liquid. The filter must hold back solid particles but be fine enough to allow liquid to pass through. Figure 3.4A shows a laboratory apparatus for filtering through filter paper. Figure 3.4B shows a faster method: filtering under reduced pressure. In Topic 15.3, you will see how important filtration is in the purification of our drinking water.

3.5 ▶ Centrifuging

Biochemists have developed methods of growing bacteria as a source of **high-protein food**. When bacteria are allowed to grow in a warm solution of nutrients, they multiply so fast that they can double in mass every 20 minutes. From time to time, the harvest of bacteria is separated from the nutrient solution. Bacteria are too small to be separated by filtration. They are so small that they are **suspended** (spread out) in the liquid. They do not settle to the bottom as heavier particles would do, and they do not dissolve to form a solution. They can be separated by the method of **centrifuging** (centrifugation). The suspension of bacteria and liquid is placed inside a centrifuge and spun at high speed. The motion causes bacteria to separate from the suspension and sink to the bottom of the centrifuge tubes. Figure 3.5A shows a small laboratory centrifuge.

1 The suspension is poured into a glass tube inside the centrifuge

2 Another tube is used to balance the first

3 As the centrifuge spins, solid particles settle to the bottom of the tube

4 The solid forms a compacted mass at the bottom of the tube. The liquid is decanted (poured off) from the centrifuge tube to leave the solid behind

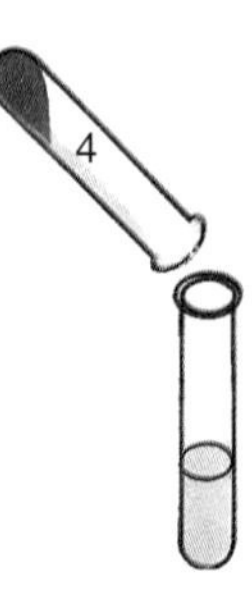

Figure 3.5A ▲ Centrifuging a suspension

SUMMARY

- Filtration will separate a solid from a liquid.
- Centrifuging will separate a suspended solid from solution.

CHECKPOINT

1. Describe how you would obtain both the substances in the following mixtures of solids.
   (a) A and B: Both A and B are soluble in hot water, but only A is soluble in cold water.
   (b) C and D: Neither C nor D is soluble in water. C is soluble in ethanol, but D is not.
2. In Trinidad and some other countries, tar trickles out of the ground. It is a valuable resource. The tar is mixed with sand and gravel. Suggest how tar may be separated from these substances.
3. Blood consists of blood cells and a liquid called plasma. When a sample of blood is taken from a person, the blood cells slowly settle to the bottom. How can the separation of blood cells from plasma be speeded up?
4. Suggest how you could obtain the following:
   (a) iron filings from a mixture of iron filings and sand,
   (b) wax from a mixture of wax and sand,
   (c) sand and gravel from a mixture of both,
   (d) rice and salt from a mixture of both.

3.6 Distillation

Sometimes you need to separate a solvent from a solution. In some parts of the world, drinking water is obtained from sea water. The method of **distillation** is employed. Figure 3.6A shows a laboratory scale distillation apparatus. The processes that take place are:

- in the distillation flask, **vaporisation**: liquid ➔ vapour
- in the condenser, **condensation**: vapour ➔ liquid.

Vaporisation followed by condensation is called **distillation**.

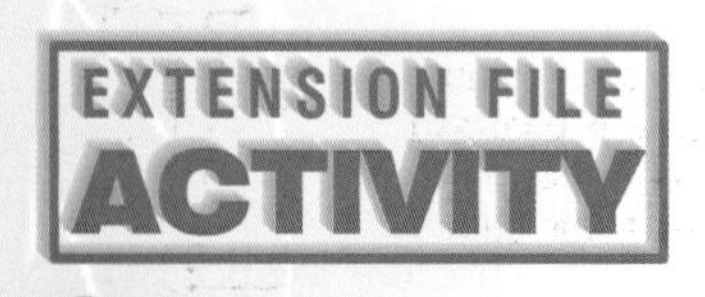

SUMMARY

Distillation is used to separate a solvent from a solution.

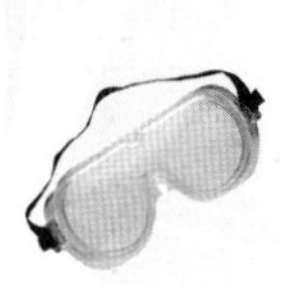

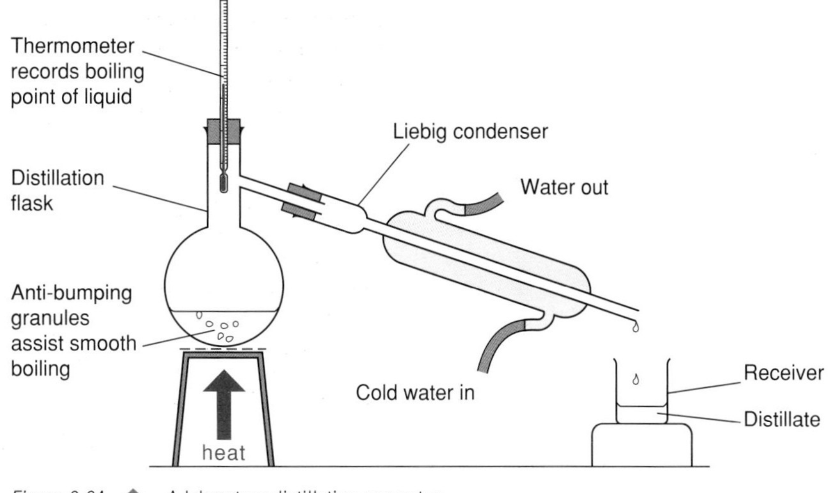

Figure 3.6A ▲ A laboratory distillation apparatus

CHECKPOINT

1. Why do some countries in the Persian Gulf obtain drinking water by distilling sea water? You will need to consider (a) the other sources of water and (b) the cost of fuel for heating the still.
2. Imagine that you are cast away on a desert island. You have two ways of obtaining drinking water. One is by separating pure water from sea water. The other is by collecting dew.
   (a) Think of a way of obtaining pure water from sea water. You do not have a proper distillation apparatus, but you have some matches and an empty petrol tin which were in the lifeboat. You find wood, bamboo canes, palm trees and coconuts on the island. Make a sketch of your design. With other members of your class, make a display of your sketches.
   (b) Every night on the island there is a heavy dew. How can you collect some of this dew? You have a sheet of plastic from the lifeboat. Again, sketch your idea. Then you can make another class display.
   (c) Write a letter to a 10-year-old, telling how you survived the shipwreck and how you obtained drinking water until you were rescued.

3.7 ▶ Separating liquids

Using a separating funnel to separate immiscible liquids

When some liquids are added to each other, they do not mix: they form separate layers. They are said to be **immiscible**. They can be separated by using a separating funnel.

1 The mixture of immiscible liquids is poured in. It settles into two layers (or more) as the liquids do not mix.

2 The tap is opened to let the bottom layer run into a receiver.

3 The tap is closed and the receiver is changed. The tap is opened to let the top layer run out.

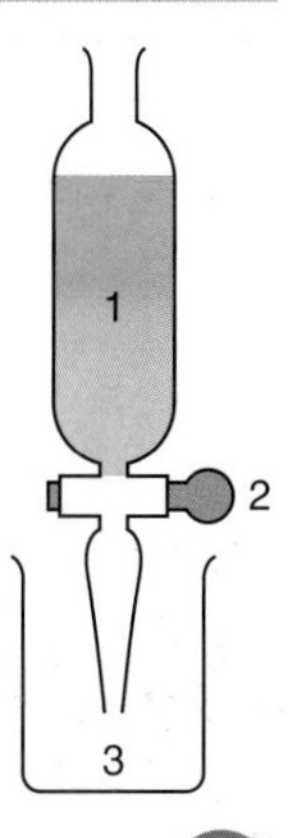

Figure 3.7A ▶ A separating funnel

SUMMARY

Immiscible liquids can be separated by means of a separating funnel.

3.8 Fractional distillation

The figure shows a solution of liquids being separated by fractional distillation by means of their different boiling points.

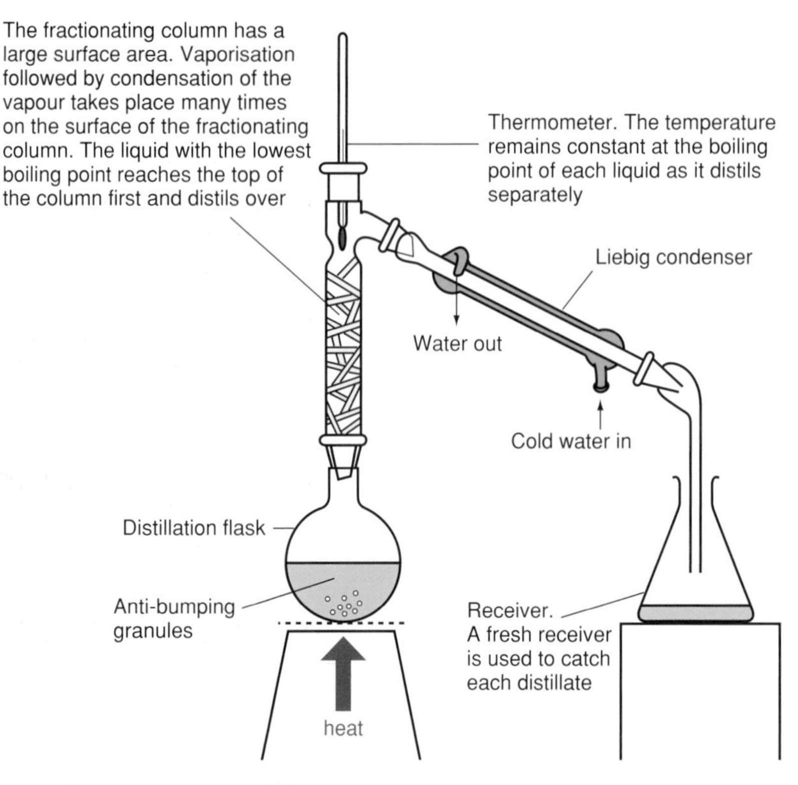

Figure 3.8A Fractional distillation

When liquids dissolve in one another to form a solution, instead of forming separate layers, they are described as **miscible**. Distillation can be used to separate a mixture of miscible liquids. Whisky manufacturers want to separate ethanol (alcohol) from water in a solution of these two liquids. They are able to do this because the two liquids have different boiling points: ethanol boils at 78 °C, and water boils at 100 °C. Figure 3.8A shows a laboratory apparatus which will separate a mixture of ethanol and water into its parts or **fractions**. The process is called **fractional distillation**. The temperature rises to 78 °C and then stays constant while all the ethanol distils over. When the temperature starts to rise again, the receiver is changed. At 100 °C, water starts to distil, and a fresh receiver is put into position to collect it.

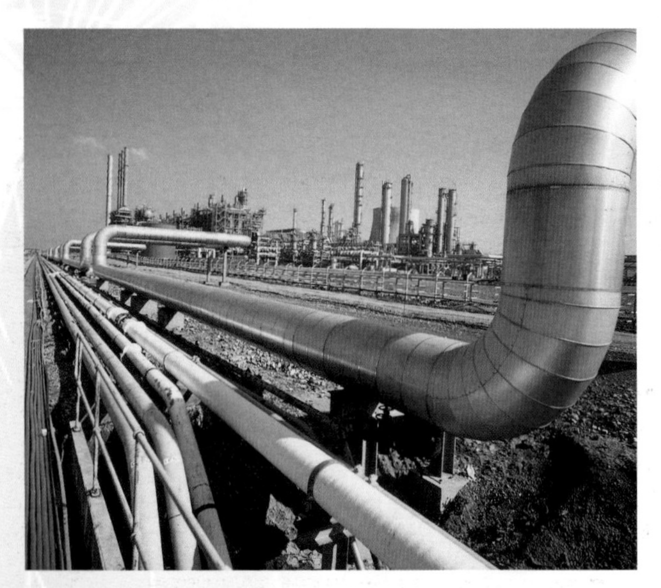

Figure 3.8B An industrial distillation plant

Continuous fractional distillation

Crude petroleum oil is not a very useful substance. By fractional distillation, it can be separated into a number of very useful products (see Figure 3.8C). In the petroleum industry, fractional distillation is made to run continuously (non-stop). Crude oil is fed in continuously, and the separated fractions are run off from the still continuously. These fractions are not pure substances. Each is a mixture of substances with similar boiling points. The fractions with low boiling points are collected from the top of the fractionating column. Fractions with high boiling points are collected from the bottom of the column.

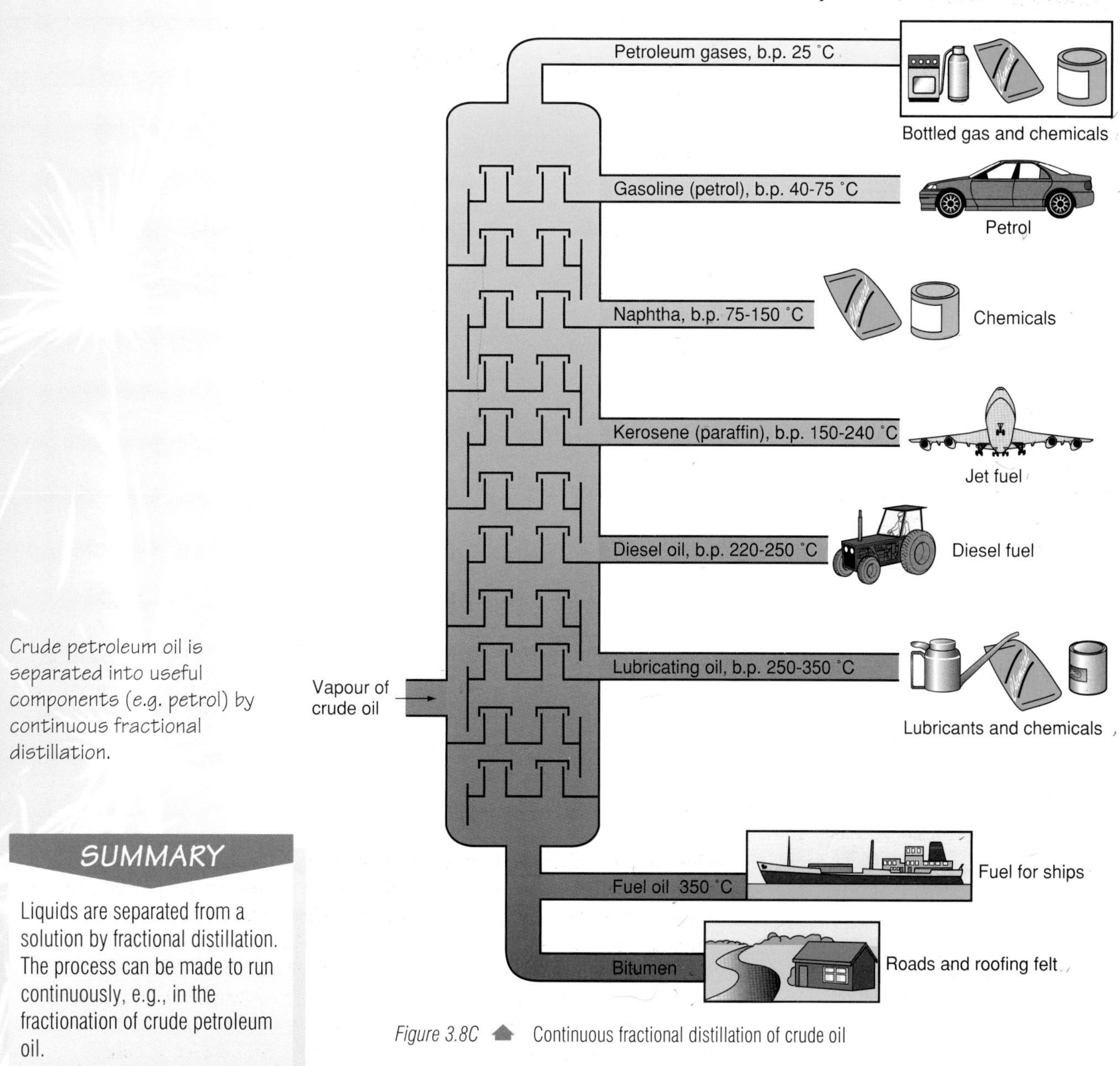

Crude petroleum oil is separated into useful components (e.g. petrol) by continuous fractional distillation.

Figure 3.8C ▲ Continuous fractional distillation of crude oil

SUMMARY

Liquids are separated from a solution by fractional distillation. The process can be made to run continuously, e.g., in the fractionation of crude petroleum oil.

CHECKPOINT

1. Your 12-year-old sister wants to know what you have been doing in your science lessons. Write a technical report, in words which she can understand, explaining *why* the petroleum industry distils crude oil and *how* fractional distillation works.
2. Your sister pours vinegar into the bottle of cooking oil by mistake. She asks you to help her to put things right. How could you separate the two liquids?
3. In Hammond Innes' novel, *Ice Station Zero*, a vehicle breaks down because the villains have put sugar in the petrol tank. Explain how you could separate the sugar from the petrol.
4. After a collision at sea, thousands of litres of oil escape from a tanker. A salvage ship sucks up a layer of oil mixed with sea water from the surface of the sea.
   (a) Describe how you could separate the oil from the sea water.
   (b) Say why it is important to be able to do this.

3.9 Chromatography

Science at work

Computers help to solve crimes. Materials found at the scene of a crime can be analysed by chromatography. The results can be stored on computer as a database to help to solve similar crimes.

Frequently, chemists want to **analyse** a mixture. (**Analysis** of a mixture means finding out which substances are present in it.) Chemists may want to find out which dyes and preservatives have been added to a food substance. They may want to find out whether there are any harmful substances present in drinking water. Chromatography is one method of separating the solutes in a solution. Figure 3.9A shows **paper chromatography**. When a drop of solution is applied to the chromatography paper, the paper adsorbs the solutes, that is, binds them to its surface. As the solvent rises through the paper, the solvent competes with the paper for the solutes. Some solutes stay put; others dissolve in the solvent and travel in it up the paper. A solute which is very soluble in the solvent travels through the paper faster than a solute which is only slightly soluble. When the solvent reaches the top of the paper, the process is stopped. Different solutes have travelled different distances. The result is a **chromatogram**.

KEY SCIENTIST

Analytical chemistry is an important branch of chemistry. A career as an analytical chemist could take you into many different fields. You could be analysing the air we breathe or the water we drink. You could be testing processed foods to find out what colourings and flavour-enhancers have been added to them. You could be in the police force. Many crimes are solved by chemists analysing materials found at the scene of the crime.

1 A drop of solution is touched on to the chromatography paper. The solvent evaporates. A spot of solute remains.

2 The chromatography paper hangs from a glass rod. It must not touch the sides of the beaker.

3 The spot of solute must be above the level of the solvent in the beaker.

4 The solvent front. The solvent has travelled up the paper to this level.

5 Spots of different substances present in the solute. As the substances travel through the paper at different speeds, they become separated. The result is a chromatogram.

Figure 3.9A ▲ Paper chromatography

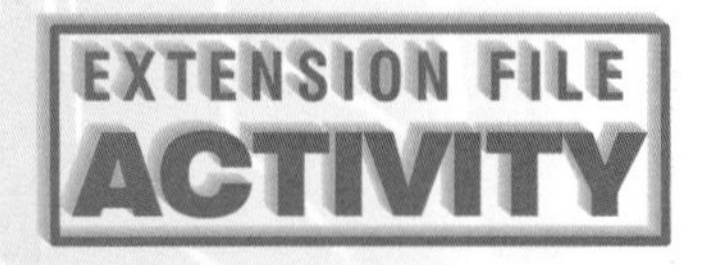

Analytical chemists use many techniques. One of them is chromatography.

Many solvents are used in chromatography. Ethanol (alcohol), ethanoic acid (vinegar) and propanone (a solvent often called acetone) are common. A chemist has to experiment to find out which solvent gives a good separation of the solutes. With a solvent other than water, a closed container should be used so that the chromatography paper is surrounded by the vapour of the solvent (see Figure 3.9B).

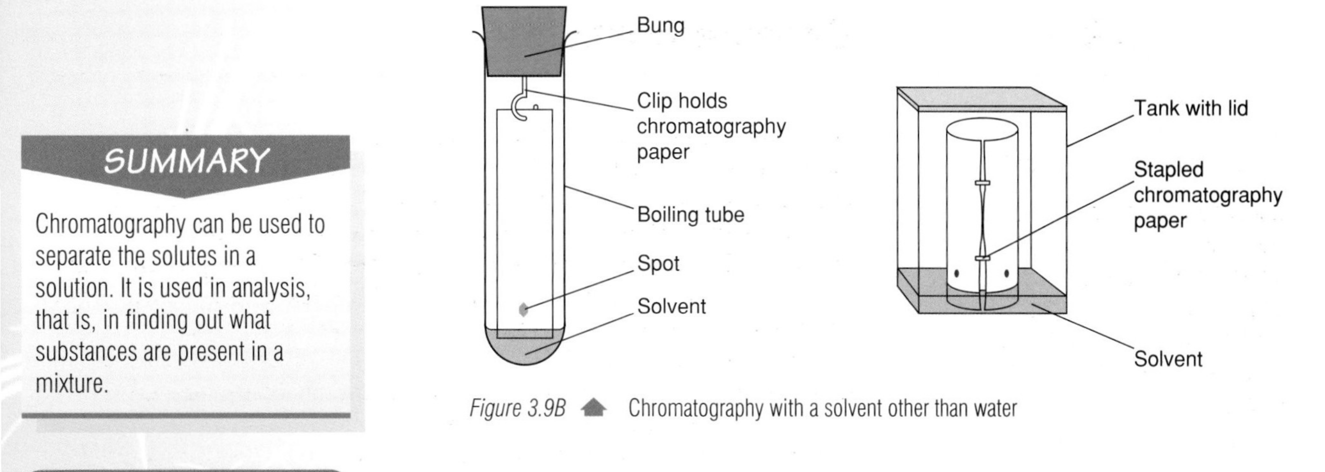

Figure 3.9B ▲ Chromatography with a solvent other than water

SUMMARY

Chromatography can be used to separate the solutes in a solution. It is used in analysis, that is, in finding out what substances are present in a mixture.

CHECKPOINT

▶ **1** A chemist is asked to find out what substances are present in two mixtures, M_1 and M_2. He makes a chromatogram of the two mixtures and a chromatogram of some substances which he suspects may be present. The figure below shows his results.
Can you interpret his results? What substances are present in M_1 and M_2?

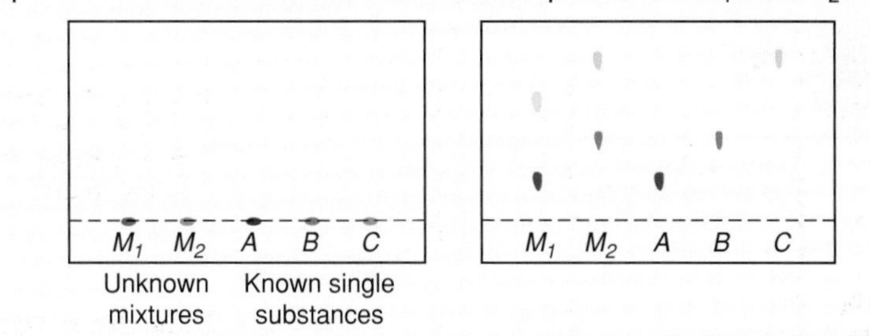

▶ **2** Imagine that you are a detective investigating forged bank notes. Your investigations take you to a house where you find a printing press and some inks. Describe how you would find out whether these inks are the same as those used to make the bank notes.

▶ **3** Dr Ecksplor discovers a strangely coloured orchid in Brazil. Professor Seeker finds an orchid of the same colour in Ecuador. The two scientists wonder whether the two flowers contain the same pigment. They feel sure that the pigment does not dissolve in water.

(a) Why do they feel sure that the pigment does not dissolve in water?

(b) Describe an experiment which they could do to find out whether the two orchids contain the same pigment.

▶ **4** An analytical chemist has the task of finding out whether the red colouring in a new food product contains any dyes which are not permitted in foods. She runs a chromatogram on the food colouring and on some of the permitted red food dyes. The figure opposite shows her results. What conclusion can you draw from these chromatograms?

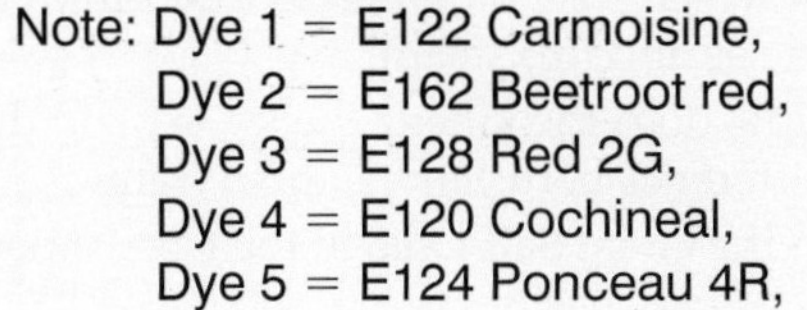

Note: Dye 1 = E122 Carmoisine,
Dye 2 = E162 Beetroot red,
Dye 3 = E128 Red 2G,
Dye 4 = E120 Cochineal,
Dye 5 = E124 Ponceau 4R,
Dye 6 = E160 Capsorubin

Dyes which are allowed in foods are given **E numbers**.

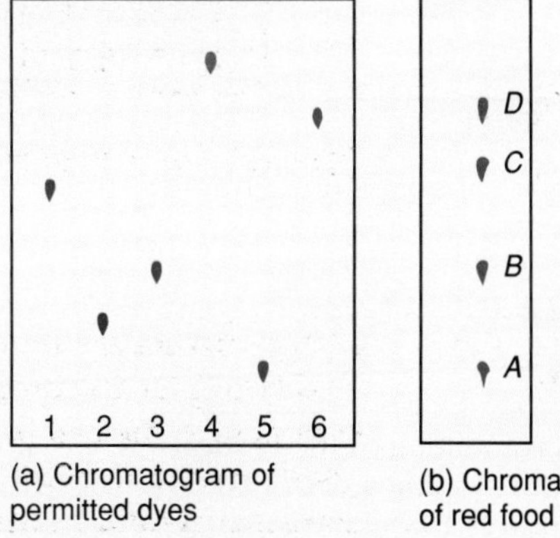

(a) Chromatogram of permitted dyes

(b) Chromatogram of red food colouring

Theme Questions

1 A solid X melts at 58 °C. Which of the following substances could be X?
A Chloroethanoic acid, m.p. 63 °C
B Diphenylamine, m.p. 53 °C
C Ethanamide, m.p. 82 °C
D Trichloroethanoic acid, m.p. 58 °C

2 An impure sample of a solid Y melts at 77 °C. Which of these solids could be Y?
E Dibromobenzene, m.p. 87 °C
F 1, 4-Dinitrobenzene, m.p. 72 °C
G 3-Nitrophenol, m.p. 97 °C
H Propanamide, m.p. 81 °C

3 An impure sample of liquid Z boils at 180 °C. Which of these liquids could be Z?
I Benzoic acid, b.p. 249 °C
J Butanoic acid, b.p. 164 °C
K Hexanoic acid, b.p. 205 °C
L Methanoic acid, b.p. 101 °C

4 Which of the substances listed below are (a) solid (b) liquid (c) gaseous at room temperature?

Pure substance	A	B	C	D	E	F
Melting point (°C)	8	−92	41	63	−111	−30
Boiling point (°C)	101	−21	182	189	11	172

5 The graph shows how the temperature of a substance rises as it is heated. At A, the substance is a solid.
(a) Say what happens:
(i) between A and B,
(ii) between B and C,
(iii) between C and D,
(iv) between D and E,
(v) between E and F.
(b) Name the temperatures T_1 and T_2.

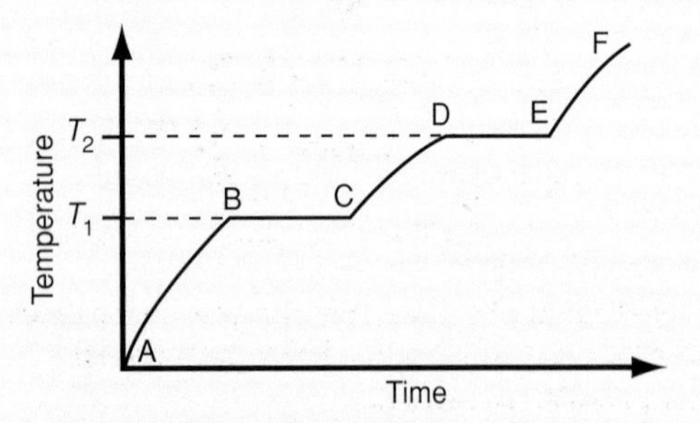

6 The table gives the solubility of potassium nitrate at various temperatures.

Temperature (°C)	0	10	20	40	60
Solubility (g per 100 g of water)	13	21	32	64	110

(a) On graph paper, plot solubility (on the vertical axis) against temperature (on the horizontal axis). Draw a smooth curve through the points. Use your graph to answer (b) and (c).
(b) What is the solubility of potassium nitrate at 30 °C?
(c) At what temperature is the solubility of potassium nitrate 85 g/100 g?
(d) A 100 g mass of water is saturated with potassium nitrate at 60 °C and cooled at 30 °C. What happens?

7 The diagram shows what happens when you breathe.

Inhalation

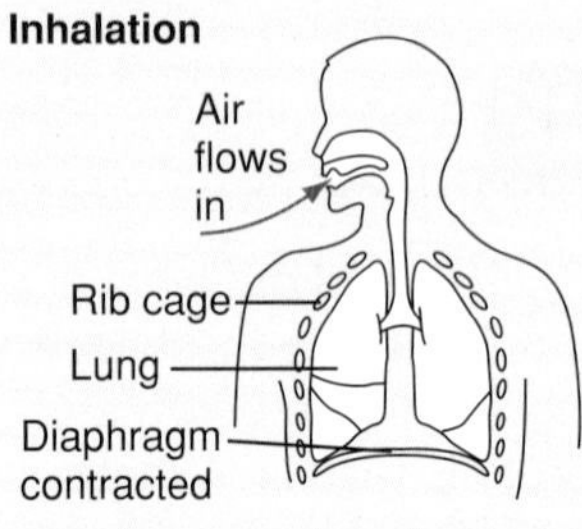

The diaphragm contracts and flattens so increasing the volume of the chest cavity.
What does this do to the air pressure in the cavity? Why does it cause air to flow into the lungs?

Exhalation

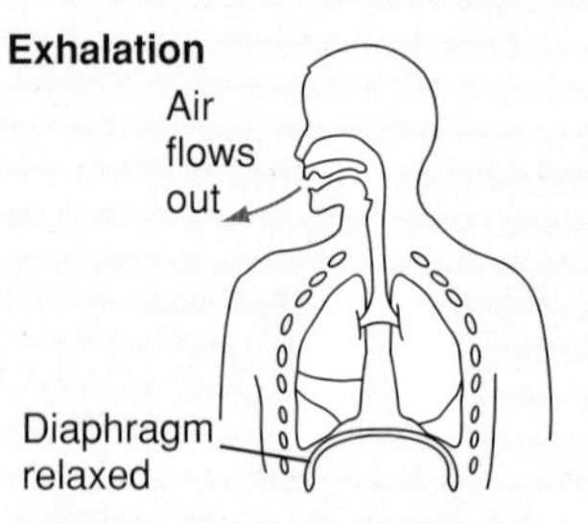

The diaphragm relaxes and pushes up into the chest cavity.
What does this do to the air pressure in the cavity? Why does it causes air to flow out of the lungs?

8 A king in medieval times asked his goldsmith to make him a new crown, nothing fancy, just plain with no jewels. He gave the goldsmith 2.00 kg of gold. When the crown arrived, the king had it weighed. It was exactly 2.00 kg in mass. He tried it on. The crown did not look exactly the right colour, and the king wondered whether the goldsmith had kept back some of the gold and alloyed the rest with a cheaper metal to make up the mass. What could the king do to find out whether the crown was pure gold? Describe the measurements which he would have to make.

9 (a) In which state of matter is water in (i) the sea (ii) icicles and (iii) steam?
(b) How does the kinetic theory explain the difference between the three states?
(c) In the Great Salt Lake in the USA, a person can float very easily.
(i) Explain why the high concentration of salt makes it easier to float.
(ii) Describe an experiment which you could do to find out what mass of salt is present in 100 cm^3 of the lake water.

10 (a) What sort of mixture can be separated by (i) filtration, (ii) chromatography, (iii) centrifuging?

(b) A mixture contains two liquids, A and B. A boils at 75 °C, and B boils at 95 °C. Draw an apparatus you could use to separate A and B. Label the drawing.

11 The police are investigating a case of poison pen letters. The police chemist makes chromatograms from the ink in the letters and the ink in the pens of three suspects. The diagram below shows her results. What conclusion can you reach?

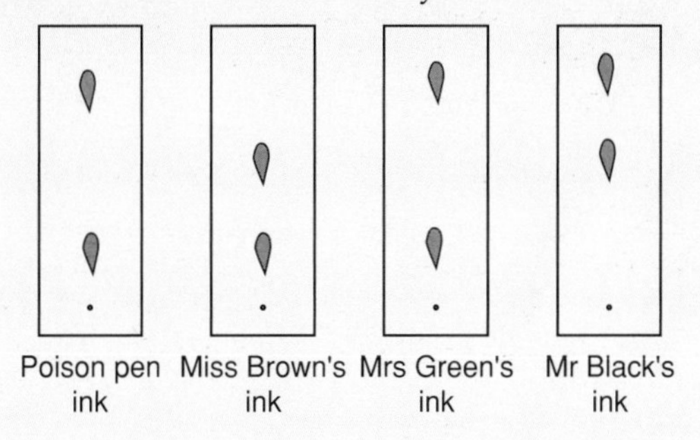

12 Use the information in the table. Say how you could separate:

(a) aluminium and cobalt
(b) chromium and polyurethane.

Substance (in powder form)	*Solubility in water*	*Solubility in ethanol*	*Magnetic properties*
Aluminium	Insoluble	Insoluble	None
Cobalt	Insoluble	Insoluble	Magnetic
Chromium	Insoluble	Insoluble	None
Polyurethane	Insoluble	Soluble	None

13 Select the correct endings to the sentences.

(a) A solid can be purified by crystallisation because
  (i) it dissolves in cold water but not in hot water
  (ii) it is insoluble in hot and cold water
  (iii) it is very soluble in hot and cold water
  (iv) it is more soluble in hot water than in cold water
  (v) it is more soluble in cold water than in hot water.

(b) Two liquids can be separated by distillation if
  (i) they have different densities
  (ii) they form two layers
  (iii) they have different boiling points
  (iv) their boiling points are the same
  (v) they have different freezing points.

(c) A solvent can be separated from a solution because
  (i) the particles of the solvent (liquid) are smaller than those of the solute (dissolved substance)
  (ii) the solvent is less dense than the solute
  (iii) the solvent has a higher freezing point than the solute
  (iv) the solvent evaporates more easily than the solute
  (v) the solvent condenses more easily than the solute.

14 (a) Materials are classified as solid, liquid or gas according to their properties. For each state give *two* typical properties. [6]

(b) The melting and boiling points of six substances are given below.

Substance	*Melting point/°C*	*Boiling point/°C*
Nitrogen	−210	−196
Carbon disulphide	−112	46
Ammonia	−78	−34
Bromine	−7	59
Phosphorus	44	280
Mercury(II) chloride	276	302

(Room temperature is taken as 20°C.)

  (i) Which *element* is a solid at room temperature? [1]
  (ii) Which *compound* is a liquid at room temperature? [1]
  (iii) Which *compound* is a gas at room temperature? [1]
  (iv) Which *element* will condense when cooled to room temperature from 100 °C? [1]
  (v) Which *compound* will freeze first on cooling from room temperature to a very low temperature? [1]
  (vi) Which of the six substances is a liquid over the widest range of temperature? [1]
  (vii) Draw, in two boxes, how the particles are arranged in the following substances at room temperature: [4]
  (I) bromine (II) ammonia

(c) Using the ideas of the kinetic theory explain why
  (i) a metal expands on heating [3]
  (ii) a gas changes to a liquid on cooling [3]
  (iii) a sample of water left in a dish at room temperature will evaporate over a period of time. [3]

(d) A student used the apparatus below to observe smoke particles.

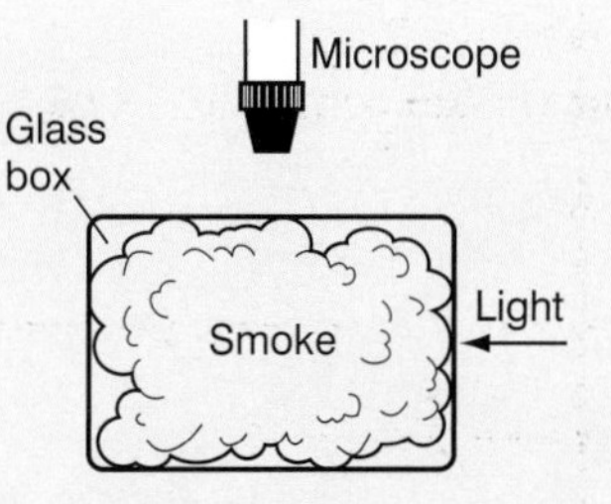

The student observed that the smoke particles moved randomly with a jerky, haphazard movement. [3]

  (i) Explain the student's observations.
  (ii) What is the name given to this type of movement? (CCEA) [1]

The atom

'Matter is composed of atoms.'

This is what John Dalton said in 1808. What a simple statement this appears to be! Yet the complex developments that followed from this statement fill the whole of physics and chemistry.

What are atoms?

How many different kinds are there?

How do atoms differ from even smaller particles?

You will find some of the answers to these questions in Theme B.

Topic 4 Elements and compounds

4.1 Silicon

FIRST THOUGHTS

Elements; they're elementary, they are simple substances, and there are 106 of them – all different. You have already met many useful elements, like silicon and gold.

The tiny chip in the mighty micro

Before 1950, a computer was a massive combination of circuits and valves which took up a whole room. Nowadays, a microcomputer the size of a typewriter can do the same job as the old-style computer. Microcomputers can be fitted in aeroplanes and spacecraft. The size and weight of the old-style computers made this impossible. Many people own a personal computer, a PC, to streamline jobs such as budgeting their expenses. The change in computer size has been brought about by the use of **silicon chips**. Figure 4.1A shows an electronic circuit built on to the surface of a silicon chip. Such circuits are very reliable because they are less affected by age, moisture and vibration than the old-style circuits.

The photograph shows how the size of some silicon chips compares with the eye of a needle.

Figure 4.1A ▲ An integrated circuit built on the surface of a silicon chip

www.keyscience.co.uk

Silicon is an **element**. An element is a pure substance which cannot be split up into simpler substances. Elements are classified as metallic elements and non-metallic elements. Silicon is a non-metallic element. Most elements are either electrical conductors (substances which allow an electric current to flow through them) or electrical insulators (substances which do not allow an electric current to pass through them). Silicon is an unusual element in being a **semiconductor**: its behaviour is between that of a conductor and that of an insulator.

Silicon is a semiconductor; treatment with other elements allows it to conduct electricity more easily. This is how electronic circuits are created on a silicon chip.

The first step in making a silicon chip is to slice large crystals of silicon into wafers. Then tiny areas of the wafer are treated with other elements

Figure 4.1B ▲ An electron diffraction micrograph of silicon

which make silicon become an electrical conductor. The result is the creation of a thousand electronic circuits on each wafer. A chip 0.5 to 1.0 cm across contains thousands of tiny electrical switches called **transistors**. They allow on-off electric signals, which are the basis of computers, to occur at very high speeds.

Number crunching

Very difficult and long calculations can be done on a computer. Computers operate so rapidly that they can solve in minutes problems that would take weeks to compute by hand. This is why computers are used to obtain fast and accurate weather forecasts. Earth scientists use computers to record the readings of their **seismometers** all over the world. The use of computers to process information is called **information technology**.

The development of silicon chips has revolutionised the computer industry. Computers are now used in every kind of job: in business as word-processors and for keeping accounts and records; in science for recording measurements and doing calculations and for collecting and organising information in a database. The internet is an international network of computers

Tedious jobs

Keeping track of records is a job for a computer. Debiting and crediting accounts, keeping a list of the stock in a shop or factory and such jobs are handled easily by a computer because a computer can repeat the same procedure over and over without error.

Planning ahead

Using computers can help businesses to plan ahead. They can try out various plans and see how each will affect profits. In a similar way, computers can be used to try out various approaches to environmental problems to see how each approach will affect the animal and plant populations.

Data-logging

Computers are used in science for recording and storing measurements. A sensor which measures pH, temperature, humidity, etc., can be connected to a microcomputer. The readings are recorded and displayed as a table or as a chart or as a graph. This can be useful for taking readings over a long period of time, e.g. monitoring water pollution, monitoring weather conditions, recording the temperature in the core of a nuclear reactor, recording the light emitted by a distant star.

Computer programs are used as sources of information on a multitude of topics; such programs are called **databases**. In forensic science, materials found at the scene of a crime are analysed. This information can be fed into a computer to become part of a database which could help the police to solve a similar crime.

It's a fact!

Plans for the car of the future include a radar set and a computer to tell the driver how far ahead the next car is. It will make fast driving much safer.

Science at work

Jobs which once used to be done laboriously by hand are now done in a fraction of the time by computer. Computers have taken the drudgery out of a lot of jobs. They have also contributed to safety in aeroplanes and in industrial plants. The use of microcomputers has created jobs in the manufacture of computers (hardware) and the writing of computer programs which tell the computer what to do (software). The new jobs are technical jobs. Skilled people are needed to fill them.

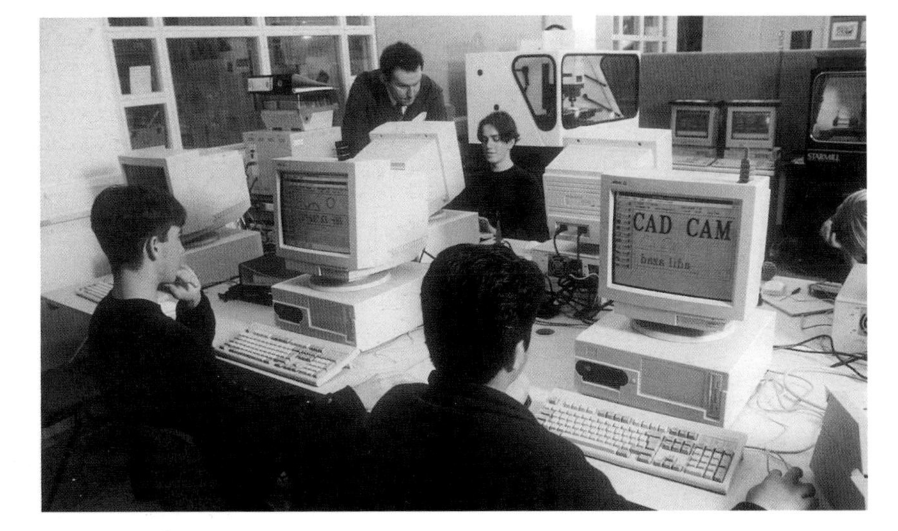

Figure 4.1C ▲ Using computers

Word processing

Computers with a word processor program are used in place of typewriters. Some programs enable the user to do 'desktop publishing'. This is used to present reports for business and scientific purposes in a neat, clear, easy-to-read format.

The internet

The **internet** is an international computer network. It consists of a core of permanently linked computers. To join the internet, you need to connect your computer to any one of these core computers. Once you are connected, 'on line', your computer can exchange information with every other computer on the internet, including those on other continents. You can send and receive messages across the world in seconds, using internet electronic mail, **e mail**. You can find information on any subject under the Sun. All this communication is made possible by computers

Your science lessons

You can use microcomputers in your science lessons for various purposes:

- extracting information from databases or making your own database,
- using a spreadsheet to display results from an experiment in a table or graph,
- using computer programs to test your ideas,
- recording readings from sensors and displaying and analysing the results.

SUMMARY

The element silicon is a semiconductor. Silicon is used to make transistors (devices which allow current to flow in one direction only). Circuits built on to silicon chips are the basis of the microcomputer industry.

CHECKPOINT

1 Can you think of jobs which are done by microcomputers in;
(a) shops (b) the car industry (c) schools (d) homes?
How do you and your family benefit from the use of micros in shops and the car industry? Do you benefit from the use of micros in any other way?

2 (a) What kinds of jobs are lost through the introduction of computers?
(b) What kinds of jobs are created by the spread of computers?
(c) Can the people who lose their jobs as a result of (a) take the jobs created in (b)? If your answer is *yes*, explain why. If your answer is *no*, explain what you think could be done to solve the problem.

3 Would space travel have been possible before the age of the microcomputer? Explain your answer.

4.2 Gold

The prospector's dream element

Gold has always held a great fascination for the human race. Gold occurs **native**, that is, as the free element, not combined with other elements. For thousands of years, people have been able to collect gold dust from river beds and melt the dust particles together to form lumps of gold. We can still see some of the objects which the goldsmiths made thousands of years ago because gold never tarnishes. For centuries, people used gold coins.

Figure 4.2A ▲ Panning for gold

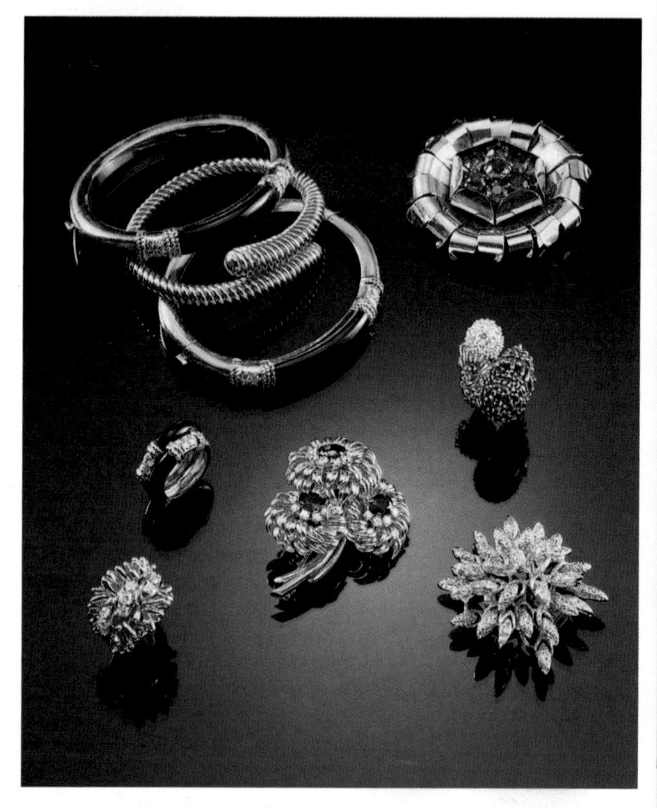

Figure 4.2B ▲ Gold jewellery

Gold is a metallic element. The shine and the ability to be worked into different shapes are characteristic of metals. Unlike gold, most metals tarnish in air. Gold is not attacked by air or water or any of the other chemicals in the environment. It is used in electrical circuits when it is essential that the circuits do not corrode. For example, in microcomputers gold wires connect silicon chips to external circuits. Spacecraft use gold connections in their electrical circuits.

SUMMARY

Gold is a metallic element. It conducts electricity and is easily worked. Unlike many other metals, gold never becomes tarnished. Gold is used for jewellery and in electrical circuits which must not corrode.

CHECKPOINT

1. Why has gold always been a favourite metal for jewellers to work with?
2. What use is made of gold in modern technology?
3. Dentists use gold to fill teeth. Why is gold a suitable metal for this job?

4.3 Copper and bronze

FIRST THOUGHTS

More metallic elements and some alloys:

- Copper, bronze and the Bronze Age
- Iron, steel and the Industrial Revolution

A step up from the Stone Age

Thousands of years ago, Stone Age humans found lumps of copper embedded in rocks. Attracted by the colour and shine of copper, they hammered it into bracelets and necklaces. Then they started to use copper to make arrowheads, spears, knives and cooking pans. They found that copper tools did not break like the stone tools they were used to. Copper knives could be ground to a sharper edge than stone, and copper dishes did not crack as pottery bowls did. The shine, the ability to be worked into different shapes and the ability to conduct heat are typical of metals. Copper is a metallic element.

Stone Age people also discovered the alloy of copper and tin called **bronze** (a mixture of metals is called an alloy). Bronze is harder than copper or tin and can be ground to a sharper edge. Bronze weapons and tools made such a difference to the way people lived that they gave their name to the Bronze Age. Thanks to the new tools hunting and farming no longer occupied all the time of all the members of the community. Some people were able to spend time on painting, making pottery and building homes. The arrival of the Bronze Age was the beginning of civilisation.

Copper is still an important metal. Copper is a good electrical conductor – a typical metal. It is easily drawn into wire. Copper wire is used in electrical circuits. Wires, cables, overhead power lines, switches and windings in electrical motors are made of copper. Half the world's production of copper (total 8 million tonnes a year) is used by electrical industries.

It's a fact!

What happened in 1886?
Three thousand miles of copper cable were laid under the Atlantic.
What for?
This was the start of the transatlantic telegraph system.

The discovery of copper and its alloy, bronze, allowed the human race to move out of the Stone Age into the Bronze Age. The science of metallurgy was born.

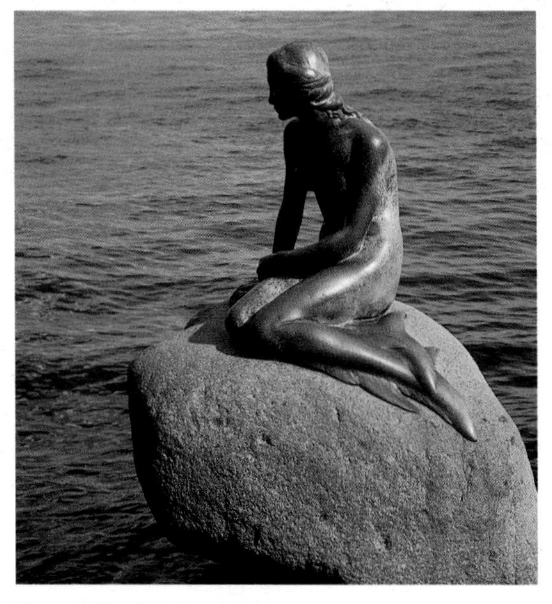

Figure 4.3A Bronze in use

CHECKPOINT

1. What advantages did copper tools have over tools made of (a) stone and (b) gold?
2. Explain why the discovery of bronze was so important that it gave its name to the Bronze Age.
3. What is the most important use of copper today?

4.4 Iron

Our most important metal

Our ancestors discovered how to extract iron from iron-bearing rocks over 3000 years ago. They used iron to make hammers, axes and knives, which did not break like stone tools and did not bend like bronze tools. Iron tools and weapons made such a difference to the way the human race lived that they gave their name to the Iron Age.

Many centuries later, iron and steel made the Industrial Revolution possible. The various types of **steel** are alloys of iron with carbon and other elements. Iron and steel are hard and strong: they can be hammered into flat blades and ground to a cutting edge. Our way of life in the twentieth century depends on machines made of iron and steel, buildings constructed on a framework of steel girders, and cars, lorries, trains, railways and ships made of steel.

Iron is a typical metallic element. An exceptional characteristic of iron is that it can be magnetised.

The discovery of iron led the human race into the Iron Age.

Metallurgists were able to make harder and sharper tools from iron than from bronze.

Steel alloys were used to make the machinery that made the Industrial Revolution possible.

Figure 4.4A ▲ A steel furnace

SUMMARY

Iron is a metallic element. Steel is an alloy. Being hard and strong, iron and steel are used for the manufacture of tools, machinery, motor vehicles, trains and ships.

4.5 Carbon

Diamond and graphite

FIRST THOUGHTS

Some non-metallic elements:

- Can brilliant diamond and greasy graphite be the same element?
- Can the killer, chlorine, save lives?

Why are diamonds so often used in engagement rings? The sparkle of diamonds comes from their ability to reflect light. The 'fire' of diamonds arises from their ability to split light into flashes of colour. The hardness of diamonds means that they can be worn without becoming scratched. Diamonds are 'for ever'. Diamond is the hardest naturally occurring substance. The only thing that can scratch a diamond is another diamond.

Figure 4.5A ▲ Diamonds

The hardness of diamond finds it many uses in industry. It is able to cut through metals, ceramics, glass, stone and concrete. Diamond-tipped saws slice silicon wafers from large crystals of silicon. As well as for cutting, diamonds are used for grinding, sharpening, etching and polishing. Oil prospectors would not be able to drill through hard rock without the help of drills studded with small diamonds (see Figure 4.5B). A 20 cm bit may be studded with 60 g of small diamonds.

Figure 4.5B ▲ Oil rig workers using a diamond studded drill

Diamond is a strange material, prized for its beauty and its usefulness. It is one form of the non-metallic element carbon.

Graphite is a shiny dark grey solid. It is so soft that it rubs off on your fingers and on paper. Pencil 'leads' contain graphite mixed with clay. Graphite is also used as a lubricant, in cars for example. Graphite is a second form of the element carbon. It is unusual in that it is the only non-metallic element which conducts electricity (see Topic 7.1).

Diamond and graphite are both pure forms of carbon. They are called **allotropes** of carbon. The existence of two or more crystalline forms of an element is called **allotropy**. Small diamonds (micron size, 10^{-6} m) can be made by heating graphite to a high temperature (1300 °C) under high pressure (60 000 atmospheres) for a few minutes. You can read more about the difference between the allotropes in Topic 4.8.

It's a fact!

Diamond knives! Where? In surgery: eye surgeons use diamond knives to remove cataracts. The edges are so sharp and so even that the surgeon can make a clean cut without tearing.

It's a fact!

Nineteenth century chemists thought graphite was a source of riches. All they had to do was to find the right conditions and graphite would change into diamond. It was more difficult than they expected. Not until the 1950s was a method found. Now 20 tonnes of manufactured diamonds are produced every year.

SUMMARY

Diamonds are beautiful jewels. Diamond is the hardest naturally occurring material. Small diamonds are used in industry for cutting, grinding and drilling.

Graphite is a slightly shiny grey solid, which conducts electricity. Graphite is used as a lubricant and as an electrical conductor. Diamond and graphite are allotropes (pure forms) of the element, carbon.

CHECKPOINT

1. How are small industrial diamonds made?
2. If the process for making industrial diamonds is carried on for a week, large gem-sized diamonds can be obtained. These large diamonds are dearer than natural diamonds. Why do you think this is so?
3. What uses are made of diamonds? What characteristics of diamond make it suitable for the uses you mention?
4. What is graphite used for? Why is graphite a suitable material for the uses you mention?

4.6 Chlorine

It's a fact!

The man who thought up the plan of using chlorine in warfare was the German chemist, Fritz Haber. Before the war, Haber won the Nobel prize for his discovery of a method of manufacturing ammonia. Haber said, 'A man belongs to the world in time of peace but to his country in time of war'. Haber's wife was so distressed by his involvement in the war that in 1916 she committed suicide.

The British scientist, Michael Faraday, was asked to develop poison warfare during the Crimean War, but he refused.

The life-saver

Chlorine is a killer. Its most infamous use was in the First World War when the German Army released cylinders of chlorine gas. The poisonous green cloud was driven by the wind into the trenches occupied by the British and French forces. Thousands of soldiers died, either choking on the gas or being shot as they retreated.

Chlorine is now used to kill germs. It is the bactericide which the water industry uses to make sure that our water supply is safe to drink. Chlorine has saved more lives than any other chemical. Before chlorine was used in the disinfection of the water supply, deaths from water-borne diseases, such as cholera and dysentery, were common. These diseases are still common in parts of the world which do not have safe water for drinking.

Chlorine is a non-metallic element. Many household bleaches and disinfectants contain chlorine.

Figure 4.6A Bleach and disinfectant

It's a fact!

In 1831, an epidemic of cholera hit London, and 50 000 people died. The cholera germs had been carried in the drinking water. That can't happen now that the water supply is disinfected with chlorine

4.7 Elements

FIRST THOUGHTS

You will have noticed that in science there is a need to classify things: to sort them into groups of similar members. Elements can be classified as metallic and non-metallic elements, with some exceptions.

Gold, copper, and iron are typical **metallic elements**. Like all metallic elements, they conduct electricity. Many of the metallic substances we use are not elements; they are **alloys**. Steel, brass, bronze, gunmetal, solder and many others are alloys. An alloy is a combination of two or more metallic elements and sometimes non-metallic elements also. Silicon, carbon and chlorine are **non-metallic elements**. Some of their characteristics are **typical** of non-metallic elements, but some are **atypical** (not typical). Diamond is atypical in being shiny; most non-metallic elements are dull. Graphite is the only non-metallic element that conducts electricity. Silicon is one of the few semiconductors of electricity.

Do you know how many elements there are, how many exist naturally, and how many are man-made?

There are 92 elements found on Earth. A further 14 elements have been made by scientists. Table 4.1 summarises the characteristics of metallic elements and non-metallic elements. You will see that there are many differences. Metallic and non-metallic elements also differ in their **chemical reactions**.

Table 4.1 Characteristics of metallic and non-metallic elements

Metallic elements	Non-metallic elements
Solids, except for mercury which is a liquid.	Solids and gases, except for bromine which is a liquid.
Hard and dense.	Most of the solid elements are softer than metals, but diamond is very hard.
A smooth metallic surface is shiny, but many metals tarnish in air, e.g. iron rusts.	Most are dull, but diamond is brilliant.
The shape can be changed by hammering: they are **malleable**. They can be pulled out into wire form: they are **ductile**.	Many solid non-metallic elements break easily when you try to change their shape. Diamond is the exception in being hard and strong.
Conduct heat, although highly polished surfaces reflect heat.	Poor thermal conductors.
Good electrical conductors.	Poor electrical conductors, except for graphite. Some are semiconductors, e.g. silicon.
Make a pleasing sound when struck: are **sonorous**.	Are not **sonorous**.

SUMMARY

Table 4.1 lists the differences between metallic and non-metallic elements. Alloys are combinations of metallic elements and sometimes non-metallic elements also.

4.8 The structures of some elements

FIRST THOUGHTS

The structure of an element means the arrangement of particles in the element. As you study this section, think about how the structure of an element affects the properties of that element.

Can you name an element which fits each of these descriptions?

(a) a solid metallic element
(b) a liquid metallic element
(c) a solid non-metallic element
(d) a gaseous non-metallic element
(e) a hard solid element
(f) a soft solid element
(g) a shiny element
(h) a dull element

How do these differences arise? The reason lies in the different arrangements of atoms in the different elements. In many elements the atoms are bonded together in groups called **molecules**.

Oxygen, chlorine and many non-metallic elements consist of individual molecules. There are strong bonds between the atoms in the molecules, but between molecules there is only a very weak attraction. The molecules move about independently, and these elements are gaseous.

Sulphur is a yellow solid. There are two forms of sulphur, which form differently shaped crystals. The crystals of **rhombic sulphur** are octahedral; those of **monoclinic sulphur** are needle-shaped. Rhombic and monoclinic sulphur are **allotropes** of sulphur: the only difference between them is the shape of their crystals (see Figure 4.8A). The reason why the crystals are shaped differently is that the sulphur molecules are packed into different arrangements in the allotropes.

Did you know?

In 1985, new allotropes of carbon were discovered. They consist of molecules containing 30–70 carbon atoms. The allotropes are known as **fullerenes**. The structure of a fullerene molecule is a closed cage of carbon atoms. The most symmetrical structure is buckminsterfullerene, C_{60}, which is a perfect sphere of carbon atoms. It is named after the geodesic domes designed by the architect Buckminster Fuller. The atoms are bonded together in 20 hexagons and 12 pentagons, which fit together like those on the surface of a football. Fullerenes have been nicknamed 'bucky balls'.

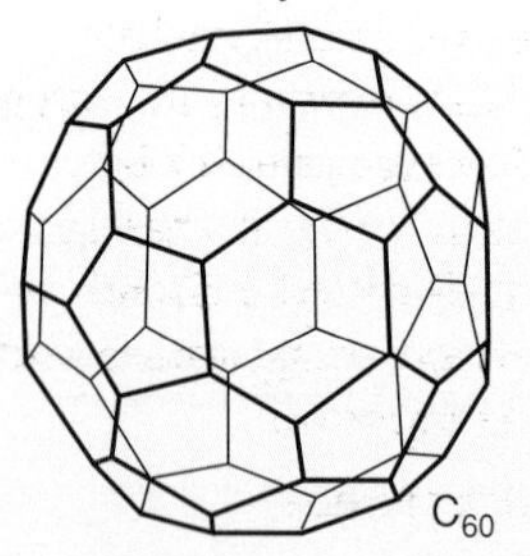

Scientists believe they will find important applications as semiconductors, superconductors, lubricants, catalysts and in batteries.

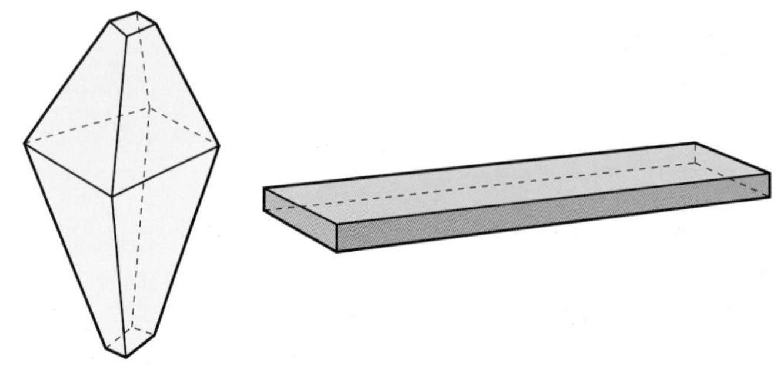

Figure 4.8A Allotropes of sulphur (×20)

Figure 4.8B shows a model of diamond. The structure is described as **macromolecular** or **giant molecular**. You can see that the carbon atoms form a regular arrangement. A crystal of diamond contains millions of carbon atoms arranged in this way. Every carbon atom is joined by chemical bonds to four other carbon atoms. It is very difficult to break this structure. The macromolecular structure is the source of diamond's hardness.

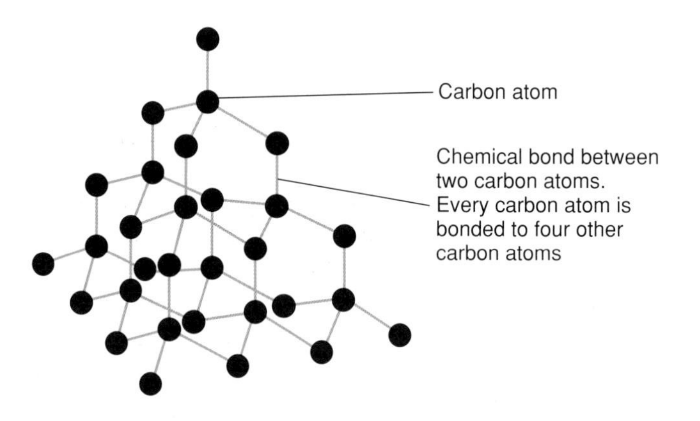

Figure 4.8B The arrangement of carbon atoms in diamonds

Figure 4.8C shows a model of graphite. Like diamond, graphite is a crystalline solid with a macromolecular structure. Graphite has a **layer structure**. Within each layer, the carbon atoms are joined by chemical bonds. Between layers, there are only weak forces of attraction. These weak forces allow one layer to slide over the next layer. This is why graphite is soft and rubs off on your fingers. The structure of graphite enables it to be used as a lubricant and in pencil 'leads' to mark paper.

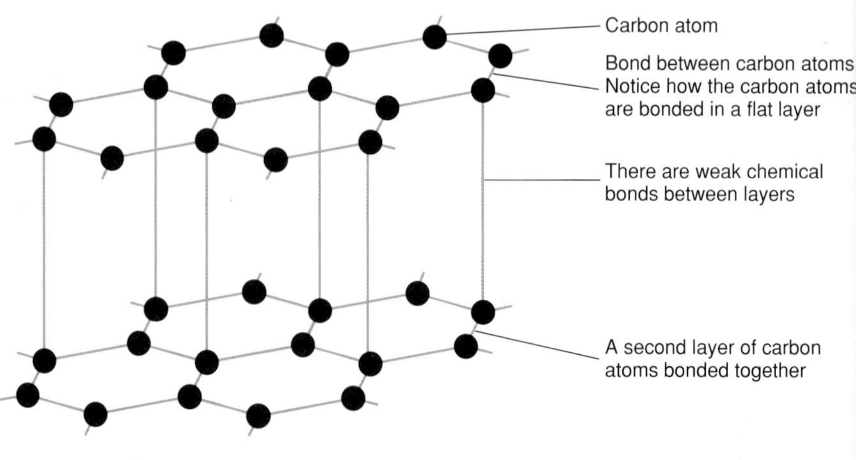

Figure 4.8C The arrangement of carbon atoms in graphite

SUMMARY

The carbon atoms in diamond are joined by chemical bonds to form a giant molecular structure.

CHECKPOINT

1. (a) What kind of atoms are there in a diamond?
   (b) What kind of structure do the atoms form?
   (c) Why does this structure make diamond a hard substance?
   (d) What uses of diamond depend on its hardness?
   (e) Why are diamonds often chosen for engagement rings?
2. (a) Explain how its structure makes graphite less hard than diamond.
   (b) Why are diamond and graphite called allotropes?
   (c) What is graphite used for?
3. Say how the allotropes differ and how they resemble one another.

4.9 Compounds

FIRST THOUGHTS

Most of the substances you see around you are not elements: they are compounds. Compounds can be made from elements by chemical reactions.

The Verey distress rocket contains the metallic element magnesium. When it is heated – when the fuse is lit – magnesium burns in the oxygen of the air. It burns with a brilliant white flame. A white powder is formed. This powder is the compound, magnesium oxide. A change which results in the formation of a new substance is called a **chemical reaction**. A chemical reaction has taken place between magnesium and oxygen. The elements have combined to form a compound.

Magnesium + Oxygen → Magnesium oxide
Element + Element → Compound

A compound is a pure substance which contains two or more elements chemically combined. A compound of oxygen and one other element is called an **oxide**. Making a compound from its elements is called **synthesis**.

Have you ever burned magnesium ribbon? You were synthesising magnesium oxide.

Chemical reactions can synthesise compounds, and chemical reactions can also decompose (split up) compounds. Some compounds can be decomposed into their elements by heat. An example is silver oxide (see Figure 4.9A). The chemical reaction that takes place is

Silver oxide $\xrightarrow{\text{Heat}}$ Silver + Oxygen
Compound   Elements

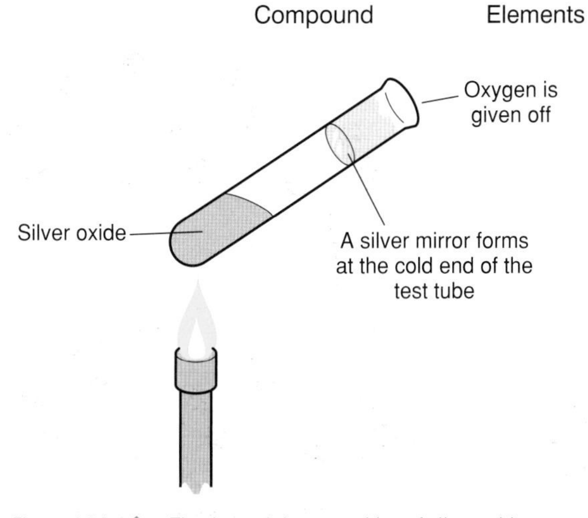

Figure 4.9A The thermal decomposition of silver oxide

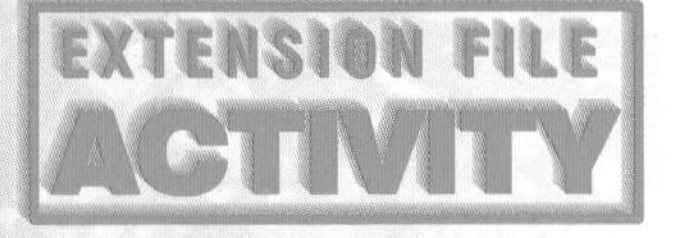

Splitting up a compound by heat is called **thermal decomposition**.

Some compounds can be decomposed into their elements by the passage of a direct electric current. An example is sodium chloride (common salt) which is a compound of sodium and chlorine. A compound of chlorine with one other element is called a **chloride**.

Some compounds can be decomposed by heat, e.g. silver oxide.

Some can be decomposed by a direct electric current, e.g. sodium chloride (when molten) and water.

Sodium chloride Compound —(Pass a direct electric current through the molten compound)→ Sodium + Chlorine Elements

The chemical reaction that occurs when a compound is split up by means of electricity is called **electrolysis**.

Water is a compound. It can be electrolysed to give the elements hydrogen and oxygen (see Figure 4.9B).

Water is a compound of hydrogen and oxygen. You could call it hydrogen oxide. It is possible to make water by a chemical reaction between hydrogen and oxygen.

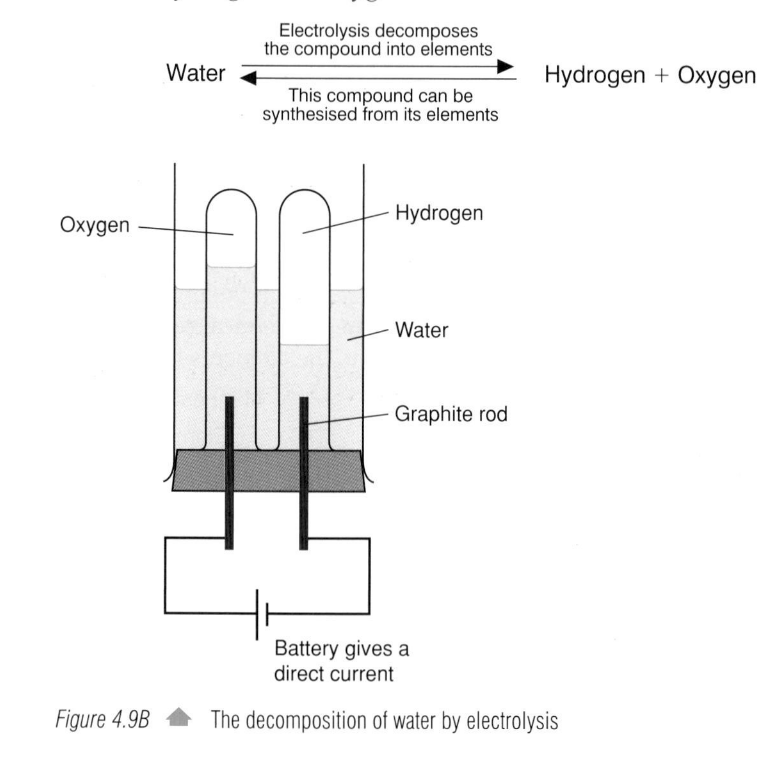

Figure 4.9B ▲ The decomposition of water by electrolysis

SUMMARY

Elements combine to form compounds. Some compounds can be split up by heat in thermal decomposition. Some compounds are decomposed by electrolysis (the passage of a direct electric current through the molten compound or a solution of the compound). Molten common salt can be electrolysed to give the elements sodium and chlorine. Water can be electrolysed to give the elements hydrogen and oxygen.

CHECKPOINT

1 In a chemical change, a new substance is formed. In a physical change, no new substance is formed. Say which of the following changes are chemical changes.
   (a) Evaporating water.
   (b) Electrolysing water.
   (c) Melting wax.
   (d) Cracking an egg.
   (e) Boiling an egg.

2 Which of the following brings about a chemical change?
   (a) Heating magnesium.
   (b) Heating silver oxide.
   (c) Heating sodium chloride.
   (d) Passing a direct electric current through copper wire.
   (e) Passing a direct electric current through molten salt.

4.10 Mixtures and compounds

FIRST THOUGHTS

Try to explain the difference between a mixture and a compound. Then read this section, and see whether you are right.

There are a number of differences between a compound and a mixture of elements. A mixture can contain its components in any proportions. A compound has a **fixed composition**. It always contains the same elements in the same percentages by mass. You can mix together iron filings and powdered sulphur in any proportions, from 1% iron and 99% sulphur to 99% iron and 1% sulphur. When a chemical reaction takes place between iron and sulphur, you get a compound called iron(II) sulphide. It always contains 64% iron and 36% sulphur by mass. Table 4.2 summarises the differences between mixtures and compounds.

Table 4.2 ▼ Differences between mixtures and compounds

Mixtures	Compounds
A mixture can be separated into its parts by methods such as distillation and dissolving.	A chemical reaction is needed to split a compound into simpler compounds or into its elements.
No chemical change takes place when a mixture is made.	When a compound is made, a chemical reaction takes place, and often heat is given out or taken in.
A mixture behaves in the same way as its components.	A compound does not have the characteristics of its elements. It has a new set of characteristics.
A mixture can contain its components in any proportions.	A compound always contains its elements in fixed proportions by mass; for example, calcium carbonate (marble) always contains 40% calcium, 12% carbon and 48% oxygen by mass.

SUMMARY

A compound is a pure substance which consists of two or more elements chemically combined. The components of a mixture are not chemically combined. Table 4.2 lists the differences between mixtures and compounds.

CHECKPOINT

1. Group the following into mixtures, compounds and elements: rain water, sea water, common salt, gold dust, aluminium oxide, ink, silicon, air.
2. Name an element which can be used for each of the following uses: surgical knife, pencil 'lead', wedding ring, saucepan, crowbar, thermometer, plumbing, electrical wiring, microcomputer circuit, disinfecting swimming pools, fireworks, artists' sketching material.
3. What are the differences between a mixture of iron and sulphur and the compound iron sulphide?
4. (a) What happens when you connect a piece of copper wire across the terminals of a battery? What happens when you disconnect the wire from the battery? Is the copper wire the same as before or has it changed?
   (b) What happens when you hold a piece of copper in a Bunsen flame for a minute and then switch off the Bunsen? Is the copper wire the same or different?
   (c) What type of change or changes occur in (a) and (b)?

Topic 5 Symbols, formulas, equations

5.1 Symbols

FIRST THOUGHTS

Molly had a little dog
Her dog don't bark no more
'Cos what she thought was H_2O
Was really H_2SO_4

Anon

This rhyme shows the importance of getting your formulas right!

SUMMARY

The symbol of an element is a letter or two letters which stand for one atom of the element.

For every element there is a **symbol**. For example, the symbol for sulphur is S. The letter S stands for one atom of sulphur. Sometimes, two letters are needed. The letters Si stand for one atom of silicon: the symbol for silicon is Si. The symbol of an element is a letter or two letters which stand for one atom of the element. In some cases the letters are taken from the Latin name of the element, Ag from *argentum* (silver) and Pb from *plumbum* (lead) are examples. Table 5.1 gives a short list of symbols. There is a complete list at the end of the book.

Table 5.1 The symbols of some carbon elements

Element	Symbol	Element	Symbol	Element	Symbol
Aluminium	Al	Gold	Au	Oxygen	O
Barium	Ba	Hydrogen	H	Phosphorus	P
Bromine	Br	Iodine	I	Potassium	K
Calcium	Ca	Iron	Fe	Silver	Ag
Carbon	C	Lead	Pb	Sodium	Na
Chlorine	Cl	Magnesium	Mg	Sulphur	S
Copper	Cu	Mercury	Hg	Tin	Sn
Fluorine	F	Nitrogen	N	Zinc	Zn

5.2 Formulas

Figure 5.2A A model of one molecule of carbon dioxide

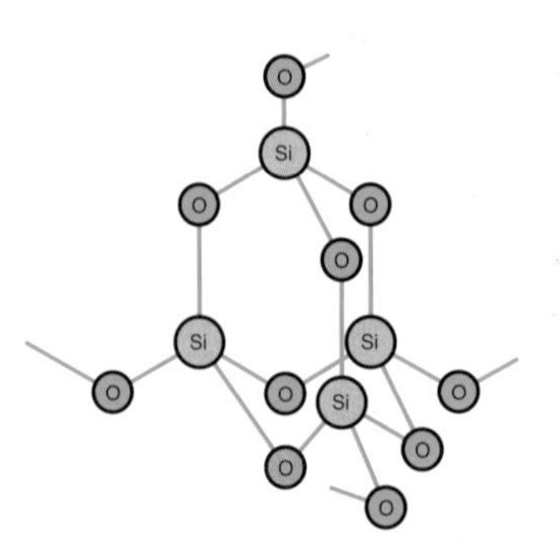

Figure 5.2B The structure of silicon(IV) oxide

www.keyscience.co.uk

For every compound there is a formula. The formula of a compound contains the symbols of the elements present and some numbers. The numbers show the ratio in which atoms are present. The compound carbon dioxide consists of molecules. Each molecule contains one atom of carbon and two atoms of oxygen. The formula of the compound is CO_2. The 2 below the line multiplies the O in front of it. To show three molecules of carbon dioxide you write $3CO_2$.

Sand is impure silicon dioxide (also called silicon(IV) oxide). It consists of macromolecules, which contain millions of atoms. There are twice as many oxygen atoms as silicon atoms in the macromolecule. The formula of silicon dioxide is therefore SiO_2.

The formulas of some of the compounds mentioned in this chapter are:

- Water, H_2O (two H atoms and one O atom; the 2 multiplies the H in front of it).
- Sodium chloride, NaCl (one Na; one Cl).
- Silver oxide, Ag_2O (two Ag; one O).
- Iron(II) sulphide, FeS (one Fe; one S).

Don't be one of those people who say that formulas are difficult!
H_2O = two H atoms + one O atom
H_2SO_4 = two H atoms + one S atom + four O atoms
Is that difficult?

The formula for aluminium oxide is Al_2O_3. This tells you that the compound contains two aluminium atoms for every three oxygen atoms. The numbers below the line multiply the symbols immediately in front of them.

The formula for calcium hydroxide is $Ca(OH)_2$. The 2 multiplies the symbols in the brackets. There are 2 oxygen atoms, 2 hydrogen atoms and 1 calcium atom. To write $4Ca(OH)_2$ means that the whole of the formula is multiplied by 4. It means 4Ca, 8O and 8H atoms. Table 5.2 lists the formulas of some common compounds.

Table 5.2 The formulas of some common compounds

Compound	Formula
Water	H_2O
Carbon monoxide	CO
Carbon dioxide	CO_2
Sulphur dioxide	SO_2
Hydrogen chloride	HCl
Hydrochloric acid	HCl(aq)
Sulphuric acid	$H_2SO_4(aq)$
Nitric acid	$HNO_3(aq)$
Sodium hydroxide	NaOH
Sodium chloride	NaCl
Sodium sulphate	Na_2SO_4
Sodium nitrate	$NaNO_3$
Sodium carbonate	Na_2CO_3
Sodium hydrogencarbonate	$NaHCO_3$
Calcium oxide	CaO
Calcium hydroxide	$Ca(OH)_2$
Calcium chloride	$CaCl_2$
Calcium sulphate	$CaSO_4$
Calcium carbonate	$CaCO_3$
Calcium hydrogencarbonate	$Ca(HCO_3)_2$
Copper(II) oxide	CuO
Copper(II) sulphate	$CuSO_4$
Aluminium chloride	$AlCl_3$
Aluminium oxide	Al_2O_3
Ammonia	NH_3
Ammonium chloride	NH_4Cl
Ammonium sulphate	$(NH_4)_2SO_4$

SUMMARY

The formula of a compound is a set of symbols and numbers. The symbols show which elements are present in the compound. The numbers give the ratio in which the atoms of different elements are present.

5.3 Valency

Some atoms can form only one chemical bond. They can therefore combine with only one other atom. Elements with such atoms are said to have a **valency** of one. Hydrogen has a valency of one, and chlorine has a valency of one.

H—

An atom of hydrogen can form one bond
Hydrogen has a valency of 1

Cl—

An atom of chlorine can form one bond
Chlorine has a valency of 1

You need to know how many bonds an atom of an element can form. Then the formulas of its compounds can be worked out.

—O—

One atom of oxygen can form two bonds
Valency of oxygen = 2

—N— (with one bond upwards)

One atom of nitrogen can form three bonds
Valency of nitrogen = 3

—C— (with one bond upwards and one downwards)

One atom of carbon can form four bonds
Valency of carbon = 4

The formula of a compound depends on the valencies of the elements in the compound. When hydrogen combines with chlorine, oxygen, nitrogen and carbon, the compounds formed have the formulas:

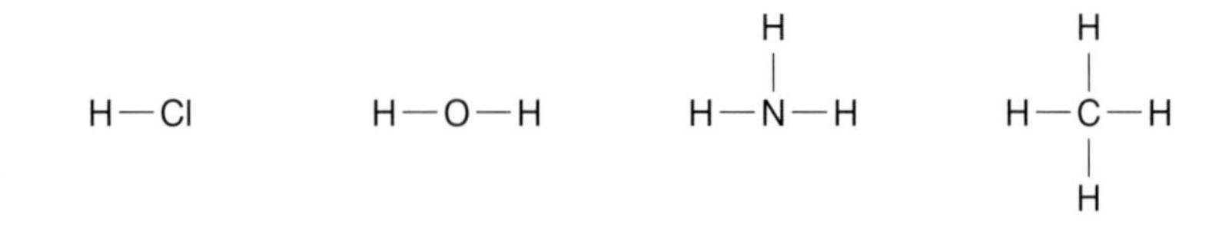

These are the compounds hydrogen chloride (HCl), water (H_2O), ammonia (NH_3), and methane (CH_4). They have different formulas because the valencies of the elements Cl, O, N and C are different.

The photographs show molecules of the compounds formed with hydrogen
by chlorine with 1 bond,
by oxygen with 2 bonds,
by nitrogen with 3 bonds
and by carbon with 4 bonds.

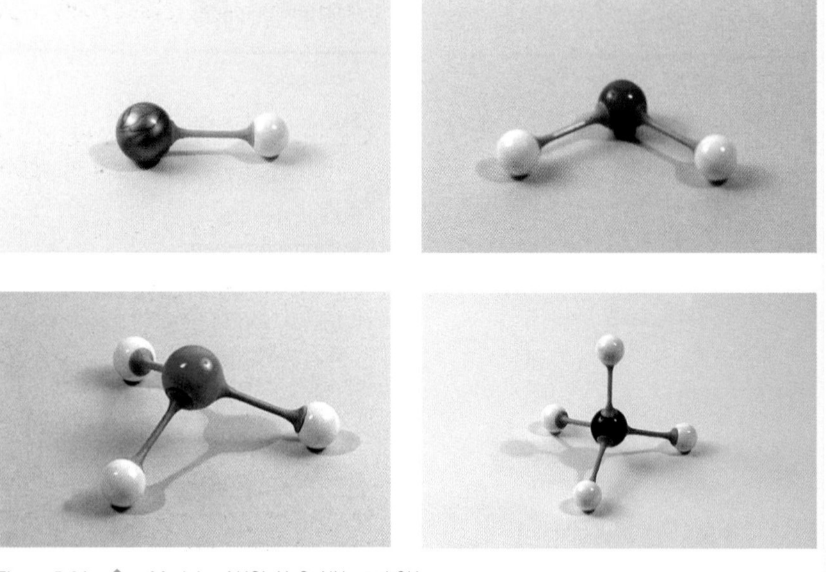

Figure 5.3A Models of HCl, H_2O, NH_3 and CH_4

Some elements have more than one valency. Sulphur, for example, forms compounds in which it has a valency of 2, e.g. H_2S, compounds in which it has a valency of 4, e.g. SCl_4, and compounds in which it has a valency of 6, e.g. SF_6.

SUMMARY

The formula of a compound depends on the valencies of the elements in the compound.

5.4 Equations

FIRST THOUGHTS

An equation tells what happens in a chemical reaction.
A word equation gives the names of the reactants and the products.
A chemical equation gives their symbols and formulas.

In a chemical reaction, the starting materials, the **reactants**, are changed into new substances, the **products**. The atoms present in the reactants are not changed in any way, but the bonds between the atoms change. Chemical bonds are broken, and new chemical bonds are made. The atoms enter into new arrangements as the products are formed. Symbols and formulas give us a nice way of showing what happens in a chemical reaction. We call this way of describing a chemical reaction a **chemical equation**.

Example 1 Copper and sulphur combine to form copper sulphide. Writing a **word equation** for the reaction

Copper + Sulphur → Copper sulphide

The arrow stands for **form**.

Writing the symbols for the elements and the formula for the compound gives the chemical equation

$$Cu + S \rightarrow CuS$$

Adding the state symbols

$$Cu(s) + S(s) \rightarrow CuS(s)$$

Why is this called an equation? The two sides are equal. On the left hand side, we have one atom of copper and one atom of sulphur; on the right hand side, we have one atom of copper and one atom of sulphur combined as copper sulphide. The atoms on the left hand side and the atoms on the right hand side are the same in kind and in number.

Example 2 Calcium carbonate decomposes when heated to give calcium oxide and carbon dioxide. The word equation is

Calcium carbonate → Calcium oxide + Carbon dioxide

The chemical equation is

$$CaCO_3(s) \rightarrow CaO(s) + CO_2(g)$$

Example 3 Carbon burns in oxygen to form the gas carbon dioxide. The word equation is

Carbon + Oxygen → Carbon dioxide

We must use the formula O_2 for oxygen because oxygen consists of molecules which contain two oxygen atoms. The chemical equation is

$$C(s) + O_2(g) \rightarrow CO_2(g)$$

SUMMARY

How to write a balanced chemical equation;

- Write the word equation.
- Put in the symbols of the elements and the formulas of the compounds.
- Add the state symbols.
- Balance the equation. Do this by multiplying symbols or formulas.
  Never change a formula.
- Check again;
  no. of atoms of each element on LHS = no. of atoms of each element on RHS

Example 4 Magnesium burns in oxygen to form the solid magnesium oxide.

$$\text{Magnesium} + \text{Oxygen} \rightarrow \text{Magnesium oxide}$$

$$Mg(s) + O_2(g) \rightarrow MgO(s)$$

There is something wrong here! The two sides are not equal. The left hand side has two atoms of oxygen; the right hand side has only one. Multiplying MgO by 2 on the right hand side should fix it

$$Mg(s) + O_2(g) \rightarrow 2MgO(s)$$

Now there are two oxygen atoms on both sides, but there are two magnesium atoms on the right hand side and only one on the left hand side. Multiply Mg by 2

$$2Mg(s) + O_2(g) \rightarrow 2MgO(s)$$

The equation is now **a balanced chemical equation**. Check up. On the left hand side, number of Mg atoms = 2; number of O atoms = 2. On the right hand side, number of Mg atoms = 2; number of O atoms = 2. The equation is balanced.

CHECKPOINT

1. Refer to the table of elements at the end of the book (i.e. the Periodic Table). Write down the names and symbols of the elements with atomic numbers 8, 10,20,24,38,47,50,80,82. Say what each of these elements is used for. (The term 'atomic number' will be explained in Topic 6.)
2. Give meanings of the state symbols (s), (l), (g), (aq). (See Topic 1.1 if you need to revise.)
3. Try writing balanced chemical equations for the following reactions.
   (a) Zinc and sulphur combine to form zinc sulphide.
   (b) Copper reacts with oxygen to form copper(II) oxide.
   (c) Sulphur and oxygen form sulphur dioxide.
   (d) Magnesium carbonate decomposes to form magnesium oxide and carbon dioxide.
   (e) Hydrogen and copper(II) oxide form copper and water.
   (f) Carbon and carbon dioxide react to form carbon monoxide.
   (g) Magnesium reacts with sulphuric acid to form hydrogen and magnesium sulphate.
   (h) Calcium reacts with water to form hydrogen and a solution of calcium hydroxide.
   (i) Zinc reacts with steam to form hydrogen and zinc oxide.
   (j) Aluminium and chlorine react to form aluminium chloride.
4. Write balanced chemical equations for the following reactions.
   (a) Zinc and sulphur combine to form zinc sulphide, ZnS
   (b) Copper and chlorine combine to form copper(II) chloride, $CuCl_2$
   (c) Sulphur burns in oxygen to form sulphur dioxide, SO_2
   (d) Magnesium carbonate decomposes to form magnesium oxide and carbon dioxide
   (e) Calcium burns in oxygen to form calcium oxide.
5. How many atoms are present in the following?
   (a) $CaCl_2$
   (b) $3CaCl_2$
   (c) $Cu(OH)_2$
   (d) $5Cu(OH)_2$
   (e) H_2SO_4
   (f) $3H_2SO_4$
   (g) $2NaNO_3$
   (h) $3Cu(NO_3)_2$

Topic 6 Inside the atom

6.1 Becquerel's key

FIRST THOUGHTS

This topic will give you a glimpse of the fascinating story of the discovery of radioactivity.

In 1896, a French physicist, Henri Becquerel, left some wrapped photographic plates in a drawer. When he developed the plates, he found the image of a key. The plates were 'fogged' (partly exposed). The areas of the plates which had not been exposed were in the shape of a key. Looking in the drawer, Becquerel found a key and a packet containing some uranium compounds. He did some further tests before coming to a strange conclusion. He argued that some unknown rays, of a type never met before, were coming from the uranium compounds. The mysterious rays passed through the wrapper and fogged the photographic plates. Where the key lay over the plates, the rays could not penetrate, and the image formed on the plates.

Figure 6.1A Becquerel's key

A young research worker called Marie Curie took up the problem in 1898. She found that this strange effect happened with all uranium compounds. It depended only on the amount of uranium present in the compound and not on which compound she used. Madame Curie realised that this ability to give off rays must belong to the 'atoms' of uranium. It must be a completely new type of change, different from the chemical reactions of uranium salts. This was a revolutionary new idea. Marie Curie called the ability of uranium atoms to give off rays **radioactivity**.

Marie Curie's husband, Pierre, joined in her research into this brand new branch of science. Together, they discovered two new radioactive elements. They called one **polonium**, after Madame Curie's native country, Poland. They called the second **radium**, meaning 'giver of rays'. Its salts glowed in the dark.

Many scientists puzzled over the question of why the atoms of these elements, uranium, polonium and radium, give off the rays which Marie Curie named radioactivity. The person who came up with an explanation was the British physicist, Lord Rutherford. In 1902 he suggested that radioactivity is caused by atoms splitting up. This was another revolutionary idea. The word 'atom' comes from the Greek word for 'cannot be divided'. When the British chemist John Dalton put forward his Atomic Theory in 1808, he said that atoms cannot be created or destroyed or split. Lord Rutherford's idea was proved by experiment to be correct. We know now that many elements have atoms which are unstable and split up into smaller atoms.

KEY SCIENTIST

The Curies worked for four years in a cold, ill-equipped shed at the University of Paris. From a tonne of ore from the uranium mine, Madame Curie extracted a tenth of a gram of uranium. The Curies published their work in research papers and exchanged information with leading scientists in Europe. A year later, they were awarded the Nobel prize, the highest prize for scientific achievement. Pierre Curie died in a road accident. Marie Curie went on with their work and won a second Nobel prize. She died in middle-age from leukaemia, a disease of the blood cells. This was caused by the radioactive materials she worked with.

6.2 Protons, neutrons and electrons

FIRST THOUGHTS

Atoms are made up of even smaller particles: protons, neutrons and electrons. As you read this section try to visualise how these particles are arranged inside the atom.

www.keyscience.co.uk

The work of Marie and Pierre Curie, Rutherford and other scientists showed that atoms are made up of smaller particles. These **subatomic particles** differ in mass and in electrical charge. They are called **protons**, **neutrons** and **electrons** (see Table 6.1).

Table 6.1 Sub-atomic particles

Particle	Mass (in atomic mass units)	Charge
Proton	1	$+e$
Neutron	1	0
Electron	0.0005	$-e$

Protons and neutrons both have the same mass. We call this mass one **atomic mass unit**, one u ($1.000\,u = 1.67 \times 10^{-27}\,kg$). The mass of an atom depends on the number of protons and neutrons it contains. The electrons in an atom contribute very little to its mass. The number of protons and neutrons together is called the **mass number**.

Electrons carry a fixed quantity of negative electric charge. This quantity is usually written as $-e$. A proton carries a fixed charge equal and opposite to that of the electron. The charge on a proton can be written as $+e$. Neutrons are uncharged particles. Whole atoms are uncharged because the number of electrons in an atom is the same as the number of protons. The number of protons (which is also the number of electrons) is called either the **atomic number** or the **proton number**. You can see that

$$\text{Number of neutrons} = \text{Mass number} - \text{Atomic (proton) number}$$

For example, an atom of potassium has a mass of 39 u and an atomic (proton) number of 19. The number of electrons is 19, the same as the number of protons. The number of neutrons in the atom is

$$39 - 19 = 20$$

KEY SCIENTIST

Rutherford had an easier life in science than the Curies. He arrived in Britain from New Zealand in 1895. By the age of 28, he was a professor. He worked in the universities of Montreal in Canada, Manchester and Cambridge. He was knighted in 1914 and made Lord Rutherford of Nelson in 1931. His co-worker Otto Hahn described him as 'a very jolly man'. Many stories were told about Rutherford. He would whistle 'Onward, Christian soldiers' when the research work was going well and 'Fight the good fight' when difficulties had to be overcome.

Relative atomic mass

The lightest of atoms is an atom of hydrogen. It consists of one proton and one electron. Chemists compared the masses of other atoms with that of a hydrogen atom. They use **relative atomic mass**. The relative atomic mass, A_r, of calcium is 40. This means that one calcium atom is 40 times as heavy as one atom of hydrogen.

$$\text{Relative atomic mass of an element} = \frac{\text{Mass of one atom of the element}}{\text{Mass of one atom of hydrogen}}$$

CHECKPOINT

1 Some relative atomic masses are:

$A_r(H) = 1$, $A_r(He) = 4$, $A_r(C) = 12$, $A_r(O) = 16$, $A_r(Ca) = 40$.

Copy and complete the following sentences.

(a) A calcium atom is ______ times as heavy as an atom of hydrogen.
(b) A calcium atom is ______ times as heavy as an atom of helium.
(c) One carbon atom has the same mass as ______ helium atoms.
(d) ______ helium atoms have the same mass as two oxygen atoms.
(e) Two calcium atoms have the same mass as ______ oxygen atoms.

2 Element E has atomic number 9 and mass number 19. Say how many protons, neutrons and electrons are present in one atom of E.

3 State (i) the atomic number and (ii) the mass number of:

(a) an atom with 17 protons and 18 neutrons,
(b) an atom with 27 protons and 32 neutrons,
(c) an atom with 50 protons and 69 neutrons.

6.3 The arrangement of particles in the atom

It's a fact!

If the nucleus of an atom was the size of a cricket ball, the nearest electron would be in the stands.

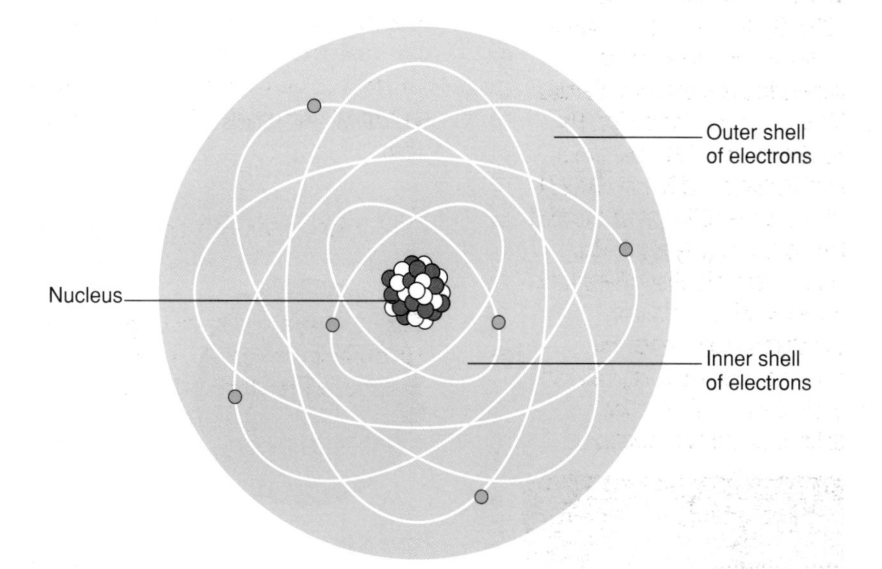

Figure 6.3A ▲ The structure of an atom

Lord Rutherford showed, in 1914, that most of the volume of an atom is space. Only protons and electrons were known in 1914; the neutron had not yet been discovered. Rutherford pictured the massive particles, the protons, occupying a tiny volume in the centre of the atom. Rutherford called this the **nucleus**. We now know that the nucleus contains neutrons as well as protons. The electrons occupy the space outside the nucleus. The nucleus is minute in volume compared with the volume of the atom.

The electrons of an atom are in constant motion. They move round and round the nucleus in paths called **orbits**. The electrons in orbits close to the nucleus have less energy than electrons in orbits distant from the nucleus.

6.4 How are the electrons arranged?

Figure 6.4A illustrates the electrons of an atom in their orbits. The orbits are grouped together in **shells**. A shell is a group of orbits with similar energy. The shells distant from the nucleus have more energy than those close to the nucleus. Each shell can hold up to a certain number of electrons. In any atom, the maximum number of electrons in the outermost group of orbits is eight.

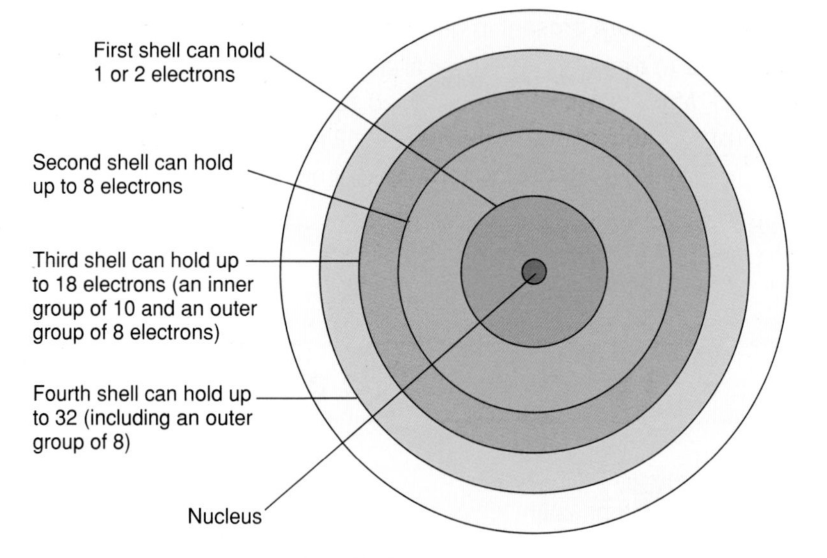

Figure 6.4A Shells of electron orbits in an atom

Electrons move round the nucleus in paths called orbits. A group of orbits of similar energy is called a shell.

Electrons need more energy to move in distant orbits than in orbits close to the nucleus.

The number of electrons in an atom = atomic number

The atomic number of an element tells you the number of electrons in an atom of the element. The electrons fill the innermost orbits in the atom first. An atom of oxygen has 8 electrons. Two electrons enter the first shell, which is then full. The other 6 electrons go into the second shell (see Figure 6.4B).

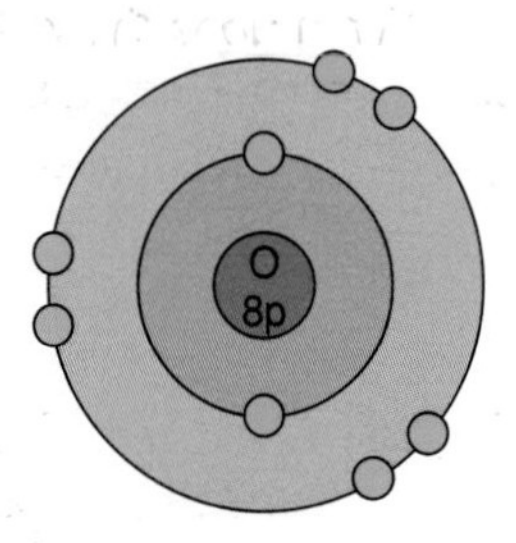

Figure 6.4B The arrangement of electrons in the oxygen atom

The diagram shows the arrangement of electrons in the different shells of the oxygen atom and the sodium atom. These arrangements are called electronic configurations.

An atom of sodium has atomic number 11. The first shell is filled by 2 electrons, the second shell is filled by 8 electrons, and 1 electron occupies the third shell. The arrangement of electrons can be written as (2.8.1). It is called the **electron configuration** of sodium. Table 6.2 gives the electron configurations of the first 20 elements.

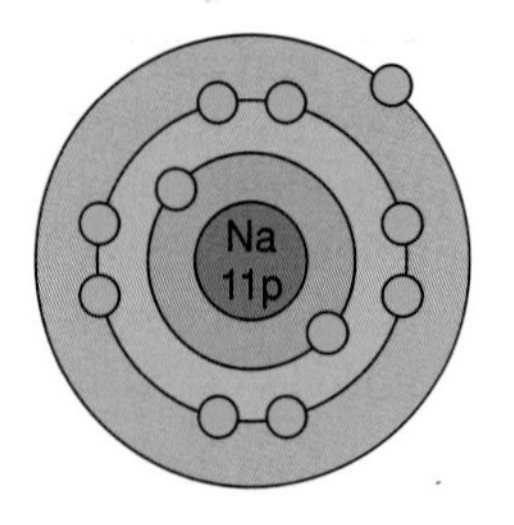

Figure 6.4C The electron configuration of sodium

SUMMARY

The atoms of all elements are made up of three kinds of particles. These are:

- protons, of mass 1 u and electric charge $+e$
- neutrons, of mass 1 u, uncharged
- electrons, of mass 0.0005 u and electric charge $-e$.

The protons and neutrons make up the nucleus at the centre of the atom. The electrons circle the nucleus in orbits. Groups of orbits with the same energy are called shells. The 1st shell can hold 2 electrons; the 2nd shell can hold 8 electrons; the 3rd shell can hold 18 electrons. The arrangement of electrons in an atom is called the electron configuration.

Table 6.2 Electron configurations of the atoms of the first twenty elements

Element	Symbol	Atomic (proton) number	Number of electrons in ... 1st shell	2nd shell	3rd shell	4th shell	Electron configuration
Hydrogen	H	1	1				1
Helium	He	2	2				2
Lithium	Li	3	2	1			2.1
Beryllium	Be	4	2	2			2.2
Boron	B	5	2	3			2.3
Carbon	C	6	2	4			2.4
Nitrogen	N	7	2	5			2.5
Oxygen	O	8	2	6			2.6
Fluorine	F	9	2	7			2.7
Neon	Ne	10	2	8			2.8
Sodium	Na	11	2	8	1		2.8.1
Magnesium	Mg	12	2	8	2		2.8.2
Aluminium	Al	13	2	8	3		2.8.3
Silicon	Si	14	2	8	4		2.8.4
Phosphorus	P	15	2	8	5		2.8.5
Sulphur	S	16	2	8	6		2.8.6
Chlorine	Cl	17	2	8	7		2.8.7
Argon	Ar	18	2	8	8		2.8.8
Potassium	K	19	2	8	8	1	2.8.8.1
Calcium	Ca	20	2	8	8	2	2.8.8.2

CHECKPOINT

1. State the meanings of the following terms: electron, proton, neutron, orbit, shell, electron configuration.
2. State the type of electric charge in (a) a proton (b) a neutron (c) an electron.
3. Silicon has the electron configuration (2.8.4). What does this tell you about the arrangement of electrons in the atom? Sketch the arrangement. (See Figures 6.4B and 6.4C for help.)
4. Sketch the arrangement of electrons in the atoms of (a) He (b) C (c) F (d) Al (e) Mg. (See Table 6.2 for atomic numbers.)
5. Copy this table, and fill in the missing numbers.

Particle	Mass number	Atomic number	Number of... protons	neutrons	electrons
Nitrogen atom	14	7	–	–	–
Sodium atom	23	–	–	–	11
Potassium atom	39	–	19	–	–
Uranium atom	235	92	–	–	–

6.5 Isotopes

FIRST THOUGHTS

There are two sorts of chlorine atom and three sorts of hydrogen atom. You can find out the difference between them in this section.

Atoms of the same element all contain the same number of protons, but the number of neutrons may be different. Forms of an element which differ in the number of neutrons in the atom are called **isotopes**. For example, the element chlorine, with relative atomic mass 35.5, consists of two kinds of atom with different mass numbers.

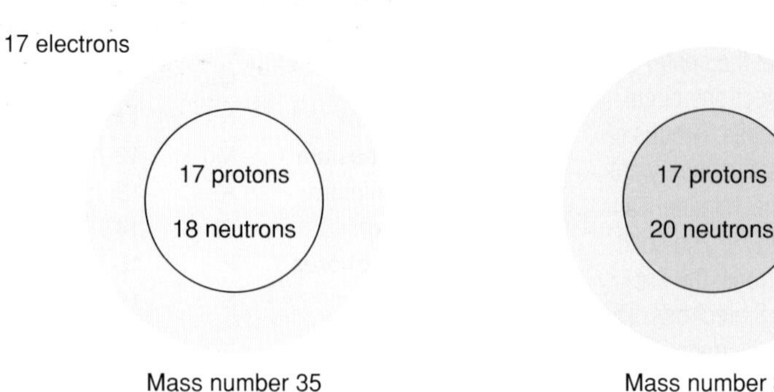

Figure 6.5A ▲ The isotopes of chlorine

Since the chemical reactions of an atom depend on its electrons, all chlorine atoms react in the same way. The number of neutrons in the nucleus does not affect chemical reactions. The different forms of chlorine are isotopes. Their chemical reactions are the same. In any sample of chlorine, there are three chlorine atoms with mass 35 u for each chlorine atom with mass 37 u so the average atomic mass is

$$\frac{(3 \times 35) + 37}{4} = 35.5\,\text{u}$$

This is why the relative atomic mass of chlorine is 35.5.

Isotopes are shown as:

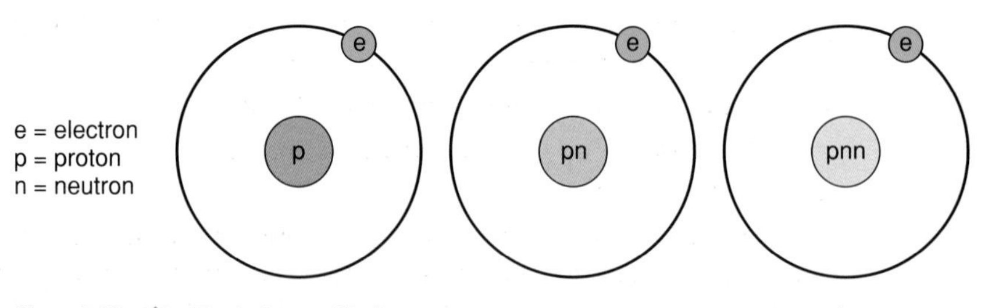

Figure 6.5B ▲ The isotopes of hydrogen

The isotopes of chlorine are written as $^{35}_{17}Cl$ and $^{37}_{17}Cl$. The isotopes of hydrogen are $^{1}_{1}H$, $^{2}_{1}H$ and $^{3}_{1}H$. They are often referred to as hydrogen-1, hydrogen-2 and hydrogen-3. Hydrogen-2 is also called deuterium, and hydrogen-3 is also called tritium. Carbon has three isotopes, carbon-12, carbon-13 and carbon-14.

SUMMARY

Isotopes are atoms of the same element which differ in the number of neutrons. They contain the same number of protons (and therefore the same number of electrons).

KEY SCIENTISTS

Otto Hahn was a German scientist. In 1939, he experimented with firing a beam of neutrons at atoms of the isotope uranium-235. He was attempting to make a new element with atoms larger than uranium-235. However the results were not what he expected: the evidence suggested that smaller atoms had been produced. This was a puzzle, and Hahn wrote to his former co-worker Lise Meitner, describing his results. She had been forced to flee from Germany to Sweden to escape the persecution of Jews in Nazi Germany. Meitner was able to suggest an explanation of Hahn's results in terms of nuclear fission – atom splitting. A neutron collides with a nucleus of uranium-235 and splits it into two smaller atoms and two neutrons, with the release of energy. The energy released in nuclear fission is enormous. Our nuclear power stations use this reaction to produce electricity.

CHECKPOINT

1 Hydrogen, deuterium and tritium are isotopes.

(a) Copy and complete this sentence.

Isotopes are _____ of an element which contain the same number of _____ and _____ but different numbers of _____.

(b) Copy and complete the table.

	Hydrogen	*Deuterium*	*Tritium*
Atomic number			
Mass number			

(c) Write the formula of the compound formed when deuterium reacts with oxygen.

(d) Explain why isotopes have the same chemical reactions.

2 Write the symbol with mass number and atomic number (as above) for each of the following isotopes.

(a) oxygen with 8 protons and 8 neutrons

(b) argon with 18 protons and 22 neutrons

(c) bromine with 35 protons and 45 neutrons

(d) chromium with 24 protons and 32 neutrons

3 Two of the atoms described below have similar chemical properties. Which two are they?

Atom X contains 9 protons and 10 neutrons.

Atom Y contains 13 protons and 14 neutrons.

Atom Z contains 17 protons and 18 neutrons.

Explain your answer.

4 Strontium-90 is a radioactive isotope formed in nuclear reactors. It can be accumulated in the human body because it follows the same chemical pathway through the body as another element X which is essential for health. After referring to the position of strontium in the Periodic Table, say which element you think is X.

5 Sodium has the electron arrangement (2.8.1).

(a) Draw and label a diagram to show how the protons, neutrons and electrons are arranged in a sodium atom.

(b) Explain why sodium is electrically uncharged.

Topic 7 Ions

7.1 Which substances conduct electricity?

FIRST THOUGHTS

Some elements and compounds conduct electricity. You can find out how they do it in this topic.

Electrical wires are often made of the metal copper. Copper is an **electrical conductor**: it allows electricity to pass through it. *Are all metals electrical conductors? Are there substances other than metals that conduct?* Figure 7.1A. shows how you can test a solid to find out whether it conducts electricity.

The diagrams show circuits which allow electricity to pass and light a bulb …

… if the solid conducts

… if the liquid conducts.

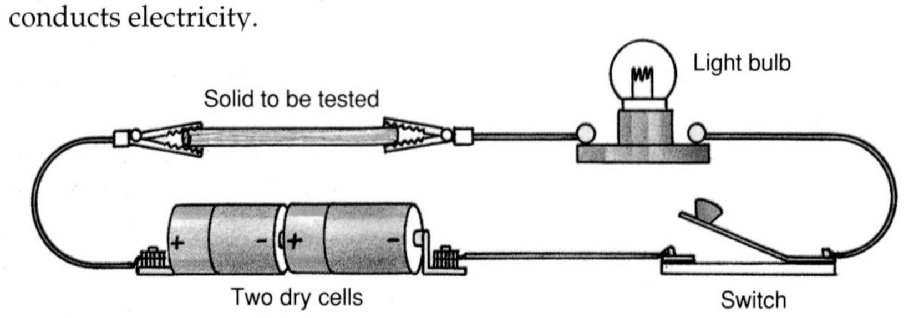

Figure 7.1A A testing circuit. If the solid conducts electricity, the bulb lights

Figure 7.1B shows a beaker of liquid. The two graphite rods in the liquid are **electrodes**: they can conduct a direct electric current into and out of the liquid. *Draw a* ***circuit diagram*** *like Figure 7.1A showing how you could test the liquid to see whether it conducts electricity.*

You should start your study of this topic by doing some experiments on conduction. *Are your results like those in Table 7.1?*

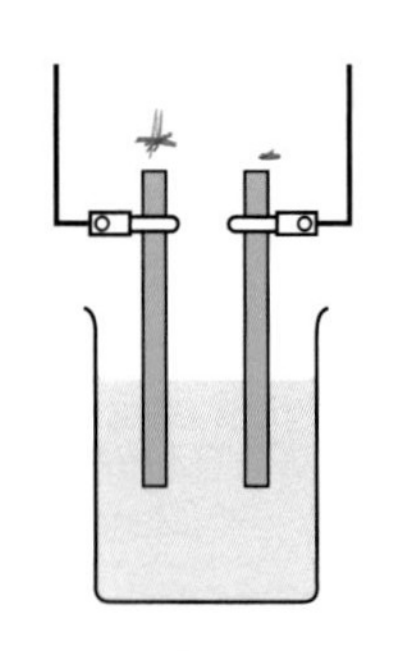

Figure 7.1B Test this liquid

Table 7.1 Electrical conductors

Solids	Liquids
Metallic elements	Mercury, the liquid metal
Alloys (mixtures of metals)	Solutions of acids, bases and salts
Graphite (a form of carbon)	Molten salts
(Solid compounds do not conduct.)	(Liquids such as ethanol and sugar solution do not conduct.)

SUMMARY

- **Solids.** Metallic elements, alloys and graphite (a form of carbon) conduct electricity.
- **Liquids.** Solutions of acids, alkalis and salts conduct electricity.

7.2 Why does electricity make some substances move?

www.keyscience.co.uk

Figure 7.2A shows an apparatus that can be used to study the movement of coloured substances in an electric field.

The following observations were made with this apparatus and a number of coloured salts, and explanations are given.

1 When a current is passed for 15 minutes with crystals of potassium manganate(VII) present, the purple colour can be seen moving slowly towards the positive electrode.

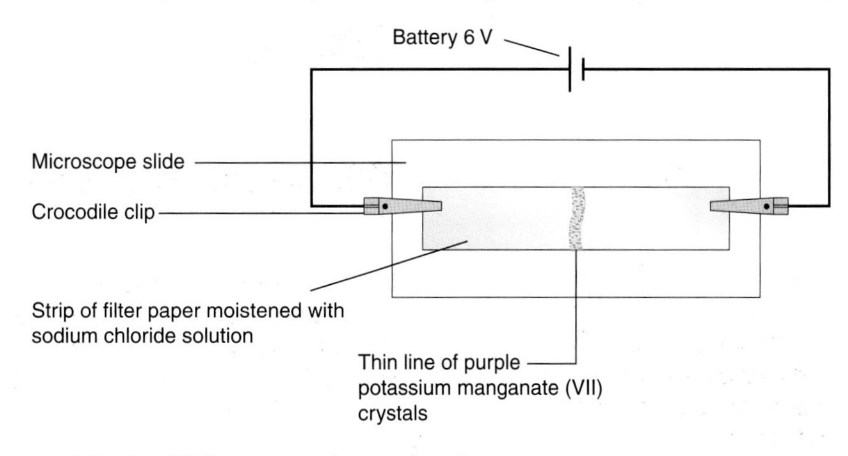

Figure 7.2A Will the coloured substances move?

Explanation: The movement of purple colour towards the positive electrode is due to the movement of purple particles with a negative charge. In fact, potassium manganate(VII) contains colourless potassium ions, K^+, and purple manganate(VII) ions, MnO_4^-.

2 When the experiment is repeated using a thin line of blue copper sulphate-5-water crystals, a blue colour moves towards the negative electrode.

Explanation: This is evidence that copper(II) sulphate contains blue positively charged copper ions, Cu^{2+}, and colourless sulphate ions, SO_4^{2-}.

EXTENSION FILE ACTIVITY

3 With potassium chromate(VI), a yellow colour moves towards the positive electrode.

Explanation: Potassium chromate(VI) consists of colourless, positive potassium ions, K^+, and yellow negative chromate(VI) ions, CrO_4^-.

4 With potassium dichromate(VI), an orange colour moves towards the positive electrode.

Explanation: Potassium dichromate(VI) consists of colourless, positive potassium ions, K^+, and orange negative dichromate(VI) ions, $Cr_2O_7^{2-}$.

5 With iron(II) sulphate, a green colour moves towards the negative electrode.

Explanation: In this case, iron(II) sulphate consists of green, positive iron(II) ions, Fe^{2+}, and colourless negative sulphate ions, SO_4^{2-}.

7.3 Molten solids and electricity

When a solid, e.g. a **metal**, conducts electricity, the current is carried by electrons. The battery forces the electrons through the conductor. The metal may change, for example it may become hot, but when the current stops flowing, the metal is just the same as before. When a metal conducts electricity, no chemical reaction occurs. When a molten salt conducts electricity, chemical changes occur, and new substances are formed.

Figure 7.3A shows an experiment to find out what happens when a direct electric current passes through a molten **salt**. A salt is a compound of a metallic element with a non-metallic element or elements. Lead(II)

When a metal conducts electricity no permanent change occurs.

Bromine and lead(II) bromide

TOXIC

When a molten salt conducts electricity the salt is decomposed – electrolysed – by the current. The diagram shows how molten lead(II) bromide can be electrolysed.

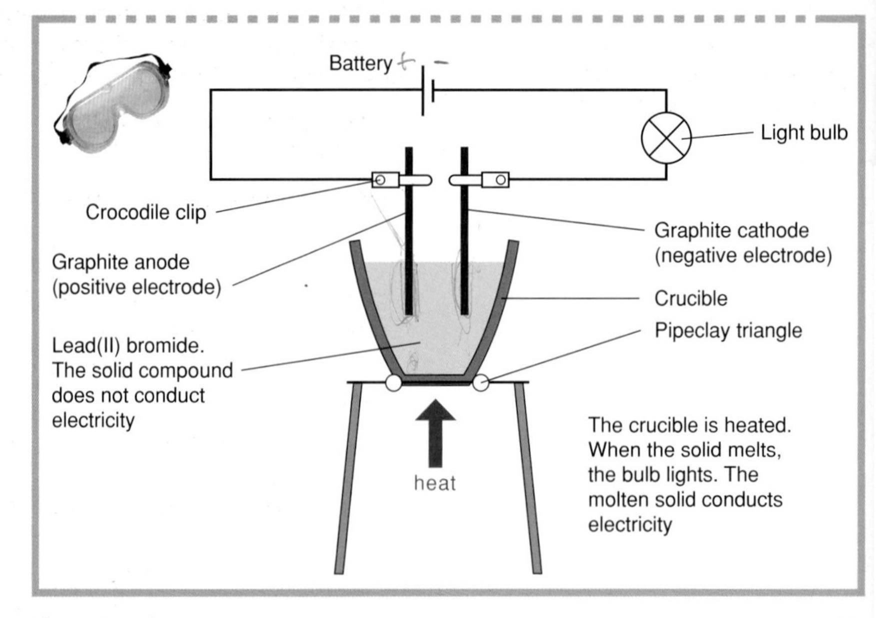

Figure 7.3A Passing a direct electric current through lead(II) bromide. This experiment should be done in a fume cupboard. (TAKE CARE: do not inhale the bromine vapour.)

Which electrode is which?
Ano**D**e
AD → ADD → + → positive
The anode is positive . . .
and the cathode is negative.

bromide is a salt with a fairly low melting point. The container through which the current passes is called a **cell**. The rods which conduct electricity into and out of the cell are called **electrodes**. The electrode connected to the positive terminal of the battery is called the **anode**. The electrode connected to the negative terminal is called the **cathode**. The electrodes are usually made of elements such as platinum and graphite, which do not react with electrolytes.

When the salt melts, the bulb lights, showing that the molten salt conducts electricity. At the positive electrode (anode), bromine can be detected. It is a non-metallic element, a reddish-brown vapour with a very penetrating smell. (TAKE CARE: do not inhale bromine vapour.) At the negative electrode (cathode), lead is formed. After cooling, a layer of lead can be seen on the cathode.

The experiment shows that lead(II) bromide has been split up by the electric current. It has been **electrolysed**. Compounds which conduct electricity are called **electrolytes**. Remember that all substances consist of particles (see Theme A, Topic 2). Since bromine goes only to the positive electrode, it follows that bromine particles have a negative charge. Since lead appears at the negative electrode only, it follows that lead particles have a positive charge. These charged particles are called **ions**. Positive ions are called **cations** because they travel towards the cathode. Negative ions are called **anions** because they travel towards the anode.

How do ions differ from atoms? A bromide ion, Br^-, differs from a bromine atom, Br, in having one more electron. The extra electron gives it a negative charge.

$$\text{Bromine atom} + \text{Electron} \rightarrow \text{Bromide ion}$$
$$Br + e^- \rightarrow Br^-$$

A lead(II) ion differs from a lead atom by having two fewer electrons. It therefore has a double positive charge, Pb^{2+}.

$$\text{Lead atom} \rightarrow \text{Lead ion} + 2\ \text{Electrons}$$
$$Pb \rightarrow Pb^{2+} + 2e^-$$

Check that you know the meaning of:
- cell
- electrode
- anode
- cathode
- anion
- cation
- electrolysis
- electrolyte.

At the electrodes ions are discharged. Electrons travel through the external circuit from the positive electrode (anode) to the negative electrode (cathode).

What happens when the ions reach the electrodes? They are **discharged**: they lose their charge. The positive electrode takes electrons from bromide ions so that they become bromine atoms.

Bromide ion ➔ Bromine atom + Electron (taken by positive electrode)

$Br^-(l) \rightarrow Br(g) + e^-$

Then bromine atoms pair up to form bromine molecules:

$$2Br(g) \rightarrow Br_2(g)$$

The negative electrode gives electrons to the positively charged lead ions so that they become lead atoms.

Lead ion + 2 Electrons (taken from the negative electrode ➔ Lead atom

$Pb^{2+}(l) + 2e^- \rightarrow Pb(l)$

The electrons which are supplied to the anode by the discharge of bromine ions travel round the external circuit to the cathode. At the cathode, they combine with lead(II) ions (see Figure 7.3B).

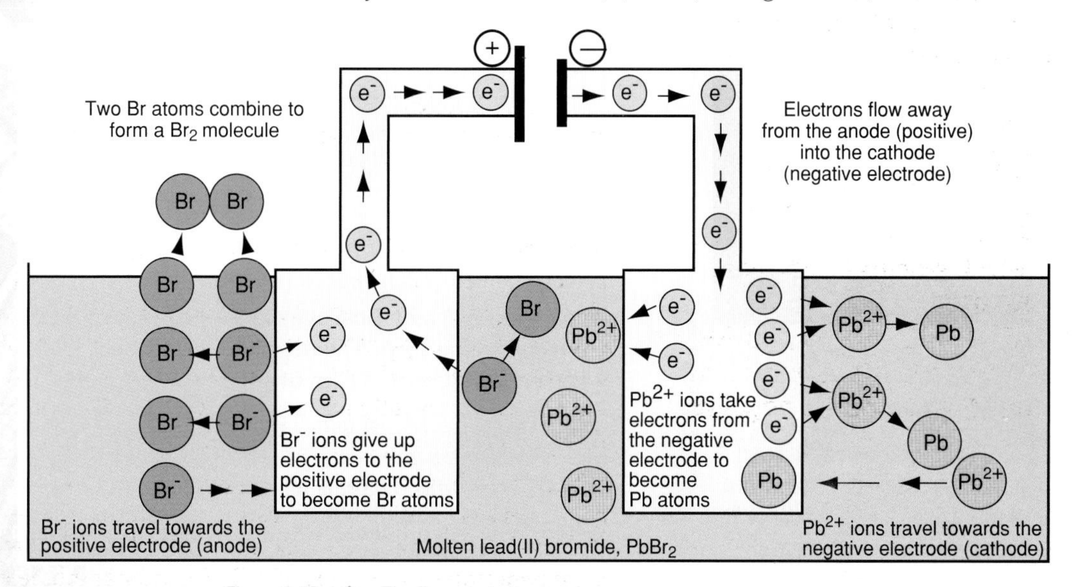

Figure 7.3B ▲ The flow of electrons

SUMMARY

Some compounds conduct electric current when they are molten. As they do they are electrolysed, that is, split up by the current. The explanation of electrolysis is that these compounds are composed of positive and negative ions.

Lead(II) bromide is an uncharged substance because there are two Br^- ions for each Pb^{2+} ions. The formula is $PbBr_2$. *Why does solid lead(II) bromide not conduct electricity?*

In the solid salt, the ions cannot move. They are fixed in a rigid three-dimensional structure. In the molten solid, the ions can move and make their way to the electrodes.

You will have noticed that lead(II) ions have two units of charge, whereas bromide ions have a single unit of charge. By experiment, it is possible to find out what charge an ion carries. Table 7.2 shows the results of such experiments.

Table 7.2 ▼ Some common ions

Positive ions			Negative ions	
+1	+2	+3	−1	−2
Hydrogen, H^+	Copper, Cu^{2+}	Aluminium, Al^{3+}	Bromide, Br^-	Oxide, O^{2-}
Sodium, Na^+	Iron(II), Fe^{2+}	Iron(III), Fe^{3+}	Chloride, Cl^-	Sulphide, S^{2-}
Potassium, K^+	Lead(II), Pb^{2+}		Iodide, I^-	Carbonate, CO_3^{2-}
	Magnesium, Mg^{2+}		Hydroxide, OH^-	Sulphate, SO_4^{2-}
	Zinc, Zn^{2+}		Nitrate, NO_3^-	

7.4 Solutions and conduction

FIRST THOUGHTS

The best way to begin this topic is by doing experiments to find out what types of solutions are electrolysed.

KEY SCIENTIST

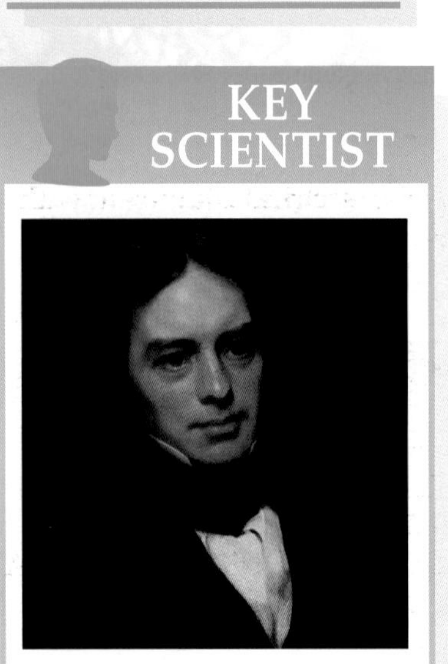

The scientist who did the first work on electrolysis was Michael Faraday. He was born in 1791, the son of a blacksmith. He received only a very basic education. When he was apprenticed to a London book binder, books on chemistry and physics came into his hands. He found them fascinating and was led to study science in his spare time. He attended a series of lectures given by Sir Humphry Davy, the director of the Royal Institution. Faraday wrote up the lectures and illustrated them with careful diagrams. He sent the notebook to Davy, asking for some kind of work in the laboratory. Davy agreed, and Faraday soon became a capable research assistant. On Davy's retirement, Faraday became director of the Royal Institution. Faraday is famous for his discovery of electromagnetic induction and for his work on the chemical effects of electric currents. In 1834, he put forward the theory that electrolysis could be explained by the existence of charged particles of matter. We now call them *ions*.

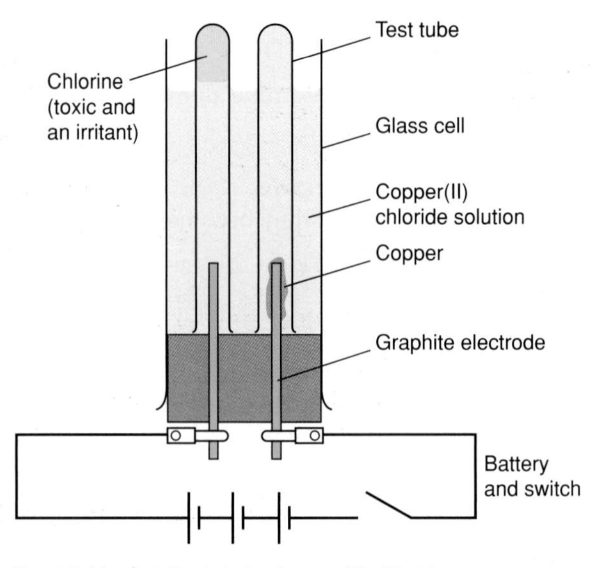

Figure 7.4A ▲ Electrolysis of copper(II) chloride

In Figure 7.4A a solution of the salt copper(II) chloride is being electrolysed. At the positive electrode (anode), bubbles of the gas chlorine can be seen. The negative electrode (cathode), becomes coated with a reddish brown film of copper.

Look at Figure 7.4A. At which electrode is copper deposited? Which kind of charge must copper ions carry? At which electrode is chlorine given off? Which kind of charge must chloride ions carry?

When they reach the electrodes, the ions are discharged.

At the negative electrode

Copper(II) ion + 2 Electrons (taken from the cathode) → Copper atom

$$Cu^{2+}(aq) + 2e^- \rightarrow Cu(s)$$

At the positive electrode

Chloride ion → Chlorine atom + Electron (given up at the anode)

$$Cl^-(aq) \rightarrow Cl(g) + e^-$$

Pairs of chlorine atoms then join to form molecules.

$$2Cl(g) \rightarrow Cl_2(g)$$

The electrons given to the anode by the discharge of chloride ions travel round the external circuit to the cathode. At the cathode, they combine with copper(II) ions.

To make sure you understand it, draw a diagram showing what happens to the ions and electrons in the electrolysis of copper(II) chloride solution. You can refer to Figure 7.3B for help.

Solid copper(II) chloride does not conduct electricity, but a solution of the salt in water does conduct. It is not the water in the solution that makes it conduct: experiment shows that water is a very poor electrical conductor. *What is the reason for the difference in behaviour between the solid and the solution?* In a solid, the ions are fixed in position, held together by strong attractive forces between positive and negative ions. In a solution of a salt, the ions are free to move (see Figure 7.4B).

The ions... which is which?
How to remember:
Current carries copper cations to cathode
Cations travel to the **cat**hode...
Anions travel to the **an**ode.

Water is a non-conductor because it is not ionic

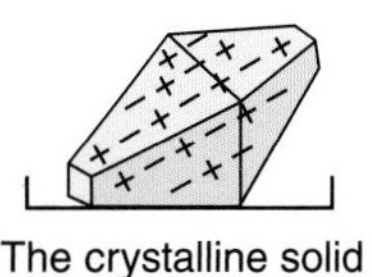
The crystalline solid is a non-conductor because the ions are held in a rigid structure

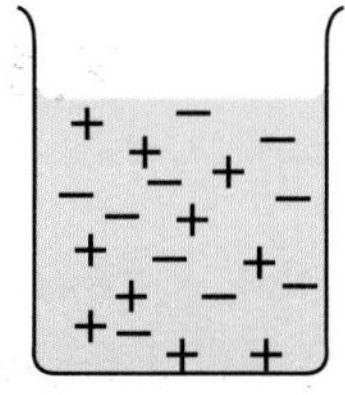
The solution is a good conductor because the ions are now free to move

Figure 7.4B ▲ The ions must be free to move

SUMMARY

When a direct electric current is passed through solutions of some compounds, they are electrolysed, that is, split up by the current. Chemical changes occur in the electrolyte, and, as a result, new substances are formed.

Compounds which are electrolytes consist of positively and negatively charged particles called ions. In molten **ionic** solids and in solutions, the ions are free to move.

Safety matters

Tap water conducts electricity. Never handle electrical equipment with wet hands. If the equipment is faulty, you run a much bigger risk of getting a lethal shock if you have wet hands.

Non-electrolytes and weak electrolytes

Some liquids and solutions do not conduct electricity. It follows that these **non-electrolytes** consist of molecules, not ions. Some substances conduct electricity to a very slight extent. They are **weak electrolytes**. These compounds consist mainly of molecules, which do not conduct. A small fraction of the molecules split up to form ions which do conduct. Such compounds are **partially ionised**.

CHECKPOINT

1 A sodium atom can be written as $^{23}_{11}Na$.
The number of protons in a sodium atom is _____ and the number of electrons is _____. The charge on a sodium atom is therefore _____.
A sodium ion can be written as $^{23}_{11}Na^+$
The number of protons in a sodium ion is _____. The charge on the ion is _____ and the number of electrons is therefore _____.
The word equation for the formation of a sodium ion is:
sodium atom _____ electron _____ sodium atom
The symbol equation is: _______________

2 An atom of chlorine can be shown as $^{35}_{17}Cl$.
(a) How many (i) protons and (ii) electrons are there in a chlorine atom?
(b) What is the overall charge on a chlorine atom?
A chloride ion can be written as $^{35}_{17}Cl^-$
(c) What is the overall charge on a chloride ion?
(d) How many (i) protons and (ii) electrons does it contain?
Copy and complete the word equation for the formation of a chloride ion:
chlorine atom _____ electron _____ chloride ion
The symbol equation is: _______________

3 (a) Divide the following list into (i) electrical conductors, (ii) non-conductors.
copper, ethanol (alcohol), mercury, limewater, sugar solution, molten copper chloride, distilled water, sodium chloride crystals, molten sodium chloride, sodium chloride solution, dilute sulphuric acid

4 Explain the words: electrolysis, electrolyte, electrode, anode, cathode, ion, anion, cation.

5 Why are sodium chloride crystals not able to conduct electricity?

7.5 More examples of electrolysis

FIRST THOUGHTS

Sometimes the products formed when solutions are electrolysed are difficult to predict, but there are rules to help you.

Sodium chloride solution

When molten sodium chloride (common salt) is electrolysed, the products are sodium and chlorine. When the aqueous solution of sodium chloride is electrolysed, the products are hydrogen (at the negative electrode) and chlorine (at the positive electrode). To explain how this happens, we have to think about the water present in the solution. Water consists of molecules, but a very small fraction of the molecules ionise into hydrogen ions and hydroxide ions.

$$\text{Water} \rightarrow \text{Hydrogen ions} + \text{Hydroxide ions}$$
$$H_2O(l) \rightarrow H^+(aq) + OH^-(aq)$$

Hydrogen ions are attracted to the negative electrode as well as sodium ions. Sodium ions are more stable than hydrogen ions. It is easier for the negative electrode to give an electron to a hydrogen ion than it is for it to give an electron to a sodium ion. Sodium ions remain in solution while hydrogen ions are discharged to form hydrogen atoms. These atoms join in pairs to form hydrogen molecules.

$$H^+(aq) + e^- \rightarrow H(g)$$
$$2H(g) \rightarrow H_2(g)$$

Can you explain how these products are formed?

- hydrogen and chlorine when sodium chloride solution is electrolysed
- copper and oxygen when copper(II) sulphate solution is electrolysed

Although the concentration of hydrogen ions in the solution is very low, it is kept topped up by the ionisation of more water molecules.

At the positive electrode, there are chloride ions and also hydroxide ions. The hydroxide ions have come, in very low concentration, from the ionisation of water molecules. Chloride ions are discharged while hydroxide ions remain in solution.

Copper(II) sulphate solution

When copper(II) sulphate is electrolysed, copper is deposited on the cathode, and oxygen is evolved at the anode. The oxygen comes from the water in the solution. At the positive electrode, there are hydroxide ions, OH^-, as well as sulphate ions, SO_4^{2-}. The hydroxide ions have come in low concentration from the ionisation of water molecules (see above). It is easier for the positive electrode to take electrons away from hydroxide ions than from sulphate ions, and hydroxide ions are discharged. The OH groups which are formed exist for only a fraction of a second before rearranging to give oxygen and water.

$$OH^-(aq) \rightarrow OH(aq) + e^-$$
$$4OH(aq) \rightarrow 2H_2O(l) + O_2(g)$$

Although only a tiny fraction of water molecules is ionised, once hydroxide ions have been discharged, more water molecules ionise to replace them with fresh hydroxide ions.

Dilute sulphuric acid

Figure 7.5A shows the electrolysis of dilute sulphuric acid.

The electrolysis of dilute sulphuric acid produces hydrogen and oxygen.

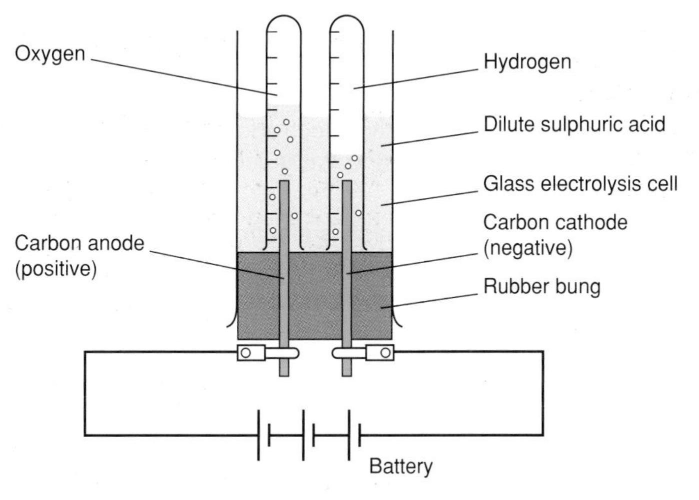

Figure 7.5A ▲ The electrolysis of dilute sulphuric acid

7.6 Which ions are discharged?

You will have seen that some ions are easier to discharge at an electrode than others.

Cations

Some metals are more **reactive** than others. The ions of very reactive metals are difficult to discharge. Sodium is a very reactive metal. When it reacts, sodium atoms form sodium ions. It is difficult to force a sodium ion to accept an electron and turn back into a sodium atom. Hydrogen ions are discharged in preference to sodium ions. The ions of less reactive metals, such as copper and lead, are easy to discharge.

Figure 7.6A ▲ Sodium ions do not want to accept electrons

Anions

Sulphate ions and nitrate ions are very difficult to discharge. When solutions of sulphates and nitrates are electrolysed, hydroxide ions are discharged instead, and oxygen is evolved.

SUMMARY

Water is ionised to a very small extent into $H^+(aq)$ ions and $OH^-(aq)$ ions. These ions are discharged when some aqueous solutions are electrolysed. The ions of very reactive metals, e.g. Na^+, are difficult to discharge. In aqueous solution, hydrogen ions are discharged instead. Hydrogen is evolved at the negative electrode. The anions SO_4^{2-} and NO_3^- are difficult to discharge. In aqueous solution, OH^- ions are discharged instead, and oxygen is evolved at the positive electrode.

CHECKPOINT

1 Copy and complete this passage.

When molten sodium chloride is electrolysed, _____ is formed at the positive electrode and _____ is formed at the negative electrode.

When aqueous sodium chloride is electrolysed, _____ is formed at the positive electrode and _____ is formed at the negative electrode. The reaction that takes place at the positive electrode is _____________________ and the reaction that takes place at the negative electrode is _____________________.

2 A solution of potassium bromide is electrolysed. At one electrode a colourless gas is given off. At the other a brown vapour appears.

(a) At which electrode does the brown vapour appear? Which ions have been discharged?

(b) Which gas is produced at the other electrode? Which ions have been discharged?

(c) Write equations for the discharge of the ions at each electrode.

7.7 Applications of electrolysis

FIRST THOUGHTS

Electrolysis is useful for:

- electroplating,
- extracting metals from their ores,
- manufacturing chemicals.

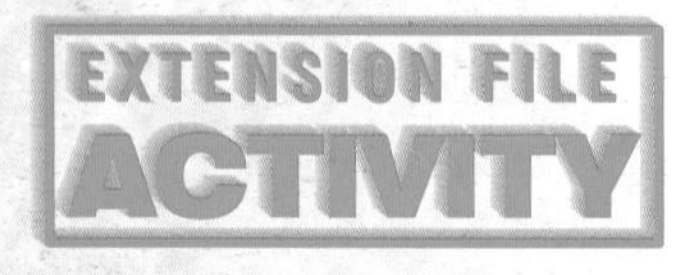

Science at work

Do you know what happens when a music group makes a recording? Sound waves activate a machine which cuts a groove in a soft plastic disc. This disc is coated with graphite so that it will conduct. Then it is electroplated with a thick layer of metal. The metal plating is prised off the plastic disc and used to stamp out copies of the original.

Electroplating

Some metals are prized for their strength and others for their beauty. Beautiful metals like silver and gold are costly. Often, objects made from less expensive metals are given a coating of silver or gold. The coating layer must stick well to the surface. It must be even and, to limit the cost, it should be thin. Depositing the metal by electrolysis is ideal. The technique is called **electroplating**.

Electroplating is used for protection as well as for decoration. You may have a bicycle with chromium-plated handlebars. The layer of chromium protects the steel underneath from rusting. Chromium does not stick well to steel. Steel is first electroplated with copper, which adheres well to steel, then it is nickel-plated and finally chromium-plated. The result is an attractive bright surface which does not corrode.

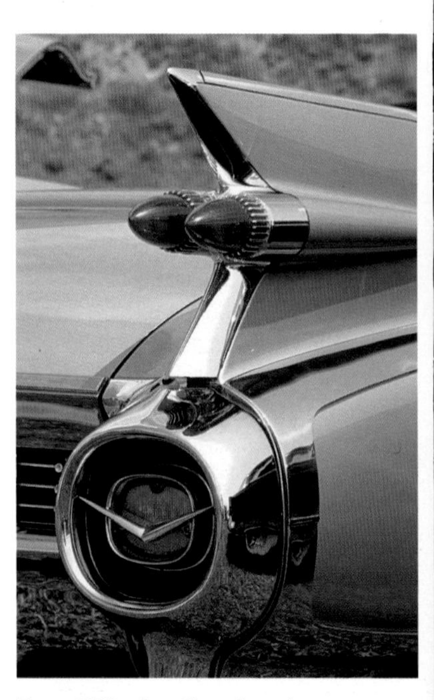

Figure 7.7A ▲ Chromium plate

The rusting of iron is a serious problem. One solution to the problem is to coat iron with a metal which does not corrode. Food cans must not rust. They are made of iron coated with tin. Tin is an unreactive metal, and the juices in foods do not react with it.

Electroplating is used to give a thin even layer of tin. A layer of zinc is applied to iron in the manufacture of 'galvanised' iron. Often, electroplating is employed.

SUMMARY

Electroplating is used to coat a cheap metal with an expensive metal. It is used to coat a metal like iron, which rusts, with a metal which does not corrode, e.g. tin or zinc.

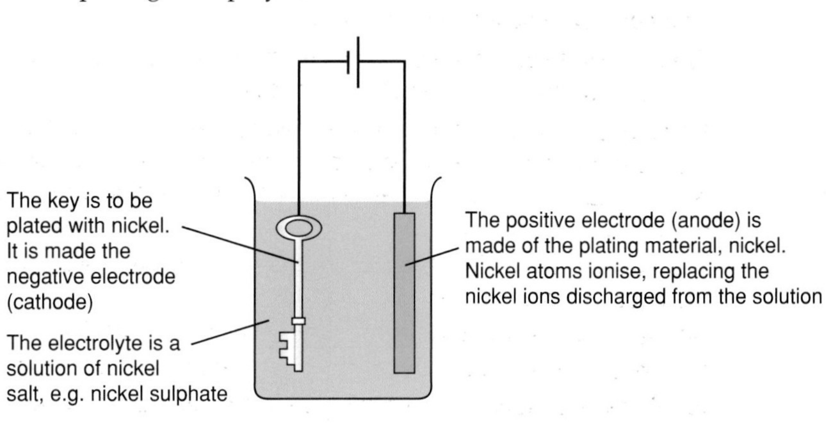

Figure 7.7B ▲ Electroplating

Extraction of metals from their ores

Reactive metals are difficult to extract from their ores. For some of them, electrolysis is the only method which works. Sodium is obtained by the electrolysis of molten dry sodium chloride, and aluminium is obtained by the electrolysis of molten aluminium oxide.

Electrolysis can also be used as a method of purifying metals. Pure copper is obtained by an electrolytic method.

SUMMARY

Electrolysis is used in the extraction of some metals from their ores, e.g. sodium, aluminium, copper.

Manufacture of sodium hydroxide

The three important chemicals, sodium hydroxide, chlorine and hydrogen are all obtained from the plentiful starting material, common salt. The method of manufacture is to electrolyse a solution of sodium chloride (common salt) in a diaphragm cell.

Figure 7.7C ▲ A room of diaphragm cells for the electrolysis of sodium chloride

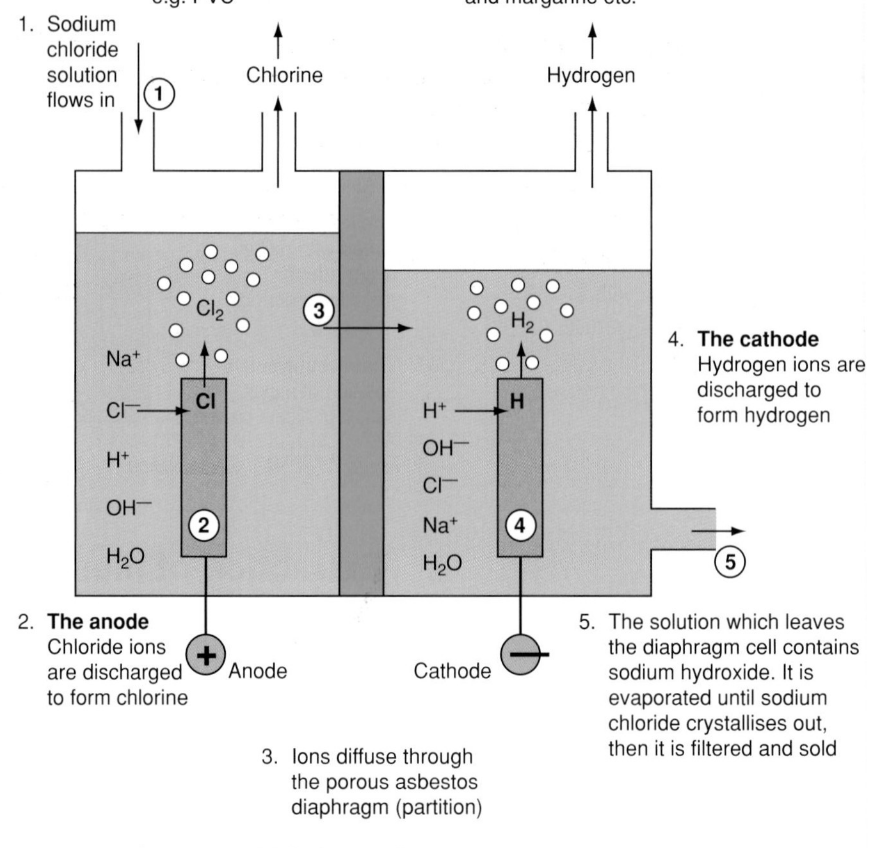

Figure 7.7D An industrial diaphragm cell

SUMMARY

The electrolysis of brine to give sodium hydroxide, chlorine and hydrogen is an important industrial process. It is carried out in the diaphragm cell.

Anodising aluminium

Aluminium is a metal which resists corrosion extremely well. After many years of use, aluminium can be recycled for the manufacture of new products. The reason is that a thin layer of aluminium oxide forms on the outside of the metal as soon as it is exposed to rthis air and protects it from further attack. The oxide layer can be made thicker and more effective by **anodisation**. A piece of aluminium is made to be the anode (the positive electrode) in a circuit such as that in Figure 7.7E. A small current is passed for a few hours through an electrolyte of dilute sulphuric acid. Oxygen forms on the surface of the aluminium and reacts to form a layer of aluminium oxide. Anodised aluminium is used for the manufacture of doors and window frames

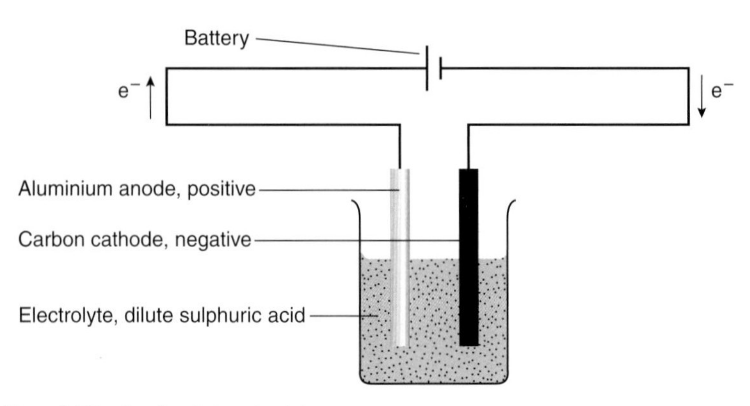

Figure 7.7E Anodising aluminium

CHECKPOINT

1 (a) You are asked to electroplate a nickel spoon with silver. Draw the apparatus and the circuit you would use. Say what the electrodes are made of and what charge they carry. Say what electrolyte you could use.

(b) Explain why silver plating is popular.

2 Both paint and chromium plating are used to protect parts of a car body.

(a) What advantages does paint have over chromium plating?

(c) What advantages does chromium plating have over paint?

(c) Why is chromium plating, rather than paint, used on the door handles?

3 Many people like gold-plated watches and jewellery.

(a) What two advantages do gold-plated articles have over cheaper metals?

(b) Why do people choose gold plate rather than pure gold?

(c) What is the advantage of solid gold over gold plate?

4 What is meant by 'anodising'? Why is this process used on aluminium?

Topic 8 The chemical bond

8.1 The ionic bond

FIRST THOUGHTS

What holds the atoms in a compound together? What holds the ions in a compound together? You can find out in this topic.

www.keyscience.co.uk

Sodium chloride

Topic 7 dealt with electrolysis. You found that some compounds are electrolytes: they conduct electricity when molten or in aqueous solution. Other compounds are non-electrolytes. Electrolysis can be explained by the theory that electrolytes consist of small charged particles called **ions**. For example, sodium chloride consists of positively charged sodium ions and negatively charged chloride ions, Na^+ Cl^-.

Sodium chloride is formed when sodium (a metallic element) burns in chlorine (a non-metallic element). During the reaction, each sodium atom loses one **electron** to become a sodium ion. Each chlorine atom gains one electron to become a chloride ion.

$$\text{Sodium atom} \rightarrow \text{Sodium ion} + \text{Electron}$$
$$Na \rightarrow Na^+ + e^-$$

$$\text{Chlorine atom} + \text{Electron} \rightarrow \text{Chloride ion}$$
$$Cl + e^- \rightarrow Cl^-$$

The noble gases are unreactive – possibly because each has a full outer shell of electrons.

A sodium atom reacts by giving one electron to leave a full outer shell.

A chlorine atom can accept one electron to attain a full outer shell.

Look at the arrangement of electrons in a sodium atom (Figure 8.1A). There is one electron more than there is in an atom of neon. Neon is one of the noble gases, helium, neon, argon, krypton and xenon (see Topic 11.3). All the noble gases have a full outer shell of 8 electrons (2 for helium). They are very unreactive elements; until 1965 they were thought to take part in no chemical reactions at all. It is believed that the lack of reactivity of the noble gases is due to the stability of the full outer shell of electrons. When atoms react they gain or lose or share electrons to attain an outer shell of 8 electrons.

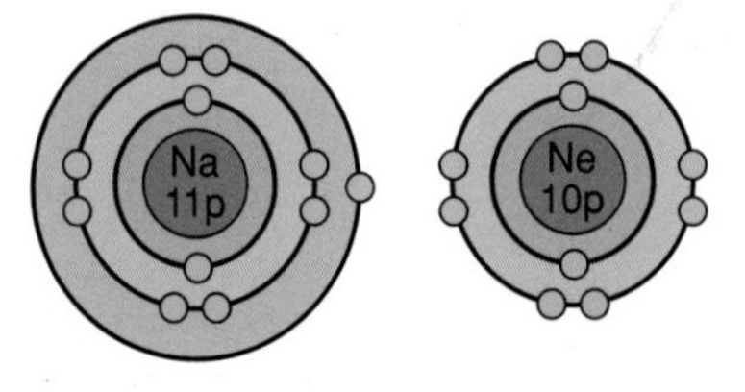

Figure 8.1A Atoms of sodium and neon

When a sodium atom loses the lone electron from its outermost shell, the outer shell that remains contains 8 electrons. A sodium atom cannot just shed an electron. The attraction between the nucleus and the electrons prevents this. It can, however, give an electron to a chlorine atom. A chlorine atom is prepared to accept an electron because this gives it the same electron arrangement as argon: a full outer shell (Figure 8.1B).

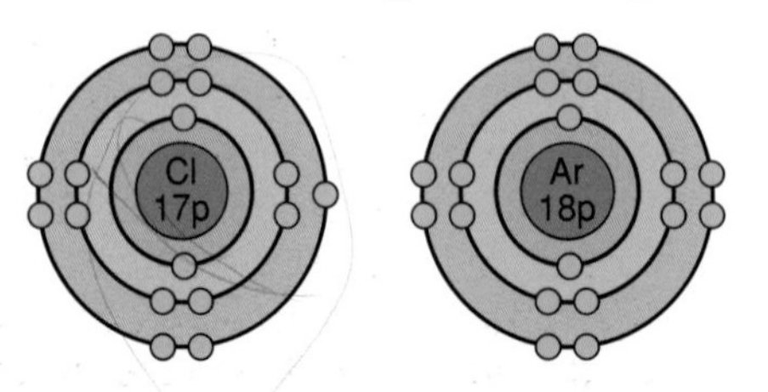

Figure 8.1B The arrangement of electrons in chlorine and argon

Figure 8.1C shows what happens when an atom of sodium gives an electron to an atom of chlorine. A full outer shell is left behind in sodium, and a full outer shell is created in chlorine. In the diagram, sodium electrons have been shown as x and chlorine electrons as o. This 'dot-and-cross diagram' does not imply that the electrons are different!

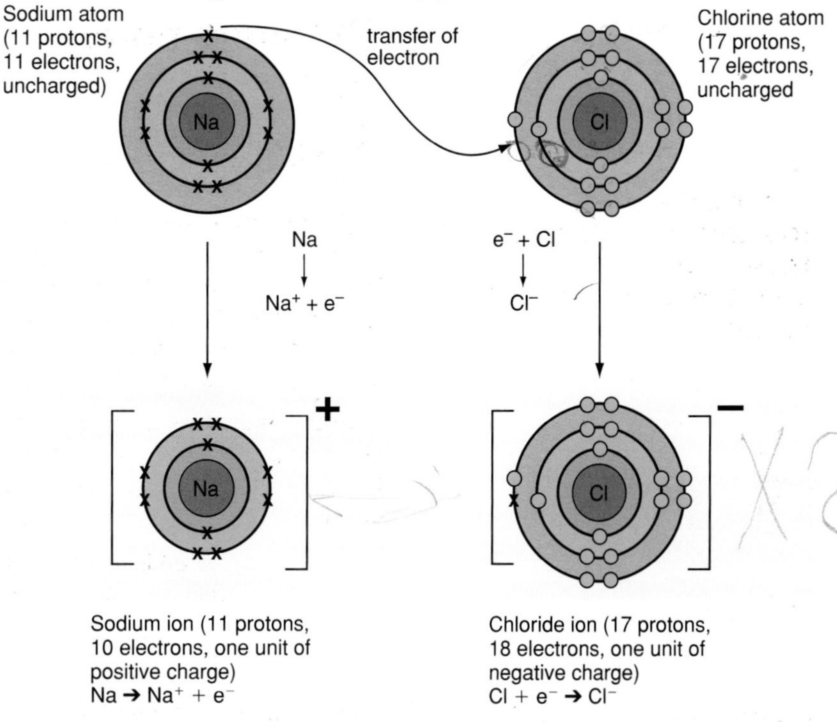

Figure 8.1C ▲ The formation of sodium chloride

Ions with opposite charges are formed. They are held together by electrostatic attraction in a crystal structure. If the solid is melted the ions move and conduct electricity.

When particles have opposite electric charges, a force of attraction exists between them. It is called **electrostatic attraction**. In sodium chloride, the sodium ions and chloride ions are held together by electrostatic attraction. The electrostatic attraction is the **chemical bond** in the compound, sodium chloride. It is called an **ionic bond** or **electrovalent bond**. Sodium chloride is an ionic or electrovalent compound. The compounds which conduct electricity when they are melted or dissolved are electrovalent compounds.

Figure 8.1D ▲ There is an attraction between sodium ions and chloride ions

A pair of ions, Na^+ Cl^-, does not exist by itself. It attracts other ions. The sodium ion attracts chloride ions, and the chloride ion attracts sodium ions. The result is a three-dimensional structure of alternate Na^+ and Cl^- ions (Figure 8.1E). This is a **crystal** of sodium chloride. The crystal is uncharged because the number of sodium ions is equal to the number of chloride ions. The forces of attraction between the ions hold them in position in the structure. Since they cannot move out of their positions, the ions cannot conduct electricity. When the salt melts, the three-dimensional structure breaks down, the ions can move towards the electrodes and the molten solid can be electrolysed. There are millions of ions in even the tiniest crystal. The structures of ionic solids, such as sodium chloride, are described as **giant ionic structures**.

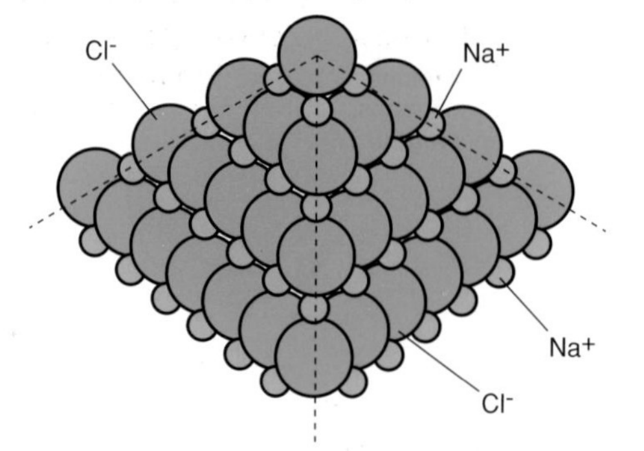

Figure 8.1E ▲ The structure of sodium chloride

SUMMARY

Sodium and chlorine react to form positive sodium ions and negative chloride ions. The electrostatic attraction between these oppositely charged ions is an ionic bond or electrovalent bond.

CHECKPOINT

1 A sodium atom has 11 electrons, 11 protons and 12 neutrons.
   (a) What is the charge on (i) an electron (ii) a proton and (iii) a neutron?
   (b) What is the overall charge on a sodium atom?
   A sodium ion has 10 electrons, 11 protons and 12 neutrons.
   (c) What is the type of charge on a sodium ion?

2 A chlorine atom has 17 protons and 17 electrons.
   (a) What is the overall charge on a chlorine atom?
   A chloride ion has 17 protons and 18 electrons.
   (b) What is the type of charge on a chloride ion?

Why is the formula of magnesium fluoride MgF_2? An Mg atom can lose 2 electrons to form an Mg^{2+} ion. An F atom can gain one electron to form an F^- ion. Therefore 2F combine with one Mg to form $Mg^{2+}\ 2F^-\ MgF_2$.

8.2 Other ionic compounds

There are thousands of ionic compounds. In general, metallic elements give away electrons to form positive ions (cations). Non-metallic elements take up electrons to form negative ions (anions). Ionic compounds are formed when metallic elements give electrons to non-metallic elements.

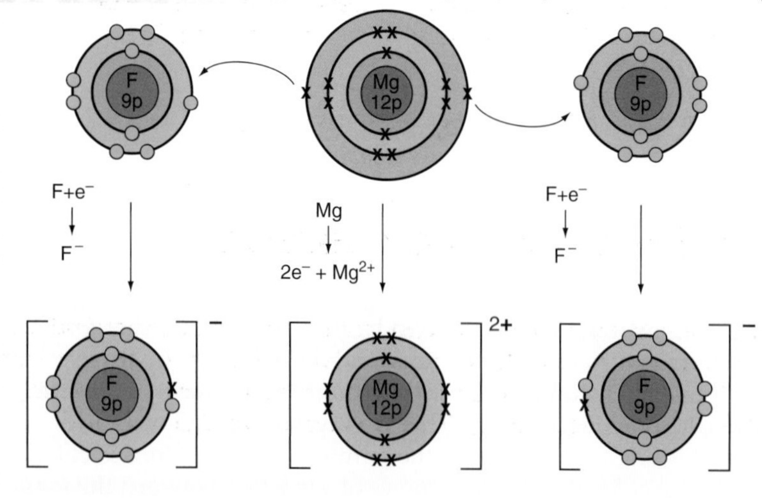

Figure 8.2A ▲ The formation of magnesium fluoride

Magnesium fluoride

Magnesium (2.8.2) has two electrons in its outermost shell. It needs to lose these to gain a stable, full outer shell of 8 electrons. Fluorine (2.7) needs to gain one electron. One magnesium atom needs two fluorine atoms to accept two electrons (see Figure 8.2A). The formula of magnesium fluoride is $Mg^{2+}2F^-$ or MgF_2.

Magnesium oxide

One atom of oxygen (2.6) can accept two electrons to give it the electron arrangement of neon. One atom of magnesium (2.8.2) therefore combines with one atom of oxygen (Figure 8.2B). The ions Mg^{2+} and O^{2-} are formed. The attraction between them is an ionic bond. The formula of magnesium oxide is $Mg^{2+}O^{2-}$ or MgO.

Why is the formula of magnesium oxide MgO? One O atom can gain 2 electrons to form O^{2-}. One O atom can therefore combine with one Mg atom to form Mg^{2+} O^{2-}, MgO.

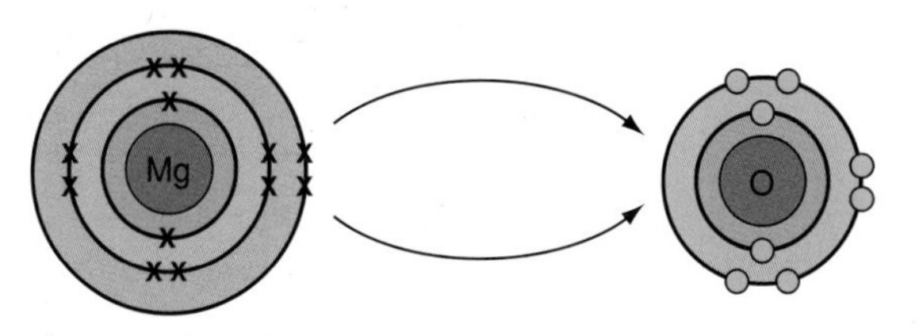

A magnesium atom gives 2 electrons to form a Mg^{2+} ion. $Mg \rightarrow Mg^{2+} + 2e^-$

An oxygen atom accepts 2 electrons to form an oxide ion, O^{2-}. $O + 2e^- \rightarrow O^{2-}$

Figure 8.2B ▲ The formation of magnesium oxide

CHECKPOINT

1. Copy and complete the following passage (lithium is an alkali metal in Group 1 and fluorine is a halogen in Group 7).

   Lithium and fluorine combine to form _____ fluoride. When this happens, each lithium atom gives _____ to a fluorine atom. Lithium ions with a _____ charge and fluoride ions with a _____ charge are formed. The _____ ions and _____ ions are held together by a strong _____ attraction. This _____ attraction is a chemical bond. This type of chemical bond is called the _____ bond or _____ bond. A _____ dimensional structure of ions is built up.

2. Sodium is a silvery-grey metal which reacts rapidly with water. Chlorine is a poisonous green gas. Sodium chloride is a white, crystalline solid, which is used as table salt. Explain how the sodium in sodium chloride differs from sodium metal. Explain how the chlorine in sodium chloride differs from chlorine gas.

3. Lithium has atomic number 3. Fluorine has atomic number 9. Draw a dot-and-cross diagram to show the bonding in lithium fluoride.

4. There are two differences between
   (a) a sodium ion, Na^+ (2.8), and a neon atom, Ne (2.8);
   (b) a chloride ion, Cl^- (2.8.8), and an argon atom, Ar (2.8.8).
   What are the differences?

5. Potassium has atomic number 19. State the number of protons and electrons in one atom of potassium. Sketch the arrangement of electrons in shells in (a) an atom of potassium (b) a potassium ion.

8.3 ▶ Formulas of ionic compounds

FIRST THOUGHTS

How do you work out the formula of an ionic compound? You balance the charges on the ions, and suddenly formulas make sense!

Ions are charged, but ionic compounds are uncharged. A compound has no overall charge because the sum of positive charges is equal to the sum of negative charges. In magnesium chloride, $MgCl_2$, every magnesium ion, Mg^{2+}, is balanced in charge by two chloride ions, $2Cl^-$.

Figure 8.3A ▲ The charges balance

This is how you can work out the formulas of electrovalent compounds.

Compound: Magnesium chloride

The ions present:	Mg^{2+} Cl^-
The charges must balance:	One Mg^{2+} ion needs two Cl^- ions.
Ions in the formula:	Mg^{2+} and $2Cl^-$ ions
The formula is:	$MgCl_2$

Compound: Sodium sulphate

The ions present:	Na^+ SO_4^{2-}
The charges must balance:	Two Na^+ are needed to balance one SO_4^{2-}
Ions in the formula:	$2Na^+$ and SO_4^{2-}
The formula is:	Na_2SO_4

Compound: Calcium hydroxide

The ions present:	Ca^{2+} OH^-
The charges must balance:	Two OH^- ions balance one Ca^{2+} ion
Ions in the formula:	Ca^{2+} and $2OH^-$
The formula is:	$Ca(OH)_2$

The brackets tell you that the 2 multiplies everything inside them. There are 2O atoms and 2H atoms, in addition to the 1Ca ion.

Have you grasped the idea? This exercise will help you to check.

Ionic compounds

Copy and complete these formulas:

Compound: Iron(II) sulphate

The ions present:	Fe^{2+} SO_4^{2-}
The charges must balance:	One Fe^{2+} ion balances __ SO_4^{2-} ions
Ions in the formula:	______ and ______
The formula is:	______

Compound: Iron(III) sulphate

The ions present:	Fe^{3+} SO_4^{2-}
The charges must balance:	__ Fe^{3+} balance __ SO_4^{2-} ions
Ions in the formula:	__ Fe^{3+} and __ SO_4^{2-}
The formula is:	$Fe_2(SO_4)_3$ The formula contains __ Fe, __ S and __ O.

What is the difference between iron(II) and iron(III) and between copper(I) and copper(II)?

You will notice that the sulphates of iron are named iron(II) sulphate and iron(III) sulphate. The roman numerals, II and III, show whether a compound contains iron(II) ions, Fe^{2+}, or iron(III) ions, Fe^{3+}.

Table 8.1 shows the symbols and charges of some ions. From this table, you can work out the formula of any compound containing these ions.

SUMMARY

The formula of an ionic compound is worked out by balancing the charges on the ions.

Table 8.1 The symbols of some common ions

Name	Symbol	Name	Symbol
Aluminium	Al^{3+}	Bromide	Br^-
Ammonium	NH_4^+	Carbonate	CO_3^{2-}
Barium	Ba^{2+}	Chloride	Cl^-
Calcium	Ca^{2+}	Hydrogencarbonate	HCO_3^-
Copper(I)	Cu^+	Hydroxide	OH^-
Copper(II)	Cu^{2+}	Iodide	I^-
Hydrogen	H^+	Nitrate	NO_3^-
Iron(II)	Fe^{2+}		
Iron(III)	Fe^{3+}	Oxide	O^{2-}
Lead(II)	Pb^{2+}	Phosphate	PO_4^{3-}
Magnesium	Mg^{2+}		
Mercury(II)	Hg^{2+}	Sulphate	SO_4^{2-}
Potassium	K^+		
Silver	Ag^+	Sulphide	S^{2-}
Sodium	Na^+		
Zinc	Zn^{2+}	Sulphite	SO_3^{2-}

CHECKPOINT

1 Write the formulas of the following ionic compounds.
(a) potassium chloride
(b) potassium sulphate
(c) ammonium chloride
(d) magnesium bromide
(e) copper(II) hydroxide
(f) zinc sulphate
(g) calcium carbonate
(h) aluminium chloride
(i) sodium hydrogencarbonate

2 Write the formulas of the following ionic compounds.
(a) sodium hydroxide
(b) calcium hydroxide
(c) iron(II) hydroxide
(d) iron(III) hydroxide
(e) aluminium oxide
(f) iron(III) oxide

3 Name the following.
(a) AgBr
(b) $AgNO_3$
(c) $Cu(NO_3)_2$
(d) $Al_2(SO_4)_3$
(e) $FeBr_2$
(f) $FeBr_3$
(g) PbO_2
(h) PbO
(i) $Zn(OH)_2$
(j) NH_4Br
(k) $Ca(HCO_3)_2$

8.4 The covalent bond

FIRST THOUGHTS

Some compounds are non-electrolytes because they consist of molecules, not ions. In this section, we think about the type of chemical bond in these molecular compounds.

Many compounds are non-electrolytes. They do not conduct electricity; therefore they cannot consist of ions. They contain a different kind of chemical bond: the covalent bond.

Chlorine Cl_2

Two atoms of chlorine combine to form a molecule Cl_2. Both chlorine atoms have the electron arrangement Cl (2.8.7). Both chlorine atoms want to gain an electron to attain a full octet. They do this by sharing electrons (see Figure 8.4A). The atoms have come close enough for the outer shells to overlap. Then the shared pair of electrons can orbit round both chlorine atoms. The shared pair of electrons is a **covalent bond**.

How to attain a full outer shell of electrons; this is the problem.

Two Cl atoms manage it by sharing a pair of electrons between them in the molecule Cl_2.

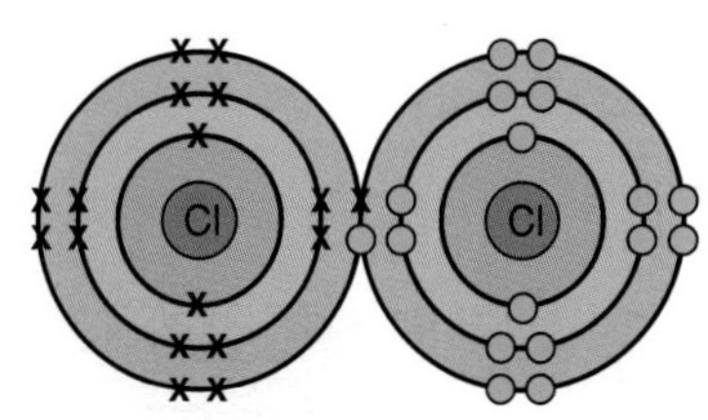

The pair of electrons ○x is shared between the 2Cl atoms. One electron comes from each Cl.

Figure 8.4A The chlorine molecule

Hydrogen chloride HCl

In hydrogen chloride HCl two electrons are shared between the hydrogen and chlorine atoms to form a covalent bond (see Figure 8.4B).

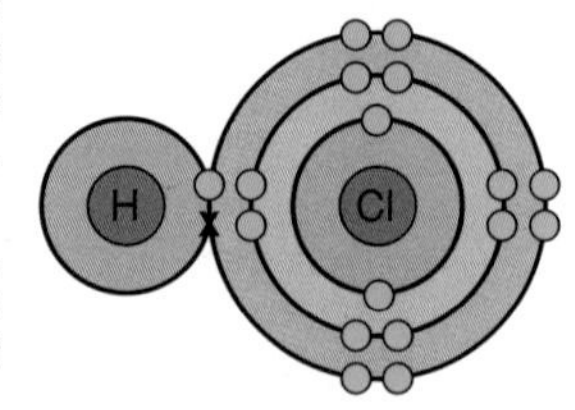

Figure 8.4B ▲ A molecule of hydrogen chloride

The shared pair of electrons is attracted to the hydrogen nucleus and to the chlorine nucleus and bonds the two nuclei together. Other covalent molecules are illustrated in Figure 8.4C.

Water H_2O

Two hydrogen atoms each share an electron with an oxygen atom. Each hydrogen atom has an outer shell of 2 electrons, and the oxygen atom has an outer shell of 8 electrons.

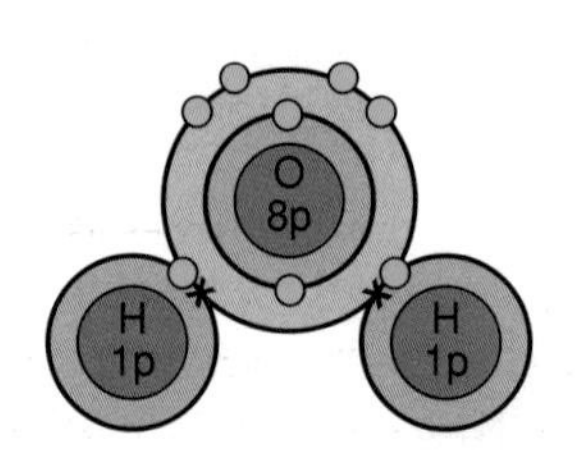

Ammonia NH_3

In ammonia, N shares 3 of its electrons, one with each of 3 H atoms. N now has 8 electrons in its outer shell (like Ne) and H has 2 electrons in its outer shell (like He).

Other examples of molecules with covalent bonds – shared pairs of electrons – are H_2O, NH_3, CH_4.

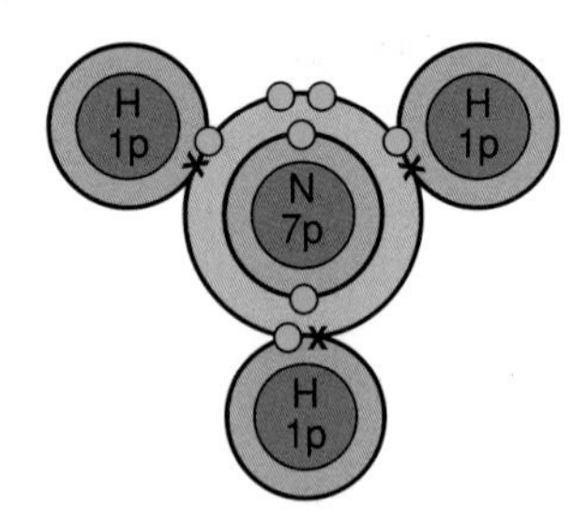

Methane CH_4

The C atom shares 4 electrons, one with each of 4 H atoms.

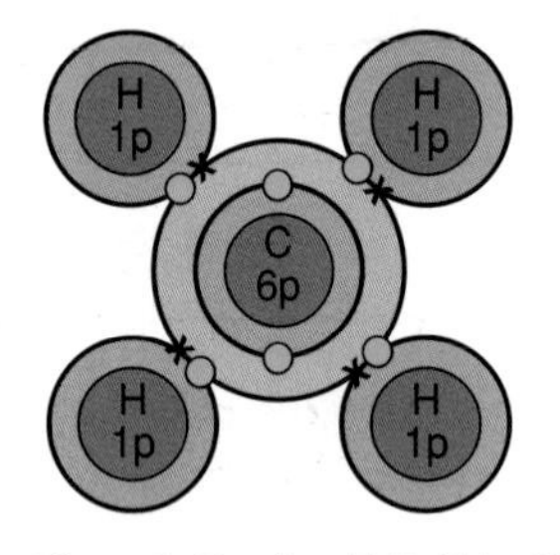

Figure 8.4C ▲ H_2O, NH_3, CH_4

Double covalent bonds

If two pairs of electrons are shared the bond is a **double bond** (see Figure 8.4D).

Carbon dioxide CO_2

There is a double covalent bond in the CO_2 molecule.

The C atom shares 2 electrons with each of the 2 O atoms. Each O atom shares 2 of its electrons with the C atom. In this way 2 pairs of electrons are shared between the C atom and each O atom. Each C=O bond is called a double bond.

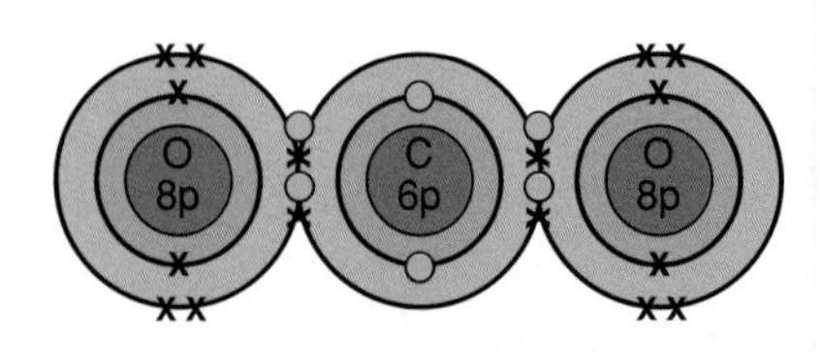

Figure 8.4D ▲ CO_2

The shapes of some covalent molecules are shown in Figure 8.4E

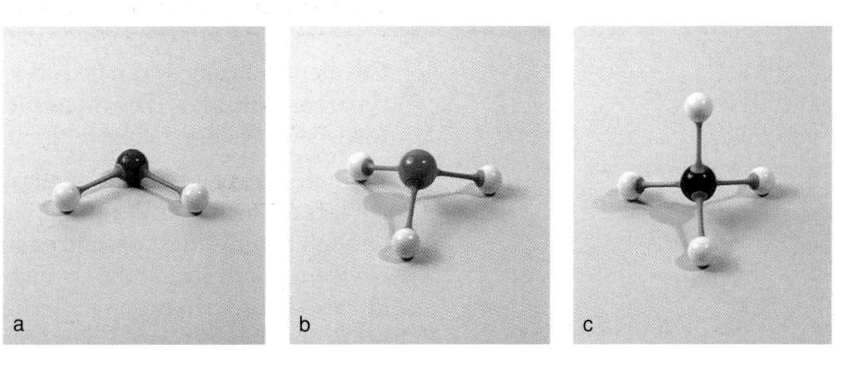

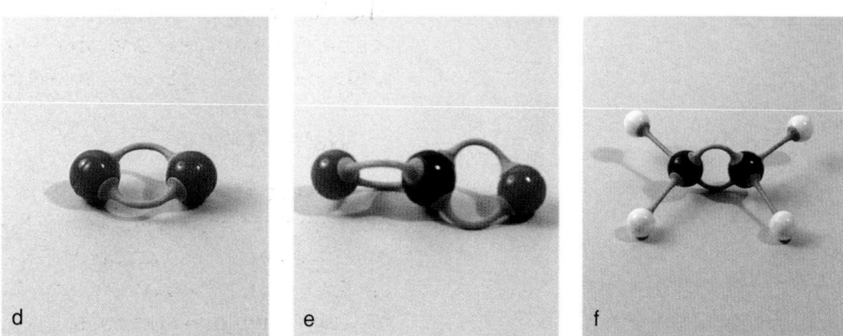

Figure 8.4E Covalent molecules (a) water, H_2O (b) ammonia, NH_3 (c) methane, CH_4 (d) oxygen, O_2 (e) carbon dioxide, CO_2 (f) ethene, C_2H_4

SUMMARY

Compounds which are non-electrolytes contain covalent bonds. Each of the atoms in a covalent bond completes its outer shell of electrons by sharing a pair of electrons. The shared electrons are attracted to both atomic nuclei and therefore bond the atoms.

CHECKPOINT

1. Sketch the arrangement of electrons in H (1) and F (2.7). Show by means of a 'dot-and-cross diagram' how a covalent bond is formed in HF.
2. Sketch the arrangement of electrons in the covalent compounds BF_3 [B (2.3), F (2.7)], PH_3 [P (2.8.5), H (1)] and H_2S [S (2.8.6), H (1)].
3. Sketch the arrangement of electrons in the covalent compound ethene, $CH_2 = CH_2$.
4. Sketch the arrangement of bonding electrons in the covalent compound carbon tetrachloride, CCl_4 (outer shells only).

8.5 Formulas of covalent compounds

SUMMARY

The formula of a covalent compound follows from a 'dot-and-cross diagram'.

The formula of a covalent compound depends on the sort of 'dot-and-cross diagram' that can be drawn for the compound. This shows the number of each type of atom in a molecule. This is the formula of the compound. Some covalent compounds you will meet on your course are sulphur dioxide, SO_2, sulphur trioxide, SO_3, ethane, C_2H_6, ethanol (alcohol), C_2H_5OH.

8.6 Forces between molecules

Covalent substances can be solids, liquids or gases. Gases consist of separate molecules. The molecules are far apart, and there are almost no forces of attraction between them.

In liquids there are forces of attraction between the molecules. In some covalent substances, the attractive forces between molecules are strong enough to make the substances solids. Ice is a solid. There are strong covalent bonds inside the covalent water molecules and weaker attractive forces between molecules (see Figure 8.6A). The forces of attraction between molecules hold them in a three-dimensional structure. It is described as a **molecular structure**.

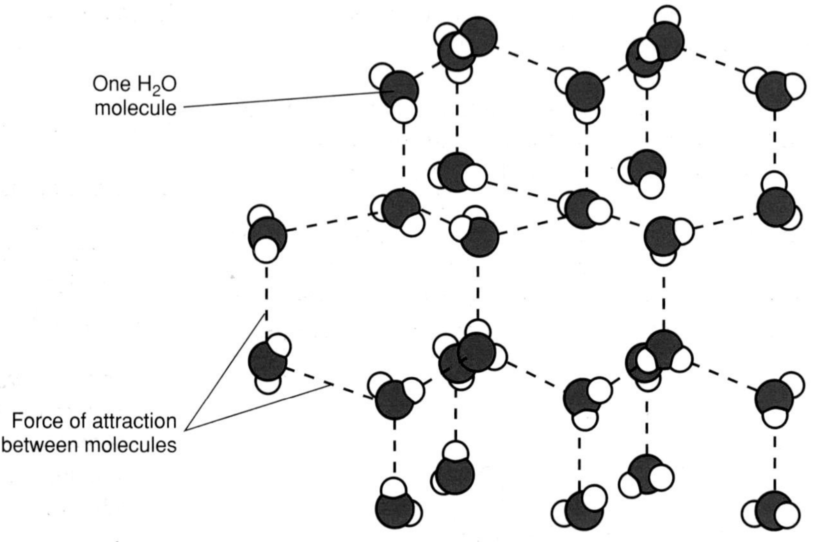

Figure 8.6A The structure of ice

Diamond is an **allotrope** of carbon. Diamond has a structure in which every carbon atom is joined by covalent bonds to four other carbon atoms (see Figure 4.8B). Millions of carbon atoms form a giant molecule. The structure is giant molecular or macromolecular.

Graphite, the other allotrope of carbon, has a layer structure (see Figure 4.8C). Strong covalent bonds join the carbon atoms within each layer. The layers are held together by weak forces of attraction.

SUMMARY

Some covalent substances are composed of individual molecules. Others are composed of larger units. These may be chains or layers or macromolecular structures.

CHECKPOINT

1. (a) Why does diamond have a high melting point?
   (b) Why is it very difficult to scratch a diamond?
   (c) Why does graphite rub off on to your hands?
2. Solid iodine consists of shiny black crystals. Iodine vapour is purple. What is the difference in chemical bonding between solid and gaseous iodine?

8.7 Ionic and covalent compounds

What difference does the type of bond make to the properties of ionic and covalent compounds?

The physical and chemical characteristics of substances depend on the type of chemical bonds in the substances.

Table 8.2 Differences between ionic and covalent substances

Ionic bonding	Covalent bonding
Ionic compounds are formed when a metallic element combines with a non-metallic element. An **ionic bond** is formed by **transfer of electrons** from one atom to another to form ions.	Atoms of non-metallic elements combine with other non-metallic elements by **sharing pairs of electrons** in their outer shells. A shared pair of electrons is a **covalent bond**.
Atoms of metallic elements form positive ions (cations), e.g. Na^+, Mg^{2+}, Al^{3+}. Atoms of non-metallic elements form negative ions (anions), e.g. O^{2-} and Cl^-.	The maximum number of covalent bonds that an atom can form is equal to the number of electrons in the outer shell. An atom may not use all its electrons in bond formation.
Ionic compounds are **electrolytes**; they conduct electricity when molten or in solution and are split up in the process – electrolysed.	Covalent compounds are **non-electrolytes**.
The strong electrostatic attraction between ions of opposite charge is an **ionic bond**. An ionic compound is composed of a giant regular structure of ions (see Figure 8.6A). This regular structure makes ionic compounds **crystalline**. The strong forces of attraction between ions make it difficult to separate the ions, and ionic compounds therefore have **high melting and boiling points**.	There are three **types of covalent substances**: (a) Many covalent substances are composed of small individual molecules with only very small forces of attraction between molecules. Such covalent substances are gases, e.g. HCl, SO_2, CO_2, CH_4.
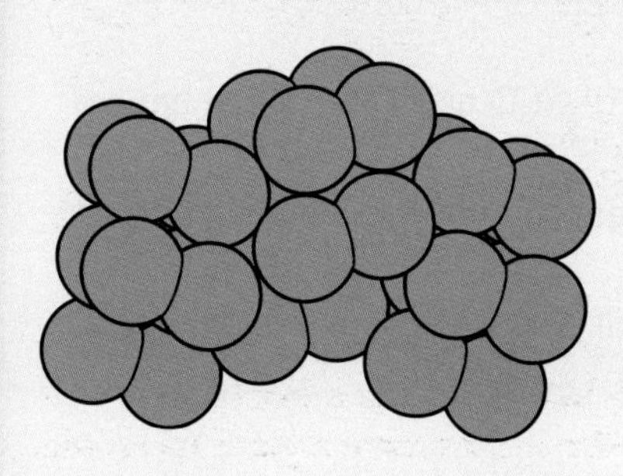 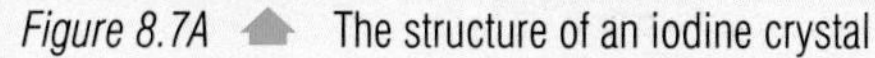 *Figure 8.7A* The structure of an iodine crystal 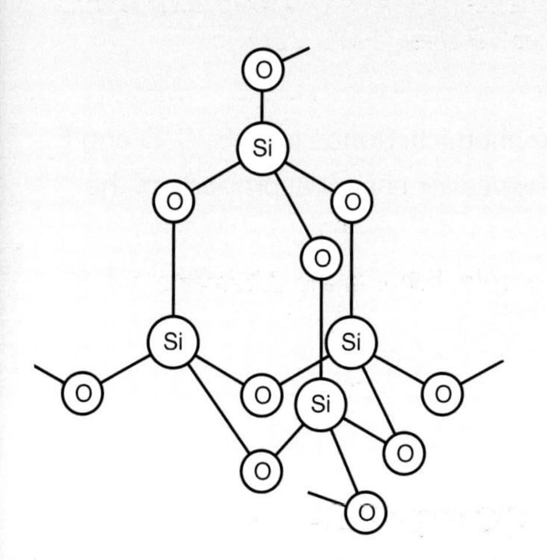  *Figure 8.7B* The structure of silicon(IV) oxide (quartz)	(b) Some covalent substances consist of small molecules with weak forces of attraction between the molecules. Some such covalent substances are low boiling point liquids, e.g. ethanol, C_2H_5OH Others are low melting point solids, e.g. iodine and solid carbon dioxide, which consist of molecular crystals (see Figure 8.7A). (c) Some covalent substances consist of giant molecules, e.g. quartz (silicon(IV) oxide, Figure 8.7B). These substances have high melting and boiling points. Atoms may link in chains or sheets, e.g. graphite (Figure 4.8C) or in 3-dimensional structures, e.g. diamond (Figure 4.8B) and quartz (Figure 8.7B). Substances with giant molecular structures have high melting and boiling points. **Organic solvents**, e.g. ethanol and propanone, have covalent bonds. They dissolve covalent compounds but not ionic compounds.

Can you tell from its properties whether a substance has ionic bonds or covalent bonds? Table 8.2 on the previous page will help you.

SUMMARY

The type of chemical bonds present, ionic or covalent, decides the properties of a compound, e.g. its physical state (s, l, g), boiling and melting points, electrical conductivity.

CHECKPOINT

1 Explain why the covalent compound CH_2Br_2 is a volatile liquid whereas the ionic compound $PbBr_2$ is a solid.

2 Explain why the compound $CHCl_3$, chloroform, has a powerful smell whereas sodium chloride, NaCl, has none.

3 Astatine, At, is an element with 7 electrons in its outer shell. Caesium, Cs, is an element with one electron in its outer shell. What type of bonding would you predict in caesium astatide? Write its formula.

4 Radium, Ra, is an element with 2 electrons in its outer shell, and iodine, I, is an element with 7 electrons in the outer shell. What type of bonding would you predict in radium iodide? Write its formula.

5 Oxygen and chlorine combine to form a number of compounds, e.g. Cl_2O and ClO_2. What type of bonding would you expect in these compounds?

6 Name covalent substances which consist of (a) individual molecules, (b) a molecular structure, (c) a giant molecular structure.

7 A girl gets some copper(II) sulphate solution on her shirt. Which will dissolve the blue stain out better, water or alcohol?

8

Solid	*State*	*Melting point*	*Does it conduct electricity?*
A	Solid	650 °C	Conducts when molten
B	Liquid	−20 °C	Does not conduct
C	Solid	700 °C	Does not conduct when molten
D	Solid	85 °C	Does not conduct when molten
E	Gas	−100 °C	Does not conduct

From the information in the table, say what you can about the chemical bonds in A, B, C, D and E.

9 Below are five statements. Give a piece of evidence (e.g. a physical or chemical property of the substance) in support of each statement.

(a) An aqueous solution of sodium chloride contains ions.

(b) Copper exists as positive ions in copper(II) sulphate solution.

(c) Ethanol (alcohol) is a covalent compound.

(d) The forces between oxygen molecules are very weak.

(e) The forces between iodine molecules are stronger than the forces between oxygen molecules.

(f) Dichromate(VI) ions are orange and negatively charged.

Theme Questions

Element	Argon	Bromine	Calcium	Carbon	Chlorine	Gold	Hydrogen	Mercury	Phosphorus	Sulphur
Metal or non-metal	N-M	N-M	M	N-M	N-M	M	N-M	M	N-M	N-M
Melting point (°C)	−189	−7	850	3730 (sublimes)	−101	1060	−259	−39	44 (white) 590 (red)	113 (rh) 119 (mono)
Boiling point (°C)	−186	59	1490	4830	−35	2970	−252	357	280	445
Density (g/cm^3)	0.0017	3.1	1.5	2.3 (gr) 3.5 (di)	0.003 017	19.3	0.000 083	13.6	1.8 (wh) 2.3 (red)	2.1 (rh) 2.0 (mono)

For questions 1 − 10 refer to the table of elements above.

1 Name the element with (a) the highest melting point (b) the lowest melting point.

2 Name the element with (a) the greatest density and (b) the lowest density.

3 Name the metal with (a) the lowest melting point and (b) the highest melting point.

4 Name the element which is liquid at room temperature and is (a) a metal (b) a non-metal.

5 Write down the names of all the elements which are (a) metals (b) non-metals.

6 Write down the names of all the elements which are (a) solids (b) liquids (c) gases.

7 Name the element which has the smallest temperature range over which it exists as a liquid.

8 Name the gaseous element which is (a) the most dense (b) the least dense.

9 Name the solid element which is (a) the most dense (b) the least dense.

10 Why are two sets of values given for (a) sulphur (b) phosphorus?

11 Copy out these equations into your book, and then balance them.

(a) $H_2O_2(aq) \rightarrow H_2O(l) + O_2(g)$

(b) $Fe(s) + O_2(g) \rightarrow Fe_3O_4(s)$

(c) $Mg(s) + N_2(g) \rightarrow Mg_3N_2(s)$

(d) $P(s) + Cl_2(g) \rightarrow PCl_3(s)$

(e) $P(s) + Cl_2(g) \rightarrow PCl_5(s)$

(f) $SO_2(g) + O_2(g) \rightarrow SO_3(g)$

(g) $Na_2O(s) + H_2O(l) \rightarrow NaOH(aq)$

(h) $KClO_3(s) \rightarrow KCl(s) + O_2(g)$

(i) $NH_3(g) + O_2(g) \rightarrow N_2(g) + H_2O(g)$

(j) $Fe(s) + H_2O(g) \rightarrow Fe_3O_4(s) + H_2(g)$

12 The element X has atomic number 11 and mass number 23. State how many protons and neutrons are present in the nucleus. Sketch the arrangement of electrons in an atom of X.

13 Atom A has atomic number 82 and mass number 204. Atom B has atomic number 80 and mass number 204. How many protons has atom A? How many neutrons has atom B? Are atoms A and B isotopes of the same element? Explain your answer.

14 The structure of one molecule of an industrial solvent, dichloromethane, is shown below.

```
      H
      |
 Cl — C — Cl
      |
      H
```

(a) How many atoms are there in one molecule of dichloromethane?

(b) How many bonds are there in one molecule of dichloromethane?

15 Describe how you could use a battery and a torch bulb to test various materials to find out whether they are electrical conductors.
Divide the following list into conductors and non-conductors:
silver, steel, polythene, PVC, brass, candle wax, lubricating oil, dilute sulphuric acid, petrol, alcohol, sodium hydroxide solution, sugar solution, limewater

16 (a) Name three types of substance that conduct electricity.

(b) Why can molten salts conduct electricity while solid salts cannot?

(c) Why can solutions of salts conduct electricity while pure water cannot?

(d) Explain why metal articles can be electroplated but plastic articles cannot.

17 Name the substances A to H in the table. Write equations for the discharges of the ions which form these substances.

Electrolyte	*Anode*	*Cathode*
Copper(II) chloride solution	A	B
Sodium chloride solution	C	D
Dilute sulphuric acid	E	F
Sodium hydroxide solution	G	H

18 Your aunt runs a business making souvenirs. She asks your advice on how to electroplate a batch of small brass medallions with copper. Describe how this could be done. Draw a diagram of the apparatus and the circuit she could use.

19 In the electrolysis of potassium bromide solution, what element is formed (a) at the positive electrode (b) at the negative electrode? Why does the solution around the negative electrode become alkaline?

20 (a) Explain why the ionic compound $PbBr_2$ melts at a higher temperature than the covalent compound CH_2Br_2.

(b) You can smell the compound $CHCl_3$, chloroform. You cannot smell the compound KCl. What difference in chemical bonding is responsible for the difference?

21 (a) Two experiments were set up as shown.

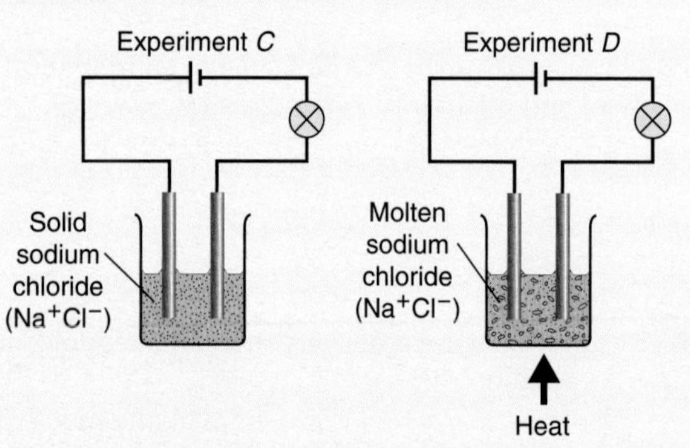

(i) Give *two* observations which would be seen only in Experiment *D*. [2]

(ii) Explain why in Experiment *C* no changes would be seen. [2]

(b) Another **electrolysis** experiment used an aqueous solution of copper chloride.

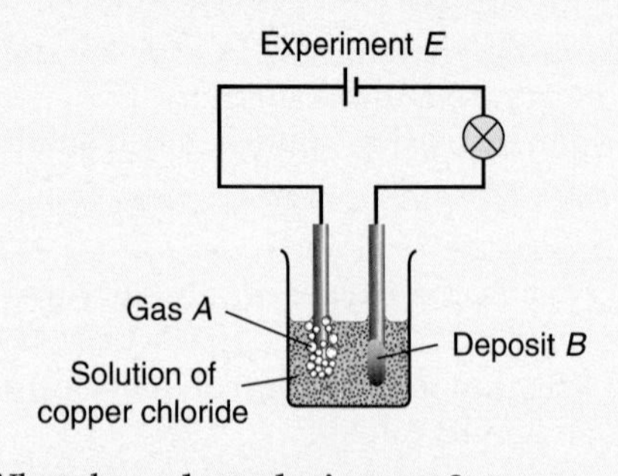

(i) What does **electrolysis** mean? [2]

(ii) Name the gas *A* and the deposit *B*. [2]

(c) Give *one* industrial use of electrolysis. [1]

AQA(SEG) 1999

22 The diagrams below show two different forms of carbon.

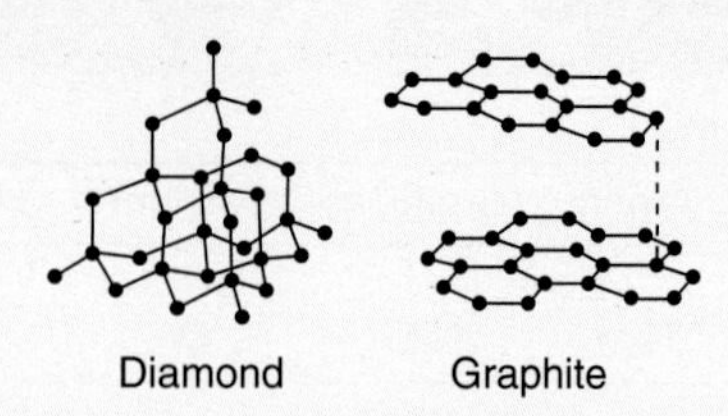

(i) Give the term used for different forms of the same element in the same physical state. [1]

(ii) Graphite conducts electricity.

I. Explain how this property is related to its structure. [2]

II. Give *one* use of graphite which relies on it being an electrical conductor. [1]

(iii) I. Explain how the hardness and high melting point of diamond are related to its structure. [2]

II. Give *one* use of diamond which relies on it being hard and having a high melting point. [1] (WJEC)

23 A student investigated the **electrolysis** of lead bromide.

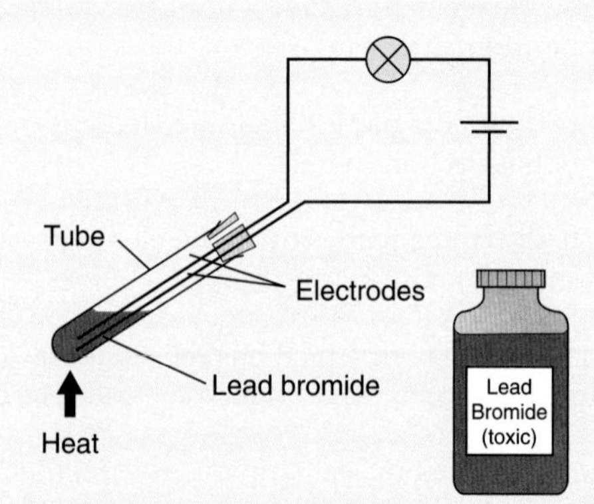

Lead bromide was placed in the tube and the circuit was switched on. The light bulb did not light up.

The tube was heated and soon the bulb lit up. The observations are shown in the table.

Positive electrode	*Negative electrode*
Red–brown gas	Silver liquid

(a) What is meant by **electrolysis**? [2]

(b) Why did the lead bromide conduct electricity when the tube was heated? [1]

(c) Name the substances formed at the: positive electrode; negative electrode. [2]

(d) Suggest *one* safety precaution that should be taken during this investigation. [1]

AQA(SEG) 2000

24 (a) Lithium, atomic number 3, reacts with oxygen, atomic number 8, to form lithium oxide.

(i) Give the electronic structures of the two elements. [2]

(ii) Explain, by means of a diagram or otherwise, the electronic changes that take place during the formation of lithium oxide. [4]

(iii) Give the chemical formula for lithium oxide. [1]

(b) Oxygen, atomic number 8, forms a compound with hydrogen, atomic number number 1, called water.

By means of a labelled diagram, show how the atoms of oxygen and hydrogen are bonded together in water. [2] (WJEC)

Theme C

Patterns of behaviour

Metals

Non-metals

Transition metals

1	2											3	4	5	6	7	0
			H														He
Li	Be											B	C	N	O	F	Ne
Na	Mg											Al	Si	P	S	Cl	Ar
K	Ca	Sc	Ti	V	Cr	Mn	Fe	Co	Ni	Cu	Zn	Ga	Ge	As	Se	Br	Kr
Rb	Sr	Y	Zr	Nb	Mo	Tc	Ru	Rh	Pd	Ag	Cd	In	Sn	Sb	Te	I	Xe
Cs	Ba	La	Hf	Ta	W	Re	Os	Ir	Pt	Au	Hg	Tl	Pb	Bi	Po	At	Rn

There are thousands of chemicals and hundreds of thousands of chemical reactions. Fortunately for us there is order, not chaos, in the universe. There are patterns of behaviour. All acids have reactions in common, bases have reactions in common, and acids and bases react together. Oxidising agents and reducing agents have complementary properties, and they react together in oxidation-reduction reactions.

The Periodic Table is a remarkable pattern of behaviour. The properties of all the elements, 105 of them, fall into place in this excellent system of classification.

Topic 9 Acids and bases

9.1 Acids

FIRST THOUGHTS

Can you say what the elements in each of these sets of materials have in common?

List A: lime juice, vitamin C, stomach juice, vinegar

List B: toothpaste, household ammonia, Milk of Magnesia, lime

If you have remembered, well done! You will learn more about acids and bases in this topic.

Cells in the linings of our stomachs produce hydrochloric acid. This is a powerful acid. If a piece of zinc was dropped into hydrochloric acid of the concentration which the stomach contains, it would dissolve. In the stomach, hydrochloric acid works to kill bacteria which are present in foods and to soften foods. It also helps in **digestion** by providing the right conditions for the enzyme pepsin to begin the digestion of proteins.

Sometimes the stomach produces too much acid. Then the result is the pain of 'acid indigestion' and 'heartburn'. The many products which are sold as 'antacids' and indigestion remedies all contain compounds that can react with hydrochloric acid to neutralise (counteract) the excess of acid. In this chapter, we shall look at the properties (characteristics) of acids and the substances which react with them.

Do not taste laboratory acids

A sour taste usually shows that a substance contains an acid. The word 'acid' comes from the Latin word for 'sour'. Vinegar contains ethanoic acid, sour milk contains lactic acid, lemons contain citric acid and rancid butter contains butanoic acid. For centuries, chemists have been able to extract acids such as these from animal and plant material. They call these acids **organic acids**. As chemistry has advanced, chemists have found ways of making sulphuric acid, hydrochloric acid, nitric acid and other acids from minerals. They call these acids **mineral acids**. Mineral acids in general react much more rapidly than organic acids. We describe mineral acids as **strong acids** and organic acids as **weak acids**. Table 9.1 lists some common acids.

This symbol means **corrosive**. Concentrated acids always carry this symbol.

Solutions of acids (and other substances) can be **dilute solutions** or **concentrated solutions**. A dilute solution contains a small amount of acid per litre of solution. A concentrated solution contains a large amount of acid per litre of solution. Solutions of acids are always called, say, dilute hydrochloric acid or concentrated hydrochloric acid. Concentrated acids are very corrosive, and you need to know which type of solution you are dealing with.

www.keyscience.co.uk

The table lists

- organic acids, many of which are weak acids
- mineral acids, many of which are strong acids

Table 9.1 Some common acids

Acid	Strong or weak	Where you find it
Ascorbic acid	Weak	In fruits. Also called Vitamin C
Carbonic acid	Weak	In fizzy drinks: these contain the gas carbon dioxide which reacts with water to form carbonic acid.
Citric acid	Weak	In fruit juices, e.g. lemon juice
Ethanoic acid	Weak	In vinegar
Hydrochloric acid	Strong	In digestive juices in the stomach; also used for cleaning ('pickling') metals before they are coated
Lactic acid	Weak	In sour milk
Nitric acid	Strong	Used for making fertilisers and explosives
Phosphoric acid	Strong	In anti-rust paint; used for making fertilisers
Sulphuric acid	Strong	In car batteries; used for making fertilisers

Check that you know the difference between

- a strong acid and a concentrated acid
- a weak acid and a dilute acid

Figure 9.1A ▲ Some common acids

Figure 9.1B ▲ The acid taste

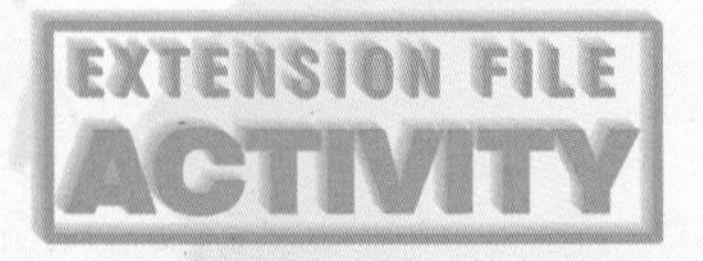

SUMMARY

Acids
- have a sour taste,
- change the colour of indicators,
- react with many metals to give hydrogen and a salt,
- react with carbonates and hydrogencarbonates to give carbon dioxide, a salt and water,
- react with bases (see Topic 9.2).

What do acids do?

1 **Acids have a sour taste.**
You will know the taste of lemon juice (which contains citric acid) and vinegar (which contains ethanoic acid). **Do not taste any of the strong acids.**

2 **Acids change the colour of substances called indicators.**
For example, acids turn blue litmus red (see Table 9.4).

3 **Acids react with many metals to produce hydrogen and a salt of the metal.**
Hydrogen is an element. It is a colourless, odourless (without smell) gas and is the least dense of all the elements. With air, hydrogen forms an explosive mixture, which is the basis of a test for hydrogen. If you put a lighted splint into hydrogen, you hear an explosive 'pop' (see Figure 9.1C).

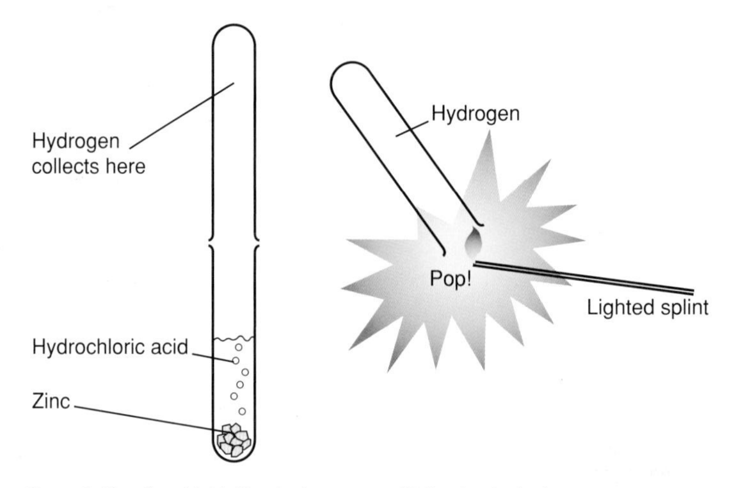

Figure 9.1C ▲ (a) Making hydrogen (b) Testing for hydrogen

Some metals, e.g. copper, react very slowly with acids. Some metals, e.g. sodium, react dangerously fast. Much heat is given out, and the hydrogen which is formed may explode. Examples of metals which react at a safe speed are magnesium, zinc and iron.

Zinc + Sulphuric acid → Hydrogen + Zinc sulphate

$Zn(s) + H_2SO_4(aq) \rightarrow H_2(g) + ZnSO_4(aq)$

In the salt, the metal replaces the hydrogen of the acid. For example when magnesium reacts with hydrochloric acid, $HCl(aq)$, hydrogen and the salt magnesium chloride, $MgCl_2$, are formed.

4 **Acids react with carbonates to give carbon dioxide, a salt and water.**
Acid indigestion is caused by an excess of hydrochloric acid in the stomach. Some indigestion tablets contain magnesium carbonate. The reaction that takes place when magnesium carbonate reaches the stomach is

Magnesium carbonate + Hydrochloric acid → Carbon dioxide + Magnesium chloride + Water

$MgCO_3(s) + 2HCl(aq) \rightarrow CO_2(g) + MgCl_2(aq) + H_2O(l)$

Other anti-acid tablets contain sodium hydrogencarbonate. In this case, the reaction that happens in the stomach is

Sodium hydrogencarbonate + Hydrochloric acid → Carbon dioxide + Sodium chloride + Water

$NaHCO_3(s) + HCl(aq) \rightarrow CO_2(g) + NaCl(aq) + H_2O(l)$

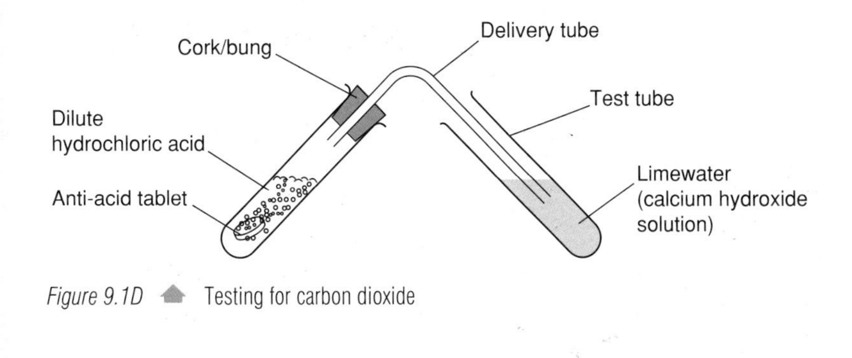

Figure 9.1D Testing for carbon dioxide

5 Acids neutralise bases
(see Topic 9.4.)

What is an acid?

A Swedish chemist called Svante Arrhenius explained why different acids have so much in common. He put forward the theory that aqueous solutions of all acids contain hydrogen ions, $H^+(aq)$, in high concentration. By 'high' concentration, he meant a concentration much higher than that in water. This theory explains many reactions of acids, including **neutralisation, electrolysis of solutions** of acids and the reaction of metals with acids

$$\text{Metal atoms} + \text{Hydrogen ions} \rightarrow \text{Metal ions} + \text{Hydrogen molecules}$$
$$M(s) + 2H^+(aq) \rightarrow M^{2+}(aq) + H_2(g)$$

Arrhenius gave this definition of an acid:

> An acid is a substance that releases hydrogen ions when dissolved in water.

A solution of a strong acid has a much higher concentration of hydrogen ions than a solution of a weak acid. A solution of a strong acid therefore reacts much more rapidly than a solution of a weak acid.

SUMMARY

Different acids have similar reactions. The reason is that solutions of all acids contain hydrogen ions in high concentration. The hydrogen ions are responsible for the typical reactions of acids.

CHECKPOINT

1. Where in the kitchen could you find
   (a) a weak acid with a pleasant taste?
   (b) a weak acid with an unpleasant taste?
   (c) a weak acid with a very sour taste?
2. What is the difference between a concentrated solution of a weak acid and a dilute solution of a strong acid?
3. What is the difference between a concentrated solution of sulphuric acid and a dilute solution of sulphuric acid? Why do road tankers of sulphuric acid carry the sign shown above?
4. *Toffee Recipe 1* Boil sugar and water with a little butter until the mixture thickens. Pour into a greased tray to set.
   Toffee Recipe 2 Repeat Recipe 1. When the mixture thickens, add vinegar and 'bicarbonate of soda' (sodium hydrogencarbonate). Pour into a greased tray to set.
   One of these recipes gives solid toffee. The other gives a honeycomb of toffee containing bubbles of gas. Which recipe gives the honeycomb? Why?

5 Rosie and Luke visited the underground caves in the limestone rock at Wookey Hole in Somerset. A guide told them that the chemical name for limestone is calcium carbonate. The children took a small piece of limestone rock to school and did an experiment with it. They put a piece of rock in a beaker and added dilute hydrochloric acid. Immediately, bubbles of gas were given off. (See the figures opposite.)

(a) Why must you always wear safety glasses when working with acids?

(b) What is the name of the gas in the bubbles?

(c) What chemical test could they do to 'prove' what the gas was?

Rosie lit a candle, and then tilted the beaker so that gas poured on to it (see lower figure). The candle went out.

(d) What two things about the gas did this experiment tell the children?

(e) What could the gas be used for?

Their teacher prepared a gas jar of the gas. Luke added distilled water and shook the gas jar. Then he added litmus solution.

(f) What colour did the indicator turn? (See Table 9.1 and *What do acids do?* for help.)

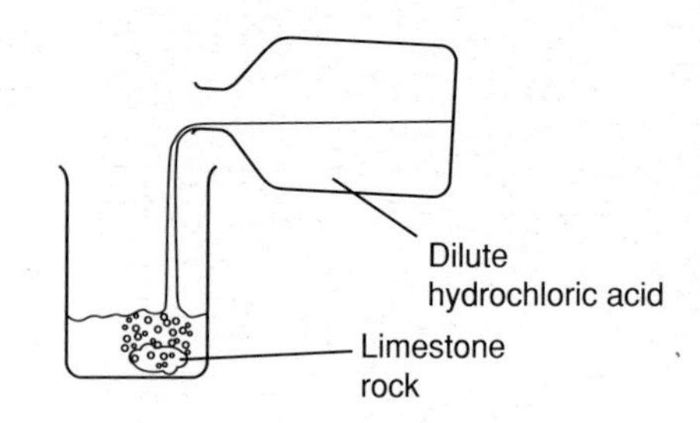

(a) Adding acid to limestone

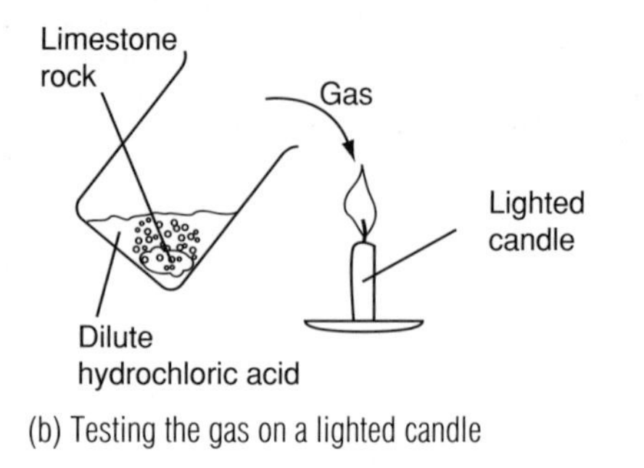

(b) Testing the gas on a lighted candle

9.2 Bases

Figure 9.2A Spreading lime

Bases are substances that neutralise (counteract) acids. Figure 9.2A shows a farmer spreading the base called 'lime' (calcium hydroxide) on a field. Some soils are too acidic to grow good crops. 'Liming' neutralises some of the acid and increases the **soil fertility**.

The product of the reaction between an acid and a base is a neutral substance, neither an acid nor a base, which is called a **salt**. Lime (calcium hydroxide) is a base. It reacts with nitric acid in the soil to form the salt calcium nitrate and water

Calcium hydroxide + Nitric acid → Calcium nitrate + Water

A definition of a base is:

> A base is a substance that reacts with an acid to form a salt and water only.

Acid + Base → Salt + Water

Soluble bases are called **alkalis**. Sodium hydroxide is an alkali; it reacts with an acid to form a salt and water.

Sodium hydroxide + Hydrochloric acid → Sodium chloride + Water

$NaOH(aq) + HCl(aq) \rightarrow NaCl(aq) + H_2O(l)$

Can you explain these terms:

- acid,
- base,
- salt,
- alkali?

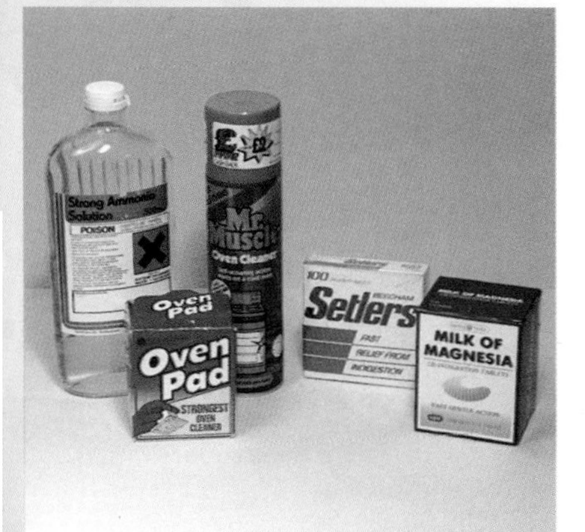

Figure 9.2B ▲ Some common bases

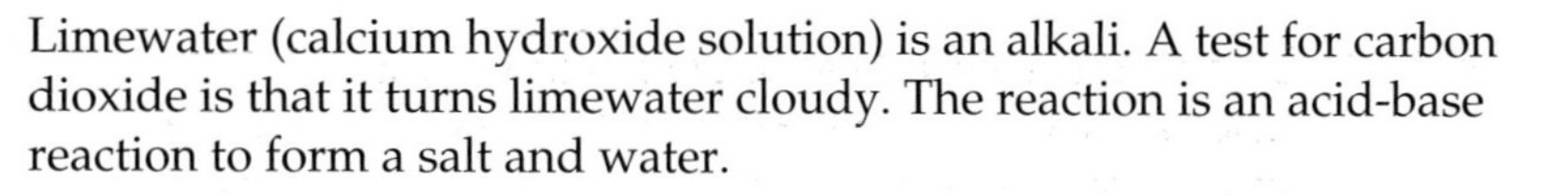

Limewater (calcium hydroxide solution) is an alkali. A test for carbon dioxide is that it turns limewater cloudy. The reaction is an acid-base reaction to form a salt and water.

$$\text{Carbon dioxide} + \text{Calcium hydroxide} \rightarrow \text{Calcium carbonate} + \text{Water}$$
$$CO_2(g) + Ca(OH)_2(aq) \rightarrow CaCO_3(s) + H_2O(l)$$

The salt, calcium carbonate, appears as a cloud of insoluble white powder.

Table 9.2 ▼ Some common bases

State	Strong or weak	Where you find it
Ammonia	Weak	In cleaning fluids for use as a degreasing agent; also used in the manufacture of fertilisers
Calcium hydroxide	Strong	Used to treat soil which is too acidic
Calcium oxide	Strong	Used in the manufacture of cement, mortar and concrete
Magnesium hydroxide	Strong	In anti-acid indigestion tablets and Milk of Magnesia
Sodium hydroxide	Strong	In oven cleaners as a degreasing agent; also used in soap manufacture

Table 9.2 lists some common bases. Different bases have a number of reactions in common.

1 Bases neutralise acids (see previous page).
2 Soluble bases can change the colour of **indicators**, e.g. turn red litmus blue (see Table 9.4).
3 Soluble bases feel soapy to your skin. The reason is that soluble bases convert some of the oil in your skin into **soap**. Some household cleaning solutions, e.g. ammonia solution, use soluble bases as degreasing agents. They convert oil and grease into soluble soaps which are easily washed away.
4 A solution of an alkali in water contains hydroxide ions, $OH^-(aq)$. This solution will react with a solution of a metal salt. Most metal hydroxides are insoluble. When a solution of an alkali is added to a solution of a metal salt, an insoluble metal hydroxide is **precipitated** from solution. (A precipitate is a solid which forms when two liquids are mixed.) For example

SUMMARY

Bases neutralise acids to form salts. Soluble bases are called alkalis. Metal oxides and hydroxides are bases. Alkalis change the colours of indicators. They are degreasing agents, and have a soapy 'feel'. Solutions of alkalis contain hydroxide ions, $OH^-(aq)$ in high concentration.

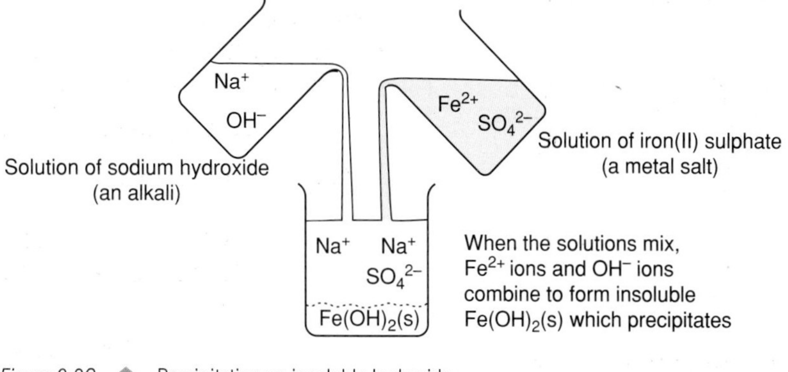

Figure 9.2C ▲ Precipitating an insoluble hydroxide

Iron(II) Sulphate (solution)	+	Sodium hydroxide (solution)	→	Iron(II) hydroxide (precipitate)	+	Sodium sulphate (solution)

$$FeSO_4(aq) + 2NaOH(aq) \rightarrow Fe(OH)_2(s) + Na_2SO_4(aq)$$

A solution of a strong base contains a higher concentration of hydroxide ions than a solution of a weak base.

Check that you know the difference between
- a strong base and a weak base
- a weak base and a dilute base
- a base and an alkali.

Table 9.3 Examples of bases

Metal oxides	*Metal hydroxides*	*Alkalis (soluble bases)*
Copper(II) oxide, CuO	Sodium hydroxide, $NaOH$	Sodium hydroxide, $NaOH$
Zinc oxide, ZnO	Magnesium hydroxide, $Mg(OH)_2$	Potassium hydroxide, KOH
		Calcium hydroxide, $Ca(OH)_2$
(*Most metal oxides and hydroxides are insoluble*)		Ammonia solution, $NH_3(aq)$

Ionic equations

An ionic equation shows the essentials.

There is another way of writing the equation of the reaction between iron(II) sulphate and sodium hydroxide. The reaction is the combination of iron(II) ions and hydroxide ions. The sodium ions and sulphate ions take no part in the reaction, and the equation can be written without them.

$$Fe^{2+}(aq) + 2OH^-(aq) \rightarrow Fe(OH)_2(s)$$

This is called an **ionic equation**.

CHECKPOINT

1 (a) Write the names of the four common alkalis.
(b) Write the names of four insoluble bases.

2 Four bottles of solution are standing on a shelf in the prep room. Their labels have come off and are lying on the floor. They read, Copper(II) sulphate, Iron(II) sulphate, Iron(III) sulphate and Zinc sulphate. The lab assistant knows that some insoluble metal hydroxides are coloured (see the table below).

Hydroxide	*Formula*	*Colour*
Copper(II) hydroxide	$Cu(OH)_2(s)$	Blue
Iron(II) hydroxide	$Fe(OH)_2(s)$	Green
Iron(III) hydroxide	$Fe(OH)_3(s)$	Rust
Magnesium hydroxide	$Mg(OH)_2(s)$	White
Zinc hydroxide	$Zn(OH)_2(s)$	White

She adds sodium hydroxide solution to a sample of each solution. Her results are:
Bottle 1 White precipitate
Bottle 2 Rust-coloured precipitate
Bottle 3 Blue precipitate
Bottle 4 Green precipitate
Say which label should be stuck on each bottle.

3 Write the formulas for the bases: calcium oxide, copper(II) oxide, zinc oxide, magnesium hydroxide, iron(II) hydroxide, iron(III) hydroxide.

9.3 ▶ Summarising the reactions of acids and bases

Figure 9.3A summarises the reactions of hydrochloric acid, a typical strong acid.

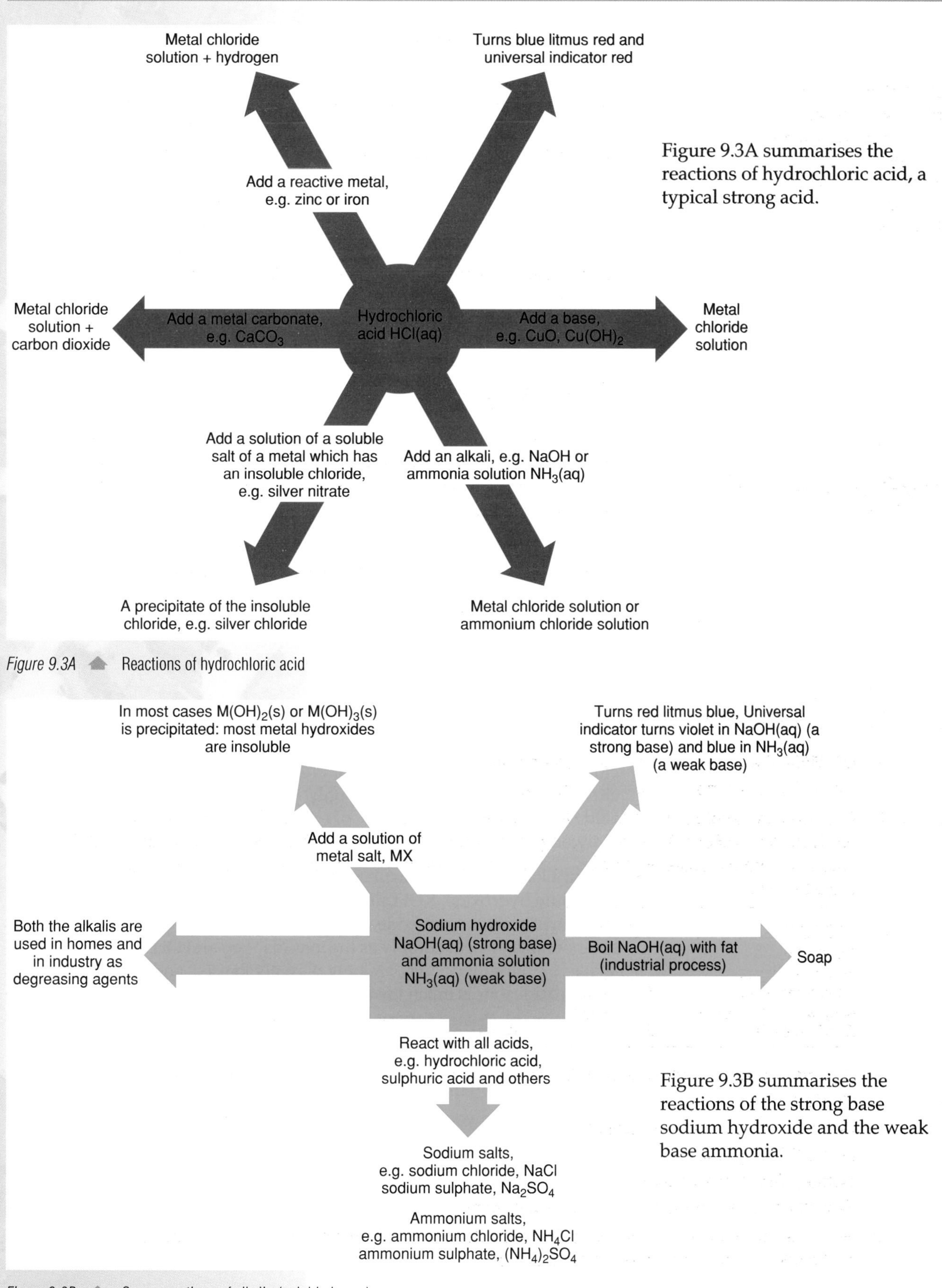

Figure 9.3A ▲ Reactions of hydrochloric acid

Figure 9.3B summarises the reactions of the strong base sodium hydroxide and the weak base ammonia.

Figure 9.3B ▲ Some reactions of alkalis (soluble bases)

The role of water

Water must be present for a compound to act as an acid or an alkali. Sherbert consists of sodium hydrogencarbonate and tartaric acid. Put a little on your tongue, and it will start to fizz. The reason for the fizz is that bubbles of carbon dioxide are formed by the reaction between tartaric acid and sodium hydrogencarbonate:

Hydrogen ions from acid	+	Hydrogencarbonate ions	→	Carbon dioxide	+	Water
$H^+(aq)$	+	$HCO_3^-(s)$	→	$CO_2(g)$	+	$H_2O(l)$

Why doesn't the sherbert fizz in the bag? There is no water present. Water on your tongue started the reaction. Tartaric acid cannot act as an acid without the presence of water.

Hydrogen chloride dissolves in water to form hydrochloric acid:

$$HCl(g) + aq \rightarrow HCl(aq)$$

When hydrogen chloride dissolves in the anhydrous (without water) covalent solvent methylbenzene the solution formed is not acidic. The solution does not react with magnesium to give hydrogen or with carbonates to give carbon dioxide. On the addition of water to the solution, hydrogen chloride passes from the methylbenzene layer into the water layer to form hydrochloric acid, which has all the usual reactions, e.g. with metals and with carbonates. The reaction that happens when hydrogen chloride dissolves in water is

$$HCl(g) + aq \rightarrow H^+(aq) + Cl^-(aq)$$

The molecule of HCl cannot split to release hydrogen ions unless there is a substance ready to accept hydrogen ions, e.g. water.

Weak acids and bases

Strong acids, e.g. hydrochloric acid, sulphuric acid and nitric acid, are completely ionised in solution. A solution of hydrochloric acid contains no molecules of HCl; it is completely in the form of hydrogen ions and chloride ions, $H^+(aq)$ and $Cl^-(aq)$. Weak acids, such as ethanoic acid, citric acid and carbonic acid, are only ionised to a small extent. Ethanoic acid in solution consists mainly of ethanoic acid molecules, CH_3CO_2H. Only one percent of molecules are ionised as $H^+(aq)$ and $CH_3CO_2^-(aq)$.

Strong alkalis (soluble bases), e.g. sodium hydroxide, NaOH, and potassium hydroxide, KOH, are completely ionised in solution. In a solution of sodium hydroxide, there are no molecules of NaOH; the compound is present entirely as the ions $Na^+(aq)$ and $OH^-(aq)$. Weak alkalis, e.g. ammonia, NH_3, are only partially ionised, and the concentration of hydroxide ions is much lower than in a solution of a strong base.

CHECKPOINT

1. Give an example of (a) a concentrated solution of a weak acid (b) a dilute solution of a strong acid.
2. Benzoic acid is a crystalline solid. How could you show that benzoic acid must be dissolved in water before it acts as an acid?
3. You are given solutions of sodium hydroxide and ammonia of the same concentration. How could you find out which solution has the higher concentration of hydroxide ions? Which solution would it be?
4. You are given solutions of acids P and Q of the same concentration. How can you find out which is the stronger acid?

9.4 Neutralisation

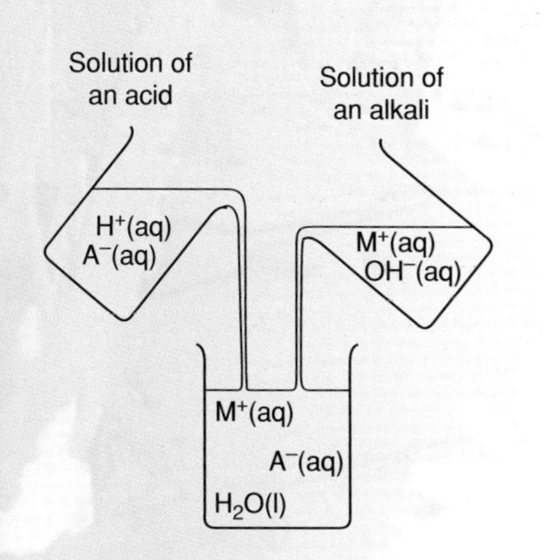

H⁺(aq) ions and OH⁻(aq) ions have combined to form water molecules, $H_2O(l)$

The solution is a solution of the salt MA

Figure 9.4A Acid + alkali → salt + water

What takes place when an acid neutralises a soluble base? An example is the reaction between hydrochloric acid and the alkali (soluble base) sodium hydroxide solution.

Hydrochloric acid	+	Sodium hydroxide	→	Sodium chloride	+	Water
HCl(aq)	+	NaOH(aq)	→	NaCl(aq)	+	$H_2O(l)$
acid	+	*alkali*	→	*salt*	+	*water*

When the solutions of acid and alkali are mixed, hydrogen ions, $H^+(aq)$, and hydroxide ions, $OH^-(aq)$, combine to form water molecules.

$$H^+(aq) + OH^-(aq) \rightarrow H_2O(l)$$

Sodium ions, $Na^+(aq)$, and chloride ions, $Cl^-(aq)$, remain in the solution, which becomes a solution of sodium chloride. If you evaporate the solution, you obtain solid sodium chloride.

> Neutralisation is the combination of hydrogen ions from an acid and hydroxide ions from a base to form water molecules. In the process, a salt is formed.

What happens when an acid neutralises an insoluble base? An example is the reaction between sulphuric acid and copper(II) oxide

Sulphuric acid	+	Copper(II) oxide	→	Copper(II) sulphate	+	Water
$H_2SO_4(aq)$	+	CuO(s)	→	$CuSO_4(aq)$	+	$H_2O(l)$
acid	+	*base*	→	*salt*	+	*water*

Hydrogen ions and oxide ions, O^{2-}, combine to form water

$$2H^+(aq) + O^{2-}(s) \rightarrow H_2O(l)$$

The resulting solution contains copper(II) ions and sulphate ions: it is a solution of copper(II) sulphate. If you evaporate it, you will obtain copper(II) sulphate crystals.

Compounds used as **anti-acids** include:
magnesium hydroxide, $Mg(OH)_2$, used in remedies for acid indigestion,
aluminium hydroxide, $Al(OH)_3$, used in indigestion tablets,
magnesium carbonate, $MgCO_3$, used in indigestion remedies,
calcium carbonate, $CaCO_3$, used on soil to neutralise excess acidity,
sodium hydrogencarbonate, $NaHCO_3$, used in indigestion remedies.

SUMMARY

Neutralisation is the combination of hydrogen ions (from an acid) with hydroxide ions (from an alkali or an insoluble base) or oxide ions (from an insoluble base) to form water and a salt.

CHECKPOINT

1. Bee stings hurt because bees inject acid into the skin. Wasp stings hurt because wasps inject alkali into the skin.
   Your little brother is stung by a bee. You are in charge. What do you use to treat the sting: 'bicarbonate of soda' (sodium hydrogencarbonate), calamine lotion (zinc carbonate), vinegar (ethanoic acid) or Milk of Magnesia (magnesium carbonate)? Would you use the same treatment for a wasp sting? Give reasons for your answers.
2. 'Acid drops' which you buy from a sweet shop contain citric acid.
   - Put an acid drop in your mouth.
   - Put some baking soda (sodium hydrogencarbonate) on your hand.
   - Lick some of the baking soda into your mouth.

   What happens to the taste of the acid drop? Why does this happen?

9.5 Indicators and pH

Table 9.4 The colours of some common indicators

indicator	Acidic colour	Neutral colour	Alkaline colour
Litmus	Red	Purple	Blue
Phenolphthalein	Colourless	Colourless	Pink
Methyl orange	Red	Yellow	Yellow

Universal indicator turns different colours in strongly acidic and weakly acidic solutions. It can also distinguish between strongly basic and weakly basic solutions (see Figure 9.5A). Each universal indicator colour is given a **pH number**. The pH number measures the acidity or alkalinity of the solution. For a neutral solution, pH = 7; for an acidic solution, pH < 7; for an alkaline solution, pH > 7.

SUMMARY

In general, mineral acids are stronger than organic acids. Some bases are stronger than others. Ammonia is a weak base. The pH of a solution is a measure of its acidity or alkalinity. Universal indicator turns different colours in solutions of different pH.

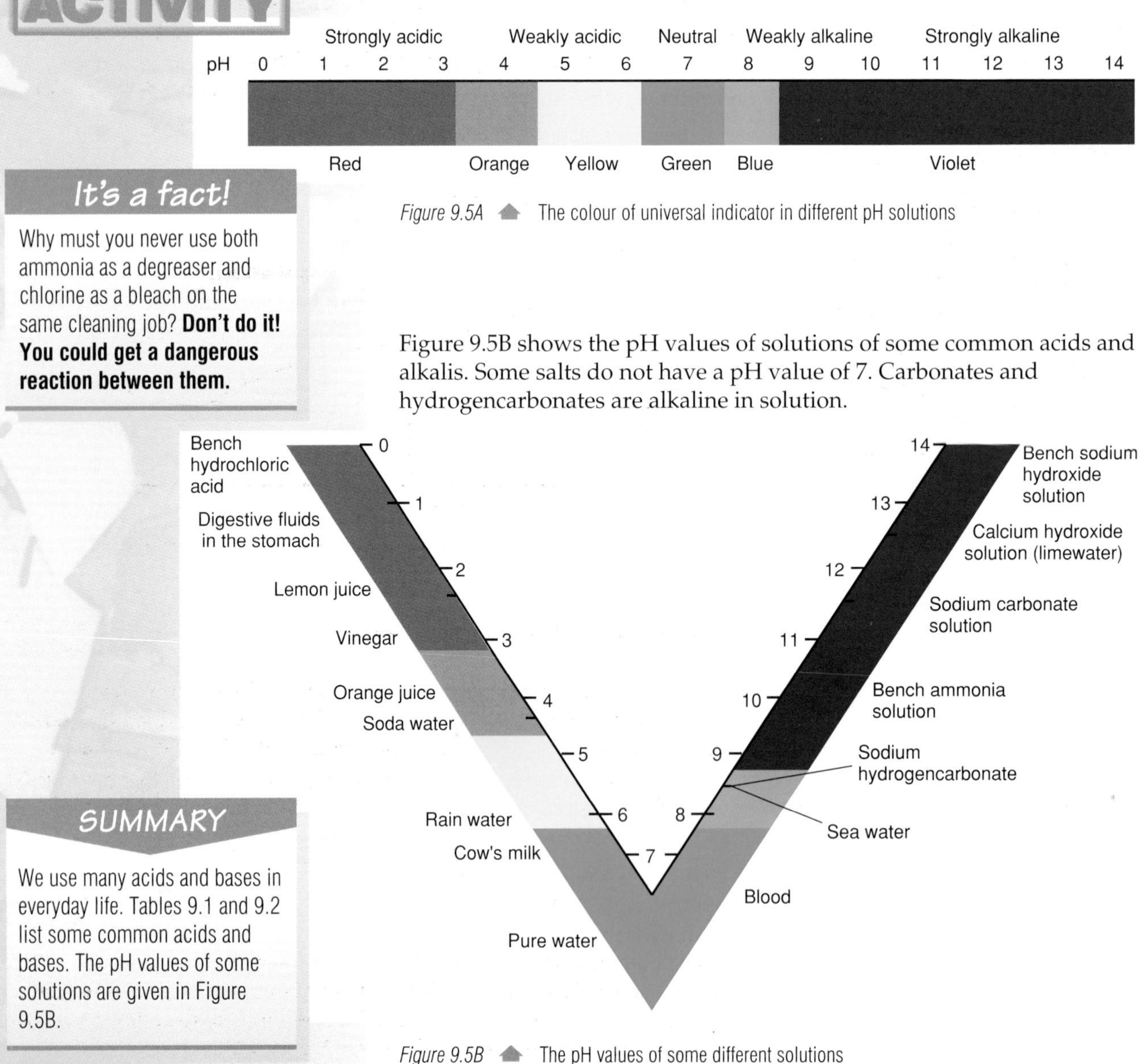

Figure 9.5A The colour of universal indicator in different pH solutions

It's a fact!

Why must you never use both ammonia as a degreaser and chlorine as a bleach on the same cleaning job? **Don't do it! You could get a dangerous reaction between them.**

Figure 9.5B shows the pH values of solutions of some common acids and alkalis. Some salts do not have a pH value of 7. Carbonates and hydrogencarbonates are alkaline in solution.

Figure 9.5B The pH values of some different solutions

SUMMARY

We use many acids and bases in everyday life. Tables 9.1 and 9.2 list some common acids and bases. The pH values of some solutions are given in Figure 9.5B.

9.6 Titration

Burette

Graduation marks

Tap – when open it lets liquid flow out of the burette

0 10 20 30 40 50

Figure 9.6A A burette

Neutralisation can be carried out gradually. When an acid is added slowly to an alkali the pH of the solution gradually changes. It starts at pH 9–14 and gradually falls to pH 7 when the amount of acid added has exactly neutralised the amount of alkali. As more acid is added, the pH falls further. If the neutralisation is carried out gradually in a measured way, it is called a **titration**. Titration is the gradual addition of one reactant to another in a measured way. A method of adding acid in a measured way is to use a **burette**. This is a graduated tube with a tap near the bottom. The graduation marks enable you to read what volume of acid has been added.

Figure 9.6B A pH meter

pH

The course of a neutralisation titration can be followed by measuring the changing pH as acid is added. An instrument called a pH meter (Figure 9.6B) is immersed in the alkali. It measures the pH after the addition of a measured volume of acid. The pH is plotted against the volume of acid added to give a graph like that in Figure 9.6C. A computer with a data-logging program linked to a pH meter will plot such a graph automatically.

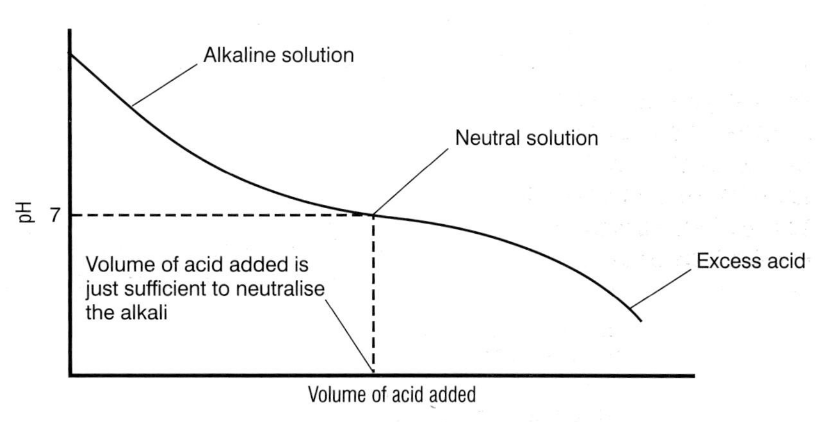

Figure 9.6C Change in pH during a titration plotted against volume of acid added

Colour of the indicator

The colour of an indicator changes with pH. If the alkali contains universal indicator the colour will be violet. As acid is added the colour changes through blue to the neutral colour green. If addition of acid continues the colour of the indicator changes to yellow, orange and then red (see Figure 9.5A).

Science at work

A pH sensor and computer can be used to record and display changes in pH.

Temperature

Heat is liberated during neutralisation; the reaction is exothermic. With a thermometer placed in the solution, temperature readings can be taken. The maximum temperature recorded corresponds to the addition of just enough acid to neutralise the alkali; see Figure 9.6D.

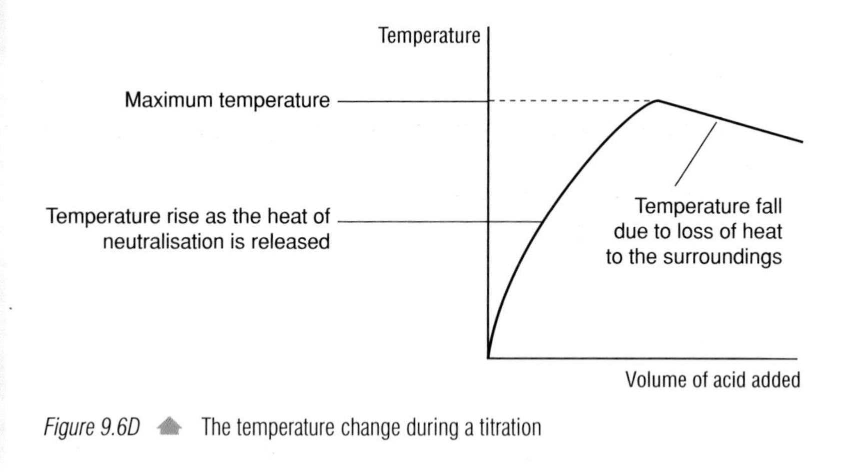

Figure 9.6D ▲ The temperature change during a titration

Conductance

The neutralisation reaction is

Hydrogen ions (from acid)	+	Hydroxide ions (from alkali)	→	Water
$H^+(aq)$	+	$OH^-(aq)$	→	$H_2O(l)$

When acid is run into a solution of an alkali, hydrogen ions from the acid remove hydroxide ions from the solution of alkali. As a result the ability of the alkaline solution to conduct electricity – the conductance of the solution – falls.

A conductance meter placed in the solution measures the current which is able to flow through the solution. The way in which conductance changes during a titration is shown in Figure 9.6E.

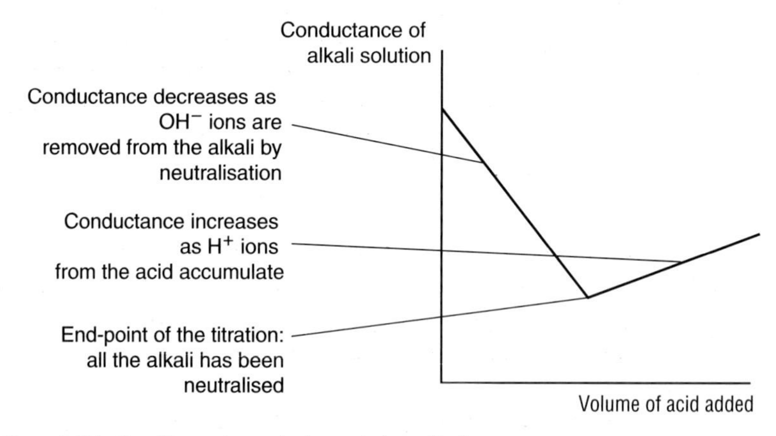

Figure 9.6E ▲ Change in conductance during a titration

Titration can be carried out by adding alkali to acid. It is however usual to have the acid in the burette because grease is used to lubricate the tap and this is saponified by alkali. Oxidation–reduction reactions can also be studied by following the course of titration. You will learn more about titration in Topic 26.11.

SUMMARY

The course of a neutralisation reaction can be followed by measuring
- the change in pH
- the change in colour of an indicator
- the rise in temperature
- the fall in conductance.

CHECKPOINT

1. Say whether these substances are strongly acidic, weakly acidic, neutral, weakly basic or strongly basic:
   (a) cabbage juice, pH = 5.0
   (b) pickled cabbage, pH = 3.0
   (c) milk, pH = 6.5
   (d) tonic water, pH = 8.2
   (e) washing soda, pH = 11.5
   (f) saliva, pH = 7.0
   (g) blood, pH = 7.4

2. You know what it feels like to be stung by a nettle. Is the substance which the nettle injects an acid or an alkali or neither? Think up a method of extracting some of the substance from a nettle and testing to see whether the extract is acidic or alkaline or neutral. If your teacher approves of your plan, try it out.

3. Figure 9.5A shows the colours of universal indicator. What colour would you expect universal indicator to be when added to each of the following? (a) distilled water (b) lemon juice (c) household ammonia (d) battery acid (e) oven cleaner

4.

Liquid	*pH value*	*Reaction with acid*
A	1.0	None
B	8.5	Produces a salt, carbon dioxide and water
C	8.5	Produces a salt and water
D	13	Produces a salt and water

Uncle Harry is suffering from acid indigestion. Explain to him why it would not be a good idea to drink either Liquid A or Liquid D. Would you advise him to take Liquid B or Liquid C? Explain your choice.

5. The oven sprays which are sold for cleaning greasy ovens contain a concentrated solution of sodium hydroxide.
Why does sodium hydroxide clean the greasy oven?
Why does sodium hydroxide work better than ammonia?
What two safety precautions should you take to protect yourself when using an oven spray?
Why do domestic cleaning fluids contain ammonia, rather than sodium hydroxide?
Why do soap manufacturers use sodium hydroxide, rather than ammonia?

6.

Crop	*Wheat*	*Potatoes*	*Sugar beet*
pH	6	9	7

The table shows the most suitable values of pH for growing some crops.
   (a) Which crop grows best in (i) an acidic soil and (ii) an alkaline soil?
   (b) A farmer wanted to grow sugar beet. On testing, he found that his soil had a pH of 5. Name a substance which could be added to the soil to make its pH more suitable for growing sugar beet.
   (c) How does the substance you mention in (b) act on the soil to change its pH?

7. Describe a test which you could do in the laboratory to show that citric acid is a weaker acid than sulphuric acid.

8. Pair them up. Give the pH of each of the solutions listed.

Solution		*pH*
1 Ethanoic acid	A	7.0
2 Sodium chloride	B	1.0
3 Sulphuric acid	C	5.0
4 Ammonia	D	13.0
5 Sodium hydroxide	E	9.0

Topic 10 Salts

10.1 Sodium chloride

FIRST THOUGHTS

Sodium chloride, NaCl, is the salt which we call 'common salt' or simply 'salt'. The average human body contains about 250 g of sodium chloride.

www.keyscience.co.uk

Sodium chloride is essential for life: it enables muscles to contract, it enables nerves to conduct nerve impulses, it regulates osmosis and it is converted into hydrochloric acid, which helps digestion to take place in the stomach. Deprived of sodium chloride, the body goes into convulsions; then paralysis and death may follow. When we sweat, we lose both water and sodium chloride. We also **excrete** sodium chloride in urine. Our kidneys control the quantity of sodium chloride which we excrete. If we eat too much salt, our kidneys excrete sodium chloride; if we eat too little salt, our kidneys excrete water but no sodium chloride.

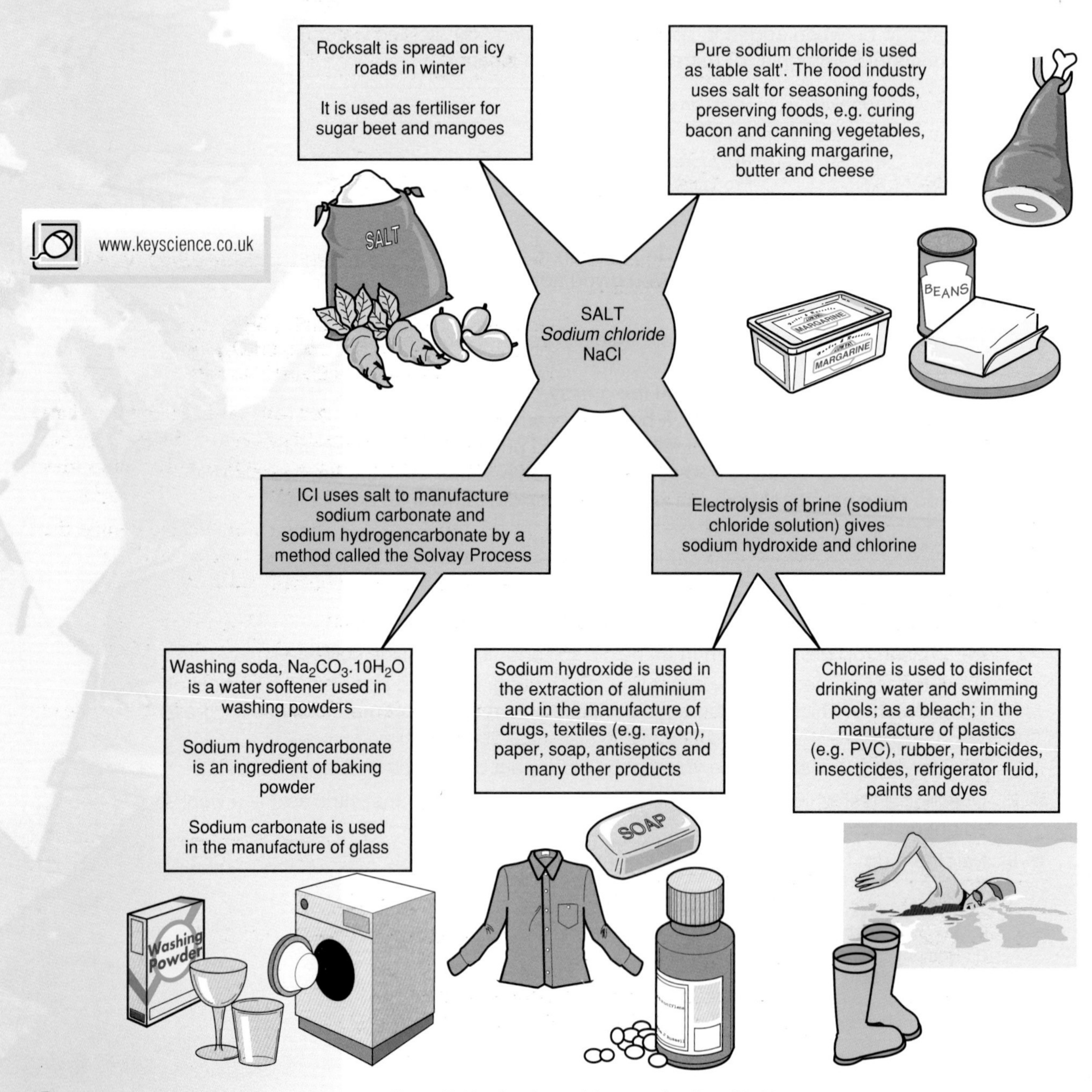

Figure 10.1A ▲ Some of the uses of sodium chloride

10.2 Salts of some common acids

It's a fact!

A French army under Napoleon I invaded Russia in 1812. The harsh Russian winter forced the French army to retreat. Salt starvation was one of the hardships which Napoleon's soldiers endured on their retreat from Moscow. Shortage of salt lowered their resistance to disease and epidemics spread; it prevented their wounds from healing and led to infection and death.

Many salts can be made by neutralising acids. In a salt, the hydrogen ions in the acid have been replaced by metal ions or by the ammonium ion. Salts of hydrochloric acid are called **chlorides**. Salts of sulphuric acid are called **sulphates**. Salts of nitric acid are called **nitrates**. See Table 10.1.

Table 10.1 Salts of some common acids

Acids	Salts
Hydrochloric acid HCl(aq)	Sodium chloride, NaCl Calcium chloride, $CaCl_2$ Iron(II) chloride, $FeCl_2$ Ammonium chloride, NH_4Cl
Sulphuric acid H_2SO_4(aq)	Sodium sulphate, Na_2SO_4 Zinc sulphate, $ZnSO_4$ Iron(III) sulphate, $Fe_2(SO_4)_3$
Nitric acid HNO_3	Sodium nitrate, $NaNO_3$ Copper(II) nitrate, $Cu(NO_3)_2$

10.3 Water of crystallisation

The crystals of some salts contain water. Such salts are called **hydrates**. Examples are:

Copper(II) sulphate-5-water, $CuSO_4.5H_2O$, *blue*
Cobalt(II) chloride-6-water, $CoCl_2.6H_2O$, *pink*
Iron(II) sulphate-7-water, $FeSO_4.7H_2O$, *green*

The water present in the crystals of hydrated salts gives them their shape and their colour. It is called **water of crystallisation**. The formula of the hydrate shows the proportion of water in the crystal, e.g. five molecules of water to each pair of copper ions and sulphate ions: $CuSO_4.5H_2O$.

When blue crystals of copper(II) sulphate-5-water are heated gently, the water of crystallisation is driven off. The white powdery solid that remains is anhydrous (without water) copper(II) sulphate.

Copper(II) sulphate-5-water	→	Water	+	Copper(II) sulphate
(blue crystals)	→	(vapour)	+	(white powder, *anhydrous*)
$CuSO_4.5H_2O$(s)	→	$5H_2O$(g)	+	$CuSO_4$(s)

If copper(II) sulphate crystals are left in the air, they slowly lose some or all of their water of crystallisation.

SUMMARY

- Some salts occur naturally, but most of them must be made by the chemical industry.
- Some salts crystallise with water of crystallisation. This gives the crystals their shape and, in some cases, their colour.

CHECKPOINT

1. Name the salts with the following formulas:
   NaI, NH_4NO_3, KBr, $BaCO_3$, Na_2SO_4, $ZnCl_2$, $CrCl_3$, $NiSO_4$, $CaCl_2$, $MgCl_2$.
2. Write formulas for the salts: potassium nitrate, potassium bromide, iron(II) chloride, iron(II) sulphate, iron(III) bromide, ammonium chloride.
3. State the number of oxygen atoms in each of these formulas:
   (a) Pb_3O_4 (b) $3Al_2O_3$ (c) $Fe(OH)_3$ (d) $MgSO_4.7H_2O$ (e) $Co(NO_3)_3.9H_2O$

4 The table shows how samples of four substances change in mass when they are left in the air.

Substance	*Mass of sample fresh from bottle (g)*	*Mass of sample after 1 week in the air (g)*
A	13.10	13.21
B	15.25	15.25
C	11.95	5.01

Which of the three substances in the table (a) loses water of crystallisation on standing (b) absorbs water from the air and (c) is unchanged by exposure to air?

5 *Barbara:* Bother! The holes in the salt cellar are blocked again. I wonder why that happens.
Razwan: It happens because sodium chloride absorbs water from the air.
Gwynneth: My mum puts a few grains of rice in the salt cellar. That stops it getting clogged.
(a) Explain why the absorption of water by sodium chloride will block the salt cellar.
(b) Explain how rice grains stop the salt cellar clogging.
(c) Describe an experiment which you could do to show that what you say in (b) is correct.

6 When a tin of biscuits is left open, do the biscuits absorb water vapour from the air or give out water vapour? Describe an experiment which you could do to find out.

10.4 Useful salts

FIRST THOUGHTS

Salts – who needs them? Farmers, doctors, dentists, industrial manufacturers, photographers: they all use salts in their work.

Agricultural uses

NPK fertilisers are mixtures of salts. They contain ammonium nitrate and ammonium sulphate to supply nitrogen to the soil. They contain phosphates to supply phosphorus and potassium chloride as a source of potassium.

Potato blight used to be a serious problem. It destroyed the potato crop in Ireland in 1846. Thousands of people starved and many more were forced to leave the country in search of food. Now, the fungus which causes potato blight can be killed by spraying with a solution of a simple salt, copper(II) sulphate. This fungicide is also used on vines to protect the grape harvest.

Figure 10.4A NPK fertiliser

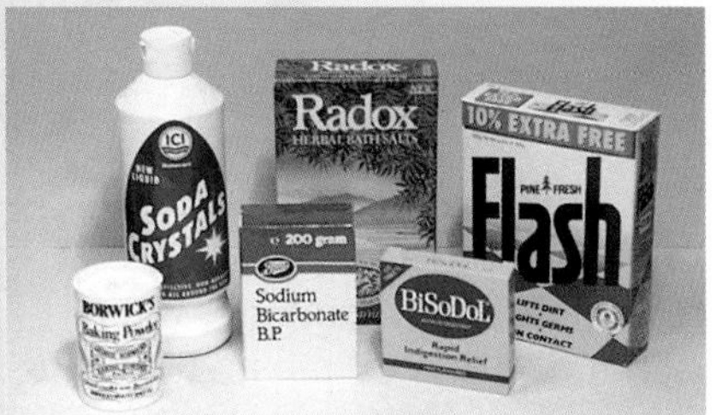

Figure 10.4B ▲ Washing soda, bath salts, baking powder

Domestic uses

Sodium carbonate-10-water is known as 'washing soda'. It is used as a **water softener**. It is an ingredient of washing powders and is also sold as bath salts.

Sodium hydrogencarbonate is known as 'baking soda'. It is added to self-raising flour. When heated at a moderate oven temperature, it decomposes.

Sodium hydrogencarbonate	→	Sodium carbonate	+	Carbon dioxide	+	Steam
$2NaHCO_3(s)$	→	$Na_2CO_3(s)$	+	$CO_2(g)$	+	$H_2O(g)$

The carbon dioxide and steam which are formed make bread and cakes 'rise'.

Medical uses

Figure 10.4C shows someone's leg being set in a plaster cast. Plaster of Paris is the salt calcium sulphate-$\frac{1}{2}$-water, $CaSO_4.\frac{1}{2}H_2O$. It is made by heating calcium sulphate-2-water, $CaSO_4.2H_2O$, which is mined. When plaster of Paris is mixed with water, it combines and sets to form a strong 'plaster cast'. It is also used for plastering walls.

People who are always tired and lacking in energy may be suffering from anaemia. This illness is caused by a shortage of haemoglobin (the red pigment which contains iron) in the blood. It can be cured by taking iron compounds in the diet. The 'iron tablets' which anaemic people may be given often contain iron(II) sulphate-7-water.

Patients who are suspected of having a stomach ulcer may be given a 'barium meal'. It contains the salt barium sulphate. After a barium meal, an X-ray photograph of the body shows the path taken by the salt. Being large, barium ions show up well in X-ray photographs.

Science at work

Liquid crystals are a form of matter which has a regular structure like a solid and which flows like a liquid. Many liquid crystals change their colour with the temperature. A recent invention uses the behaviour of liquid crystals to diagnose appendicitis. A thin plastic film coated with a suitable liquid crystal is placed on the patient's abdomen. Since an inflamed appendix produces heat, it shows up immediately as a change in the colour of the film. Appendicitis can be difficult to diagnose by other methods because the pain is often felt some distance from the appendix. Can you think of some further uses of liquid crystal temperature strips?

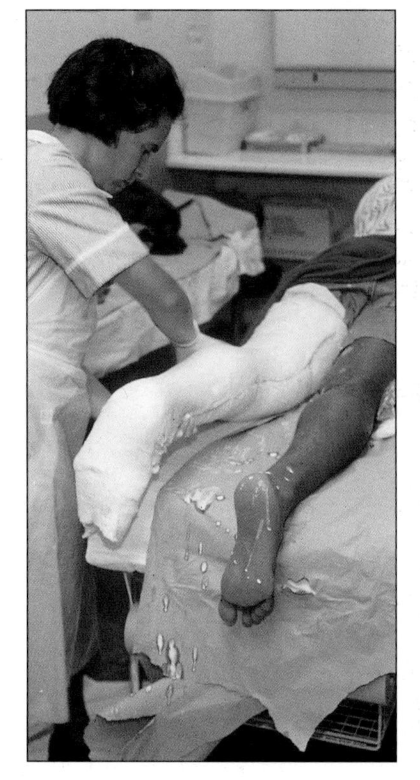

Figure 10.4C ▲ A patient having his leg set in a plaster cast

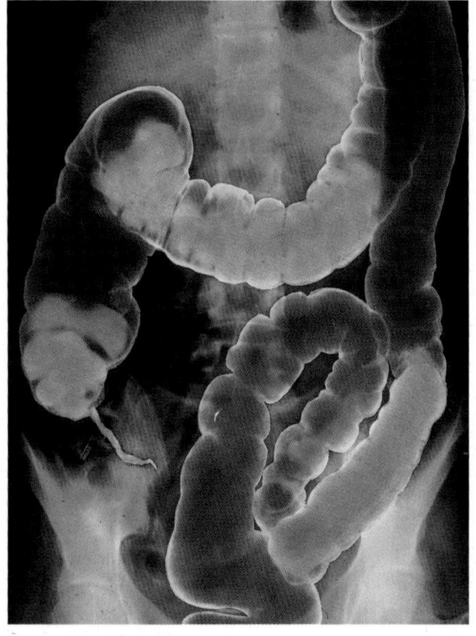

Figure 10.4D ▲ X-ray photograph of a patient after consuming a barium meal

KEY SCIENTIST

The fluoride story started in 1901 when Fred McKay began work as a dentist in Colorado, USA. Many of his patients had mottled teeth or dark brown stains on their teeth. He could find no information about this condition and called it 'Colorado Brown Stain'. He became convinced that the stain was caused by something in the water. After Oakley in Idaho changed its source of drinking water, the children there no longer developed mottled or stained teeth. The drinking water in a number of towns was analysed and showed that teeth were affected when there were over 2 p.p.m (parts per million) of fluoride in the drinking water. It affected children's teeth more than adults' teeth. McKay also noticed that people with 'Colorado Brown Stain' had less dental decay than other patients!

Fluorides for healthy teeth

Why do dentists recommend that people use toothpastes containing the salt calcium fluoride? Tooth enamel reacts with calcium fluoride to form a harder enamel which is better at resisting attack by mouth acids. To make sure that everyone gets protection from tooth decay, many water companies add a small amount of calcium fluoride to drinking water. The concentration of fluoride ions must not rise above 1 p.p.m. Spending 10p per person each year on fluoridation saves the National Health Service £10 per person each year in dentistry. Some people are violently opposed to the fluoridation of water supplies. This is because they are worried about the effects of drinking too much fluoride. Worldwide, 250 million people drink fluoridated water.

Industrial uses

Sodium chloride has many uses in industry; see Figure 10.1A.

Photography

Silver bromide, AgBr, is the salt which is used in black and white photography. Light affects silver bromide. A photographic film is a piece of celluloid covered with a thin layer of gelatin containing silver bromide. When you take a photograph, light falls on to some areas of the film. In exposed areas of film, light converts some silver ions, Ag^+, into silver atoms, which are black. These black atoms form a *latent image* of the object.

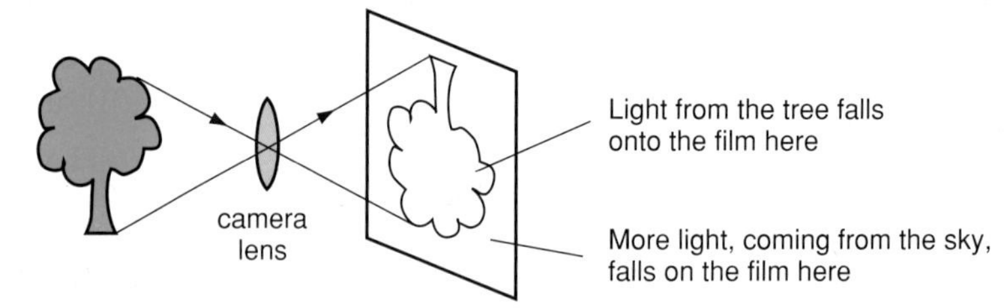

1. The film is **exposed**. The light converts some silver ions, Ag^+, into silver atoms, which are black. The more light falls on the film, the more silver ions are converted. These black atoms form a **latent image** of the tree ('latent' means hidden)

2. The film is **developed**. It is placed in a 'developer' solution which continues the process started by light, converting more silver ions in the areas affected by light into silver atoms
3. The film is placed into a solution of **fixer** which removes unchanged silver bromide
4. The result is a **negative**. The tree appears light against a dark background
5. To make a **print** the negative is placed on a piece of light-sensitive printing paper and exposed to light. The pattern of dark and light areas in the print is the reverse of that of the negative

Figure 10.4E ▲ Black and white photography

Some of these useful salts are mined, e.g. sodium chloride, magnesium sulphate and calcium sulphate. Others must be made by the chemical industry, e.g. silver bromide and copper(II) sulphate.

SUMMARY

Salts are used
- in the home, e.g. for cleaning and in baking
- in medicine, e.g. plaster casts, 'iron' tablets, 'barium meals'
- in dentistry, e.g. fluorides to protect against decay
- in photography – silver bromide and 'fixer' and 'developer'.

Some salts occur naturally; others do not. Methods of making salts are therefore important.

10.5 ▸ Methods of making salts

FIRST THOUGHTS

Since salts are so useful, chemists have found methods of making them. This topic describes the methods.

SUMMARY

Soluble salts are made by the reactions:
- Acid + Reactive metal → Salt + Hydrogen
- Acid + Metal oxide → Salt + Water
- Acid + Metal carbonate → Salt + Water + Carbon dioxide

In these three methods, an excess of the solid reactant is used, and unreacted solid is removed by filtration.

The method which you choose for making a salt depends on whether it is soluble or insoluble. Soluble salts are made by neutralising an acid. Insoluble salts are made by adding two solutions. Table 10.2 summarises the facts about the solubility of salts.

Table 10.2 ▼ Soluble and insoluble salts

Salts	*Soluble*	*Insoluble*
Chlorides	Most are soluble	Silver chloride Lead(II) chloride
Sulphates	Most are soluble	Barium sulphate Calcium sulphate Lead(II) sulphate
Nitrates	All are soluble	None
Carbonates	Sodium and potassium carbonates	Most are insoluble
Ethanoates	All are soluble	None
Sodium salts	All are soluble	None
Potassium salts	All are soluble	None
Ammonium salts	All are soluble	None

Methods for making soluble salts

An acid is neutralised by adding a metal, a solid base or a solid metal carbonate or a solution of an alkali.

Method 1: Acid + Metal → Salt + Hydrogen
Method 2: Acid + Metal oxide → Salt + Water
Method 3: Acid + Metal carbonate → Salt + Water + Carbon dioxide
Method 4: Acid + Alkali → Salt + Water

The practical details of Methods 1–3 are as follows.

Step one Add an excess (more than enough) of the solid to the acid (see Figure 10.5A).

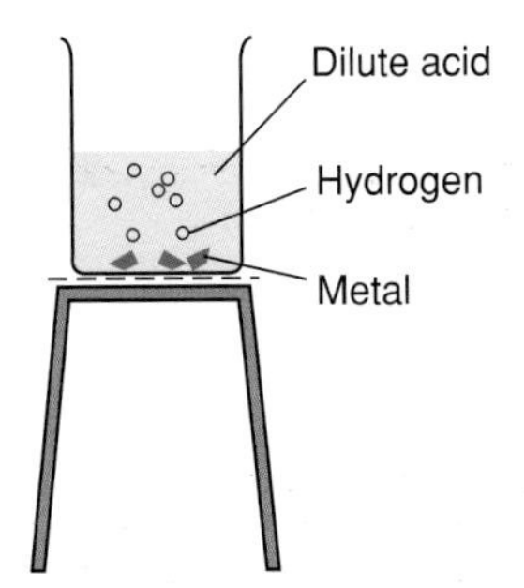

Method 1
Warm the acid.
Switch off the Bunsen.
Add an excess of the metal to the acid.
Wait until no more hydrogen is evolved.
The reaction is then complete.

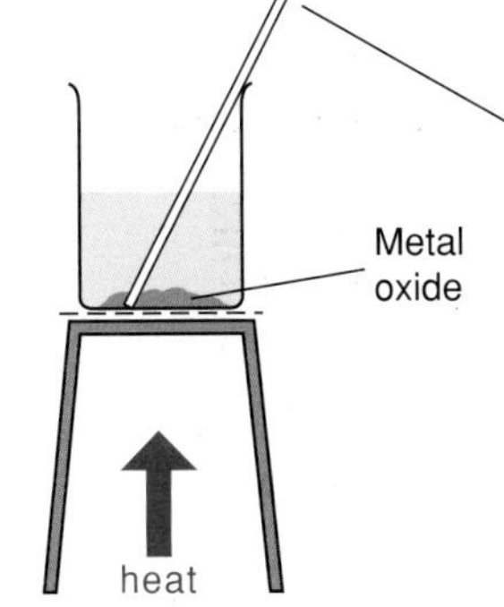

Method 2
Add an excess of the metal oxide to the acid. Wait until the solution no longer turns blue litmus red. The reaction is then over.

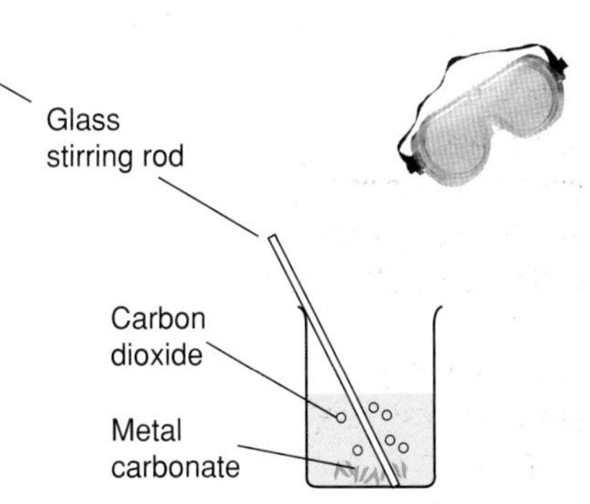

Method 3
Add an excess of the metal carbonate to the acid. Wait until no more carbon dioxide is evolved. The reaction is then over.

Figure 10.5A ▲ Adding an excess of the solid reactant to the acid

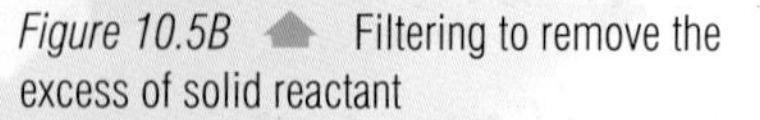

Step two Filter to remove the excess of solid (Figure 10.5B).

Step three Gently evaporate the filtrate (Figure 10.5C).

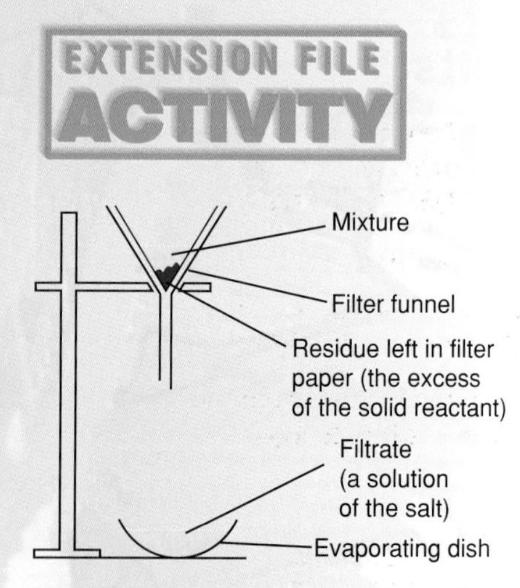

Figure 10.5B ▲ Filtering to remove the excess of solid reactant

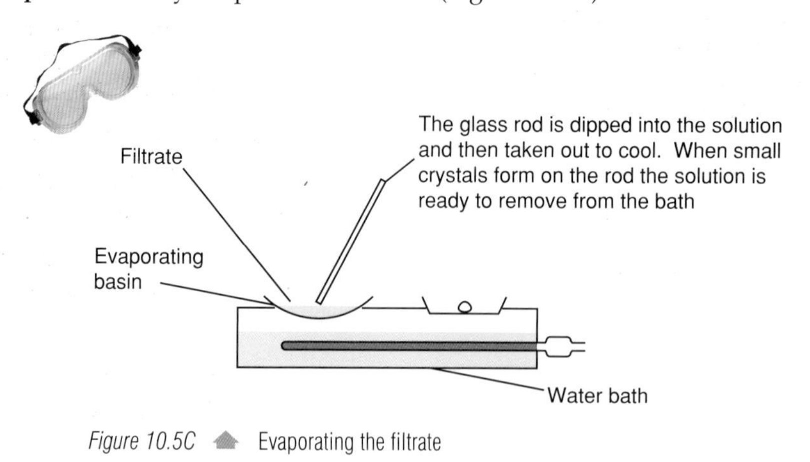

Figure 10.5C ▲ Evaporating the filtrate

Step four As the solution cools, crystals of the salt form. Separate the crystals from the solution by filtration. Use a little distilled water to wash the crystals in the filter funnel. Then leave the crystals to dry.

Acid + alkali

Method 4 Reaction between an acid and an alkali

Ammonia is an alkali. The preparation of ammonium salts is important because they are used as fertilisers. A laboratory method for making ammonium sulphate is shown in Figure 10.5D.

1 Add the ammonia solution to the dilute sulphuric acid, stirring constantly

2 From time to time, remove a drop of acid on the end of a glass rod. Spot it onto a strip of indicator paper on a white tile. When the test shows that the solution has become akaline, stop adding ammonia. You have now added an excess of ammonia

3 Evaporate the solution in a fume cupboard until it begins to crystallise. Leave it to stand. Filter to obtain crystals of ammonium sulphate

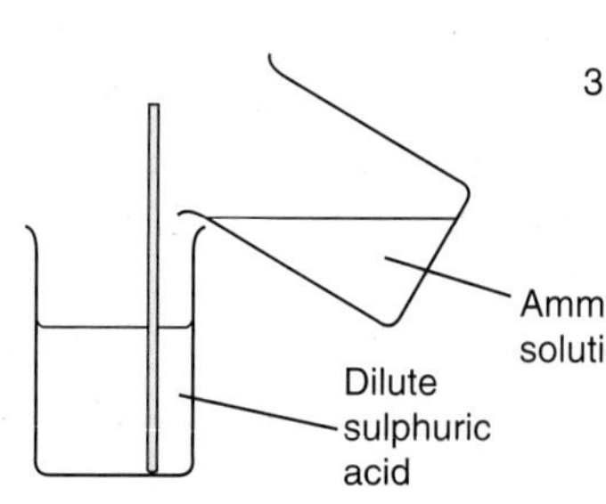

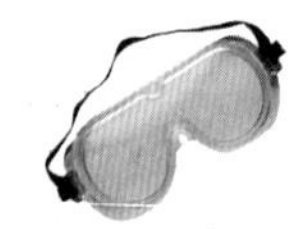

Figure 10.5D ▲ Making ammonium sulphate

SUMMARY

Soluble salts can be made by the reaction:
Acid + Alkali → Salt + Water

This method is used for ammonium salts. An excess of ammonia is added. The excess is removed when the solution is evaporated.

Method for insoluble salts

Insoluble salts are made by mixing two solutions. For example, you can make the insoluble salt barium sulphate by mixing a solution of the soluble salt barium chloride with a solution of the soluble salt sodium sulphate. When the two solutions are mixed, the insoluble salt barium sulphate is precipitated (thrown out of solution). What do you think remains in solution? The method is called **precipitation**.

SUMMARY

Insoluble salts are made by precipitation. A solution of a soluble salt of the metal is added to a solution of a soluble salt of the acid.

Remember

- All nitrates are soluble.
- All sodium, potassium and ammonium salts are soluble.

Precipitation

The equation for the reaction is

Barium chloride	+	Sodium sulphate	→	Barium sulphate	+	Sodium chloride
$BaCl_2(aq)$	+	$Na_2SO_4(aq)$	→	$BaSO_4(s)$	+	$2NaCl(aq)$

An ionic equation can be written for the reaction

Barium ions	+	Sulphate ions	→	Barium sulphate
$Ba^{2+}(aq)$	+	$SO_4^{2-}(aq)$	→	$BaSO_4(s)$

The ionic equation shows only the ions which take part in the precipitation reaction: the barium ions and sulphate ions.

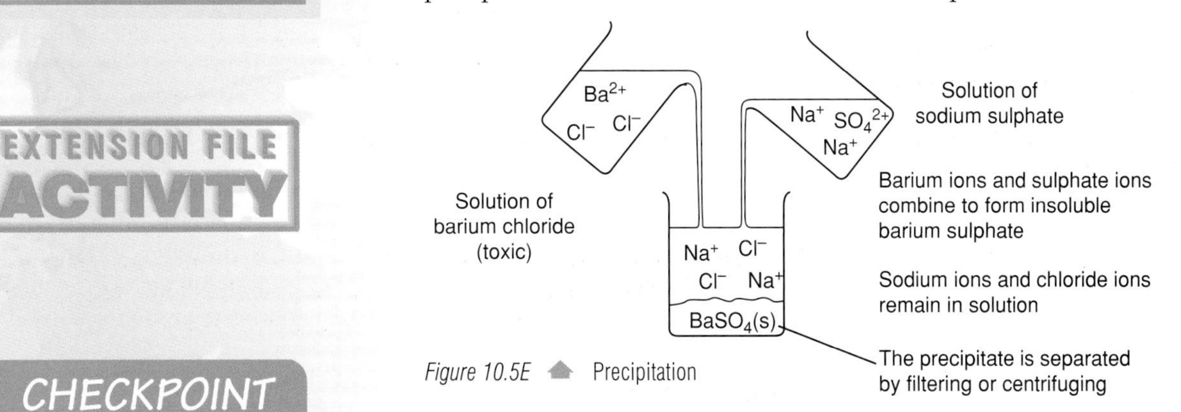

Figure 10.5E ▲ Precipitation

CHECKPOINT

1 Refer to Figure 10.5A.

(a) How do you know that all the acid has been used up
   (i) in the reaction with a metal?
   (ii) in the reaction with a metal oxide?
   (iii) in the reaction with a metal carbonate?

(b) Why is it important to make sure that all the acid is used up?

(c) If some acid were left unneutralised in Step 1, what would happen to it in Step 3? How would it affect the crystals of salt formed in Step 4?

(d) Why is it easier to remove an excess of base than an excess of acid?

2 Complete the following word equations

(a) magnesium + sulphuric acid → _____ sulphate + _____

(b) zinc oxide + hydrochloric acid → _____ chloride + _____

(c) calcium carbonate + hydrochloric acid → _____ + _____ + _____

(d) nickel oxide + hydrochloric acid → _____ + _____

(e) chromium oxide + sulphuric acid → _____ + _____

(f) magnesium oxide + nitric acid → _____ nitrate + _____

3 Write chemical equations for the reactions (a), (b) and (c) in Question 2.

4 Refer to Table 10.2 for solubility

(a) Strontium sulphate, $SrSO_4$, is insoluble. Which soluble strontium salt and which soluble sulphate could you use to make strontium sulphate?

(b) What would you do to obtain a dry specimen of strontium sulphate?

(c) Write a word equation for the reaction.

(d) Write an ionic equation for the reaction (the strontium ion is Sr^{2+}).

5 Barium carbonate, $BaCO_3$, is insoluble.

(a) Name two solutions which you could mix to give a precipitate of barium carbonate.

(b) Say what you would do to obtain barium carbonate from the mixture.

(c) Write a word equation for the reaction.

(d) Write an ionic equation (the barium ion is Ba^{2+}).

Topic 11 The Periodic Table

11.1 Arranging elements in order

FIRST THOUGHTS

The physical and chemical properties of elements and the arrangement of electrons in atoms of the elements; they all fall into place in the Periodic Table.

The atom is composed of protons, neutrons and electrons; see Topic 6.2. The number of protons in an atom is the same as the number of electrons and is called the **atomic number**. Electrons are arranged in groups at different distances from the nucleus called shells; see Topic 6.4. The arrangement of electrons in the atom of an element is called the **electron configuration**. For example calcium with 20 electrons in the atom has the electron configuration 2.8.8.2, meaning that there are 2 electrons in the first shell (nearest to the nucleus), 8 in the second shell, 8 in the third shell and 2 electrons in the fourth shell (furthest away from the nucleus). In this topic we shall see how the electron configurations of elements tie in with their chemical reactions. Some interesting patterns emerge when the elements are taken in the order of their atomic numbers and then arranged in rows. A new row is started after each noble gas (see Table 11.1). The arrangement is called the Periodic Table. You will have seen copies of the Periodic Table on the walls of chemistry laboratories. It simplifies the job of learning about the chemical elements.

Check that you remember the meanings of these terms:

- atomic number
- shell of electrons
- electron configuration.

If not, look at Topic 6 again!

Table 11.1 A section of the Periodic Table

	Group 1	Group 2	Group 3	Group 4	Group 5	Group 6	Group 7	Group 0
Period 1	H (1)							He (2)
Period 2	Li (2.1)	Be (2.2)	B (2.3)	C (2.4)	N (2.5)	O (2.6)	F (2.7)	Ne (2.8)
Period 3	Na (2.8.1)	Mg (2.8.2)	Al (2.8.3)	Si (2.8.4)	P (2.8.5)	S (2.8.6)	Cl (2.8.7)	Ar (2.8.8)
Period 4	K (2.8.8.1)	Ca (2.8.8.2)						

www.keyscience.co.uk

You can see that the arrangement in Table 11.1 has the following features:

- The elements are listed in order of increasing atomic number.
- The eight vertical columns are called **groups**. Elements which have the same number of electrons in the outermost shell fall into the same group of the Periodic Table.
- The noble gases are in Group 0. For the rest of the elements, the group number is the number of electrons in the outermost shell.
- The horizontal rows are called **periods**. The first period contains only hydrogen and helium. The second period contains the elements lithium to neon. The third period contains the elements sodium to argon.

The complete Periodic Table is shown in Figure 11.1A and at the back of this book.

The Periodic Table lists the elements in order of a certain property. What is it?

What can you say about

- all the members of a group of the Periodic Table?
- all the members of a period of the Periodic Table?

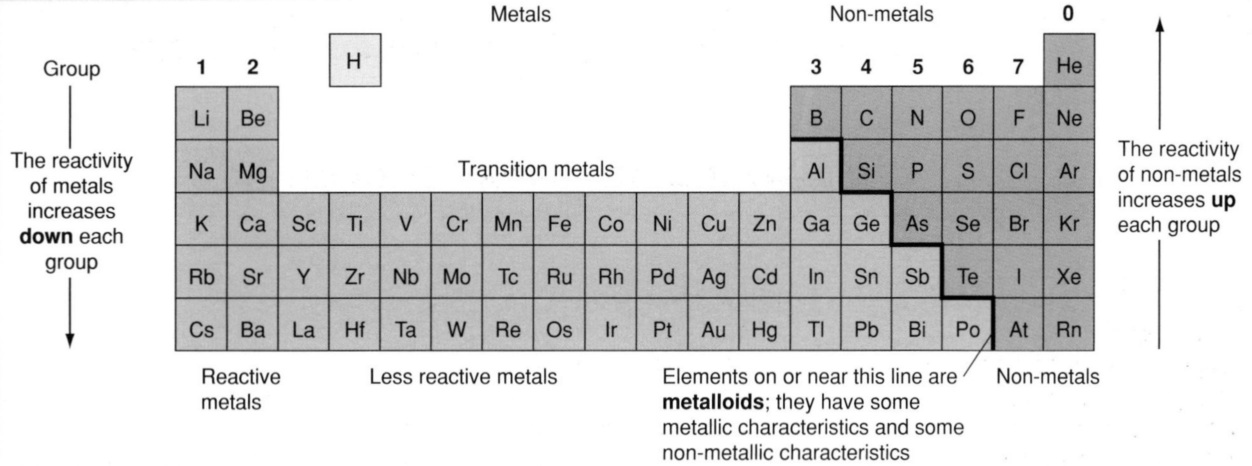

Metals — Non-metals

Group — The reactivity of metals increases **down** each group

1	2											3	4	5	6	7	0
			H														He
Li	Be											B	C	N	O	F	Ne
Na	Mg											Al	Si	P	S	Cl	Ar
K	Ca	Sc	Ti	V	Cr	Mn	Fe	Co	Ni	Cu	Zn	Ga	Ge	As	Se	Br	Kr
Rb	Sr	Y	Zr	Nb	Mo	Tc	Ru	Rh	Pd	Ag	Cd	In	Sn	Sb	Te	I	Xe
Cs	Ba	La	Hf	Ta	W	Re	Os	Ir	Pt	Au	Hg	Tl	Pb	Bi	Po	At	Rn

Transition metals

The reactivity of non-metals increases **up** each group

Reactive metals — Less reactive metals — Elements on or near this line are **metalloids**; they have some metallic characteristics and some non-metallic characteristics — Non-metals

Figure 11.1A ▲ The Periodic Table

11.2 Patterns in the Periodic Table

Could you list the differences between the physical properties of metallic and non-metallic elements? If not, refer to Table 4.1.

Patterns can be seen in the arrangement of the elements in the Periodic Table. The metallic elements are on the left-hand side and in the centre block and the non-metallic elements on the right-hand side.

Table 11.2 summarises the differences between the chemical properties of metallic and non-metallic elements.

Table 11.2 ▼ Chemical properties of metallic and non-metallic elements

Metallic elements	Non-metallic elements
Metals which are high in the reactivity series react with dilute acids to give hydrogen and a salt of the metal	Non-metallic elements do not react with dilute acids
Metallic elements form positive ions, e.g. Na^{+}, Zn^{2+}, Fe^{3+}	Non-metallic elements form negative ions, e.g. Cl^{-}, O^{2-}
Metal oxides and hydroxides are bases, e.g. Na_2O, CaO, NaOH. If they dissolve in water they give alkaline solutions, e.g. NaOH	Many oxides are acids and dissolve in water to give acidic solutions, e.g. CO_2, SO_2. Some oxides are neutral and insoluble, e.g. CO
The chlorides of metals are ionic crystalline solids, e.g. NaCl	The chlorides of the non-metals are covalent liquids or gases, e.g. HCl(g), CCl_4(l)

SUMMARY

Metallic and non-metallic elements differ in:
- their reactions with acids,
- the acid–base nature of their oxides,
- the nature of their chlorides,
- the type of ions they form.

Metals

Table 11.2 compares
- the alkali metals in Group 1,
- the alkaline earths in Group 2 and
- the transition metals in between Groups 2 and 3.

The reactive metals are at the left-hand side of the table, less reactive metallic elements in the middle block and non-metallic elements at the right-hand side.

The differences between the metals in Group 1, those in Group 2 and the transition metals are summarised in Table 11.3 (M stands for the symbol of a metallic element). Dilute sulphuric acid reacts with metals in the same way as dilute hydrochloric acid. Dilute nitric acid is an oxidising agent and attacks metals, e.g. copper, which are not sufficiently reactive to react with other dilute acids.

Table 11.3 Some reactions of metals

Group 1 • The alkali metals

Element	**Symbol**	**Reaction with ...**				**Trend**
		air	**water**	**non-metallic elements**	**dilute hydrochloric acid**	
Lithium Sodium Potassium Rubidium Caesium	Li Na K Rb Cs	All burn vigorously to form an oxide of formula M_2O (M=symbol of metal)	All are stored under oil. They react vigorously with cold water to give hydrogen and the hydroxide MOH. The hydroxides are all strong alkalis.	All combine with non-metals to form salts (and oxides). The salts are crystalline ionic solids. The alkali metals are the cations (positive ions) in the salts.	The reaction is dangerously violent.	The vigour of all these reactions increases down the group.

Group 2 • The alkaline earths

Element	**Symbol**	**Reaction with ...**			**Trend**
		air	**water**	**dilute hydrochloric acid**	
Beryllium Magnesium Calcium Strontium Barium	Be Mg Ca Sr Ba	Burn to form the strongly basic oxides, MO, which are sparingly soluble or insoluble.	Be reacts very slowly. Mg burns in steam. Ca, Sr, Ba react readily to form hydrogen and the alkali $M(OH)_2$.	React readily to give hydrogen and a salt, e.g. $MgCl_2$.	The vigour of all these reactions increases down the group. Group 2 elements are less reactive than Group 1.

Transition metals • The block of elements between Group 2 and Group 3

Element	**Symbol**	**Reaction with ...**			**Trend**
		air	**water**	**dilute hydrochloric acid**	
Iron Zinc	Fe Zn	When heated, form oxides without burning. The oxides and hydroxides are weaker bases than those of Groups 1 and 2 and are insoluble.	Iron rusts slowly. Iron and zinc react with steam to form hydrogen and the oxide.	Iron and zinc react to give hydrogen and a salt.	Transition metals are less reactive than Groups 1 and 2. In general, their compounds are coloured. They are used as catalysts.
Copper	Cu		Copper does not react.	Copper does not react.	

Table 11.4 Physical properties of Group 1

Element	**Symbol**	**m.p. (°C)**	**b.p. (°C)**	**Density (g/cm^3)**
Lithium	Li	180	1336	0.53
Sodium	Na	98	883	0.97
Potassium	K	64	759	0.86
Rubidium	Rb	39	700	1.53
Caesium	Cs	29	690	1.9

SUMMARY

The alkali metals of Group 1 are very reactive.
Their oxides are strong bases.
Their hydroxides are strong alkalis.
Reactivity increases with the size of the atom (down the group).
The alkaline earths of Group 2 are less reactive than Group 1.
Reactivity increases with the size of the atom.

The alkali metals

From Table 11.3, you can see how the Periodic Table makes it easier to learn about all the elements. Look at the elements in Group 1: lithium, sodium, potassium, rubidium and caesium. They are all very reactive metals. Their oxides and hydroxides have the general formulas M_2O and MOH (where M is the symbol of the metallic element) and are strongly basic. The oxides and hydroxides dissolve in water to give strongly alkaline solutions. The reactivity of the alkali metals increases as you pass down the group. If you know these facts, you do not need to learn the properties of all the metals separately. If you know the properties of sodium, you can predict those of potassium and lithium. The Periodic Table saves you from having to learn the properties of 106 elements separately!

The alkaline earths

The metals in Group 2 are less reactive than those in Group 1. They form basic oxides and hydroxides with the general formulas MO and $M(OH)_2$. Their oxides and hydroxides are either sparingly soluble or insoluble. The reactivity of the alkaline earths increases as you pass down the group. Again, if you know the chemical reactions of one element, you can predict the reactions of other elements in the group.

The transition metals

The transition metals in the block between Group 2 and Group 3 are a set of similar metallic elements. They are less reactive than those in Groups 1 and 2. Their oxides and hydroxides are less strongly basic and are insoluble.

Physical properties

Transition metals are hard and dense. They are good conductors of heat and electricity (e.g. copper wire in electrical circuits). Their melting points, boiling points and heats of melting are all higher than those for Group 1 and Group 2 metals. These are all a measure of the strength of the metallic bond. Iron, cobalt and nickel are strongly magnetic. Some of the other transition metals are weakly magnetic.

EXTENSION FILE
ASSIGNMENT

Extraction

Transition metals are less reactive than those in Groups 1 and 2. They can be extracted from their ores by means of chemical reducing agents. The method of extraction follows the steps listed below.

1 The ore is concentrated, e.g. by flotation.
2 Sulphide ores are roasted to convert them into oxides.
3 The oxide is reduced by heating with coke to form the metal and carbon monoxide.
4 The metal is purified.

SUMMARY

Transition metals compared with Groups 1 and 2 are
- less reactive
- harder, denser and stronger
- use more than one valency (except zinc)
- have coloured ions (except zinc)
- form oxo-ions, e.g. MnO_4^-
- are catalysts.

Valency

Transition metals characteristically use more than one valency.

Table 11.5 Some transition metals, ions and compounds

Metal	Ions	Compounds
Copper	Cu^+, Cu^{2+}	Cu_2O, $CuSO_4$
Iron	Fe^{2+}, Fe^{3+}	$FeSO_4$, $Fe_2(SO_4)_3$
Cobalt	Co^{2+}, Co^{3+}	$CoCl_2$, $CoCl_3$
Chromium	Cr^{2+}, Cr^{3+} $Cr_2O_7^{2-}$	$CrCl_2$, $CrCl_3$ $K_2Cr_2O_7$
Manganese	Mn^{2+}, MnO_4^-	$MnSO_4$, MnO_2, $KMnO_4$

Colour

Transition metals have coloured ions, except for zinc (see Table 11.6).

In addition to forming simple cations, transitions metals form oxo-ions in combination with oxygen. Examples are:

chromate(VI), CrO_4^{2-} (yellow), dichromate(VI), $Cr_2O_7^{2-}$ (orange)
manganate(VII), MnO_4^- (purple), zincate, ZnO_2^{2-} (colourless)

Table 11.6 The colours of some ions

Ion	Colour
Cu^{2+}(aq)	blue
Fe^{2+}(aq)	green
Fe^{3+}(aq)	rust
Ni^{2+}(aq)	green
Mn^{2+}(aq)	pink
Cr^{3+}(aq)	blue
Zn^{2+}(aq)	colourless

Catalysis

Many transition metals and their compounds are important catalysts. Examples are:

- iron and iron(III) oxide in the Haber process for making ammonia (Topic 22.2)
- platinum in the oxidation of ammonia during the manufacture of nitric acid (Topic 22.3)
- vanadium(V) oxide in the Contact process for making sulphuric acid (Topic 22.4)
- nickel in the hydrogenation of oils to form fats (Topic 28.1).

Silicon and germanium

Silicon and germanium are semiconductors.

On the borderline between metals and non-metals in the Periodic Table (Figure 11.1A) are silicon and germanium. These elements are semiconductors, intermediate between metals, which are electrical conductors, and non-metals, which are non-conductors of electricity. They are vital to the computer industry.

The halogens

The elements in Group 7 are fluorine, chlorine, bromine and iodine. These elements are a set of very reactive non-metallic elements (see Table 11.7). They are called **the halogens** because they react with metals to form salts. (Greek: halogen = salt-former)

Chlorine is a bactericide, used in the purification of drinking water and in swimming pools.

Iodine is used as an antiseptic for dressing wounds.

Table 11.7 The halogens

Element	Symbol	m.p. (°C)	b.p. (°C)	Colour	Reaction with metals to form salts	Reaction with hydrogen to form the compound HX	
Fluorine	F	−223	−188	Pale yellow	Dangerously reactive	Explodes	Reactivity decreases down the group ↓
Chlorine	Cl	−103	−35	Yellow-green	Readily combines to form chlorides.	Explodes in sunlight	
Bromine	Br	−7	59	Red-brown	Combines when heated to form bromides	Reacts when heated	
Iodine	I	114	184	Purple-black	Combines when heated to form iodides. The halogens are the anions (negative ions) in their salts.	Reaction is not complete	

Reactions of the halogens

1. Sodium burns in halogens vigorously to form the halide, sodium fluoride, NaF, sodium chloride, NaCl, sodium bromide, NaBr, or sodium iodide, NaI. The reaction with fluorine is dangerously explosive.
2. When heated, iron reacts vigorously with chlorine to form iron(III) chloride, $FeCl_3$. With bromine the product is iron(III) bromide, $FeBr_3$. With iodine the reaction is less vigorous, and the product is iron(II) iodide, FeI_2. Iodine is a less powerful oxidising agent than the other halogens; see Topic 12.5.
3. The halogens are oxidising agents; see Topic 12.5.
4. The halogen halides (hydrogen fluoride, hydrogen chloride, hydrogen bromide and hydrogen iodide) are gases. They are covalent compounds. They dissolve in and react with water to form solutions of strong acids. The acids consist of hydrogen ions and halide ions, e.g.

Hydrogen cloride + Water → Hydrochloric acid

$$HCl(g) + aq \rightarrow H^+(aq) + Cl^-(aq)$$

SUMMARY

The halogens in Group 7 are a reactive group of non-metallic elements.
They react with metals to form salts.
Reactivity decreases with the size of the atom (down the group).
The halogens are oxidising agents.
Hydrogen halides react with the water to form strong acids.

SUMMARY

The noble gases in Group 0 are very unreactive.

Each atom of a noble gas has a full outer shell of electrons

The noble gases

The elements in group 0 are helium, neon, argon, krypton, xenon and radon. These elements are called the noble gases. They are present in air (see Topic 14). The noble gases exist as single atoms, e.g. He, Ne. Their atoms do not combine in pairs to form molecules as do the atoms of most gaseous elements, e.g. O_2, H_2. For a long time, no-one was able to make the noble gases take part in any chemical reactions. In 1960, however, two of them, krypton and xenon, were made to combine with the very reactive element, fluorine. *Why are the noble gases so exceptionally unreactive?* Chemists came to the conclusion that it is the full outer shell of electrons that makes the noble gases unreactive.

CHECKPOINT

1. Which of the alkali metals (a) float on water? (b) can be cut with a knife made of iron? (c) melt at the temperature of boiling water?
   - (d) Why would it be very dangerous to put an alkali metal into boiling water?
   - (e) What is another name for sodium chloride?
   - (f) What would you expect rubidium chloride to look like?
   - (g) The alkali metals are kept under oil to protect them from the air. Which substances in the air would attack them?
   - (h) What pattern can you see in (i) the melting points (ii) the boiling points of the alkali metals?
2. (a) Which of the halogens is (i) a liquid and (ii) a solid at room temperature (20 °C)?
   - (b) Why is fluorine not studied in school laboratories?
   - (c) Write the formulas for (i) sodium iodide (ii) potassium fluoride.
3. Magnesium chloride, $MgCl_2$, is a solid of high melting point and carbon tetrachloride, CCl_4, is a volatile liquid. Explain how the differences in chemical bonding account for these differences.
4. Choose from the elements: Na, Mg, Al, Si, P, S, Cl, Ar.
   - (a) List the elements that react readily with cold water to form alkaline solutions.
   - (b) List the elements that form sulphates.
   - (c) Name the elements which exist as molecules containing (i) 1 atom (ii) 2 atoms.
   - (d) Which element has both metallic and non-metallic properties?
5. The elements sodium and potassium have the electron arrangements Na (2.8.1), K (2.8.8.1). How does this explain the similarity in their reactions?

11.3 Electron configuration and chemical reactions

Group 0: helium, neon, argon, krypton and xenon

The noble gases have a full outer shell of electrons, e.g. helium (2), neon (2.8) and argon (2.8.8). For many years no reactions of the noble gases were known, until in 1960 krypton and xenon were made to combine with fluorine. They exist as single atoms: their atoms do not combine to form molecules as do the atoms of other gaseous elements, e.g. O_2, N_2.

SUMMARY

The chemical properties of elements depend on the electron configurations of their atoms. The unreactive noble gases all have a full outer shell of 8 electrons.

The reactive alkali metals have a single electron in the outer shell.

The halogens are reactive non-metallic elements which have 7 electrons in the outer shell: they are 1 electron short of a full shell.

Group 1: lithium, sodium, potassium, rubidium, caesium

The alkali metals (see Table 11.3) all have one electron in the outer shell, e.g. lithium (2.1), sodium (2.8.1) and potassium (2.8.8.1). They form ionic compounds. The outer electrons are involved in the formation of ionic bonds; see Topic 8.1–2. This is why it is believed that the similar electron configuration is the reason why the metals behave in a very similar way.

Group 2: beryllium, magnesium, calcium, strontium, barium

The alkaline earth metals (see Table 11.3) all have two electrons in the outer shell, e.g. beryllium (2.2), magnesium (2.8.2) and calcium (2.8.8.2). It is the outer electrons that are involved in the formation of bonds. This is why it is believed that the similar electron configuration is the reason why the metals all behave in a very similar way.

Group 7: fluorine, chlorine, bromine, iodine

The halogens all have 7 electrons in the outer shell, e.g. fluorine (2.7) and chlorine (2.8.7). They all have atoms that want to acquire an extra electron to form a halide ion and form an ionic compound. This is why the halogens all behave in a similar way; see Table 11.7.

11.4 Ions in the Periodic Table

The Periodic Table helps when it comes to remembering the charge on an ion.

Investigating ions

This exercise will show you how useful the Periodic Table is. One of the things it helps you to remember is the charge on an ion.

Take the number of the group in the Periodic Table to which an element belongs. How is this related to the charge on the ions of the element?

Element	*Symbol for ion*	*Group of Periodic Table*
Sodium	Na^{+}	
Potassium	K^{+}	
Calcium	Ca^{2+}	
Magnesium	Mg^{2+}	
Aluminium	Al^{3+}	
Nitrogen	N^{3-}	
Oxygen	O^{2-}	
Sulphur	S^{2-}	
Chlorine	Cl^{-}	
Bromine	Br^{-}	

- Copy the table. Fill in the number of the Group in the Periodic Table to which each of the elements belongs.
- What is the connection between the charge on a cation and the number of the Group in the Periodic Table to which the element belongs?
- What is the connection between the charge on an anion and the number of the Group in the Periodic Table to which the element belongs?
- What would you expect to be the charge on (a) a barium ion (Group 2) and (b) a fluoride ion (Group 7)?

The charge on the ions formed by an element is called the valency of that element. Magnesium form Mg^{2+} ions: magnesium has a valency of 2. Sodium forms Na^{+} ions: sodium has a valency of 1. Oxygen forms O^{2-} ions: oxygen has a valency of 2. Iodine forms I^{-} ions: iodine has a valency of 1. Some elements have a variable valency. Iron forms Fe^{2+} ions and Fe^{3+} ions: iron has a valency of 2 in some compounds and 3 in others.

Non-metallic elements often combine with oxygen to form anions, e.g. sulphate, SO_4^{2-} and nitrate, NO_3^{-}.

- **Metallic elements**

Charge on cation = Number of the Group in the Periodic Table to which the element belongs = Valency of element

- **Non-metallic elements**

Charge on anion = 8 − Number of Group in Periodic Table to which the element belongs = Valency of element

11.5 The history of the Periodic Table

FIRST THOUGHTS

John Newlands had the idea of the Periodic Table, but it didn't catch on. Dmitri Mendeleev took over and improved on Newlands' work. He was able to predict the properties of new elements that had not yet been discovered

The person who has the credit for drawing up the Periodic Table is a Russian scientist called Dmitri Mendeleev. He extended the work of a British chemist called John Newlands. In 1864, 63 elements were known to Newlands. He arranged them in order of relative atomic mass. When he started a new row with every eighth element, he saw that elements with similar properties fell into vertical groups. He spoke of 'the regular periodic repetition of elements with similar properties'. This gave rise to the name 'periodic table'. Sometimes, there were misfits in Newlands' table, for example, iron did not seem to belong where he put it with oxygen and sulphur. Newlands' ideas were not accepted.

Mendeleev had an idea about the troublesome misfits. He realised that many elements had yet to be discovered. Instead of slotting elements into positions where they did not fit he left gaps in the table. He expected that when further elements were discovered they would fit the gaps. He predicted that an element would be discovered to fit into the gap he had left in the table below silicon.

Table 11.8 The predictions which Mendeleev made for the undiscovered element which he called 'eka-silicon' (below silicon) compared with the properties of germanium

Mendeleev's predicted properties for eka-silicon, Ek (1871)		The properties of germanium Ge (discovered in 1886)
Appearance	Grey metal	Grey-white metal
Density	~5.5 g/cm^3	5.47 g/cm^3
Relative atomic mass	73.4	72.6
Melting point	~800 °C	958 °C
Reaction with oxygen	Forms the oxide EkO_2. The oxide EkO may also exist.	Forms the oxide GeO_2 GeO also exists

SUMMARY

In the Periodic Table, elements are arranged in order of increasing atomic (proton) number in 8 vertical groups. The horizontal rows are called periods.

By the time Mendeleev died in 1907, many of the gaps in the table had been filled by new elements. The noble gases were unknown to Mendeleev when he drew up his table. The first of them was discovered in 1894. As the noble gases were discovered, they all fell into place between the halogens and the alkali metals. They formed a new group, Group 0. This was a spectacular success for the Periodic Table.

Why did Newlands have such difficulty in getting his ideas accepted? Why did Mendeleev succeed? Although you see a copy of the Periodic Table in every chemical laboratory, it is sad that the person who set the ball rolling, Newlands, received so little credit for his novel idea. Newlands had an enormous knowledge of the chemical properties of all the elements. This knowledge enabled him to spot the similarities between groups of elements and classify them. When he spotted that lithium (the third known element), sodium (the tenth known element) and potassium (the 17th known element) were similar, he compared the repeating pattern of elements with a musical octave. This comparison caused a lot of jokes at his expense, and distracted people's minds from the chemical facts. In addition, some elements did not fit into the repeating pattern on the basis of their relative atomic masses.

Mendeleev had the brilliance to realise that some elements had still to be discovered. He left gaps in the table, rather than force elements into spaces where their properties did not fit. He used his theory to make predictions. He predicted that new elements would be discovered and their properties would fit the empty places. He predicted the properties of the missing elements. When new elements were discovered and found to agree closely with his predictions, this forced the scientific community to accept the ideas behind the Periodic Table. Newlands had to watch the same people who had made fun of his ideas praising Mendeleev's triumph.

CHECKPOINT

1 (a) When was germanium discovered?
(b) How do the properties of germanium compare with Mendeleev's prediction?
(c) How long did Mendeleev have to wait to see if he was right?
(d) In which group of the Periodic Table does germanium come?
(e) What important articles are manufactured from germanium and silicon?

2 Radium, Ra, is a radioactive element of atomic number 88 which falls below barium in Group 2. What can you predict about
(a) the nature of radium oxide
(b) the reaction of radium and water
(c) the reaction of radium with dilute hydrochloric acid?
State the physical state and type of bonding in any compounds you mention. Give the names and formulas of any compounds formed.

3 Astatine, At, is a radioactive element of atomic number 85 which follows fluorine in Group 7. What can you predict about
(a) the nature of its compound with hydrogen
(b) the reaction of astatine with sodium?
State the physical state and type of bonding in any compounds you mention. Give the names and formulas of any compounds formed.

Topic 12 Oxidation-reduction reactions

12.1 Oxidation and reduction

FIRST THOUGHTS

The rocket launch shown in Figure 12.1A takes place thanks to oxidation–reduction reactions. Kerosene burns in oxygen to give the power required for lift-off. In stage 2 and stage 3 of the rocket's journey into space, hydrogen burns in oxygen. These are oxidation–reduction reactions.

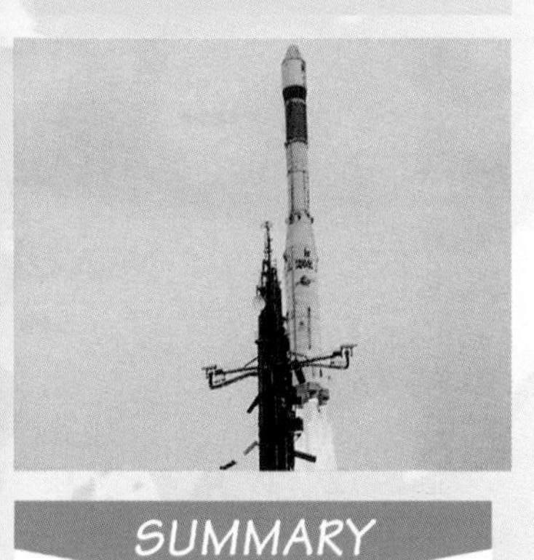

You have already met both chemical reactions that are described as oxidation reactions and also reduction reactions. In this topic, we shall take a closer look at some of these reactions.

Example 1:

Copper + Oxygen ➔ Copper(II) oxide

$2Cu(s) + O_2(g) \rightarrow CuO(s)$

In this reaction, copper has gained oxygen, and we say that copper has been **oxidised**. This reaction is an **oxidation**, and oxygen is an **oxidising agent**.

Example 2:

Lead(II) oxide + Hydrogen ➔ Lead + Water

$PbO(s) + H_2(g) \rightarrow Pb(s) + H_2O(l)$

In this reaction, lead(II) oxide has lost oxygen, and we say that lead(II) oxide has been **reduced**. Hydrogen has taken oxygen from another substance, and we therefore describe hydrogen as a **reducing agent**. The reaction is a **reduction**. On the other hand, notice that hydrogen has gained oxygen: hydrogen has been oxidised. You can also describe this reaction as an oxidation. Since lead(II) oxide has given oxygen to another substance, in this reaction lead(II) oxide is an oxidising agent. Oxidation and reduction are occurring together. One reactant is the oxidising agent and the other reactant is the reducing agent.

This is reduction (PbO ➔ Pb)

$PbO(s) + H_2(g) \rightarrow Pb(s) + H_2O(l)$

This is oxidation (H_2 ➔ H_2O)

Oxidising agent (PbO) Reducing agent (H_2)

Example 3:

Zinc oxide + Carbon ➔ Zinc + Carbon monoxide

$ZnO(s) + C(s) \rightarrow Zn(s) + CO(g)$

Again, you see in this example that oxidation and reduction are occurring together. This is always true, however many examples you consider. Since oxidation never occurs without reduction, it is better to call these reactions **oxidation–reduction reactions** or **redox reactions**.

SUMMARY

Oxidation is the gain of oxygen or loss of hydrogen by a substance. Reduction is the loss of oxygen or gain of hydrogen by a substance. An oxidising agent gives oxygen to or takes hydrogen from another substance. A reducing agent takes oxygen from or gives hydrogen to another substance.

CHECKPOINT

1 State which is the oxidising agent and which is the reducing agent in each of these reactions:

(a) the thermit reaction: Aluminium + Iron oxide ➔ Iron + Aluminium oxide

(b) roasting the ore tin sulphide in air: Tin sulphide + Oxygen ➔ Tin oxide + Sulphur dioxide

(c) 'smelting' tin oxide with coke: Tin oxide + Carbon ➔ Tin + Carbon monoxide

(d) roasting the ore copper sulphide in air: $2CuS(s) + 3O_2(g) \rightarrow 2CuO(s) + 2SO_2(g)$

12.2 Gain or loss of electrons

SUMMARY

Oxidation and reduction occur together in oxidation–reduction reactions or redox reactions.

Are you ready for another way of looking at oxidation–reduction reactions?

In a redox reaction

- the reducing agent gives electrons to the oxidising agent,
- and the oxidising agent accepts them.

How to remember

Which is which: oxidation or reduction – loss or gain of electrons?

OIL Oxidation is loss.

RIG Reduction is gain.

Metals are reducing agents. They react by giving electrons to e.g. the hydrogen ions of an acid or the molecules of a halogen.

www.keyscience.co.uk

It is not obvious in Example 1 in Topic 12.1 that reduction accompanies oxidation.

$$\text{Copper} + \text{Oxygen} \rightarrow \text{Copper(II) oxide}$$
$$2Cu(s) + O_2(g) \rightarrow 2CuO(s)$$

Atoms of copper, Cu, have been converted into copper(II) ions, Cu^{2+}. Molecules of oxygen, O_2, have been converted into oxide ions, O^{2-}. This means that copper atoms have lost electrons:

$$Cu(s) \rightarrow Cu^{2+}(s) + 2e^-$$

and oxygen molecules have gained electrons:

$$O_2(g) + 4e^- \rightarrow 2O^{2-}(s)$$

This picture holds for all oxidation–reduction reactions.

- The substance which is oxidised loses electrons.
- The substance which is reduced gains electrons.
- An oxidising agent accepts electrons.
- A reducing agent gives electrons.

Oxidation is the gain of oxygen or the loss of hydrogen or the loss of electrons.

Reduction is the loss of oxygen or the gain of hydrogen or the gain of electrons.

Metals are reducing agents. Metals react (see Topics 12.1, 19.3) by losing electrons to form positive ions, M^+, M^{2+} and M^{3+}. When a metal reacts with an acid:

$$\text{Metal atom} \rightarrow \text{Metal ion} + \text{electrons}$$
$$M(s) \rightarrow M^{2+}(aq) + 2e^-$$

$$\text{Hydrogen ions} + \text{electrons} \rightarrow \text{Hydrogen molecule}$$
$$2H^+(aq) + 2e^- \rightarrow H_2(g)$$

The metal atoms have supplied electrons, and hydrogen ions have accepted electrons. The metal atoms have been oxidised, and hydrogen ions have been reduced.

$$M(s) + 2H^+(aq) \rightarrow M^{2+}(aq) + H_2(g)$$

Some metals form more than one type of ion. When iron(II) compounds are converted into iron(III) compounds, Fe^{2+} ions lose electrons: they are oxidised.

$$Fe^{2+}(aq) \rightarrow Fe^{3+}(aq) + e^-$$

The oxidation is brought about by a substance which can accept electrons – an oxidising agent, e.g. chlorine. Chlorine is reduced to chloride ions.

$$Cl_2(aq) + 2e^- \rightarrow 2Cl^-(aq)$$

The complete reaction is:

$$2Fe^{2+}(aq) + Cl_2(aq) \rightarrow 2Fe^{3+}(aq) + 2Cl^-(aq)$$

12.3 Redox reactions and cells

Electrolysis cells

Redox reactions involve the gain and the loss of electrons. The reactions that happen at the electrodes in electrolysis involve either the gain or the loss of electrons. Let us look at the connection between electrode reactions and redox reactions.

In the electrolysis of copper(II) chloride, the electrode reactions are as follows.

Cathode (negative electrode): $Cu^{2+}(aq) + 2e^- \rightarrow Cu(s)$

Here copper ions, Cu^{2+}, gain electrons: they are reduced to copper atoms, Cu.

Anode (positive electrode): $Cl^-(aq) \rightarrow Cl(g) + e^-$

Here chloride ions, Cl^- give up electrons: they are oxidised to chlorine atoms, Cl. These immediately combine to form chlorine molecules, Cl_2. The cathode reaction is reduction; the anode reaction is oxidation.

SUMMARY

In an electrolysis cell, the anode reaction is oxidation, and the cathode reaction is reduction.

Chemical cells

A simple chemical cell can be made by immersing two different metals in an electrolyte and connecting them with a conducting wire. A current flows through the wire (see Figure 12.3A).

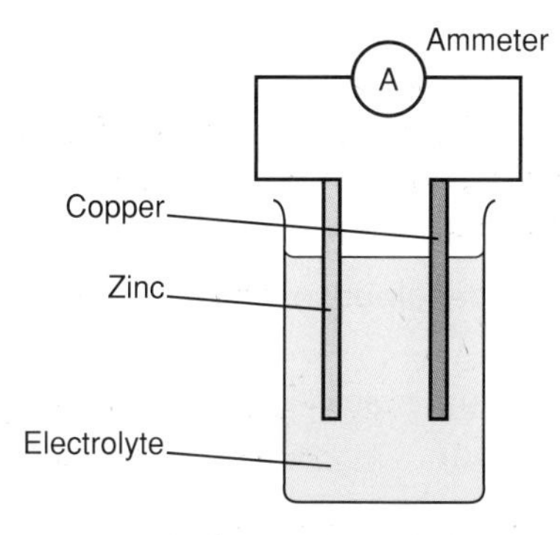

Figure 12.3A ▲ A chemical cell

SUMMARY

- The reactions which take place in chemical cells are redox reactions.
- Oxidation takes place at the anode.
- Reduction takes place at the cathode.
- Half-equations are written for the electrode reactions.
- The half-equations can be added to give the overall cell reaction.

Example 1: The zinc–copper chemical cell

The half-equations for the two electrode reactions are:

Zinc electrode: $Zn(s) \rightarrow Zn^{2+}(aq) + 2e^-$

Oxidation (loss of electrons) occurs at this electrode; therefore this is the anode.

Copper electrode: $Cu^{2+}(aq) + 2e^- \rightarrow Cu(s)$

Reduction (gain of electrons) occurs at this electrode; therefore this is the cathode. Notice that, in the chemical cell, the anode is negative and the cathode is positive.

The equation for the whole cell reaction is:

$$Zn(s) + Cu^{2+}(aq) \rightarrow Zn^{2+}(aq) + Cu(s)$$

12.4 Tests for oxidising agents and reducing agents

Test for reducing agents

Acidified potassium manganate(VII) is a powerful oxidising agent. A reducing agent will change the colour from the purple of $MnO_4^-(aq)$ to the pale pink of $Mn^{2+}(aq)$. Since potassium manganate(VII) is a strong oxidising agent, it can oxidise – and therefore test for – many reducing agents.

For example, (see Figure 12.7A),

Soak a piece of filter paper in acidified potassium manganate(VII) solution, and hold it in a stream of sulphur dioxide.

The paper turns from purple to a very pale pink.

Test for oxidising agents

Many oxidising agents will oxidise the iodide ion, I^-, to iodine, I_2. The presence of iodine can be detected because it forms a dark blue compound with starch. Potassium iodide solution and starch can therefore be used to test for oxidising agents.

$2I^-(aq) \rightarrow I_2(aq) + 2e^-$

colourless → brown, forms a dark blue compound with starch

For example,

Wet a filter paper with a solution of potassium iodide and starch, and hold it in a stream of chlorine.

The colour changes from white to dark blue.

SUMMARY

Reducing agents decolourise acidified potassium manganate(VII) solution. Oxidising agents give a dark blue colour with a solution of potassium iodide and starch.

CHECKPOINT

1 (a) Say which of the following will decolourise acidified potassium manganate(VII):
$FeSO_4(aq)$, $ZnSO_4(aq)$, $SO_2(g)$

(b) Say which of the following will turn starch-iodide paper blue:
$Br_2(aq)$, $Fe_2(SO_4)_3$, Cl_2, I_2, $HCl(aq)$

12.5 Chlorine as an oxidising agent

EXTENSION FILE ACTIVITY

The reactions of chlorine are dominated by its readiness to act as an oxidising agent: to gain electrons and form chloride ions.

Chlorine molecules + Electrons → Chloride ions

$Cl_2(aq) + 2e^- \rightarrow 2Cl^-(aq)$

Water treatment

Of major importance is the role played by chlorine in making our tap water safe to drink. The ability of chlorine to kill bacteria is due to its oxidising power.

Chlorine

TOXIC

Metals

Sodium and chlorine react to form sodium chloride. Sodium atoms are oxidised (give electrons), becoming sodium ions, Na^+. Chlorine molecules are reduced (gain electrons) to become chloride ions, Cl^-.

$$Na(s) \rightarrow Na^+(s) + e^-$$

So the full equation for sodium and chlorine is:

$$2Na(s) + Cl_2(g) \rightarrow 2Na^+(s) + 2Cl^-(s)$$

Ions of metals of variable valency

The oxidation of iron(II) salts to iron(III) salts by chlorine has been mentioned (Topic 12.2).

Halogens

The reactions of chlorine with bromides and iodides are oxidation–reduction reactions. Chlorine displaces bromine from bromides. This happens because chlorine is a stronger oxidising agent than bromine is. Chlorine takes electrons away from bromide ions: bromide ions are oxidised to bromine molecules. Chlorine molecules are reduced to chloride ions.

REDUCTION

$$Cl_2(aq) + 2Br^-(aq) \rightarrow 2Cl^-(aq) + Br_2(aq)$$

OXIDATION

Chlorine molecules are ready to accept electrons to form chloride ions.

This is the basis of chlorine's reactivity as an oxidising agent, with, e.g.
- metals
- metal ions
- other halogens.

The bromine formed can be detected more easily if a small volume of an organic solvent is added. Bromine dissolves in the organic solvent to form an orange solution.

Similarly, chlorine displaces iodine from iodides because chlorine is a stronger oxidising agent than iodine. Again, iodine can be detected readily by adding a small volume of an organic solvent. Iodine dissolves to give a purple solution.

Bromine displaces iodine from iodides because bromine is a stronger oxidising agent than iodine.

Fluorine is the most powerful oxidising agent of the halogens. It is dangerously reactive and is not used in schools and colleges. In order of oxidising power, the halogens rank:

$$F_2 > Cl_2 > Br_2 > I_2$$

EXTENSION FILE ACTIVITY

Oxidation–reduction

Oxidation is
- the gain of oxygen or
- the loss of hydrogen or
- the loss of electrons by a substance.

Reduction is
- the loss of oxygen or
- the gain of hydrogen or
- the gain of electrons by a substance.

An oxidising agent
- gives oxygen to or
- takes hydrogen from or
- takes electrons from a substance.

A reducing agent
- takes oxygen from or
- gives hydrogen to or
- gives electrons to a substance.

SUMMARY

Chlorine is an oxidising agent. It is used in water treatment to kill bacteria. It oxidises metals, metal ions, bromides and iodides. In order of oxidising power, the halogens rank $F_2 > Cl_2 > Br_2 > I_2$

12.6 Bromine from sea water

The oxidising power of the halogens decreases as you descend Group 7. Chlorine is a more powerful oxidising agent than bromine; therefore chlorine will oxidise bromide ion to bromine while chlorine is reduced to chloride ion.

$$\text{chlorine} + \text{bromide ion} \rightarrow \text{chloride ion} + \text{bromine}$$
$$Cl_2(aq) + 2Br_2(aq) \rightarrow 2Cl^-(aq) + Br_2(l)$$

Sea water is a plentiful source of bromide ion. The extraction of bromine from sea water is carried out in Anglesey, Wales and in the Dead Sea, Israel. Israel has the advantage that the sea water can be concentrated by evaporation by the sun's heat before treatment. During evaporation, sodium chloride, potassium chloride and magnesium chloride crystallise. These salts are electrolysed to give chlorine. The process is illustrated in Figure 12.6A.

The oxidising power of chlorine is used to convert bromide ions in sea water into bromine.

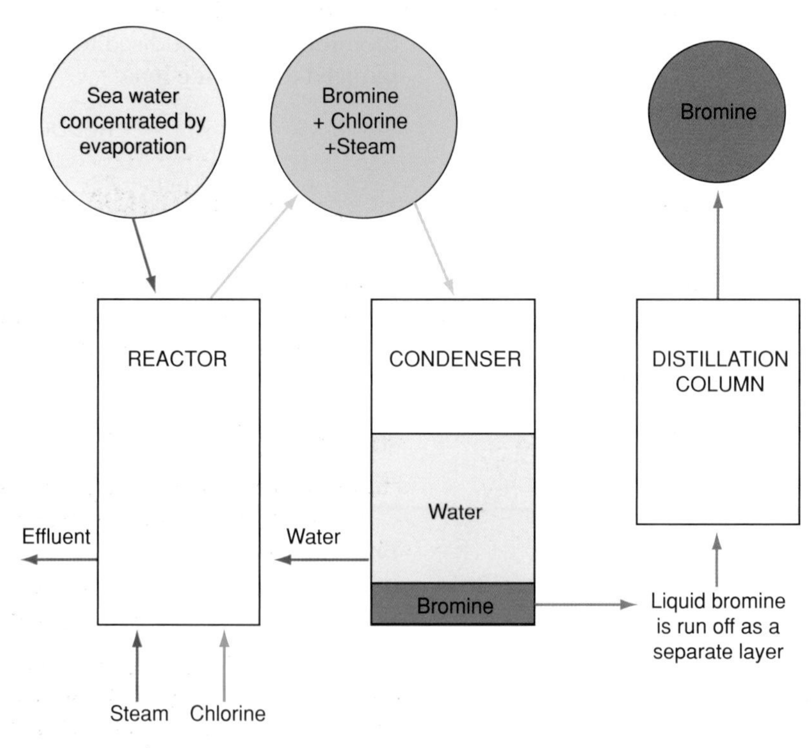

Figure 12.6A Flow diagram for the extraction of bromine from sea water

Bromine is a corrosive liquid with a poisonous vapour. Extreme care has to be taken in transporting bromine from the producer to the user.

Bromine is shipped from Israel to the petrochemicals plant at Fawley near Southampton. Bromine is a corrosive liquid which vaporises easily (b.p. 58 °C). The vapour is poisonous, with a penetrating smell, and inhaling it makes you choke. Both liquid and vapour are caustic to the skin, producing serious burns. Transporting this dangerous liquid from Israel to Britain is a difficult business, but the safety record is good. It is shipped in lead-lined steel tanks which hold several tonnes of bromine. At Fawley the tanks must be emptied. Nitrogen is led into the tank to pump bromine out through a plastic pipe into a storage cylinder of glass-lined steel. When required bromine is pumped from the storage tank through plastic pipes into the plant. A bromine detector in the area ensures the safety of operators.

CHECKPOINT

1 (a) Write a word equation for the oxidation of sodium iodide by chlorine.
(b) Write a balanced chemical equation for the reaction in (a).
(c) Write a word equation for the oxidation of potassium iodide by bromine.
(d) Write a balanced chemical equation for the reaction in (c).

2 The lab. technician has taken delivery of three bottles containing crystalline white solids. Unfortunately, their labels have come off.

Potassium chloride	**Potassium bromide**	**Potassium iodide**

The technician has to decide which label to stick on each of the three bottles. All she has to work with is a bottle of chlorine water and an organic solvent. How can she solve the problem?

3 With which one of the following pairs of reagents would a displacement reaction take place?
(a) aqueous bromine and aqueous potassium chloride
(b) aqueous bromine and aqueous sodium chloride
(c) aqueous chlorine and aqueous potassium iodide
(d) aqueous iodine and aqueous potassium bromide.

12.7 Sulphur dioxide as a reducing agent

Sulphur dioxide

TOXIC

Sulphur dioxide is a gas with a very unpleasant, penetrating smell. It is poisonous, causing congestion followed by choking, and at sufficiently high concentrations it will kill.

Sulphurous acid

Sulphur dioxide is extremely soluble in water. It 'fumes' in moist air as it reacts with water vapour to form a mist. It reacts with water to form sulphurous acid, H_2SO_3.

Sulphur dioxide + Water → Sulphurous acid

$$SO_2(g) + H_2O(l) \rightarrow H_2SO_3(aq)$$

Sulphur dioxide reacts with water to form sulphurous acid ...
... which is oxidised to sulphuric acid ...
... by oxygen in the air ...
... and by bromine water, potassium dichromate(VI) and potassium manganate(VII).

Sulphurous acid is a weak acid which forms salts called sulphites, e.g. sodium sulphite Na_2SO_3. Sulphurous acid is slowly oxidised by oxygen in the air to sulphuric acid, H_2SO_4. Sulphuric acid is formed whenever sulphurous acid accepts oxygen from an oxidising agent.

Sulphurous acid + Oxygen from the air or from an oxidising agent → Sulphuric acid

$$H_2SO_3(aq) + (O) \rightarrow H_2SO_4(aq)$$

For example,

Sulphurous acid + Bromine water → Sulphuric acid + Hydrogen bromide

$$H_2SO_3(aq) + Br_2(aq) + H_2O(l) \rightarrow H_2SO_4(aq) + 2HBr(aq)$$

Tests for sulphur dioxide

1 Sulphur dioxide reduces potassium dichromate(VI) from orange dichromate ions, $Cr_2O_7^{2-}$, to blue chromium(III) ions, Cr^{3+}, passing through the intermediate colour of green, which results from a mixture of orange ions and blue ions.

2 It reduces potassium manganate(VII) from purple manganate(VII) ions, $MnO_4^-(aq)$ to pale pink, almost colourless manganese(II) ions, Mn^{2+}.

FUME CUPBOARD

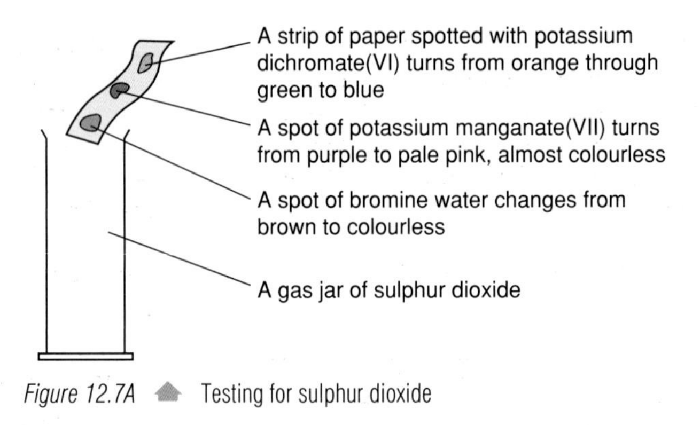

Figure 12.7A Testing for sulphur dioxide

SUMMARY

Sulphur dioxide is an acidic, poisonous gas with a pungent smell. It dissolves in water to form sulphurous acid, which is a reducing agent.

CHECKPOINT

1. Makers of home-brewed beer and wine rinse their bottles with a solution of sodium sulphite before filling them. Explain why they do this.
2. Write the symbol for the reducing agent in each of the following reactions.

   $Na(s) + 2H_2O(l) \rightarrow H_2(g) + 2NaOH(aq)$
   $O_2(g) + S(s) \rightarrow SO_2(g)$
   $2Al(s) + 3Cl_2(g) \rightarrow 2AlCl_3(s)$
   $Cu(s) + S(s) \rightarrow CuS(s)$
3. Explain why the reaction

   $2FeSO_4 + H_2SO_4 + Cl_2 \rightarrow Fe_2(SO_4)_3 + 2HCl$

   is classified as an oxidation–reduction reaction.
4. In the electrolysis of molten lead(II) bromide, the electrode processes are

   $Pb^{2+}(l) + 2e^- \rightarrow Pb(l)$
   $2Br^-(l) \rightarrow Br_2(g) + 2e^-$

   Which of these two processes is oxidation? Explain your choice.
5. Two bottles have lost their labels. One reads 'Oxidising agent' and the other 'Reducing agent'. Describe how you could test the solutions in the bottles to decide which label belongs to which bottle.

Theme Questions

1 The indicator phenolphthalein is colourless in neutral and acidic solutions and turns pink in alkaline solutions. Of the solutions A, B and C, one is acidic, one is alkaline, and one is neutral. Explain how, using only phenolphthalein, you can find out which solution is which.

2 Seven steel bars were placed in solutions of different pH for the same time. The table shows the percentage corrosion of the steel bars.

pH of solution	1	2	3	4	5	6	7
Percentage corrosion of steel bar	65	60	55	50	20	15	10

(a) On graph paper, plot the percentage corrosion against the pH of the solution.

(b) Read from your graph the percentage corrosion at pH 4.5.

3 Look at the information from the label of a bottle of concentrated orange squash.

Concentrated orange squash
Ingredients: Sugar, Water, Citric acid, Flavourings, Preservative E250, Artificial sweetener, Yellow colourings E102 and E103

(a) Explain what is meant by 'concentrated'.

(b) A sample of concentrated orange squash is mixed with water. A piece of universal indicator paper is dipped in. What colour will the paper turn?

(c) Why does using universal indicator paper give a better result than adding universal indicator solution to the orange squash?

(d) Describe a simple experiment you could do to prove that **only two** yellow substances are present in the concentrated orange squash.

4 Zinc sulphate crystals, $ZnSO_4.7H_2O$, can be made from zinc and dilute sulphuric acid by the following method.

Step 1 Add an excess of zinc to dilute sulphuric acid. Warm.

Step 2 Filter.

Step 3 Partly evaporate the solution from Step 2. Leave it to stand.

(a) Explain why an excess of zinc is used in Step 1.

(b) How can you tell when Step 1 is complete?

(c) Name the residue and the filtrate in Step 2.

(d) Explain why the solution is partly evaporated in Step 3.

(e) Would you dry the crystals by strong heating or by gentle heating? Explain your answer.

(f) Write a word equation for the reaction. Write a symbol equation.

5 What method would you use to make the insoluble salt lead(II) carbonate? Explain why you have chosen this method. Say what starting materials you would need, and say what you would do to obtain solid lead(II) carbonate. (For solubility information, see Theme C, Table 10.2)

6 Explain the following:

(a) Tea changes colour when lemon juice is added.

(b) Sodium sulphate solution is used as an antidote to poisoning by barium compounds.

(c) Washing soda takes some of the pain out of bee stings.

(d) Toothpastes containing aluminium hydroxide fight tooth decay.

(e) Calcium fluoride is added to some toothpastes.

(f) Scouring powders often contain sodium hydroxide and powdered stone.

7 A student receives the following instructions: 'Make a solution of copper sulphate by neutralising dilute sulphuric acid, using the base copper oxide.'

(a) Explain the meanings of the terms: solution, neutralising, dilute, base.

(b) Draw an apparatus which the student could use for carrying out the instructions.

(c) Describe exactly how the student should carry out the instructions.

8 The table shows the colours of some indicators.

Indicator	*pH* 1 2 3 4 5 6 7 8 9 10 11 12 13 14
Methyl orange	←Red→ ←——— Yellow ———→
Bromocresol green	←Yellow→ ←——— Blue ———→
Phenol red	←Yellow→ ←———Red———→
Phenolphthalein	←Colourless→ ←———Red———→

(a) What colour is a solution of pH 10 when a few drops of bromocresol green are added?

(b) What colour is a solution of pH 3.5 when a few drops of methyl orange are added?

(c) What colour is a solution of pH 6.5 when a few drops of phenol red are added?

(d) A solution turns yellow when either methyl orange or phenol red is added. What is the approximate pH of the solution?

(e) A solution is colourless when phenolphthalein is added and red when phenol red is added. What is the pH of the solution?

(f) A mixture of bromocresol green, phenol red and phenolphthalein is added to a solution of pH 10. What is the colour?

(g) A mixture of all four indicators is added to a strong acid. What is the colour?

9 (a) Use the Periodic Table to answer the following questions:
  (i) Give the symbols for the elements carbon and sodium.
  (ii) Give the symbol for any inert gas.
  (iii) Give the symbol for any element in Group 7 (the halogens)
  (iv) Give **one** reason why the symbol for helium is He and not H.

(b) One molecule of carbon dioxide contains one atom of carbon and two atoms of oxygen. Its formula is written as CO_2.
Write the formula of:
  (i) a molecule of sulphur dioxide, which has one atom of sulphur and two atoms of oxygen.
  (ii) a molecule of sulphuric acid, which contains two atoms of hydrogen, one atom of sulphur and four atoms of oxygen.

10 Explain why Mendeleev fitted the elements into vertical groups in his Periodic Table. Why did he fit lithium, sodium, potassium, rubidium and caesium into the same group? Why did he leave some gaps in his table? The noble gases were discovered after Mendeleev had written his Periodic Table. What do they have in common (a) in their chemical reactions, (b) in their electronic configurations and (c) in their positions in the Periodic Table?

11 (a) Copy and complete the table (refer to Figure 9.5A if necessary).

Colour of universal indicator	*pH of solution*
red	______
______	11
green	______

(b) Copy and complete the following word equation:
acid + alkali → ______ + ______
(c) Name this type of reaction.
(d) Name the products formed when the acid is dilute hydrochloric acid and the alkali is potassium hydroxide.
(e) Name the acid and alkali needed to give ammonium nitrate.

12 The graph shows the average number of fillings per child plotted against the fluoride content of the water in different regions.

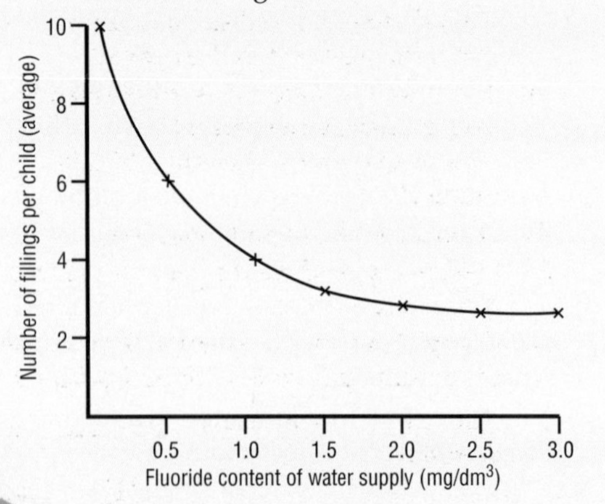

(a) Say what you can deduce from the graph about tooth decay.
(b) Give one reason why some people are in favour of adding fluoride to the water supply.
(c) Give one reason why some people are against adding fluoride to the water supply.

13 Choose from the following list a substance which fits each of the descriptions below.

fluorine hydrogen sodium hydroxide potassium chromium copper(II) sulphate potassium chloride chlorine calcium silver

(a) a halogen used to treat drinking water
(b) a transition element
(c) a compound which dissolves to form a coloured solution
(d) an alkali
(e) a metallic element which reacts vigorously with water
(f) a metallic element which does not react with water
(g) an alkaline earth
(h) the salt of a Group 1 element and a Group 7 element.

14 Refer to the Periodic Table (Figure 11.1A) and to Tables 11.3 and 11.7.
(a) Rubidium, Rb, is in Group 1.
  (i) Name two products you would expect when rubidium reacts with water.
  (ii) Describe what you would expect to see during the reaction.
(b) Iodine, I, is in Group 7. It reacts with hydrogen to form hydrogen iodide, a soluble gas.
  (i) Name a common chemical which you would expect to behave like hydrogen iodide solution.
  (ii) Name two products which you would expect when hydrogen iodide solution reacts with magnesium.
  (iii) Describe what you would expect to see during the reaction.
(c) Give the name and formula of the compound formed between rubidium and iodine.

15 This question is about the halogen elements in Group 7 of the Periodic Table. The reactivity of the elements in Group 7 decreases down the group.
(a) Write the name and symbol of the *most* reactive halogen in Group 7. [1]
(b) The table gives information about three halogens.

Halogen	*Colour*	*Melting point in °C*	*Boiling point in °C*	*State at room temperature and pressure*
Chlorine	Greenish-yellow	−101	−34	Gas
Bromine	Red	−7	60	Liquid
Iodine	Dark grey	114	185	Solid

Astatine is in Group 7 of the Periodic Table. It is below iodine.
Use the information in the table and your knowledge to answer the following.

(i) Predict the state of astatine at room temperature and pressure. [1]
(ii) Suggest a melting point for astatine. Use the data to explain your choice. [2]
(iii) Predict the colour of astatine. [1]
(iv) What are the name and formula of the compound formed by sodium and astatine? [2]

(c) The table summarises the results of reactions when halogens are added to solutions of sodium halides.

(i) Finish a copy of the table by adding a tick (✓) if a reaction takes place and a cross (✗) if a reaction does not take place. Some have been done for you. [3]

halogen added	*solutions of*		
	sodium chloride	*sodium bromide*	*sodium iodide*
bromine	✗	✗	✓
chlorine	✗		
iodine			✗

(ii) Finish the sentence by choosing the *best* word from the list.
decomposition; displacement; neutralisation.
The reaction of bromine with sodium iodide is an example of a ______ reaction. [1]
(iii) Write an equation for the reaction taking place when bromine, Br_2, is added to potassium iodide solution, KI. [2]

(d) Sodium chloride is used as a raw material for producing other sodium compounds. These include
sodium carbonate; sodium hydrogencarbonate; sodium hydroxide.
Choose *two* of these. For each one, write down a use of the sodium compound. [2] (OCR)

16 The diagram below shows an apparatus used to pass steam over burning magnesium ribbon.

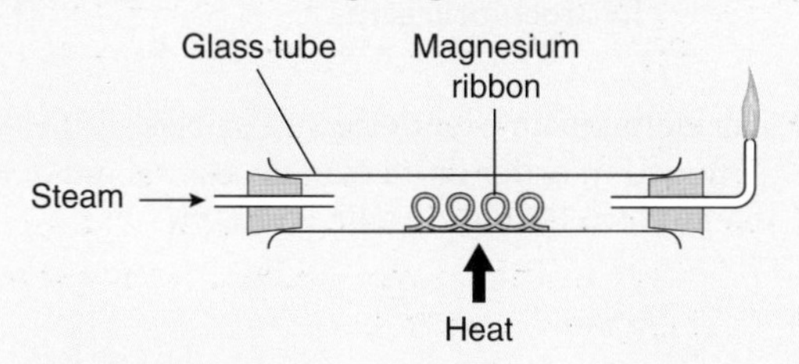

(a) Describe what you would *see* inside the glass tube when steam reacts with burning magnesium. [1]
(b) Give the name and appearance of the main product remaining in the glass tube. [2]
(c) If the gas formed in the reaction had been collected how could you show it was hydrogen? [2]
(d) Using the Periodic Table of Elements, state the Group II element with which it would be too dangerous to repeat this experiment. [2] (WJEC)

17 (a) Use the Periodic Table to give:
(i) the symbol for an atom of sulphur;
(ii) an element in the same group as sodium;
(iii) an element in group 2;
(iv) an element in group 6;
(v) the atomic number of neon;
(vi) an element in period 2. [6]

(b) Elements can be classified as metals or non-metals. Give the names of TWO non-metals. [2]
(c) State TWO properties of most metals from the following.
They are shiny; They are soft:
They break easily; They conduct electricity. [2]
(d) You have an unknown solid element.
Describe a test to find out if it is a metal. [3] (Edexcel)

18 Dilute hydrochloric acid was added slowly to dilute sodium hydroxide solution in a beaker. The graph below shows how the pH of the solution in the beaker changed as the acid was added.

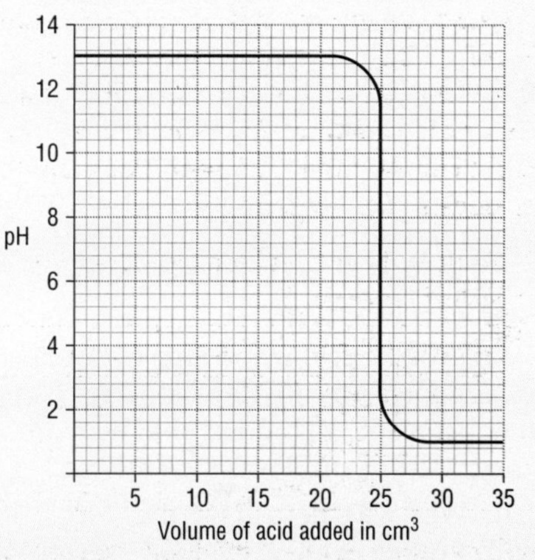

(a) What was the pH of the solution in the beaker when 30 cm³ of dilute hydrochloric acid had been added? [1]
(b) The dilute sodium hydroxide solution in the beaker contained Univeral indicator. What colour was the solution in the beaker when the following volumes of dilute hydrochloric acid had been added?
(i) 30.0 cm³ (ii) 10.0 cm³ [2]
(c) (i) What is the pH of a neutral solution? [1]
(ii) What volume of dilute hydrochloric acid was added to neutralise the sodium hydroxide solution in the beaker? [1]
(iii) The neutral solution was evaporated to dryness to leave a solid salt.
What is the name of the salt which is formed? [1]
(iv) Describe what the salt looks like. [1]
(v) State the type of bonding that is present in this salt. [1]
(vi) Complete the word equation for the reaction of sodium hydroxide with hydrochloric acid.

Sodium hydroxide + Hydrochloric acid → ______ + ______

[1] (Edexcel)

19 Bordeaux Mixture controls some fungal infections on plants.
A student wanted to make some Bordeaux Mixture.

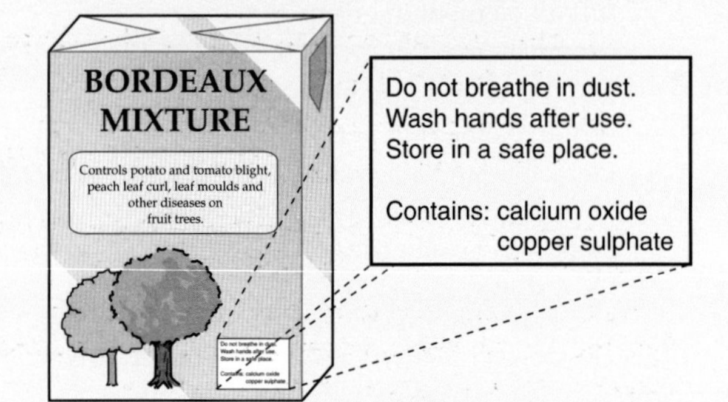

(a) The student knew that calcium oxide could be made by heating limestone. Limestone contains calcium carbonate, $CaCO_3$.
(i) Write the word equation for this reaction. [1]
(ii) What type of reaction is this? [1]

(b) The student knew that copper sulphate $CuSO_4$, could be made by the following general reaction.

Acid + Base → Salt + Water [1]

(i) What type of reaction is this?
(ii) The base used is copper oxide. Name and give the chemical formula of the acid used. [2]

(c) The student wrote about how the copper sulphate was made.
'Some of the acid was warmed. "Copper oxide was added. The mixture was stirred. More copper oxide was added until no more would react. The mixture was then filtered."
(i) Why was the acid warmed? [1]
(ii) Copper oxide was added until no more would react. Explain why. [2]
(iii) The filtration apparatus is shown.

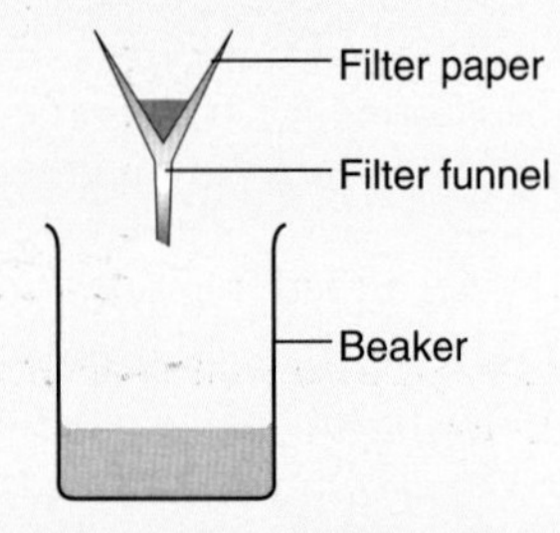

Describe and explain what happens as the mixture is filtered. [2] AQA(SEG) 1999

20 Fluorine is more reactive than chlorine.
Fluorine reacts with most elements in the Periodic Table. However, fluorine does not react with argon.
Atomic numbers: F 9; Cl 17; Ar 18.
(a) To which group of the Periodic Table do fluorine and chlorine belong? [1]
(b) (i) Give *one* use for argon. [1]
(ii) Explain why the noble gas argon is unreactive. [2]
(c) (i) Give *one* use for chlorine. [1]
(ii) Draw the electron arrangement of a chlorine atom.

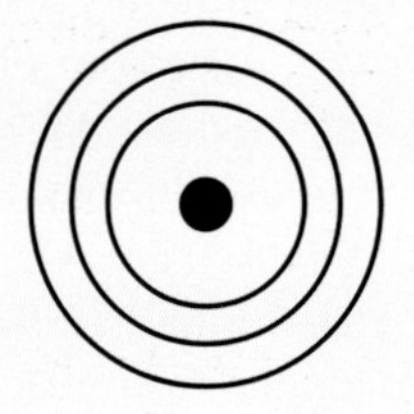

[2]

(iii) Explain why fluorine is more reaction than chlorine. [3]

(d) When chlorine is bubbled into potassium bromide solution the colourless solution becomes an orange-red colour.
(i) Complete and balance the ionic equation for this reaction.

Cl_2 + _____ → _____ + _____ [3]

(ii) Explain how the solution becomes orange-red in colour. [2] AQA(SEG) 1998

21 (a) A solution of zinc chloride can be prepared by adding excess zinc carbonate to dilute hydrochloric acid. At the end of the reaction, the remaining zinc carbonate is removed by filtration.
(i) Explain why excess zinc carbonate is used. [1]
(ii) State *one* other zinc *compound* which reacts with dilute hydrochloric acid to form zinc chloride solution. [1]

(b) Silver chloride can be made by reacting silver nitrate solution with hydrochloric acid.
(i) Write the ionic equation, including state symbols, for this reaction. [2]
(ii) Explain why pure silver chloride could NOT be made by adding silver carbonate to hydrochloric acid. [2] (Edexcel)

22 The elements in Mendeleev's Periodic Table were arranged in order of increasing atomic mass. Part of the modern Periodic Table is shown.

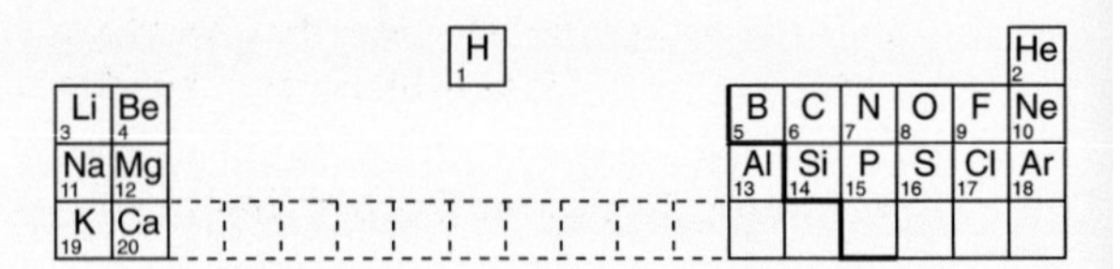

(a) Complete the following sentence by adding the missing words.
The modern Periodic Table is arranged in order of increasing .... [1]

(b) (i) Name a metal in the same group as lithium. [1]
(ii) Name a non-metal in the same period as magnesium. [1]

(c) The table contains some information about *two* elements.

Element	Symbol	Number of protons	Number of neutrons	Number of electrons
Fluorine	F	9	10	9
Chlorine	Cl	17	18	17
Chlorine	Cl	17	20	17

(i) In terms of atomic structure, give *one* feature that both these elements have in common. [1]

(ii) There are two **isotopes** of chlorine shown in the table. Explain what **isotope** means. [2]

(iii) Explain, in terms of electron arrangement, why fluorine is more reactive than chlorine. [2]

(d) Sodium reacts with chlorine to form the compound sodium chloride.

$$2Na + Cl_2 \rightarrow 2NaCl$$

Describe, in terms of electron arrangement, the type of bonding in:

(i) a molecule of chlorine. [3]

(ii) the compound sodium chloride. [4]

AQQ(SEG) 2000

23 (a) Five solutions, *A–E*, were tested with universal indicator solution, to find their pH. The results are shown below.

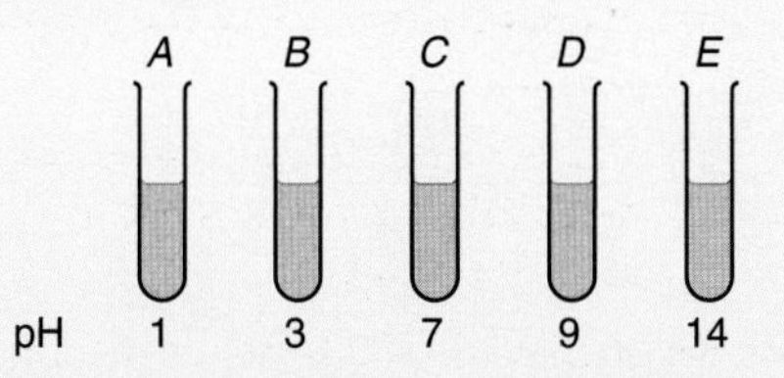

(i) Give the colour of each of the following solutions during testing: Solution *A*, Solution *E*. [2]

(ii) The five solutions were known to be ammonia solution, potassium hydroxide, sodium chloride, sulphuric acid and vinegar. Identify each of the solutions *A–E* by assigning the appropriate letter to each name below. [5]

(b) Complete the following passage by inserting the correct words/formulae in the spaces.

Acids are compounds which dissolve in water producing ______ ions which have the formula ______.

An example is nitric acid which has formula ______.

It may be neutralised by alkalis to form salts called ______.

Alkalis dissolve in water producing ______ ions which have the formula ______.

These ions combine with the ions in acids during neutralisation to form ______.

Acids react with metal carbonates to produce a gas. The name of the gas is ______. [8]

(c) Barium sulphate is an example of an insoluble salt. It is made by mixing solutions of barium chloride and magnesium sulphate. An insoluble white solid forms immediately.

(i) How would you prepare a pure dry sample of barium sulphate from the mixture? [3]

(ii) The preparation of barium sulphate is an example of a ______ reaction. [1] (CCEA)

24 Two students made some copper sulphate solution and wrote these notes about the experiment.

"We took 25 cm^3 of acid and added some green powder called copper carbonate. The mixture fizzed and we tested the gas with limewater. The limewater turned milky. All the green powder disappeared so we added some more until the fizzing stopped and some green powder remained. Then we filtered the mixture to get the copper sulphate solution."

(a) Describe ONE safety precaution which the students should take during this experiment. [1]

(b) Name the gas given off during the reaction. [1]

(c) (i) Which acid must be used to make copper sulphate?

A Hydrochloric acid
B Nitric acid
C Sulphuric acid [1]

(ii) Draw, and name, the piece of apparatus that should be used to measure 25 cm^3 of the acid. [2]

(d) Label a copy of the diagram below which shows the mixture being filtered to obtain copper sulphate solution.

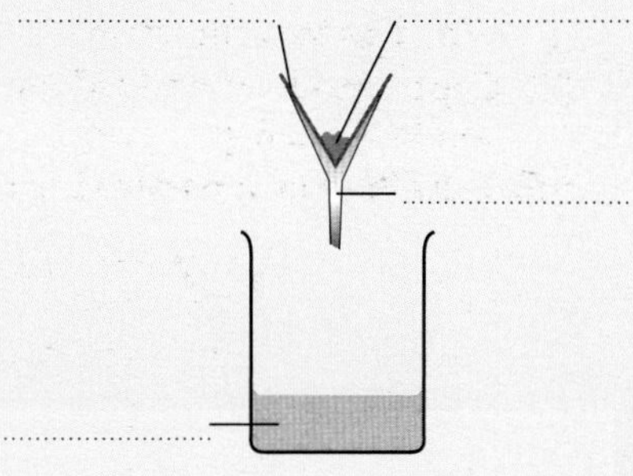

[4]

(e) Name the solute and solvent in copper sulphate solution. [2] (Edexcel)

Theme D

Planet Earth

Our planet provides us with the materials which we use to make the food, clothing, housing and all the possessions we need and prize. The plenty we enjoy has led to pollution. The air is polluted by objectionable gases, dust and smoke from our cars and factories. Rivers and lakes are polluted by waste from factories and fields. The land is spoiled by rubbish tips. People are awakening to the need to combat pollution and save our planet. Steps are being taken to reduce the emission of pollutants into the air and into rivers and lakes. A start has been made in recycling rubbish to preserve the landscape and to conserve Earth's resources.

The Earth's atmosphere has evolved slowly over billions of years to the composition which it has today. The rocks in the Earth's crust are constantly undergoing slow geological changes. The continents we know today are vastly different from those that first emerged on Earth. This theme tells the story of these changes.

13.1 ▶ Oxygen the life-saver

FIRST THOUGHTS

As you study this topic, think about the vital importance of air as a source of:

- oxygen – which supports the life of plants and animals
- carbon dioxide – which supports plant life
- nitrogen – the basis of natural and synthetic fertilisers
- the noble gases – filling our light bulbs

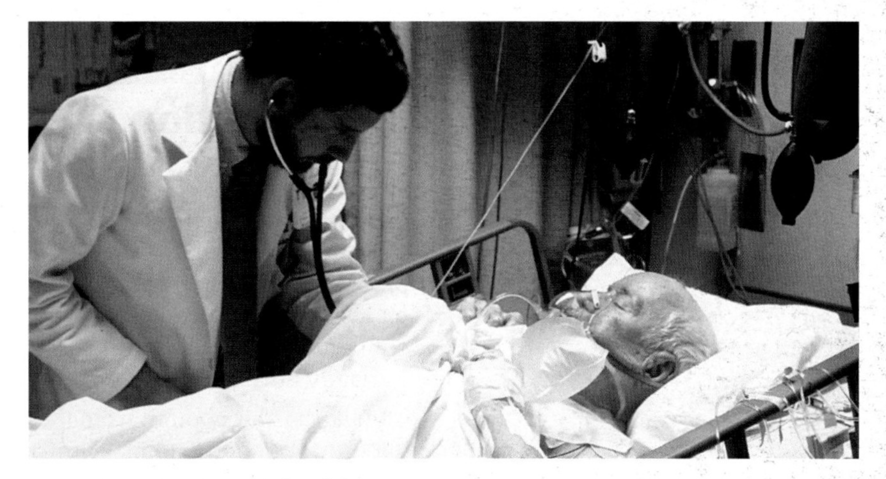

Figure 13.1A ▲ Oxygen in hospitals

Hospitals need oxygen for patients who have difficulty in breathing. Tiny premature babies often need oxygen. They may be put into an oxygen high pressure chamber. This chamber contains oxygen at 3–4 atmospheres pressure, which makes 15–20 times the normal concentration of oxygen dissolve in the blood. People who are recovering from heart attacks and strokes also benefit from being given oxygen. Patients who are having operations are given an anaesthetic mixed with oxygen.

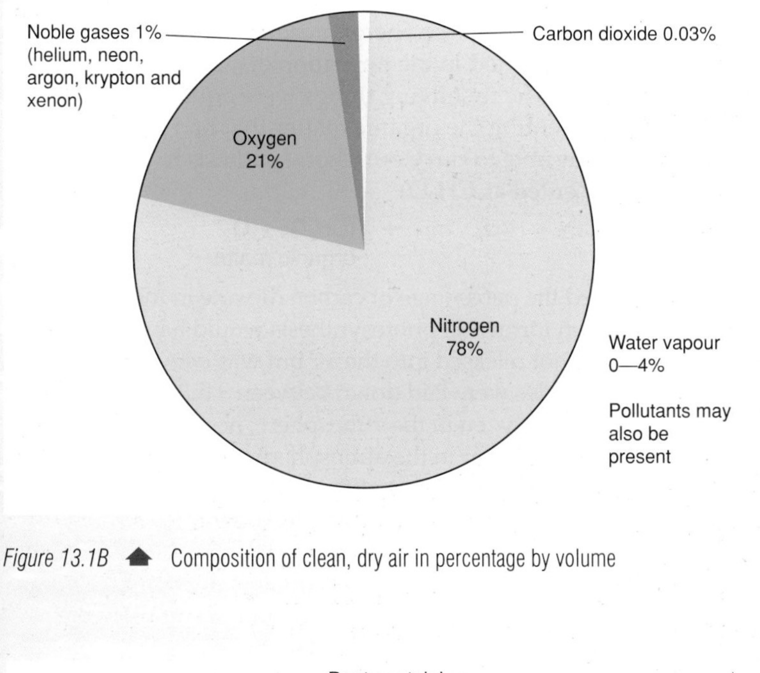

Figure 13.1B ▲ Composition of clean, dry air in percentage by volume

In normal circumstances, we can obtain all the oxygen we need from the air. Figure 13.1B shows the percentages of oxygen, nitrogen and other gases in pure, dry air. Water vapour and pollutants may also be present in air.

Figure 13.1 C shows an apparatus which can be used to measure the percentage of oxygen in air.

Figure 13.1C ▲ Finding the percentage of oxygen in air

The first gas syringe contains 100 cm^3 of dry air. The air is pushed slowly by the plunger from syringe 1 to syringe 2, across the hot copper powder. This combines with some of the oxygen in the air to form copper(II) oxide. Plunger 2 pushes air slowly back into syringe 1. This is continued until the volume of air no longer changes, showing that all the oxygen has been removed. Then the remaining air in the syringe is cooled to room temperature and measured. The results might be:

Volume of air in syringe at room temperature at start = 100 cm^3
Volume of air in syringe after repeated passes over hot copper and cooling to room temperature = 79 cm^3

Percentage of oxygen in air = $\left(\frac{21}{100}\right) \times 100\% = 21\ \%$

13.2 The evolution of Earth's atmosphere

www.keyscience.co.uk

The present-day atmosphere has evolved from a mixture of gases released from the interior of the Earth about 4.6 billion years ago. The mixture was probably similar to that released from volcanoes today: water vapour 64% by mass, carbon dioxide 24%, sulphur dioxide 10%, nitrogen 1.5%, ammonia 0.5%. This mixture has evolved into today's atmosphere which is mainly nitrogen and oxygen with only a trace of carbon dioxide.

The Earth is far enough from the Sun to allow the primordial water vapour to condense to make oceans. Carbon dioxide dissolved in the oceans in large quantities and was eventually converted into limestone beds. Since Venus is closer to the Sun than Earth is, oceans never formed and carbon dioxide remains in the atmosphere.

When living things emerged on Earth, they altered the composition of the atmosphere. The first primitive living things evolved in the ocean about 3.5 billion years ago. The atmosphere was irradiated with ultraviolet light which provided the energy needed to bring about chemical reactions that resulted in the formation of substances such as amino acids and sugars. The first living things were aquatic. They derived energy by fermenting the organic matter that had been formed. Eventually they became able to carry out photosynthesis to produce organic matter (represented as CH_2O).

$$CO_2 + H_2O + h\upsilon \rightarrow \underset{\text{organic matter}}{(CH_2O)} + O_2$$

Photosynthesis reduced the percentage of carbon dioxide in the atmosphere. The oxygen formed in photosynthesis would have poisoned primitive plants. It was not released into the air but was converted into iron oxides. Layers of iron oxides were laid down between 3 billion and 1.5 billion years ago. With no oxygen in the atmosphere, no ozone, O_3, was able to form. With no ozone layer in the atmosphere to absorb ultraviolet (UV) radiation from the Sun, the surface of the Earth was too hot to support life. However, bacteria were able to live in the ocean, protected from UV radiation by water. They eventually acquired enzymes that enabled them to use the oxygen in the atmosphere to oxidise organic matter in the ocean with the liberation of energy. The process evolved into respiration, the mechanism by which present-day organisms obtain energy.

Ammonia in the early atmosphere was removed by

- nitrifying bacteria which converted it into nitrates
- denitrifying bacteria which converted it into nitrogen.

Later, when the level of oxygen in the atmosphere rose, reaction with oxygen removed ammonia from the atmosphere.

Between 1800 and 800 million years ago, oxygen began to accumulate in the atmosphere. It enabled the ozone layer to form, shielding the Earth from excessive ultraviolet radiation and making it an environment in

SUMMARY

The composition of the air is very different from the atmosphere that originally surrounded the Earth. Changes happened gradually over millions of years. Water vapour condensed to make oceans ... carbon dioxide dissolved in the oceans ... primitive plants evolved and carried on photosynthesis ... bacteria evolved and began to use oxygen in respiration ... oxygen accumulated and the ozone layer formed ... making Earth an environment in which living things could prosper.

which living things could prosper (see § 16.11). The Earth became a welcoming environment and land plants emerged 450 million years ago, followed by land animals 400 million years ago. Living things evolved from sea-dwellers into land-dwellers.

Volcanic activity had released nitrogen since the beginning of Earth's existence. The nitrogen had built up since no living things metabolised it. The composition of the atmosphere has remained the same for about 300 million years.

13.3 How oxygen and nitrogen are obtained from air

The method used by industry to obtain oxygen and nitrogen from air is **fractional distillation** of liquid air. Air must be cooled to −200 °C before it liquefies. It is very difficult to get down to this temperature. However, one way of cooling a gas is to compress it and then allow it to expand suddenly. Figures 13.3A and 13.3B show how air is first liquefied and then distilled.

Compression of a gas causes a rise in temperature; expansion causes a cooling. This cooling effect is used to liquefy air.

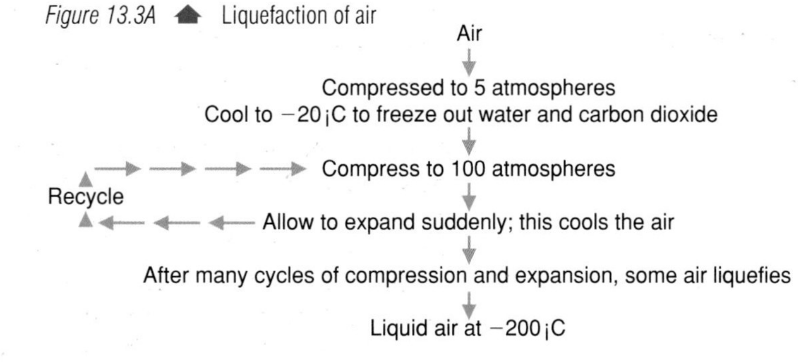

Figure 13.3A ▲ Liquefaction of air

Figure 13.3B ▲ Fractional distillation of air

Nitrogen gas (b.p. −196 °C)

The fractionating column is well insulated. The top of the column is at −190 °C and the bottom is at −200 °C

Argon gas (b.p. −186 °C)

Liquid air at −190 °C

Perforated shelves allow the ascending gases and descending liquids to mix

Liquid oxygen (b.p. −183 °C)

Oxygen, nitrogen and argon are redistilled. They are stored under pressure in strong steel cylinders. Industry finds many important uses for them.

SUMMARY

Air is a mixture of nitrogen, oxygen, noble gases, carbon dioxide, water vapour and pollutants. The fractional distillation of liquid air yields oxygen, nitrogen and argon.

CHECKPOINT

1 The top of the fractionating column in Figure 13.3B is at −190 °C. Explain why nitrogen is a gas at the top of the column but oxygen is a liquid.

2 The table lists nitrogen, oxygen and the noble gases.

(a) Which gas has
(i) the lowest boiling point and
(ii) the highest boiling point?

(b) List the gases which would still be gaseous at −200 °C.

Gas	*Boiling point* (°C)
Argon	−186
Helium	−269
Krypton	−153
Neon	−246
Nitrogen	−196
Oxygen	−183
Xenon	−108

13.4 Uses of oxygen

Oxygen is a colourless, odourless gas which is slightly soluble in water. All living things need oxygen for respiration. Breathing becomes difficult at a height of 5 km above sea level, where the air pressure is only half that at sea level. Climbers who are tackling high mountains take oxygen with them. Aeroplanes which fly at high altitude carry oxygen. Astronauts must carry oxygen, and even unmanned space flights need oxygen. Deep-sea divers carry cylinders which contain a mixture of oxygen and helium.

Industry has many uses for oxygen. Figure 13.4B shows an oxyacetylene torch being used to weld metal at a temperature of about 4000 °C. The hot flame is produced by burning the gas ethyne (formerly called acetylene) in oxygen. Substances burn faster in pure oxygen than in air.

The steel industry converts brittle cast **iron** into strong **steel**. Cast iron is brittle because it contains impurities such as carbon, sulphur and phosphorus. These impurities will burn off in a stream of oxygen. Steel plants use one tonne of oxygen for every tonne of cast iron turned into steel. Many steel plants make their own oxygen on site.

The proper treatment of sewage is important for public health. Air is used to help in the decomposition of sewage. Without this treatment, sewage would pollute many rivers and lakes. One method of treating polluted lakes and rivers to make them fit for plants and animals to live in is to pump in oxygen.

It's a fact!

The introduction of the oxy-acetylene flame brought about a revolution in metal working. Industry moved out of the blacksmith era into the twentieth century with gas-welding and flame-cutting techniques.

SUMMARY

Uses for pure oxygen are:

- treating patients who have breathing difficulties
- supporting high-altitude pilots, mountaineers, deep-sea divers and space flights
- in steel making and other industries
- treating polluted water

Figure 13.4A Astronaut using oxygen

Figure 13.4B Oxyacetylene welding being used.

CHECKPOINT

▸ 1 Explain why oxygen is used in (a) steelworks (b) metal working (c) sewage treatment (d) space flights and (e) hospitals. Say what advantage pure oxygen has over air for each purpose.

13.5 ▸ Nitrogen

FIRST THOUGHTS

So far, oxygen seems to be the important part of the air, but nitrogen has its uses too, as you will find out in this section.

Nitrogen is a colourless, odourless gas which is slightly soluble in water. It does not readily take part in chemical reactions. Many uses of nitrogen depend on its unreactive nature. Liquid nitrogen (below −196 °C) is used when an inert (chemically unreactive) refrigerant is needed. The food industry uses it for the fast freezing of foods.

Vets use the technique of artificial insemination to enable a prize bull to fertilise a large number of dairy cows. They carry the semen of the bull in a type of vacuum flask filled with liquid nitrogen.

Many foods are packed in an atmosphere of nitrogen. This prevents the oils and fats in the foods from reacting with oxygen to form rancid products. As a precaution against fire, nitrogen is used to purge oil tankers and road tankers. The silos where grain is stored are flushed out with nitrogen because dry grain is easily ignited.

Figure 13.5A ▲ Oil tanker delivering oil to a refinery. Once it is empty, the tanks will be purged with nitrogen

Figure 13.5B ▲ Grain silos

SUMMARY

Nitrogen is a rather unreactive gas. It is used to provide an inert (chemically unreactive) atmosphere.

13.6 ▸ The nitrogen cycle

Nitrogen is an essential element in **proteins**. Some plants have nodules on their roots which contain nitrogen-fixing bacteria. These bacteria **fix** gaseous nitrogen, that is, convert it into nitrogen compounds. From these nitrogen compounds, the plants can synthesise proteins. Members of the legume family, such as peas, beans and clover, have nitrogen-fixing bacteria.

Plants other than legumes synthesise proteins from nitrates. *How do nitrates get into the soil?* Nitrogen and oxygen combine in the atmosphere during lightning storms and in the engines of motor vehicles during combustion. They form nitrogen oxides (compounds of nitrogen and oxygen). These gases react with water to form nitric acid. Rain showers bring nitric acid out of the atmosphere and wash it into the ground, where it reacts with

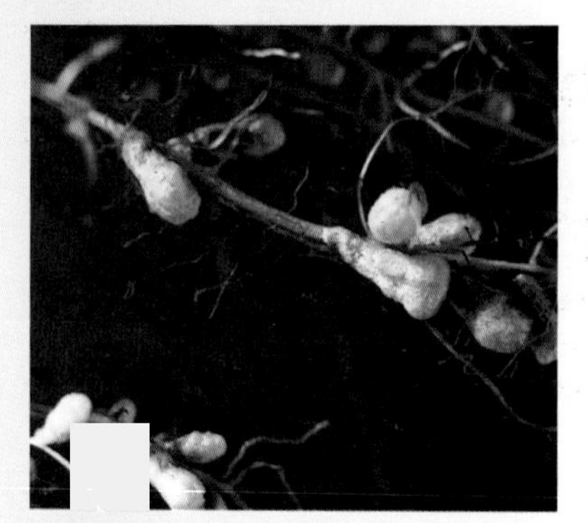

Figure 13.6A ▲ Nodules containing nitrogen-fixing bacteria on the roots of a legume

minerals to form nitrates. Plants take in these nitrates through their roots. They use them to synthesise proteins. Animals obtain the proteins they need by eating plants or by eating the flesh of other animals.

In the excreta of animals and the decay products of animals and plants, **ammonium salts** are present. Nitrifying bacteria in the soil convert ammonium salts into nitrates. Both nitrates and ammonium salts can be removed from the soil by **denitrifying bacteria**, which convert the compounds into nitrogen. To make the soil more fertile, farmers add both nitrates and ammonium salts as fertilisers. The balance of processes which put nitrogen into the air and processes which remove nitrogen from the air is called the **nitrogen cycle** (see Figure 13.6B).

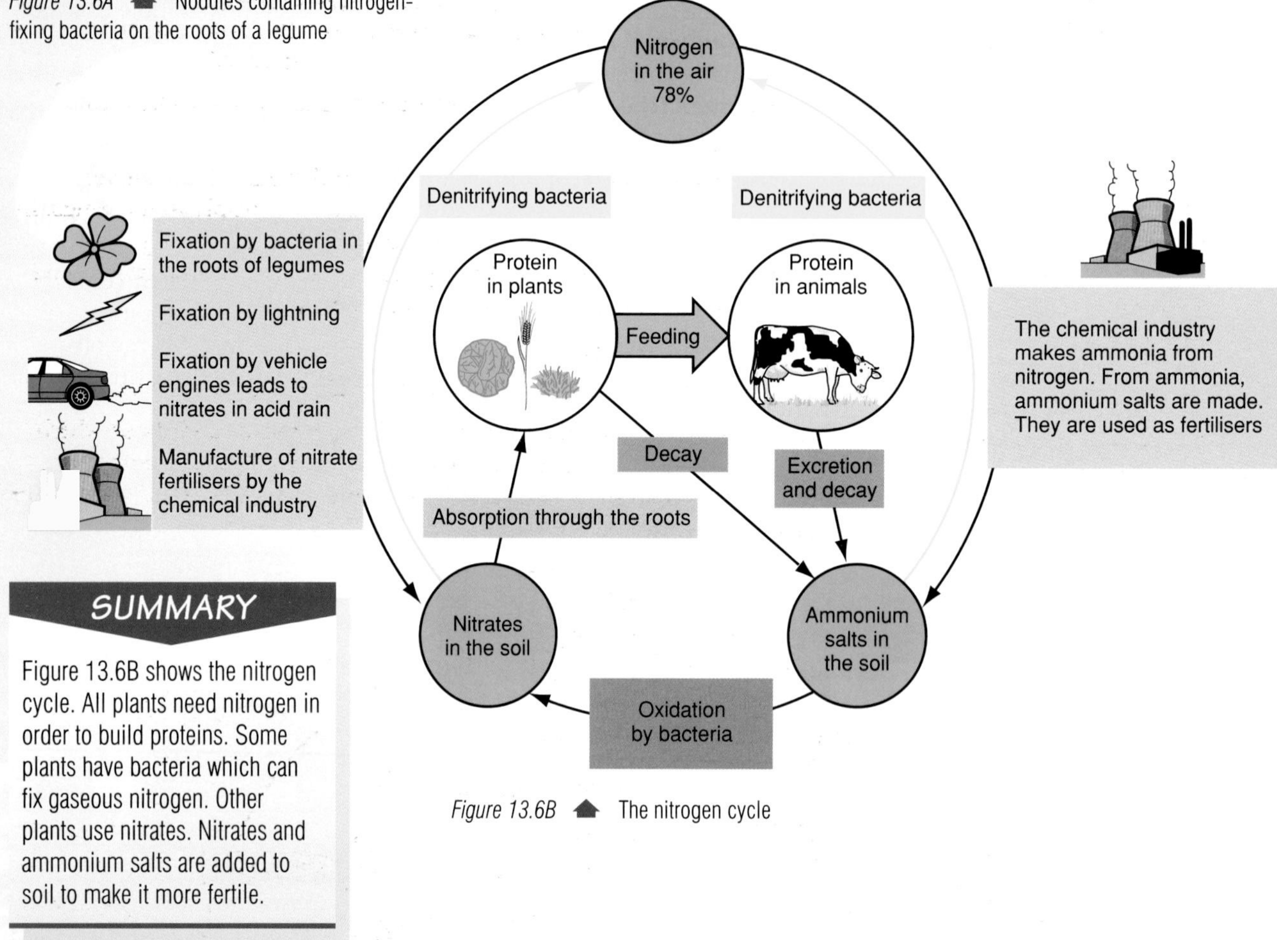

Figure 13.6B ▲ The nitrogen cycle

SUMMARY

Figure 13.6B shows the nitrogen cycle. All plants need nitrogen in order to build proteins. Some plants have bacteria which can fix gaseous nitrogen. Other plants use nitrates. Nitrates and ammonium salts are added to soil to make it more fertile.

CHECKPOINT

1. *Process A* Nitrogen-fixing bacteria convert nitrogen gas into nitrates.
   Process B Nitrifying bacteria use oxygen to convert ammonium compounds (from decaying plant and animal matter) into nitrates.
   Process C Denitrifying bacteria turn nitrates into nitrogen.
   (a) Say where *Process A* takes place.
   (b) Say what effect the presence of air in the soil will have on *Process B*.
   (c) Say what effect waterlogged soil which lacks air will have on *Process C*.
   (d) Explain why plants grow well in well-drained, aerated soil.
   (e) A farmer wants to grow a good crop of wheat without using a fertiliser. What could he plant in the field the previous year to ensure a good crop?
   (f) Explain why garden manure and compost fertilise the soil.
2. Explain why nitrogen is used in (a) food packaging (b) oil tankers (c) hospitals and (d) food storage.

13.7 Carbon dioxide and the carbon cycle

FIRST THOUGHTS

The percentage by volume of carbon dioxide in clean, dry air is only 0.03%. Perhaps you think this makes carbon dioxide sound rather unimportant. Through studying this section, you may change your mind about the importance of carbon dioxide!

SUMMARY

Figure 13.7A shows the carbon cycle. Photosynthesis is the process in which green plants use sunlight, carbon dioxide and water to make sugars and oxygen. Respiration is the process in which animals and plants oxidise carbohydrates to carbon dioxide and water with the release of energy.

Plants need carbon dioxide, and animals, including ourselves, need plants. Plants take in carbon dioxide through their leaves and water through their roots. They use these compounds to **synthesise** (build) sugars. The reaction is called **photosynthesis** (photo means light). It takes place in green leaves in the presence of sunlight. Oxygen is formed in photosynthesis.

Photosynthesis (in plants)

$$\text{Sunlight} + \text{Carbon dioxide} + \text{Water} \xrightarrow{\text{catalysed by chlorophyll in green leaves}} \text{Glucose (a sugar)} + \text{Oxygen}$$

The energy of sunlight is converted into the energy of the chemical bonds in glucose.

Animals eat foods which contain starches and sugars. They inhale (breathe in) air. Inhaled air dissolves in the blood supply to the lungs. In the cells, some of the oxygen in the dissolved air oxidises sugars to carbon dioxide and water and energy is released. This process is called **respiration**. Plants also respire to obtain energy.

Respiration (in plants and animals)

$$\text{Glucose} + \text{Oxygen} \rightarrow \text{Carbon dioxide} + \text{Water} + \text{Energy}$$

The processes which take carbon dioxide from the air and those which put carbon dioxide into the air are balanced so that the percentage of carbon dioxide in the air stays at 0.03%. This balance is called the **carbon cycle** (see Figure 13.7A).

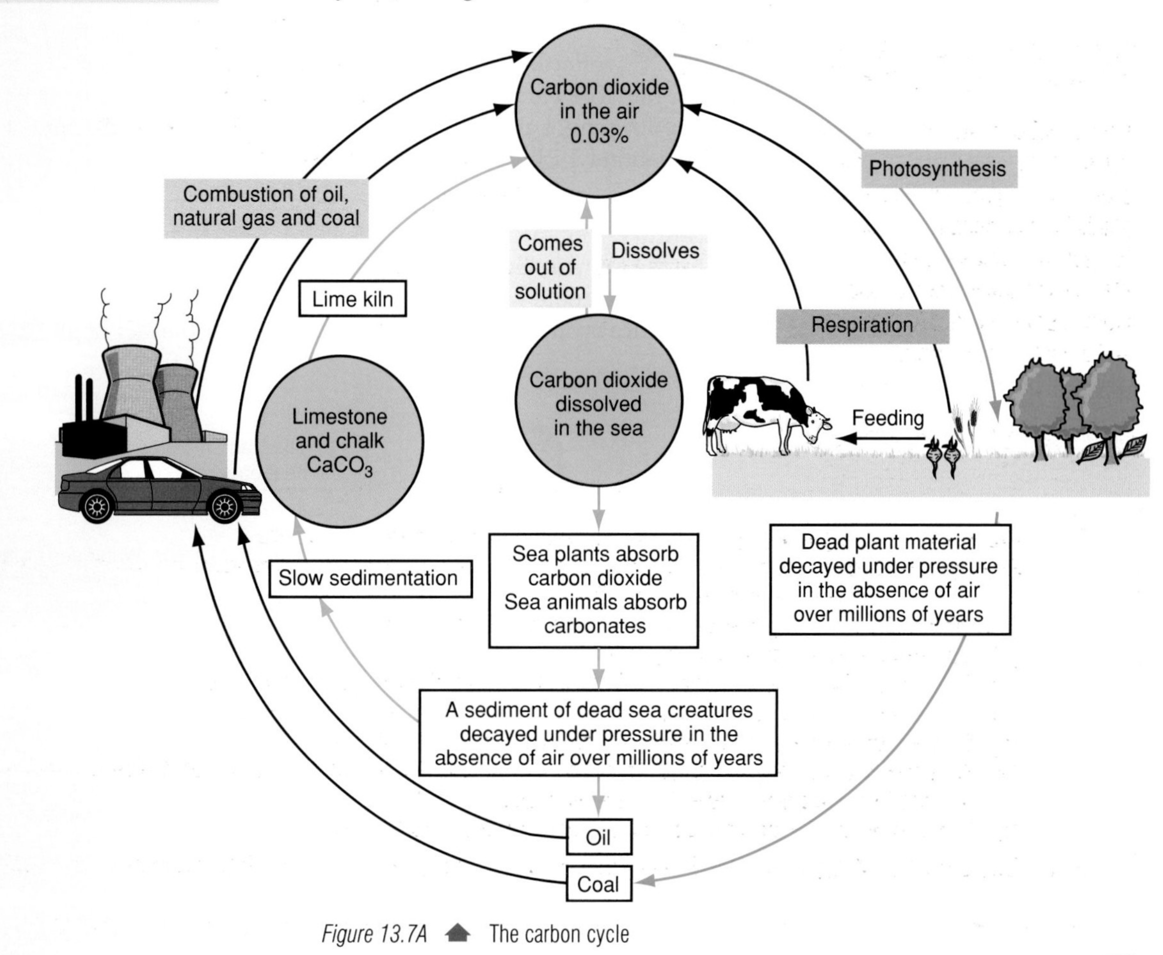

Figure 13.7A ▲ The carbon cycle

CHECKPOINT

1. Name two processes which add carbon dioxide to the atmosphere.
2. Suggest a place where you would expect the percentage of carbon dioxide in the atmosphere to be lower than average.
3. Suggest two places where you would expect the percentage of carbon dioxide in the atmosphere to be higher than average.

13.8 The greenhouse effect

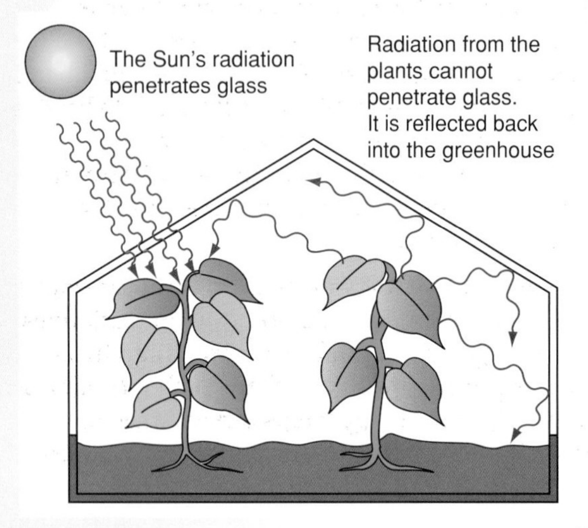

Figure 13.8A ▲ A greenhouse

The sun is so hot that it emits high-energy radiation. The Sun's rays can pass easily through the glass of a greenhouse. The plants in the greenhouse are at a much lower temperature. They send out infra-red radiation which cannot pass through the glass. The greenhouse therefore warms up (Figure 13.8A).

Radiant energy from the Sun falls on the Earth and warms it. The Earth radiates heat energy back into space as infra-red radiation. Unlike sunlight, infra-red radiation cannot travel freely through the air surrounding the Earth. Both water vapour and carbon dioxide absorb some of the infra-red radiation. Since carbon dioxide and water vapour act like the glass in a greenhouse, their warming effect is called the **greenhouse effect**. Without carbon dioxide and water vapour, the surface of the Earth would be at −18 °C instead of the average value of 15 °C which makes life on Earth possible. Most of the greenhouse effect is due to water vapour. The Earth does, however, radiate some wavelengths which water vapour cannot absorb. Carbon dioxide is able to absorb some of the radiation which water vapour lets through.

The surface of the Earth has warmed up by about 0.5 °C during the last century. The rate of warming is increasing. If the temperature continues to rise at the present rate, Earth will experience **global warming**. The Earth's climate will change and some agricultural areas will cease to produce crops and become deserts. Another danger of global warming is

It's a fact!

Worldwide, 16 thousand million tonnes of carbon dioxide are formed each year by the combustion of fossil fuels!

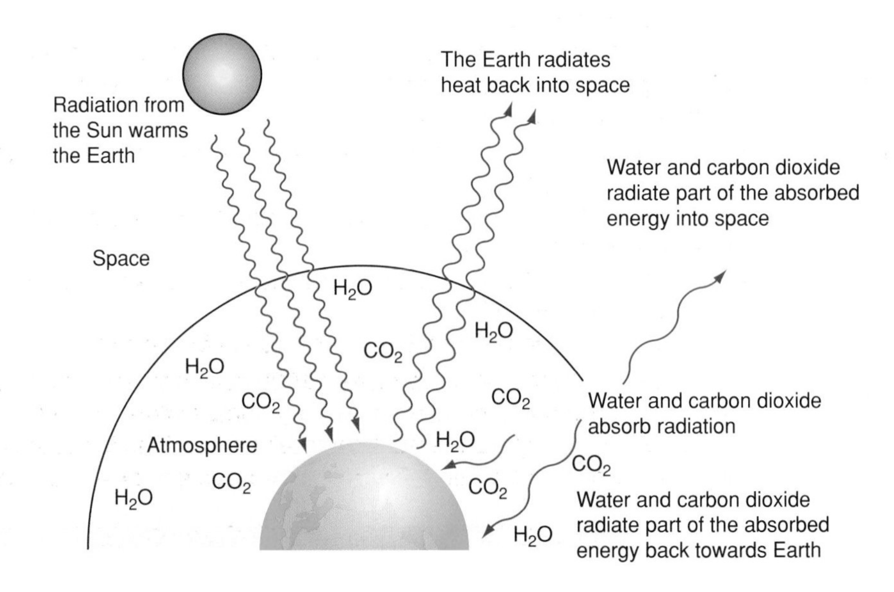

Figure 13.8B ▶ The greenhouse effect

SUMMARY

Carbon dioxide and water vapour reduce the amount of heat radiated from the Earth's surface into space and keep the Earth warm. Their action is called the greenhouse effect. The percentage of carbon dioxide in the atmosphere is increasing, and the temperature of the Earth is rising. If it continues to rise, the polar ice caps could melt.

that the temperature of the Arctic and Antarctic regions might rise above 0 °C. Then polar ice would melt slowly over the course of a century or two and flow into the oceans. The US Environmental Protection Agency predicts a rise in sea level of 0.5–1.5 m by the year 2100. The result would be that low-lying areas of land would disappear under the sea. The British Antarctic Survey of 1996 reported that an area of 8500 km^2 of the ice shelf that surrounds the continent had melted.

One reason for the increase in Earth's temperature is that we are putting too much carbon dioxide into the air. There are other greenhouse gases, e.g. methane, dinitrogen oxide and CFCs, but the concentration of carbon dioxide in the air is much higher than that of the other gases so it makes sense to focus on it. The combustion of coal and oil in power stations, factories and vehicles sends carbon dioxide into the air. The second reason is that we are felling too many trees. In South America huge areas of tropical forest have been cut down to make timber and provide land for farming. In many Asian countries forests have been cut down for firewood. The result is that worldwide there are fewer trees to take carbon dioxide from the air by photosynthesis. The percentage of carbon dioxide in the air increased by 25% between 1850 and 1990, while Earth's temperature rose by 0.5 °C. Recently the rate of increase of carbon dioxide has speeded up to 5% per year. If the concentration of carbon dioxide were to rise to twice its present level, it is predicted that the Earth's temperature would rise by up to 5 °C.

Measures can be taken to reduce carbon dioxide emission and to remove carbon dioxide from industrial exhaust gases. Alternatives to fossil fuels have been developed, e.g. hydrogen. Coal can be converted into fuels with a lower carbon content, e.g. methane and methanol, which emit less carbon dioxide for the same output of energy. Steps can be taken to remove carbon dioxide from the exhaust gases of power plants, e.g. by absorbing it in a base.

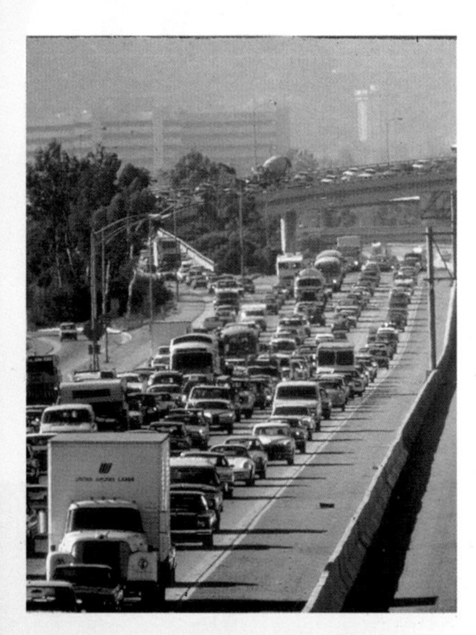

Figure 13.8C ◀ Nearly one quarter of the carbon dioxide formed by the combustion of fossil fuels comes from vehicles

CHECKPOINT

1. The amount of carbon dioxide in the atmosphere is slowly increasing.
   (a) Suggest two reasons why this is happening.
   (b) Explain why people call the effect which carbon dioxide has on the atmosphere the 'greenhouse effect'.
   (c) Why are some people worried about the greenhouse effect?
   (d) Suggest two things which could be done to stop the increase in the percentage of carbon dioxide in the atmosphere.
2. *Selima* Did you hear what Miss Sande said about the greenhouse effect making the temperature of the Earth go up?
   Joshe I don't know what she's worried about. We wouldn't be here at all if it weren't for the greenhouse effect.
   (a) What does Joshe mean by what he says? What would the Earth be like without the greenhouse effect?
   (b) Is Joshe right in thinking there is no cause for worry?
3. The burning of fossil fuels produces 16 000 million tonnes of carbon dioxide per year. Carbon dioxide is thought to increase the average temperature of the air. It is predicted that the effect of this increase in temperature will be to melt some of the ice at the North and South Poles. Describe the effects which this could have on life for people in other parts of the world.

13.9 Carbon dioxide

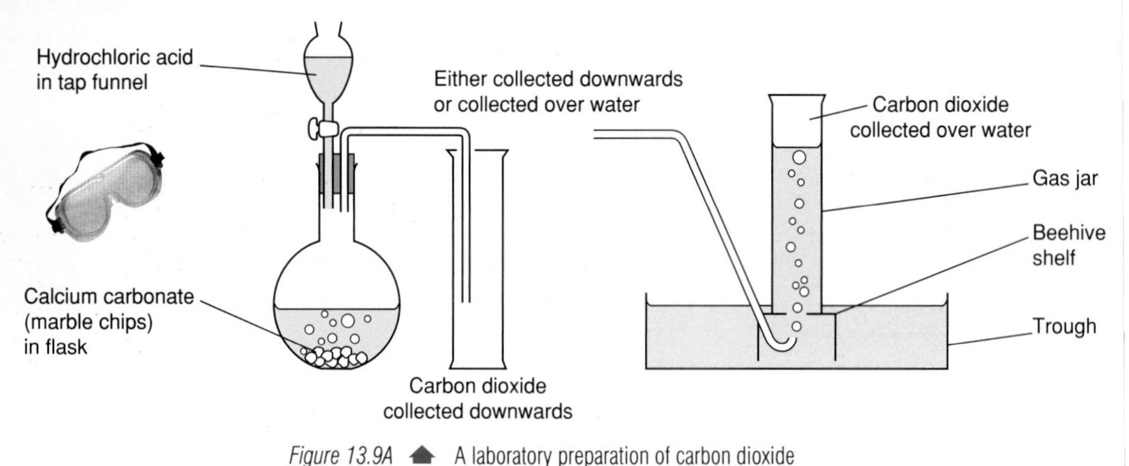

Figure 13.9A A laboratory preparation of carbon dioxide

Do not try using dilute sulphuric acid. A layer of insoluble calcium sulphate forms on the outside of each marble chip and brings the reaction to a stop.

Carbon dioxide can be made in the laboratory. Figure 13.9A shows one method. You will remember from Topic 9 that a carbonate reacts with an acid to form carbon dioxide, a salt and water. In this case

Calcium carbonate (marble chips) + Hydrochloric acid → Carbon dioxide + Calcium chloride + Water

$$CaCO_3(s) + 2HCl(aq) \rightarrow CO_2(g) + CaCl_2(aq) + H_2O(l)$$

Carbon dioxide can be collected over water. (*What does this tell you about the solubility of carbon dioxide?*) It can also be collected downwards. (*What does this tell you about the density of carbon dioxide?*)

Science at work

Pop stars sometimes use Dricold on stage. As it sublimes, it cools the air on stage. Water vapour condenses from the cold air in swirling clouds.

In water, carbon dioxide dissolves slightly to form the weak acid carbonic acid, H_2CO_3. Under pressure, the solubility increases. Many soft drinks are made by dissolving carbon dioxide under pressure and adding sugar and flavourings. When you open a bottle of fizzy drink, the pressure is released and bubbles of carbon dioxide come out of solution. People like the bubbles and the slightly acidic taste.

When carbon dioxide is cooled, it turns into a solid. This is known as **dry ice** and as **Dricold**. It is used as a refrigerant for ice cream and meat. When it warms up, dry ice sublimes (turns into a vapour without melting first).

Figure 13.9B Subliming carbon dioxide being poured from a gas jar

SUMMARY

Carbon dioxide:

- is a colourless, odourless gas
- is denser than air
- dissolves slightly in water to form carbonic acid
- does not burn
- allows few materials to burn in it

CHECKPOINT

1. Why do ice cream sellers prefer dry ice to ordinary ice?
2. Why is carbon dioxide used by the soft drinks industry?

13.10 ▸ Testing for carbon dioxide and water vapour

It's a fact!

Lake Nyos is a beautiful lake in Cameroon, West Africa. On 21 August 1986, a loud rumbling noise came from the lake. One billion cubic metres of gas burst out of the lake. The dense gas rushed downward to the village of Lower Nyos. The 1200 villagers were suffocated in their sleep. The gas was carbon dioxide. It had been formed by volcanic activity beneath the lake.

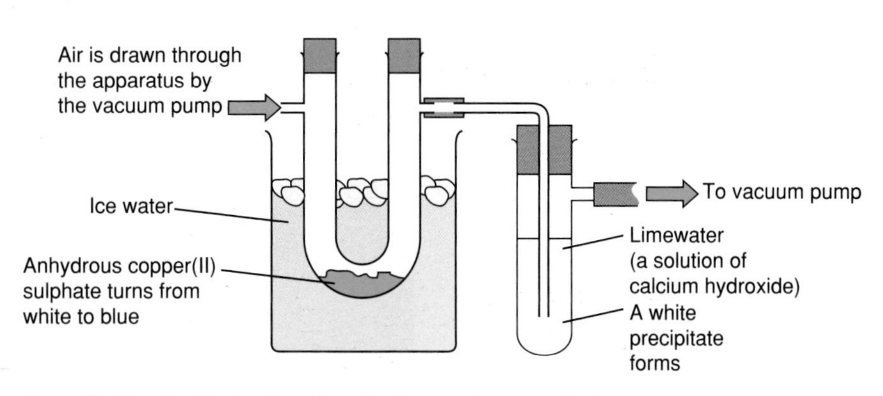

Figure 13.10A ▲ Testing for carbon dioxide and water vapour in air

Figure 13.10A shows how you can test for the presence of carbon dioxide and water vapour in air.

Test for carbon dioxide

A white precipitate is formed when carbon dioxide reacts with a solution of calcium hydroxide, **limewater**.

Carbon dioxide + Calcium hydroxide (limewater) → Calcium carbonate (white precipitate) + Water

$$CO_2(g) + Ca(OH)_2(aq) \rightarrow CaCO_3(s) + H_2O(l)$$

Test for water vapour

Water turns anhydrous copper(II) sulphate from white to blue.

Copper(II) sulphate (anhydrous, a white solid) + Water → Copper(II) sulphate-5-water (blue crystals)

$$CuSO_4(s) + 5H_2O(l) \rightarrow CuSO_4.5H_2O(s)$$

SUMMARY

The carbon dioxide in the air turns limewater cloudy. The water vapour in the air turns anhydrous copper(II) sulphate from white to blue.

CHECKPOINT

1. Malachite is a green mineral found in many rocks. How could you prove that malachite is a carbonate?
2. (a) How could you prove that whisky contains water?
   (b) How could you show that whisky is not pure water? (Don't say 'Taste it': tasting chemicals is often dangerous.)

13.11 ▸ The noble gases

Deep-sea divers have to take their oxygen with them (see Figure 13.11A). If they take air in their cylinders, nitrogen dissolves in their blood. Although the solubility of nitrogen is normally very low, at the high pressures which divers experience the solubility increases. When the divers surface, their blood can dissolve less nitrogen than it can at high pressure. Dissolved nitrogen leaves the blood to form tiny bubbles. These cause severe pains, which divers call 'the bends'. The solution to this problem is to breathe a mixture of oxygen and **helium**. Since helium

It's a fact!

Why are the noble gases so called?

They don't take part in many chemical reactions. Like the nobility of old, they don't seem to do much work.

is much less soluble than nitrogen, there is much less danger of the bends. Helium is a safe gas to use because it takes part in no chemical reactions. It is one of the **noble gases** (see Figure 13.1B).

For a long time, it seemed that the noble gases (helium, neon, argon, krypton and xenon) were unable to take part in any chemical reactions. They were called the 'inert gases'. Argon, the most abundant of them, makes up 0.09% of the air. Figure 13.3B shows how argon is obtained by the fractional distillation of liquid air.

What are 'the bends'? Why does breathing a mixture of oxygen and helium stop divers from getting 'the bends'?

Figure 13.11A ▲ A diver carries a mixture of oxygen and helium

Neon, argon, krypton and xenon are used in display lighting (see Figure 13.11B). The discharge tubes which are used as strip lights contain these gases at low pressure. When the gases conduct electricity, they glow brightly. Most electric light bulbs are filled with argon. The filament of a light bulb is so hot that it would react with other gases. Krypton and xenon are also used to fill light bulbs, and krypton is used in lasers.

The low density of helium makes it useful. It is used to fill balloons and airships.

Figure 13.11B ▲ Neon lights

SUMMARY

The noble gases take part in few chemical reactions. They are present in the air. Argon is used to fill light bulbs. Neon and other noble gases are used in illuminated signs. Helium is used to fill airships.

CHECKPOINT

1 The first airships contained hydrogen. Modern airships use helium. The table gives some information about the two gases and air.

	Hydrogen	*Helium*	*Air*
Density (g/cm³)	8.3×10^{-5}	1.66×10^{-4}	1.2×10^{-3}
Chemical reactions with air	Forms an explosive mixture with air	No known chemical reactions	

(a) What advantage does helium have over hydrogen for filling airships?

(b) What advantage does hydrogen have compared with helium?

(c) Why is air not used for 'airships'?

14.1 ▶ Blast-off

FIRST THOUGHTS

In this topic, you can find out about some of the elements and compounds that react with oxygen. Oxygen is a very reactive element, and many substances burn in it.

Why do rocket launches use pure oxygen? After all, burning a tonne of kerosene can supply only a certain amount of heat. Burning the fuel in oxygen does not produce more heat than burning it in air.

The reason is that fuels burn faster in pure oxygen and release heat faster and can therefore deliver more power.

It's a fact!

The mass of oxygen in the atmosphere is 12 hundred million million tonnes!

On 16 July 1969, ten thousand people gathered at the Kennedy Space Centre in Florida, USA. They had come to watch the spacecraft *Apollo 11* lift off on its journey to the moon. While the *Saturn* rocket stood on the launch pad, its roaring jet engines burned 450 tonnes of kerosene in 1800 tonnes of pure oxygen. The thrust from the jets shot the rocket through the lower atmosphere, trailing a jet of flame. At a height of 65 km the first stage of the rocket separated. For six minutes, the second stage burned hydrogen in pure oxygen, taking the spacecraft to a height of 185 km. Then the second stage separated. The third stage, burning hydrogen in oxygen, put *Apollo 11* into an orbit round the Earth at a speed of 28 000 km/hour. From this orbit, *Apollo 11* headed for the moon. The energy needed to lift *Apollo 11* into space came from burning fuels (kerosene and hydrogen) in pure oxygen. Fuels burns faster in oxygen than they do in air. They therefore deliver more power. We shall return to this important reaction of burning in Topic 14.7.

Figure 14.1A ▲ Rocket launch

14.2 ▶ Test for oxygen

KEY SCIENTIST

A new gas was discovered by the British chemist, Joseph Priestley, on 1 August 1786. Heating mercury in air he found that it combined with part of the air to form a solid which he called 'red calx of mercury'. When Priestley heated 'red calx of mercury', he obtained mercury and a new gas, which when he breathed it produced a 'light and easy sensation in his chest'. The gas was oxygen.

Oxygen is a colourless, odourless gas, which is only slightly soluble in water. It is neutral. Oxygen allows substances to burn in it: it is a good **supporter of combustion**. One test for oxygen is to lower a glowing wooden splint into the gas. If the splint starts to burn brightly, the gas is oxygen.

Oxygen relights a glowing splint (see Figure 14.2A).

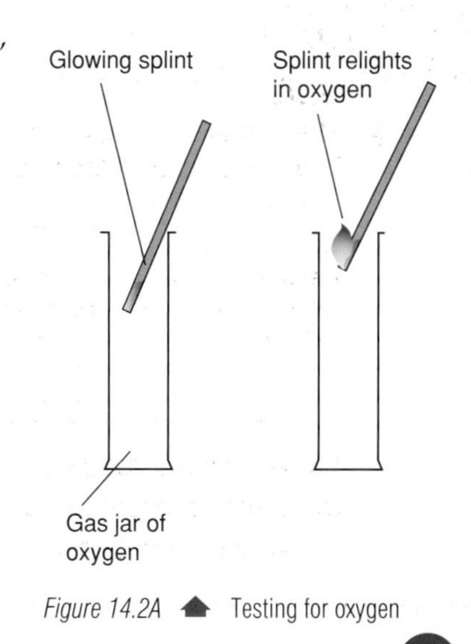

Figure 14.2A ▲ Testing for oxygen

14.3 A method of preparing oxygen in the laboratory

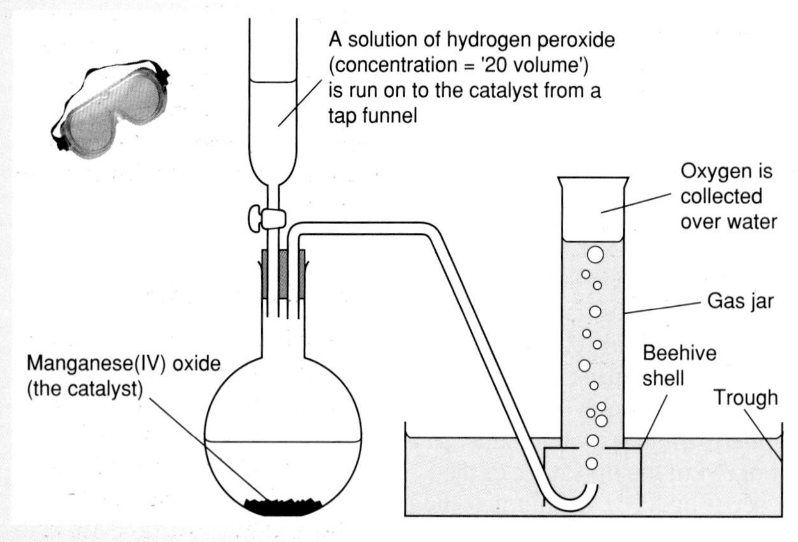

Figure 14.3A ▲ A laboratory preparation of oxygen

Hydrogen peroxide is a colourless liquid which decomposes to give oxygen and water.

Hydrogen peroxide → Oxygen + Water
$2H_2O_2(aq) \rightarrow O_2(g) + 2H_2O(l)$

Solutions of hydrogen peroxide are kept in stoppered brown bottles to slow down the rate of decomposition. If you want to speed up the formation of oxygen, you can add a **catalyst**. A substance which speeds up a reaction without being used in the reaction is called a catalyst. The catalyst manganese(IV) oxide, MnO_2, is often used.

14.4 The reactions of oxygen with some elements

FIRST THOUGHTS

Most elements combine with oxygen; many elements burn in oxygen.

Table 14.1 shows how some elements react with oxygen. The products of the reactions are **oxides**. An oxide is a compound of oxygen with one other element.

Table 14.1 ▼ How some elements react with oxygen

Element	Observation	Product	Action of product on water
Calcium (metal)	Burns with a red flame	Calcium oxide, CaO (a white solid)	Dissolves to give a strongly alkaline solution
Copper (metal)	Does not burn; turns black	Copper(II) oxide, CuO (a black solid)	Insoluble
Iron (metal)	Burns with yellow sparks	Iron oxide, Fe_3O_4 (a blue-black solid)	Insoluble
Magnesium (metal)	Burns with a bright white flame	Magnesium oxide, MgO (a white solid)	Dissolves slightly to give an alkaline solution, pH = 9
Sodium (metal)	Burns with a yellow flame	Sodium oxide, Na_2O (a yellow-white solid)	Dissolves readily to form a strongly alkaline solution, pH = 10
Carbon (non-metal)	Glows red	Carbon dioxide, CO_2 (an invisible gas)	Dissolves slightly to give a weakly acidic solution, pH = 4
Phosphorus (non-metal)	Burns with a yellow flame	Phosphorus(V) oxide, P_2O_5 (a white solid)	Dissolves to give a strongly acidic solution, pH = 2
Sulphur (non-metal)	Burns with a blue flame	Sulphur dioxide, SO_2 (a fuming gas with a choking smell)	Dissolves readily to form a strongly acidic solution, pH = 2

14.5 ▸ Oxides

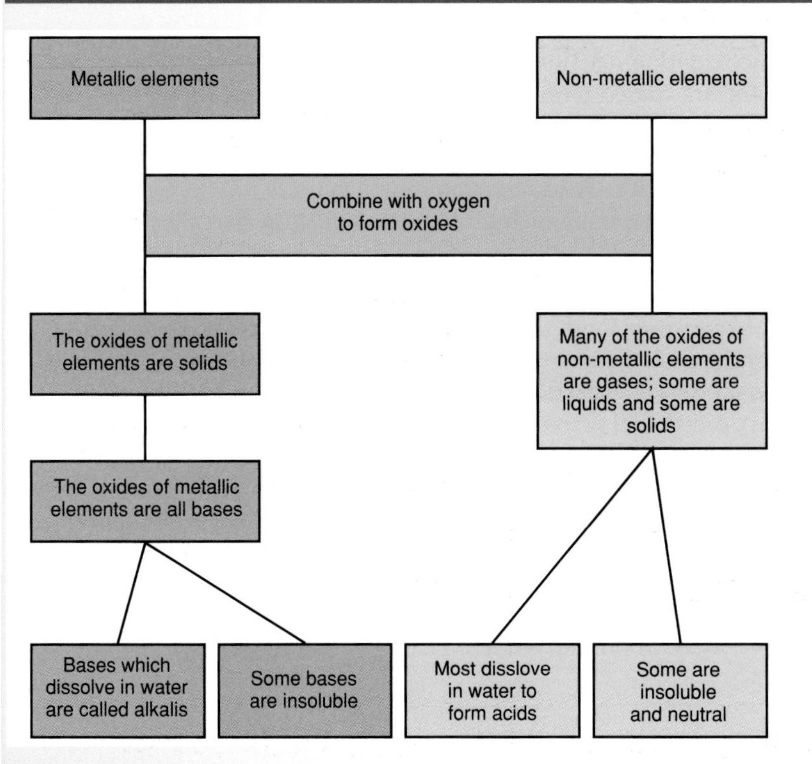

Figure 14.5A ▲ Oxides

A pattern can be seen in the characteristics of oxides. The oxides of metallic elements are **bases**. The bases which dissolve in water are called **alkalis**. Most of the oxides of non-metallic elements are acids, but some are neutral. Acids react with bases to form **salts**.

Amphoteric oxides

Some oxides and hydroxides are both basic (they react with acids to form salts) and acidic (they react with bases to form salts). Such oxides and hydroxides are **amphoteric**. They include the oxides and hydroxides of zinc, aluminium and lead.

SUMMARY

Most elements combine with oxygen to form oxides. Many elements burn in oxygen. The oxides of metals are basic. The oxides of non-metallic elements are acidic or neutral. An oxidising agent gives oxygen to another substance. A reducing agent takes oxygen from another substance.

When an element combines with oxygen, we say it has been **oxidised**. Oxygen **oxidises** copper to copper(II) oxide. This reaction is an **oxidation**.

Copper is oxidised

Copper + Oxygen → Copper(II) oxide

$2Cu(s) + O_2(g) \rightarrow 2CuO(s)$

This reaction is oxidation

The opposite of oxidation is **reduction**. When a substance loses oxygen, it is **reduced**. Copper(II) oxide can be reduced by heating it and passing hydrogen over it.

Copper(II) oxide is reduced

Copper(II) oxide + Hydrogen → Copper + Water

$CuO(s) + H_2(g) \rightarrow Cu(s) + H_2O(l)$

Hydrogen is oxidised

You can see from the equation that oxidation and reduction occur together. Copper(II) oxide is reduced to copper while hydrogen is oxidised to water. Copper(II) oxide is called an **oxidising agent** because it gives oxygen to hydrogen. Hydrogen is called a **reducing agent** because it takes oxygen from copper(II) oxide.

CHECKPOINT

▸ 1 (a) Describe two differences in the physical characteristics of metallic and non-metallic elements (see Topic 4.7 also).

(b) State two differences between the chemical reactions of metallic and non-metallic elements (see Topic 9.1 also).

▸ 2 Write word equations and balanced chemical equations for the combustion in oxygen of (a) sulphur (b) carbon (c) magnesium (d) sodium.

14.6 Combustion

SUMMARY

The combustion of fuels is an oxidation reaction. The combustion of hydrocarbon fuels is a vital source of energy in our economy. These fuels burn to form carbon dioxide and water. If the supply of air is insufficient, the combustion products include carbon monoxide (a poisonous gas) and carbon (soot).

In many oxidation reactions, energy is given out. The fireworks called 'sparklers' are coated with iron filings. When the iron is oxidised to iron oxide, you can see that energy is given out in the form of heat and light. An oxidation reaction in which energy is given out is called a combustion reaction. A combustion in which there is a flame is described as burning. Substances which undergo combustion are called fuels.

Daily, we make use of the combustion of fuels. In respiration, the **combustion of foods** provides us with energy. We use fuels to heat our homes, to cook our food, to run our cars and to generate electricity. Many of the fuels which we use are derived from petroleum oil. Petrol (used in motor vehicles), kerosene (used in aircraft and as domestic paraffin), diesel fuel (used in lorries and trains) and natural gas (used in gas cookers) are obtained from **petroleum oil**. These fuels are mixtures of **hydrocarbons**. Hydrocarbons are compounds of carbon and hydrogen only. It is important to know what products are formed when these fuels burn. Figure 14.6A shows how you can test the products of combustion of kerosene which is burned in paraffin heaters. **Do not use petrol in this apparatus**. You can burn a candle instead of kerosene. Candle wax is another hydrocarbon fuel obtained from crude oil.

As you read, check your understanding of these terms:

- oxidation,
- combustion,
- burning,
- fuel,
- respiration.

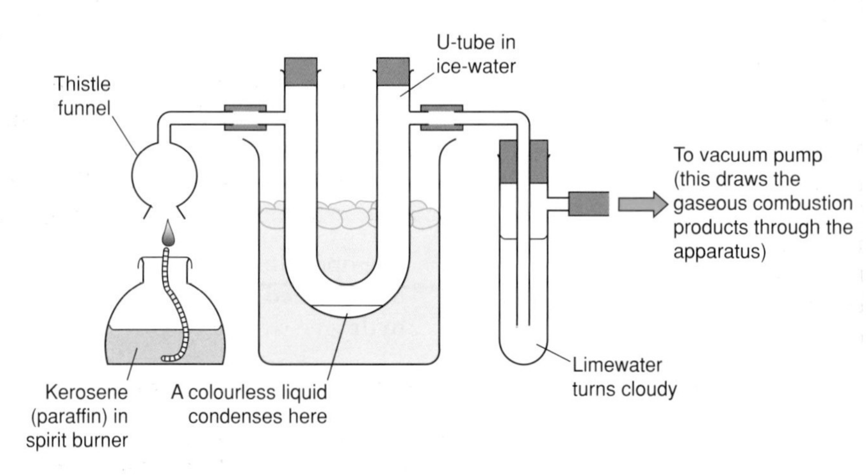

Figure 14.6A ▲ How to test the combustion products of a hydrocarbon fuel, e.g. kerosene or candle wax (*NOTE: do not use petrol*)

SUMMARY

Oxidation is the addition of oxygen to a substance.
Combustion is oxidation with the release of energy.
Burning is combustion accompanied by a flame.
Respiration is combustion which takes place in living tissues. In respiration, food materials are oxidised in the cells with the release of energy.

The combustion products are carbon dioxide and water. Other hydrocarbon fuels give the same products.

Hydrocarbon + Oxygen → Carbon dioxide + Water vapour

If you burn a candle in this apparatus, you will see a deposit of carbon (soot) in the thistle funnel. This happens when the air supply is insufficient to oxidise all the carbon in the hydrocarbon fuel to carbon dioxide, CO_2. Another product of incomplete combustion is the poisonous gas carbon monoxide, CO. Because you cannot see or smell carbon monoxide, it is doubly dangerous. Many times, people have been poisoned by carbon monoxide while running a car engine in a closed garage. The engine could not get enough oxygen for complete combustion to occur. The exhaust gases from petrol engines always contain some carbon monoxide, some unburnt hydrocarbons and some soot, in addition to the harmless products: carbon dioxide and water.

CHECKPOINT

1. (a) What type of compound is present in petrol?
   (b) What products are formed in combustion (i) if there is plenty of air and (ii) if there is a limited supply of air?
2. In February 1988 newspapers carried a report of a woman who fell asleep in front of a fire and never woke up. Later, workmen removed three buckets full of birds' nesting materials from the chimney. What do you think had caused the woman's death?
3. Why should you make sure the window is open if you use a gas heater in the bathroom?

14.7 ▶ Rusting

FIRST THOUGHTS

Rusting is a costly nuisance. Methods of slowing down the process can save a lot of money.

EXTENSION FILE ACTIVITY

These experiments explore the conditions which favour rusting. The point is that if you know the conditions that promote rusting you can avoid these conditions and prolong the life of iron objects.

Many metals become corroded by exposure to the air. The corrosion of iron and steel is called rusting. Rust is the reddish brown solid, hydrated iron(III) oxide, $Fe_2O_3.nH_2O$. (The number of water molecules, n, varies.)

Rusting is an oxidation reaction:

$$\text{Iron} + \text{Oxygen} \rightarrow \text{Iron(III) oxide}$$
$$4Fe(s) + 3O_2(g) \rightarrow 2Fe_2O_3(s)$$

Rusting is a nuisance. Cars, ships, bridges, machines and other costly items made from iron and steel rust. To prevent rusting, or at least to slow it down, saves a lot of money. Before you can prevent rusting, you first have to know what conditions speed up rusting. Figure 14.7A shows some experiments on the rusting of iron nails.

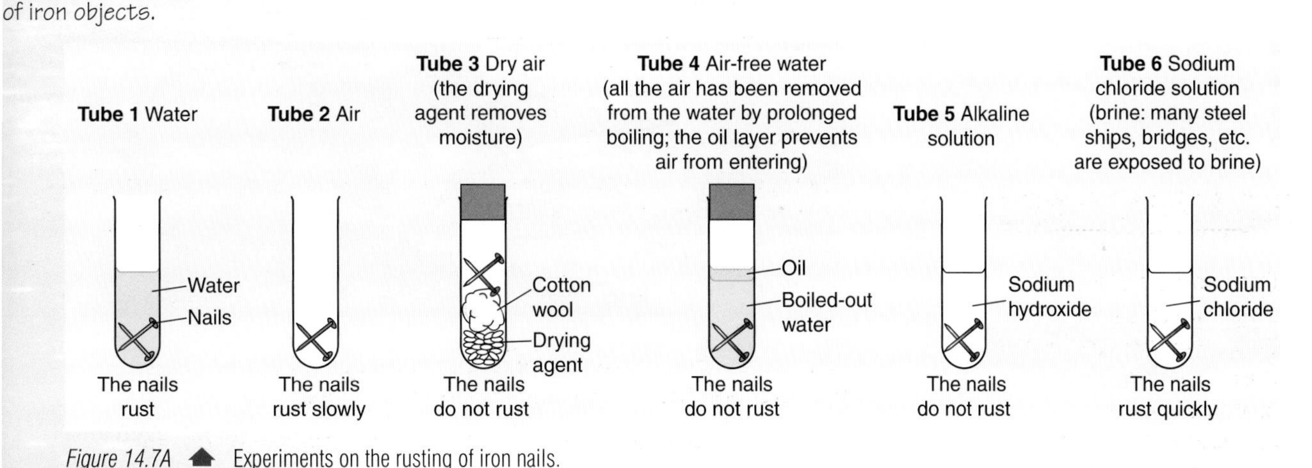

Figure 14.7A ▲ Experiments on the rusting of iron nails.

SUMMARY

Rusting is the oxidation of iron and steel to iron(III) oxide.

The experiments show that in order to rust, iron needs air and water and a trace of acid. The carbon dioxide in the air provides sufficient acidity. Rusting is accelerated by salt. Bridges and ships are exposed to brine, and cars are exposed to the salt that is spread on the roads in winter. It is obviously very important to find ways of rust-proofing these objects.

15.1 The water cycle

FIRST THOUGHTS

The first living things evolved in water. As more complex plants and animals evolved, water remained essential for life.

www.keyscience.co.uk

Where does all the rain come from? Why does the atmosphere never run out of water? Four-fifths of the world's surface is covered by water. From oceans, rivers and lakes, water evaporates into the atmosphere. Plants give out water vapour in **transpiration.** As it rises into a cooler part of the atmosphere, water vapour condenses to form clouds of tiny droplets. If the clouds are blown upward and cooled further, larger drops of water form and fall to the ground as rain (or snow). *Where does the rain go*? Rain water trickles through soil, where some is taken up by plants. The rest passes through porous rocks to become part of rivers, lakes, ground water and the sea. This chain of events is called the **water cycle** (see Figure 15.1A).

It's a fact!

A large tree can lose 300 litres of water vapour in an hour by transpiration.

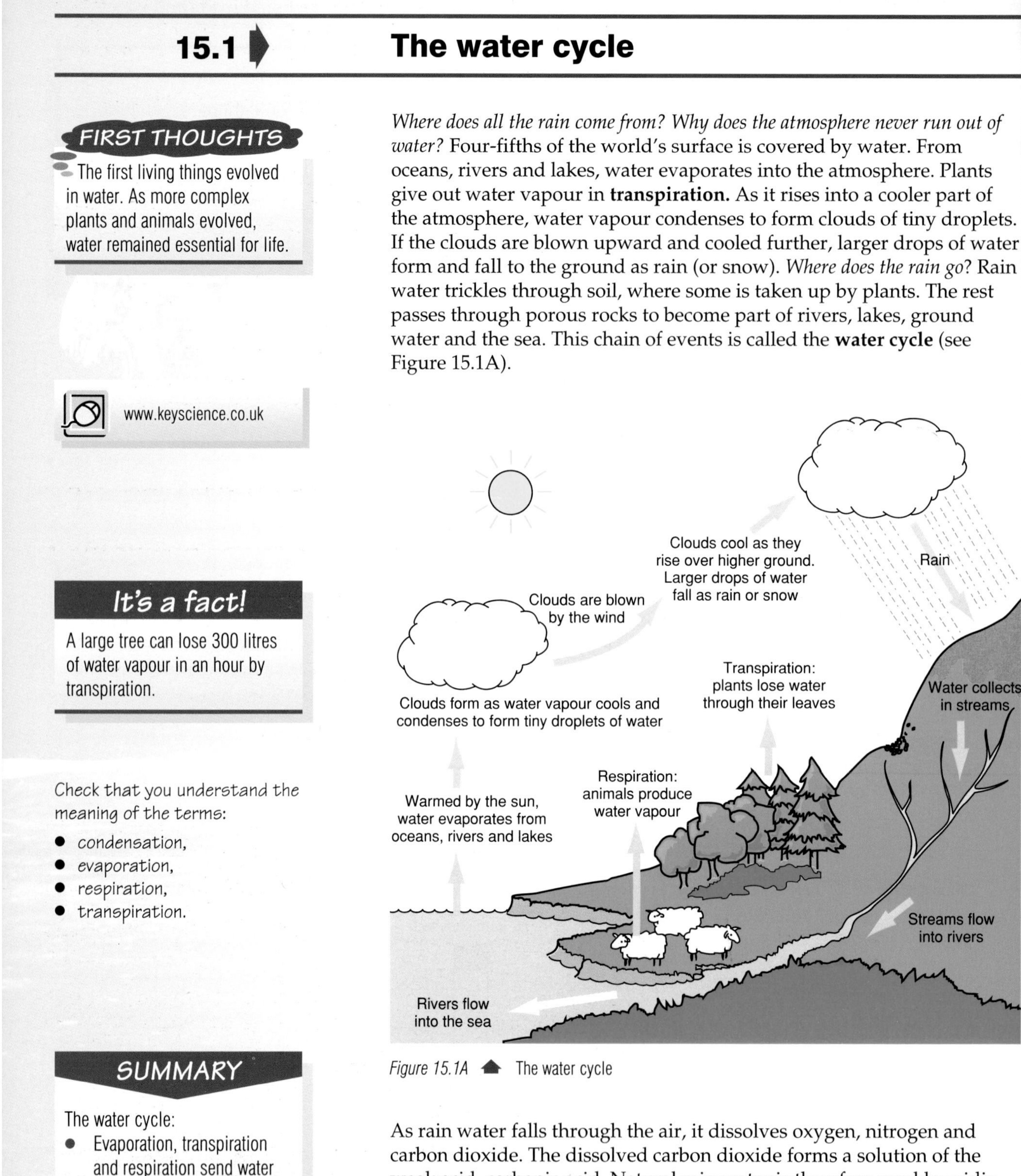

Figure 15.1A ▲ The water cycle

Check that you understand the meaning of the terms:

- condensation,
- evaporation,
- respiration,
- transpiration.

As rain water falls through the air, it dissolves oxygen, nitrogen and carbon dioxide. The dissolved carbon dioxide forms a solution of the weak acid, carbonic acid. Natural rain water is therefore weakly acidic. In regions where the air is polluted, rain water may dissolve sulphur dioxide and oxides of nitrogen, which make it strongly acidic; it is then called **acid rain**. As rain water trickles through porous rocks, it dissolves salts from the rocks. The dissolved salts are carried into the sea. When sea water evaporates, the salts remain behind.

SUMMARY

The water cycle:
- Evaporation, transpiration and respiration send water vapour into the atmosphere.
- Condensation forms clouds which return water to the Earth as rain, hail or snow.

15.2 ▸ Dissolved oxygen

SUMMARY

Air dissolves in water. The dissolved oxygen in water keeps fish alive. If too much organic matter, e.g. sewage, is discharged into a river, the dissolved oxygen is used up in oxidising the organic matter, and the fish die.

The fact that oxygen dissolves in water is vitally important. The solubility is low: water can dissolve no more than 10 g oxygen per tonne of water, that is 10 p.p.m. (parts per million). This is high enough to sustain fish and other water-living animals and plants. When the level of dissolved oxygen falls below 5 p.p.m. aquatic plants and animals start to suffer.

Water is able to purify itself of many of the pollutants which we pour into it. Bacteria which are present in water feed on plant and animal debris. These bacteria are **aerobic** (they need oxygen). They use dissolved oxygen to oxidise organic material (material from plants and animals) to harmless products, such as carbon dioxide and water. This is how the bacteria obtain the energy which they need to sustain life. If a lot of untreated sewage is discharged into a river, the dissolved oxygen is used up more rapidly than it is replaced, and the aerobic bacteria die. Then **anaerobic** bacteria (which do not need oxygen) attack the organic matter. They produce unpleasant-smelling decay products.

Some synthetic (manufactured) materials, e.g. plastics, cannot be oxidised by bacteria. These materials are non-biodegradable, and they last for a very long time in water.

15.3 ▸ Water treatment

FIRST THOUGHTS

The earliest human settlements were always beside rivers. The settlers needed water to drink and used the river to carry away their sewage and other waste. Obtaining clean water is more difficult now.

The tap water we use is taken mainly from lakes and rivers. Before we use it, we want to remove from the water

- solids in suspension
- solids in colloidal form (see Topic 15.15)
- germs
- substances with unpleasant tastes and smells
- acidic substances.

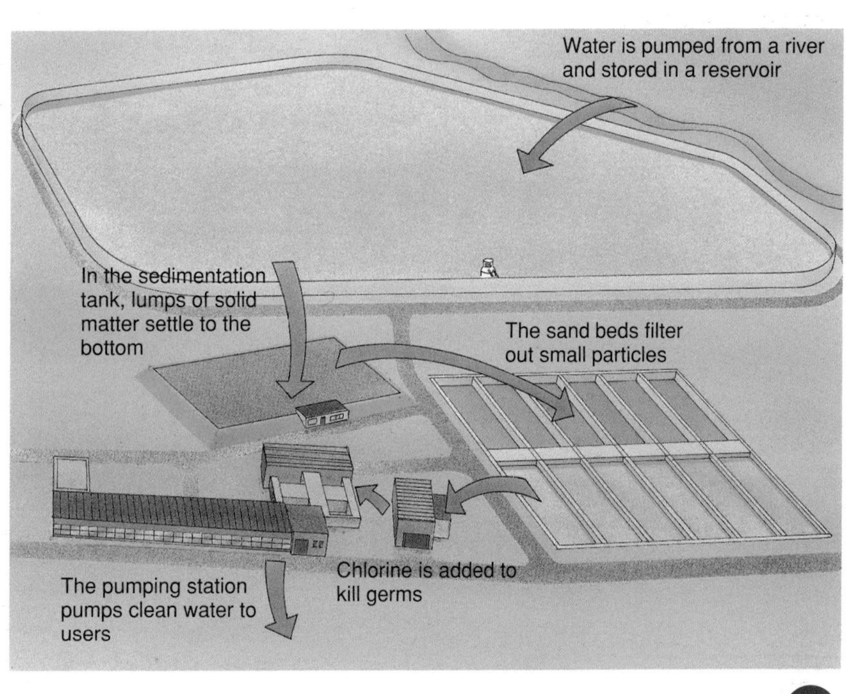

Figure 15.3A ▸ A water treatment works

Science at work

The water industry makes use of computers. Sensors detect the pH, oxygen concentration and other qualities of the water and relay the measurements to a computer. This constant monitoring enables the industry to control the quality of the water it provides.

SUMMARY

Water treatment works take water from lakes and rivers. After filtration, followed by chlorination, the water is safe to drink.

Water treatment plants (Figure 15.3A) make the water safe to drink by:

1 addition of a coagulant, e.g. aluminium sulphate, to make colloidal clay precipitate as a solid so that it can be filtered off
2 sedimentation and filtration to remove solids
3 passing through a carbon slurry to remove tastes and smells
4 addition of a lime slurry to neutralise acidity
5 addition of chlorine to kill bacteria
6 addition of sulphur dioxide to remove excess chlorine

In some areas, the water supply comes from ground water (water held underground in porous layers of rock). As rain water trickles down from the surface through porous rocks, the solid matter suspended in it is filtered out. Ground water therefore does not need the complete treatment. It is pumped out of the ground and chlorinated before use.

Iron(III) hydroxide, in colloidal form, is a problem in some water supplies. It leaves rust-coloured stains in baths and sinks and sometimes on laundry. Vegetables turn brown when boiled in the water. Tea made with the water has a dark colour and a bitter taste.

15.4 Sewage works

FIRST THOUGHTS

Rivers carry away our sewage, our industrial grime, the waste chemicals from our factories and the waste heat from our power stations. Sewage works try to ensure that rivers are not overloaded with waste.

Homes, factories, businesses and schools all discharge their used water into sewers which take it to a sewage works. There, the dirty water is purified until it is fit to be discharged into a river (see Figure 15.4A). The river dilutes the remaining pollutants and oxidises some of them. The digested sludge obtained from a sewage works can be used as a fertiliser. Raw sewage cannot be used as fertiliser because it contains harmful bacteria.

After treatment, the water is clean enough to be discharged into a river

Filter beds filled with lumps of coke. Water from the settling tanks is sprayed on to the beds through rotating metal pipes. Aerobic bacteria in the beds break down harmful substances in the water

Sewer water flows into settling tanks. Sludge, the muddy part of sewage, sinks to the bottom

The sludge is pumped into sludge digestion tanks. There, anaerobic bacteria feed on it. Methane is formed. It can be sold as a fuel. The digested sludge can be sold as a fertiliser

Figure 15.4A ▲ A sewage works

SUMMARY

Sewage works treat used water to make it clean enough to be emptied into rivers or the sea. The treatment is sedimentation followed by aerial oxidation.

15.5 ➧ Uses of water

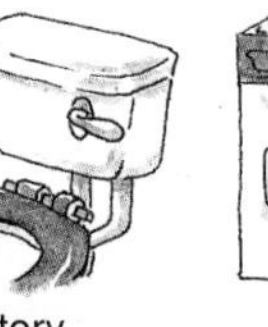

Washing and baths 55 litres | Lavatory 50 litres | Laundry 20 litres | Washing up 15 litres | Cooking 10 litres | Gardening 5 litres | Waste (dripping taps, leaking pipes) 20 litres

Figure 15.5A ▲ Water: 175 litres a day

The uses shown in Figure 15.5A show only 10% of the total amount of water you use. The other 90% is used:

- to grow your food (agricultural use),
- to make your possessions (industrial use as a solvent, for cleaning and for cooling),
- to generate electricity (used as a coolant in power stations).

The total water consumption in an industrialised country amounts to around 80 000 litres (80 tonnes) a year per person. The manufacture of:

- 1 tonne of steel uses 45 tonnes of water,
- 1 tonne of paper uses 90 tonnes of water,
- 1 tonne of nylon uses 140 tonnes of water,
- 1 tonne of bread uses 4 tonnes of water,
- 1 motor car uses 450 tonnes of water,
- 1 litre of beer uses 10 litres of water.

Water used for many industrial purposes is purified and recycled.

It's a fact!

An industrial country uses about 80 tonnes of water per person per year.

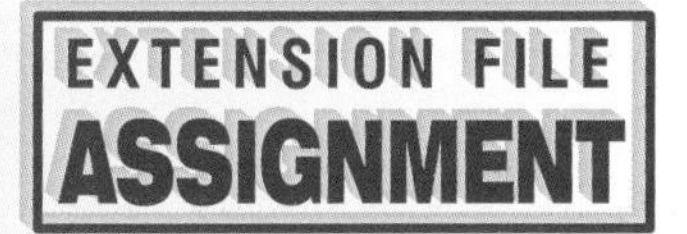

CHECKPOINT

1. Name three substances that are used to treat water before it can be drunk. Say what purpose each of them serves.
2. Why does a shortage of dissolved oxygen make water begin to smell?
3. Name two useful products that come from the digestion of sewage sludge, and state their uses.
4. You may have seen rotating metal pipes spraying water on to beds of gravel in sewage works. What is the purpose of this treatment?

15.6 ➧ Water: the compound

The one thing that everyone knows – the formula of water!

Water is a compound. When a direct electric current passes through it, water splits up: it is electrolysed. The only products formed in the electrolysis of water are the gases hydrogen and oxygen. The volume of hydrogen is twice that of oxygen. From this result, chemists have calculated that the formula for water is H_2O.

$$\text{Water} \xrightarrow{\text{electrolyse}} \text{Hydrogen} + \text{Oxygen}$$

$$2H_2O(l) \rightarrow 2H_2(g) + O_2(g)$$

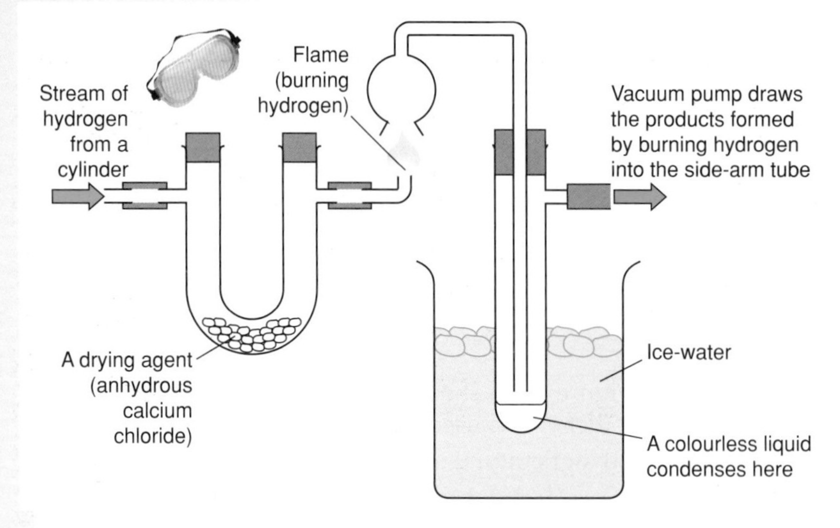

Figure 15.6A What is formed when hydrogen burns in air?

Water is the oxide of hydrogen. *Can it be made by the combination of hydrogen and oxygen?* Figure 15.6A shows an experiment to find out what forms when you burn hydrogen in air. All the air is flushed out of the apparatus before the flame is lit. Then hydrogen burns quietly. The only product is a colourless liquid. You can test this liquid to see whether it is water.

Since hydrogen burns with the evolution of much heat to form a harmless product, water, hydrogen is used as a fuel.

SUMMARY

Tests for water:

- Turns anhydrous copper(II) sulphate from white to blue.
- Turns anhydrous cobalt(II) chloride from blue to pink.

Tests for pure water:

- Boiling point = 100 °C at 1 atm
- Freezing point = 0 °C at 1 atm

Water is formed when hydrogen burns in air.

Tests for water

- Water turns white anhydrous copper(II) sulphate blue.

Copper(II) sulphate + Water → Copper(II) sulphate-5-water

$$CuSO_4(s) + 5H_2O(l) \rightarrow CuSO_4.5H_2O(s)$$

(white solid) (blue solid)

- Water turns blue anhydrous cobalt(II) chloride pink.

Cobalt(II) chloride + Water → Cobalt(II) chloride-6-water

$$CoCl_2(s) + 6H_2O(l) \rightarrow CoCl_2.6H_2O(s)$$

(blue solid) (pink solid)

Any liquid which contains water will give positive results in these tests. To find out whether a liquid is pure water, you can find its boiling point and freezing point. At 1 atm, pure water boils at 100 °C and freezes at 0 °C.

The tests show that the liquid formed when hydrogen burns in air is in fact water.

Hydrogen + Oxygen → Water

$$2H_2(g) + O_2(g) \rightarrow 2H_2O(l)$$

15.7 Pure water

SUMMARY

Water is a good solvent. The presence of a solute raises the boiling point and lowers the freezing point. Pure water is obtained by distillation.

Almost all substances dissolve in water to some extent: that is, water is a good **solvent**. Since water is such a good solvent, it is difficult to obtain pure water. Distillation is one method of purifying water. In some countries, distillation is used to obtain drinking water from sea water. The technique is called desalination (desalting). Hong Kong has a large desalination plant which has never been used because the cost of importing the oil needed to run it is so high. Saudi Arabia and Bahrain operate desalination plants. *Why do you think they need the plants and can afford to run them?*

When chemists describe water as pure, they mean that the water contains no dissolved material. This is different from what a water company means by pure water: they mean that the water contains no harmful substances. Safe drinking water contains dissolved salts. Water which contains substances that are bad for health is **polluted** water.

CHECKPOINT

These questions will enable you to revise solubility curves (Topic 1.6). Use the figure opposite to help you.

1. One kilogram of water saturated with potassium chloride is cooled from 80 °C to 20 °C. What mass of potassium chloride crystallises out?
2. One kilogram of water saturated with sodium chloride is cooled from 80 °C to 20 °C. What mass of sodium chloride crystallises out?
3. Dissolved in 100 g of water at 100 °C are 30 g of sodium chloride and 50 g of potassium chloride. What will happen when the solution is cooled to 20 °C?
4. Dissolved in 100 g of water at 80 °C are 10 g of potassium sulphate and 70 g of potassium bromide. What will happen when the solution is cooled to 20 °C?

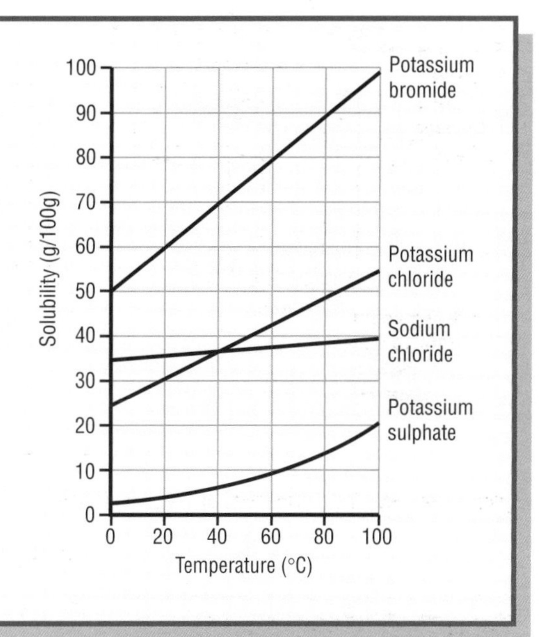

15.8 ▶ Underground caverns

In limestone regions, rain water trickles over rocks composed of calcium carbonate (limestone) and magnesium carbonate. These carbonates do not dissolve in pure water, but they react with acids. The carbon dioxide dissolved in rain water makes it weakly acidic. It reacts with the carbonate rocks to form the soluble salts, calcium hydrogencarbonate and magnesium hydrogencarbonate.

Calcium carbonate (limestone) + Water + Carbon dioxide → Calcium hydrogencarbonate solution

$$CaCO_3(s) + H_2O(l) + CO_2(g) \rightarrow Ca(HCO_3)_2(aq)$$

What caused the formation of this cavern at Wookey Hole?

A well-known reaction between calcium carbonate and a dilute acid!

Figure 15.8A ▲ A cavern at Wookey Hole (note the stalactites and stalagmites)

This chemical reaction is responsible for the formation of the underground **caves** and potholes which occur in limestone regions. Over thousands of years, large masses of carbonates have been dissolved out of the rock (see Figure 15.8A).

The reverse reaction can take place. Sometimes, in an underground cavern, a drop of water becomes isolated. With air all round it, water will evaporate. The dissolved calcium hydrogencarbonate turns into a grain of solid calcium carbonate.

Calcium hydrogencarbonate → Calcium carbonate + Water + Carbon dioxide

$$Ca(HCO_3)_2(aq) \rightarrow CaCO_3(s) + H_2O(l) + CO_2(g)$$

Slowly, more grains of calcium carbonate are deposited. Eventually, a pillar of calcium carbonate may have built up from the floor of the cavern. This is called a **stalagmite**. The same process can lead to the formation of a **stalactite** on the roof of the cavern.

SUMMARY

In limestone regions, acidic rainwater reacts with calcium carbonate to form soluble calcium hydrogencarbonate. The reverse process leads to the formation of stalactites and stalagmites.

15.9 Soaps

FIRST THOUGHTS

Have you ever tried to wash greasy hands without using soap? The problem is that grease and water do not mix. You can find out how soap solves the problem in this section.

Soaps are able to form a bridge between grease and water. They are the sodium and potassium salts of organic acids. One soap is sodium hexadecanoate, $C_{15}H_{31}CO_2Na$. A model of the soap is shown in Figure 15.9A.

(Hexadecane means sixteen. Count up. *Are there 16 carbon atoms?*)

It consists of a sodium ion and a hexadecanoate ion, which we will call a *soap* ion for short. The *soap* ion has two parts (see Figure 15.9B). The head, which is attracted to water, is a—CO_2^- group (a **carboxylate** group). The tail, which is repelled by water and attracted by grease, is a long chain of —CH_2— groups. Figure 15.9C shows how *soap* ions wash grease from your hands.

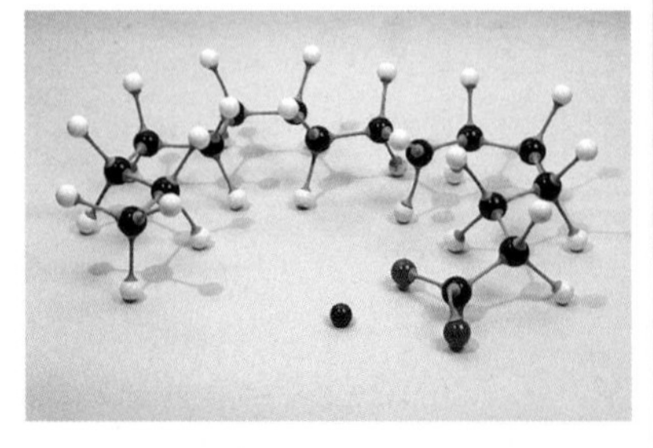

Figure 15.9A ▲ A model of the soap, sodium hexadecanoate

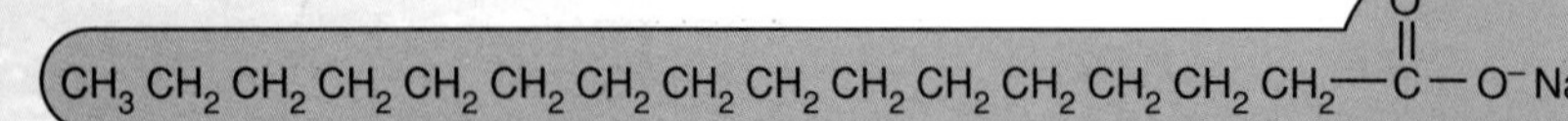

Figure 15.9B ▲ A *soap* ion

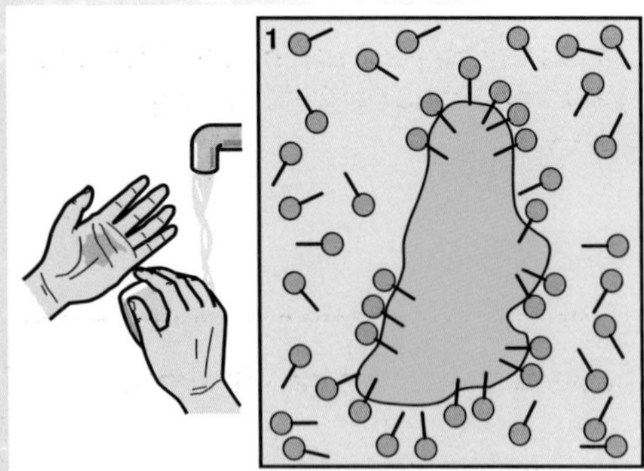

The tails of the *soap* ions begin to dissolve in the grease. The heads remain dissolved in the water

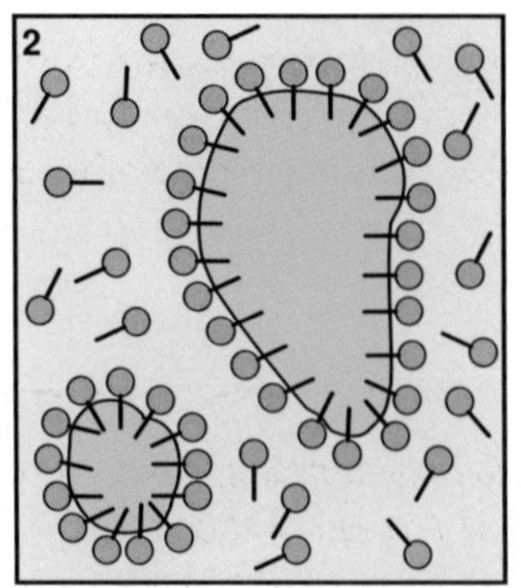

The negatively charged heads of the *soap* ions repel one another; this repulsion makes the grease break up into small droplets, which are suspended in water. The soap has **emulsified** the grease and water (made them mix)

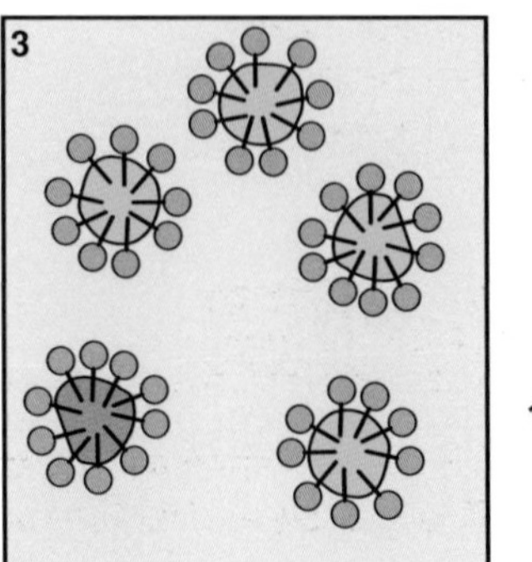

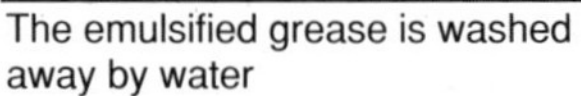
The emulsified grease is washed away by water

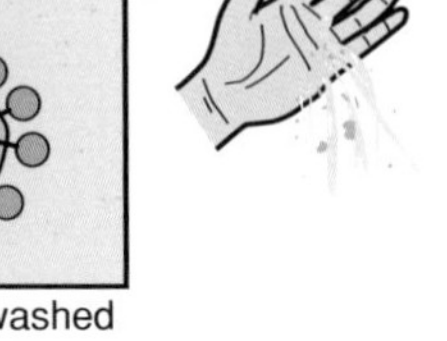

Figure 15.9C ▲ The cleansing action of soap

Manufacture of soap

Soaps are made by boiling together animal fats or vegetable oils and a strong alkali, e.g. sodium hydroxide. The reaction is called **saponification** (soap-making).

Fat (or oil) + Sodium hydroxide → Soap + Glycerol

When sodium chloride is added to the mixture, soap solidifies. The product is purified to remove alkali from it. Perfume and colouring are added before the soap is formed into bars.

SUMMARY

Soaps are able to emulsify oil and water.

15.10 Soapless detergents

Soaps are one type of detergent (cleaning agent). There is another type of detergent known as **soapless detergents**. Many washing powders and household cleaning fluids are soapless detergents. Often they are referred to simply as 'detergents'. Soapless detergents are made from petroleum oil. They are the sodium salts of sulphonic acids (see Figure 15.10A). Like soaps, they are **emulsifying agents** (see Topic 15.15).

Figure 15.10A ▲ A model of a soapless detergent (note the tail, a chain of $—CH_2—$ groups which dissolves in grease, and the head, a sulphate group, $—SO_4^-$, which dissolves in water)

EXTENSION FILE ACTIVITY

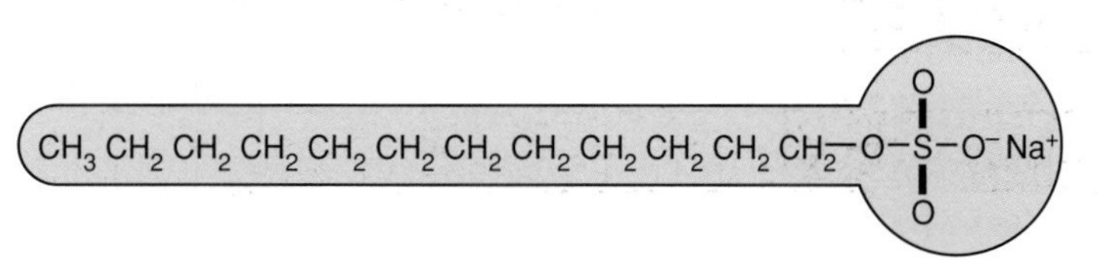

Figure 15.10B ▲ A soapless detergent

SUMMARY

Soapless detergents have many advantages over soaps and are used in washing powders. Soaps are better for use on the skin.

Soapless detergents are very good at removing oil and grease. They are too powerful for use on the skin, and the gentler action of a soap is better. Shampoos are mild detergents.

CHECKPOINT

1. Figure 15.10B shows the formula of a soapless detergent. Say which part of the ion will attach itself to grease and which part will remain in the water.
2. Explain how a soapless detergent is able to dislodge grease from dirty clothes.
3. Why is it important that the water in a washing machine is agitated?
4. Why is it important to rinse clothes well after washing them?

15.11 Bleaches and alkaline cleaners

Bleaches

Many household bleaches contain chlorine compounds, e.g. sodium chlorate(I), NaClO. They are powerful oxidising agents, killing germs and oxidising dirt. You should not use bleaches together with other cleaning agents. An acid will react with sodium chlorate(I) to liberate the poisonous gas chlorine.

Alkaline cleaners

Many household cleaners are alkalis. They react with grease and oil to form an emulsion of glycerol and soap, which can be washed away. The reaction is saponification. Sodium hydroxide, NaOH, is used in oven-cleaners; sodium carbonate, Na_2CO_3, is used in washing powders, and ammonia is used in solution as a household cleaner. You should not use ammonia together with a bleach because they can react to form poisonous chloroamines.

SUMMARY

Household bleaches are chlorine compounds. They work by oxidising dirt and germs. It is not safe to use a bleach together with an acid. Alkaline cleaners work by saponifying grease and oil. It is not safe to use ammonia and a bleach together.

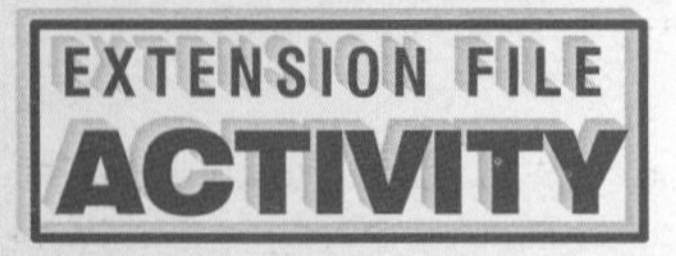

Dry cleaning

Many covalent substances do not dissolve in water but dissolve in covalent solvents such as white spirit (a petroleum fraction), ethanol and propanone. Dry cleaning is carried out with such non-aqueous, covalent solvents.

CHECKPOINT

Oven-cleaners contain sodium hydroxide. A greasy oven is wiped with a pad of oven-cleaner and left for a few minutes. The grease can then be washed off with water.

1. Explain how sodium hydroxide makes it easier to remove grease.
2. Explain why you should wear rubber gloves when you use this kind of oven-cleaner.
3. What effect does it have on the cleaning job if you warm the oven first?

15.12 Hard water and soft water

Something went wrong with the names: hard water and soft water. Should we call them difficult water and easy water – meaning difficult or easy to get a lather with soap?

In some parts of the country, the tap water is described as **hard water**. This means that it is hard to get a lather with soap. Instead of forming a lather, soap forms an insoluble scum. Water in which soap lathers easily is **soft water**. Hard water contains soluble calcium and magnesium salts. They combine with *soap* ions to form insoluble calcium and magnesium compounds. These compounds are the insoluble scum that floats on the water.

Soap ions (in solution) + Calcium ions (in solution) → Scum (insoluble solid)

SUMMARY

Hard water contains calcium ions and magnesium ions. They combine with soap ions to form an insoluble scum. Detergents work well even in hard water because their calcium and magnesium salts are soluble.

If you go on adding soap, eventually all the calcium ions and magnesium ions will be precipitated as scum. After that, the soap will be able to work as a cleaning agent.

Soapless detergents are able to work in hard water because their calcium and magnesium salts are soluble. For many purposes, people prefer soapless detergents to soaps. Sales of soapless detergents are four times as high as those of soaps.

15.13 Methods of softening hard water

Temporary hardness

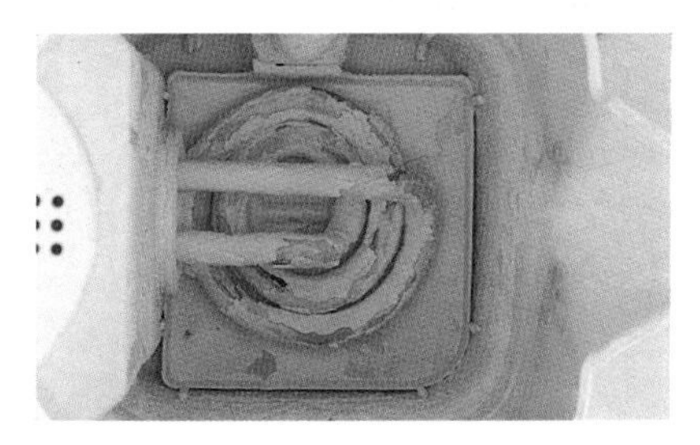

Figure 15.13A ▲ Scale on a kettle element

Hardness which can be removed by boiling is called temporary hardness. Temporarily hard water contains dissolved calcium hydrogencarbonate and magnesium hydrogencarbonate, and these compounds decompose when the water is boiled. The resulting water is soft water.

Calcium hydrogencarbonate → Calcium carbonate + Carbon dioxide + Water

$$Ca(HCO_3)_2(aq) \rightarrow CaCO_3(s) + CO_2(g) + H_2O(l)$$

A deposit of calcium carbonate and magnesium carbonate forms. This is the **scale** which is deposited in kettles and water pipes.

EXTENSION FILE ACTIVITY

EXTENSION FILE ASSIGNMENT

Permanent hardness

Hardness which cannot be removed by boiling is called permanent hardness. It is caused by dissolved chlorides and sulphates of calcium and magnesium. These compounds are not decomposed by heat.

Washing soda

Washing soda is sodium carbonate-10-water. It can soften both temporary and permanent hardness. Washing soda precipitates calcium ions and magnesium ions as insoluble carbonates.

Calcium ions + Carbonate ions → Calcium carbonate

$$Ca^{2+}(aq) + CO_3^{2-}(aq) \rightarrow CaCO_3(s)$$

Exchange resins

Ion exchange resins are substances which take ions of one kind out of aqueous solution and replace them with ions of a different kind. Permutits are manufactured ion exchange resins. They replace calcium and magnesium ions in water by sodium ions.

Calcium ions + Sodium permutit → Sodium ions + Calcium permutit

SUMMARY

- Temporary hardness is removed by boiling.
- Permanent hardness is removed by adding sodium carbonate (washing soda) or by running water through an exchange resin.

How soft is the water?

How well do these treatments for softening water work? The hardness of water before and after treatment can be compared. The number of drops of soap solution needed to give a permanent lather is a measure of the hardness of a water sample. The harder the water, the more drops of soap solution needed. Soap solution is added drop by drop to 100 cm^3 of hard water, with shaking. The number of drops needed to give a lather that lasts for 60 seconds is counted. Then 100 cm^3 of water that has been softened are taken. The number of drops of soap solution needed to give a lasting lather is found. An alternative method is to add the soap solution from a burette (see Topic 9.6) and measure the volume of soap solution added.

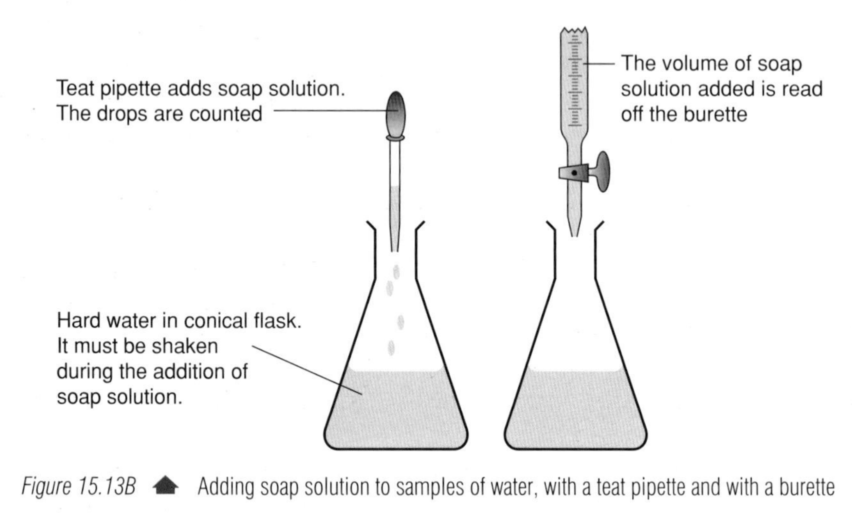

Figure 15.13B ▲ Adding soap solution to samples of water, with a teat pipette and with a burette

15.14 Advantages of hard water

SUMMARY

- Hard water is better than soft water for drinking …
  … and preferred by some industries.

Hard water has some advantages over soft water for health reasons. The **calcium** compounds in hard water strengthen bones and teeth. The calcium content is also beneficial to people with a tendency to develop heart disease.

Some industries prefer hard water. The leather industry prefers to cure leather in hard water. The brewing industry likes hard water for the taste which the dissolved salts give to the beer.

CHECKPOINT

1. Gwen washes her hair in hard water. Which kind of shampoo would you advise her to choose: a mild soapless detergent or a soap? Explain your advice.
2. (a) Explain the difference between hard and soft water.
   (b) Why is drinking hard water better for health than drinking soft water?
   (c) Which solutes make water hard? Explain how the substances you mention get into tap water.
   (d) Name a use for which soft water is preferred to hard water. Explain why.
   (e) Describe one method of softening hard water. Explain how it works.
   (f) Why are detergents preferred to soaps for use in hard water?
   (g) Why is it better to use distilled water rather than tap water in a steam iron?

▶ **3** The table gives some information on three brands of shampoo.

Brand	*Price of bottle (p)*	*Volume (cm³)*
Soffen	80	204
Sheeno	100	350
Silken	120	480

(a) Which brand is sold in the smallest bottle?

(b) Calculate what volume of shampoo (in cm^3) you get for 1 p if you buy (i) Soffen (ii) Sheeno and (iii) Silken. Say which shampoo is the cheapest.

(c) Suggest three things which a person might consider, other than price, when choosing a shampoo.

(d) Describe an experiment you could do to find out which of the shampoos is best at producing a lather. Mention any steps you would take to make sure the test was fair.

▶ **4** Some washing powders contain enzymes. Zenab decides to test whether the washing powder Biolwash, which contains an enzyme, washes better than Britewash, which does not. Zenab decides to use 1 g of washing powder in 100 cm^3 of warm water and to do her tests on squares of cotton fabric.

(a) Suggest some everyday substances which stain cloth and which would be interesting to experiment on.

(b) Describe how Zenab could do a fair test to compare the washing action of Biolwash and Britewash. What factors must be kept the same in the two experiments?

(c) Zenab finds that Biolwash washes better than Britewash on many stains. Another student, Ahmed, did his tests at 80 °C and found that Britewash gave a cleaner result than Biolwash. Can you explain the difference between Zenab's and Ahmed's results? For enzymes, see Topic 25.8.

▶ **5** A student was set the problem of finding out which softened tap water best; boiling, adding washing soda or running it through an ion exchange resin. She ran soap solution from a burette into 100 cm^3 samples of water and measured the volume of soap solution needed to give a lather which lasted for 60 seconds. These are the results:

Water sample, 100 cm³	*Volume of soap solution needed to make a lather(cm³)*
Tap water	2.20
Boiled tap water	1.10
Tap water with washing soda	0.65
Tap water treated with resin	0.30

(a) Which method of softening works the best?

(b) What is washing soda? How does it react with tap water?

(c) How does boiling affect tap water?

(d) What reactions happen when tap water is trickled through ion exchange resin?

15.15 ▶ Colloids

Sand is largely silica, SiO_2. Silica does not dissolve in water. Why is it, then, that when an aqueous solution of a silicate is acidified, silica is not precipitated? Instead there forms either a clear solution with a slight pearly sheen or a jelly-like precipitate or a solid-like gel in which liquid is trapped. The form of the product depends on the pH. The solution and the jelly-like precipitate and the gel are all **colloids**, also called **colloidal**

FIRST THOUGHTS

The bluish haze of tobacco smoke and the brilliant sunsets in deserts – what do they have in common? The answer is the scattering of light by **colloidal particles** suspended in air. In this topic you will find out what colloidal particles are.

dispersions and **colloidal suspensions**. The particles of silica are **suspended** or **dispersed** (spread) through the liquid. The solid (silica) is the **disperse phase** and the liquid (water) is the **dispersion phase** or **dispersion medium**. The silica content may be as high as 30% by mass. Measurements show that the silica particles contain between 2000 and 20 000 SiO_2 units.

How do colloids differ from solutions and suspensions?

Solutions are homogeneous (the same all through) mixtures of two or more substances. The particles of the solute (the dissolved substance) are of atomic or molecular size (about one nanometre long, 1 nm; 1 nm = 10^{-9} m).

Suspensions are heterogenous (not the same all through) mixtures. They contain relatively large particles (over 1000 nm long) of insoluble solid or liquid suspended in (scattered throughout) a liquid. In time, the particles settle out.

Colloids are heterogenous mixtures whose particles are larger than the molecules and ions which form solutions but smaller than the particles which form suspensions (between 1 nm and 1000 nm long). The particles cannot be separated by filtration through filter paper.

Colloids include:

- sols, in which solid particles are dispersed in (spread throughout) a liquid, e.g. clay, paint, starch
- foams, in which a gas is dispersed in a liquid, e.g. soap suds, whipped cream, beer foam
- emulsions, in which one liquid is dispersed in another, e.g. milk, mayonnaise, butter

Optical properties

Colloidal particles are smaller than the particles of solutes and larger than the particles of suspensions.

Colloidal particles are invisible but scatter light.

The particles of a colloid are too small to be visible, but they are large enough to scatter light. The effect resembles the scattering of light in dust-laden air. It is named the Tyndall effect after its discoverer (see Figure 15.15A).

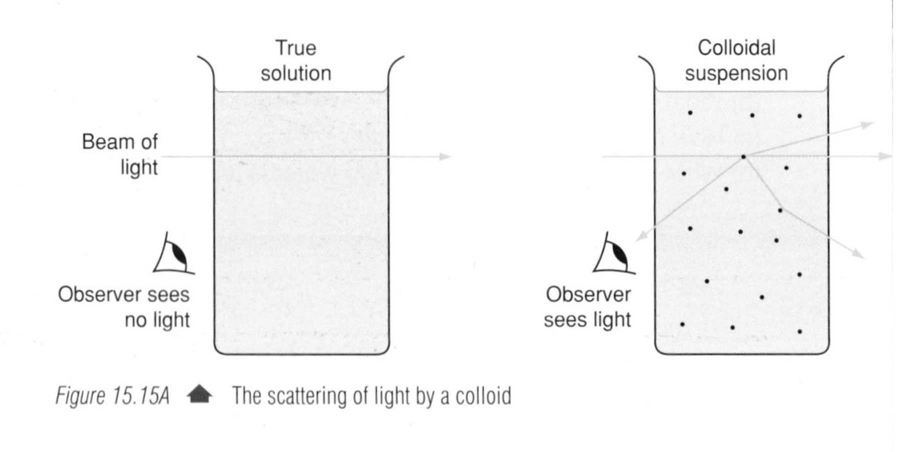

Figure 15.15A ▲ The scattering of light by a colloid

Charge

The reason colloidal particles do not clump together to form a precipitate is that they are charged. In order to precipitate a colloid, the charge on the particles must be neutralised. This can be done by adding ions of opposite charge. When water is purified for drinking, clay particles and other colloidal suspensions must be removed. The impure water is treated with, for example, aluminium sulphate. Aluminium ions, Al^{3+}, are small in size and highly charged. They neutralise the negative charges on the clay particles, which are then able to clump together and settle out of solution.

The proteins in blood are colloidal and are negatively charged. Small cuts can be treated with styptic pencils. These contain Al^{3+} or Fe^{3+} ions which neutralise the charges on the colloidal particles of protein and help the blood to clot.

Electrostatic precipitation

When gases and air are fed into an industrial process, they often contain colloidal particles. To remove these particles, **electrostatic precipitation** is used. The charge on the particles is used to attract the particles to charged metal plates (see Figure 15.15B).

Colloidal particles are charged. They can therefore be

- precipitated by neutralising the charge
- separated by electrophoresis.

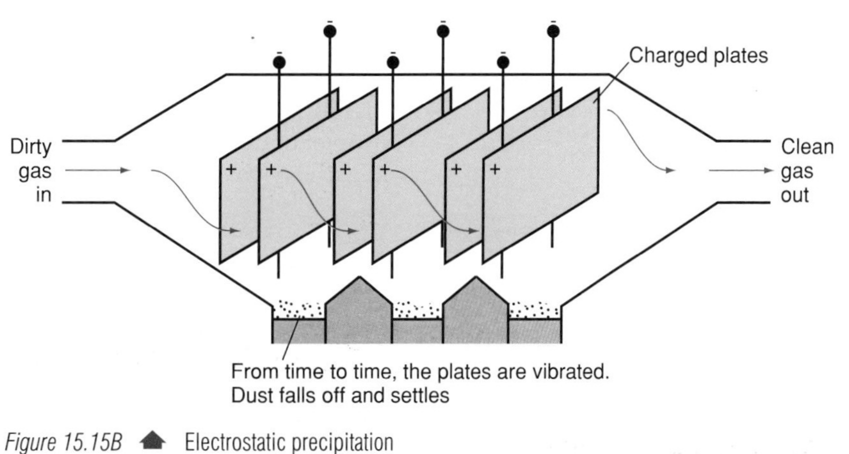

Figure 15.15B ▲ Electrostatic precipitation

Classification

Some colloids of different kinds are tabulated in Table 15.1.

Table 15.1 ▼ Some types of colloids

Dispersed phase	Dispersed medium	Type	Example
Liquid	Gas	Aerosol	Fog, mist from aerosol spray can, clouds
Solid	Gas	Aerosol	Smoke, dust-laden air
Gas	Liquid	Foam	Soap suds, whipped cream
Liquid	Liquid	Emulsion	Oil in water, milk, mayonnaise, protoplasm
Solid	Liquid	Sol	Clay, starch in water, protein in water, gelatin in water
Gas	Solid	Solid foam	Lava, pumice, styrofoam, marshmallows
Liquid	Solid	Solid emulsion	Pearl, opal, jellies, butter, cheese
Solid	Solid	Solid sol	Some gems, e.g. black diamond, coloured glass, some alloys

Electrophoresis

Colloidal particles migrate in an electric field as ions do. They are attracted to one electrode and repelled by the other. Colloidal particles of different types migrate at different rates. This movement under the influence of an electric field is called **electrophoresis**. Electrophoresis can be used to separate different substances. Blood contains a number of proteins in colloidal suspension.

Emulsifiers

Emulsifiers are able to make oil and water mix. They are important.

- in salad dressings
- as soaps
- as soapless detergents
- in paints
- in toothpastes
- in cosmetics.

An emulsion is a colloidal mixture of oil and water. The oil is dispersed through the water as a suspension of tiny drops. Examples are milk, cream, mayonnaise and many sauces. The natural tendency is for oil and water to separate. An emulsifier keeps the two together as a colloidal dispersion. An emulsifier ion has two parts. One is a polar group, with a negative charge, which is attracted to water (the water-loving group). The other is a hydrocarbon chain which is attracted to fat and oil (the fat-loving group) (see Figure 15.15C(a)). When the emulsifier is added to a mixture of oil and water, the emulsifier ions orient themselves so that the water-loving group dissolves in the water and the fat-loving group dissolves in the oil. Emulsifier ions arrange themselves round each droplet of oil (Figure 15.15C(b)). As the surface of each droplet is negatively charged, the drops repel one another and do not run together.

SUMMARY

Colloidal particles are larger than those in solutions but smaller than those in suspensions. They scatter light. They are charged, and if they lose their charge they settle out of the colloidal dispersion. Colloidal particles move under the influence of an electric field in electrophoresis.

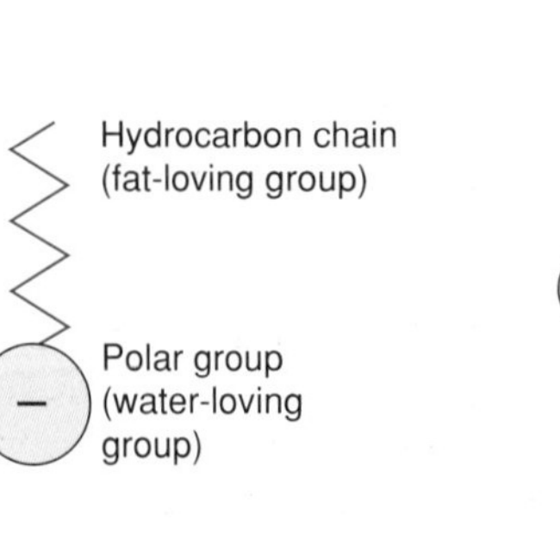

Figure 15.15C ▲ (a) An emulsifier ion; (b) A drop of oil surrounded by emulsifier ions

CHECKPOINT

1. How can you tell the difference between:
   (a) a solution and a suspension,
   (b) a suspension and a colloid,
   (c) a solution and a colloid?
2. How can solid colloidal particles be made to separate:
   (a) from a sol,
   (b) from an aerosol?
3. Explain:
   (a) why a salad dressing contains an emulsifier,
   (b) how the emulsifier works.

16.1 ▶ Smog

FIRST THOUGHTS

As you read through this topic, think about the 15 000 or 20 000 litres of air that you breathe in each day. Obviously you hope that it is clean air. Unfortunately the air that most of us breathe is polluted. In this topic, we shall look into what can be done to reduce the pollution of air.

Four thousand people died in the great London smog of December 1952. Smog is a combination of smoke and fog. Fog consists of small water droplets. It forms when warm air containing water vapour is suddenly cooled. The cool air cannot hold as much water vapour as it held when it was warm, and water condenses. When smoke combines with fog, fog prevents smoke escaping into the upper atmosphere. Smoke stays around, and we inhale it. Smoke contains particles which irritate our lungs and make us cough. Smoke also contains the gas sulphur dioxide. This gas reacts with water and oxygen to form sulphuric acid, H_2SO_4. This strong acid irritates our lungs, and they produce a lot of mucus which we cough up.

The Government did very little about the cause of smog until 1956. Then there was another killer smog. A private bill brought by a Member of Parliament (the late Mr Robert Maxwell, the newspaper owner) gained such widespread support that the Government was forced to act. The Government introduced its own bill, which became the Clean Air Act of 1956. The Act allowed local authorities to declare smokeless zones. In these zones, only low-smoke and low-sulphur fuels can be burned. The Act banned dark smoke from domestic chimneys and industrial chimneys.

16.2 ▶ The problem

All the dust and pollutants in the air pass over the sensitive tissues of our lungs. Any substance which is bad for health is called a **pollutant**. The lung diseases of cancer, bronchitis and emphysema are common illnesses in regions where air is highly polluted. From our lungs, pollutants enter our bloodstream to reach every part of our bodies. The main air pollutants are shown in Table 16.1.

Can you spot from the table the three main sources of pollutants?

Table 16.1 ▼ The main pollutants in air (Emissions are given in millions of tonnes per year in the UK.)

Pollutant	*Emission*	*Source*
Carbon monoxide, CO	100	Vehicle engines and industrial processes
Sulphur dioxide, SO_2	33	Combustion of fuels in power stations and factories
Hydrocarbons	32	Combustion of fuels in factories and vehicles
Dust	28	Combustion of fuels; mining; factories
Oxides of nitrogen, NO and NO_2	21	Vehicle engines and fuel combustion
Lead compounds	0.5	Vehicle engines

In this topic, we shall look at where these pollutants come from, what harm they do and what can be done about them.

www.keyscience.co.uk

16.3 Dispersing air pollutants

Fortunately there is a mechanism for carrying away many of our air pollutants.

The surface of the Earth absorbs energy from the Sun and warms up. The Earth warms the lower atmosphere. The air in the upper atmosphere is cooler than the air near the Earth. **Convection currents** carry warm air upwards. Cold air descends to take its place (see Figure 16.3A). In this way, the warm dirty air from factories and motor vehicles is carried upwards and spread through the vast upper atmosphere.

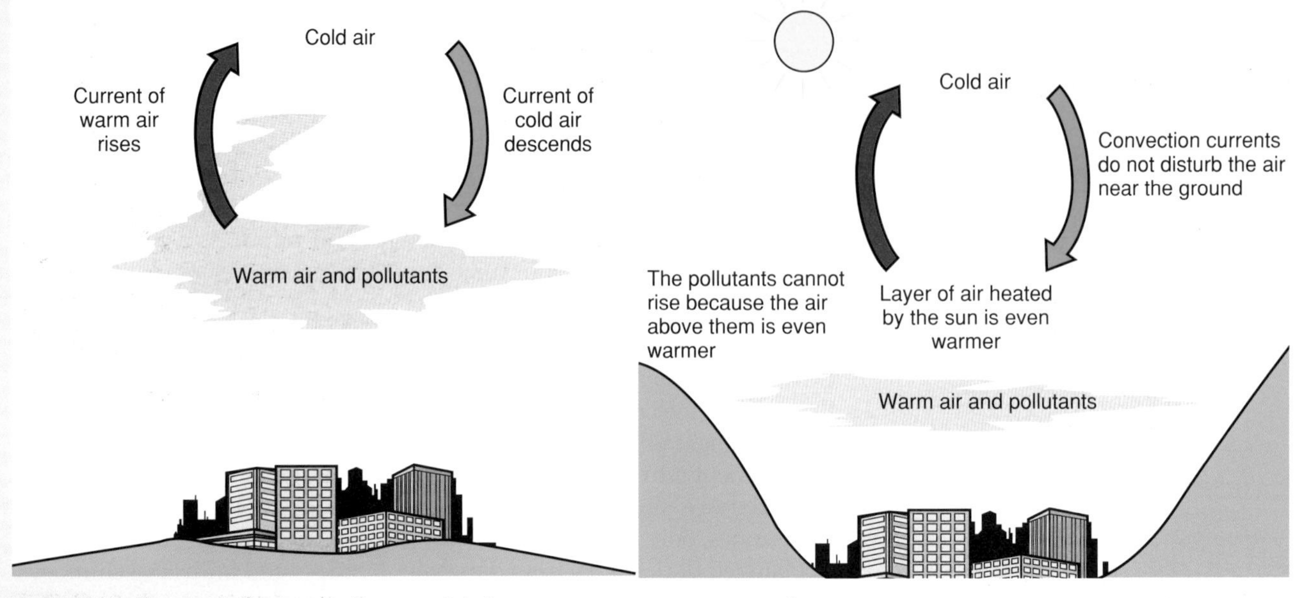

Figure 16.3A ▲ Convection currents disperse pollutants

Figure 16.3B ▲ A temperature inversion traps pollutants

A low-lying area surrounded by higher ground tends to have still air. If an area like this has a hot climate, it is possible for the Sun to warm a layer of air in the upper atmosphere (Figure 16.3B). If the Sun is very hot, this layer of air may become warmer than that near the ground. There is a **temperature inversion**. The air near the ground is no longer carried upwards and dispersed. Pollutants accumulate in the layer of still air at ground level, and the city dwellers are forced to breathe them.

SUMMARY

Pollutants are carried upwards by rising currents of warm air. A temperature inversion stops the dispersal of pollutants. Temperature inversions occur in places with a hot climate and still air.

16.4 Sulphur dioxide

The root of the sulphur dioxide problem is that coal and oil contain sulphur.

Where does sulphur dioxide come from?

Worldwide, 150 million tonnes of sulphur dioxide a year are emitted. Almost all the sulphur dioxide in the air comes from industrial sources. The emission is growing as countries become more industrialised. Half of the output of sulphur dioxide comes from the burning of coal. Most of the coal is burned in power stations. All coal contains between 0.5 and 5 per cent sulphur.

Sulphur (coal) + Oxygen (air) → Sulphur dioxide

$$S(s) + O_2(g) \rightarrow SO_2(g)$$

Industrial smelters, which obtain metals from sulphide ores, also produce tonnes of sulphur dioxide daily.

It's a fact!

In Canada, the countryside surrounding the nickel smelter in Sudbury, Ontario, was so devastated by sulphur dioxide emissions that in 1968 US astronauts practised moon-walking there before attempting the first lunar landing.

SUMMARY

Sulphur dioxide causes bronchitis and lung diseases. The Clean Air Acts have reduced the emission of sulphur dioxide from low chimneys. Factories, power stations and metal smelters send sulphur dioxide into the air. In the upper atmosphere, sulphur dioxide reacts with water to form acid rain.

What harm does sulphur dioxide do?

Sulphur dioxide is a colourless gas with a very irritating smell. Inhaling sulphur dioxide causes coughing, chest pains and shortness of breath. It is poisonous; at a level of 0.5%, it will kill. Sulphur dioxide is thought to be one of the causes of bronchitis and lung diseases.

What can be done about it?

After the Clean Air Acts of 1956 and 1968, the emission of sulphur dioxide and smoke from the chimneys of houses decreased. At the same time, the emission of sulphur dioxide and smoke from tall chimneys increased. Tall chimneys carry sulphur dioxide away from the power stations and factories which produce it (see Figure 16.4A). Unfortunately it comes down to earth again as acid rain (see Figure 16.5B).

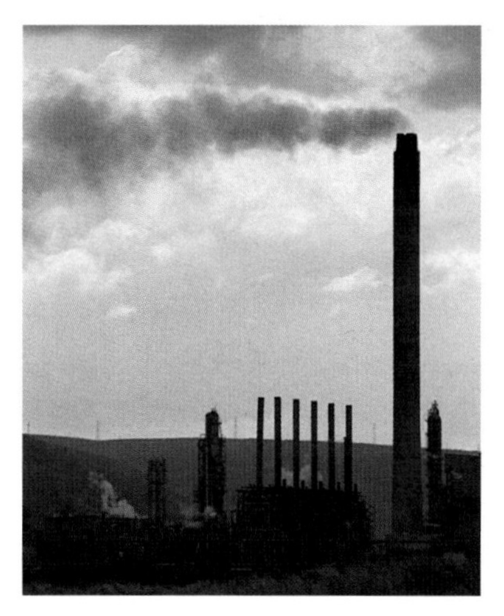

Figure 16.4A ▲ Smoking chimney

CHECKPOINT

1 The emission of sulphur dioxide from low chimneys decreased between 1955 and 1975 from 1.7 million tonnes a year to 0.6 million tonnes a year. In the same period, the emission of sulphur dioxide from tall chimneys increased from 1.4 million tonnes to 3.0 million tonnes.
   (a) Explain why there was a decrease in sulphur dioxide emission from low chimneys.
   (b) Who benefited from the decrease in emission from low chimneys?
   (c) Explain why tall chimneys are not a complete answer to the problem of sulphur dioxide emission.

16.5 Acid rain

Rain water is naturally weakly acidic. It has a pH of 5.4. Carbon dioxide from the air dissolves in it to form the weak acid, carbonic acid, H_2CO_3. What we mean by acid rain is rain which contains the strong acids, sulphuric acid and nitric acid. Acid rain has a pH between 2.4 and 5.0 (see Figure 16.5A).

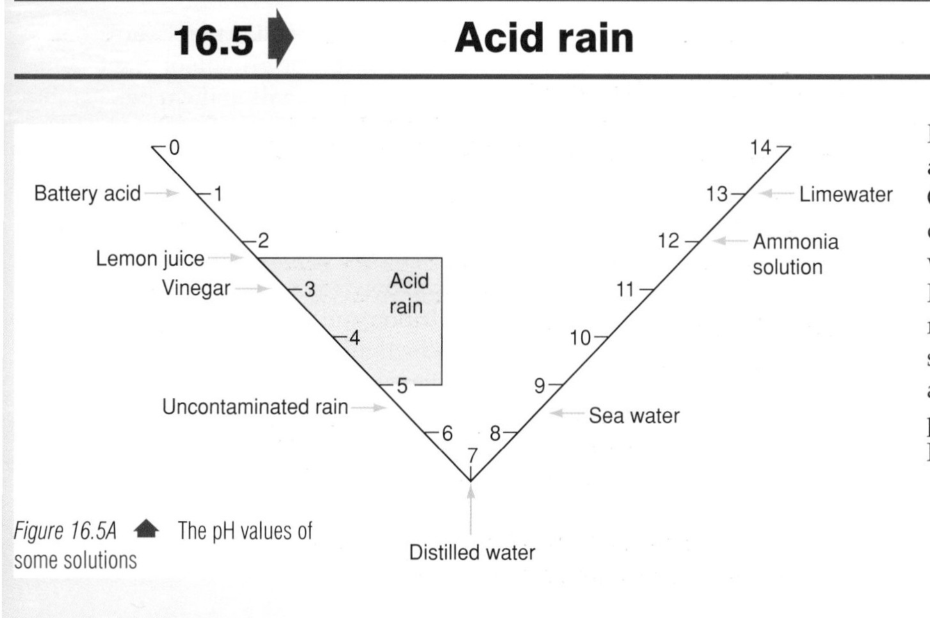

Figure 16.5A ▲ The pH values of some solutions

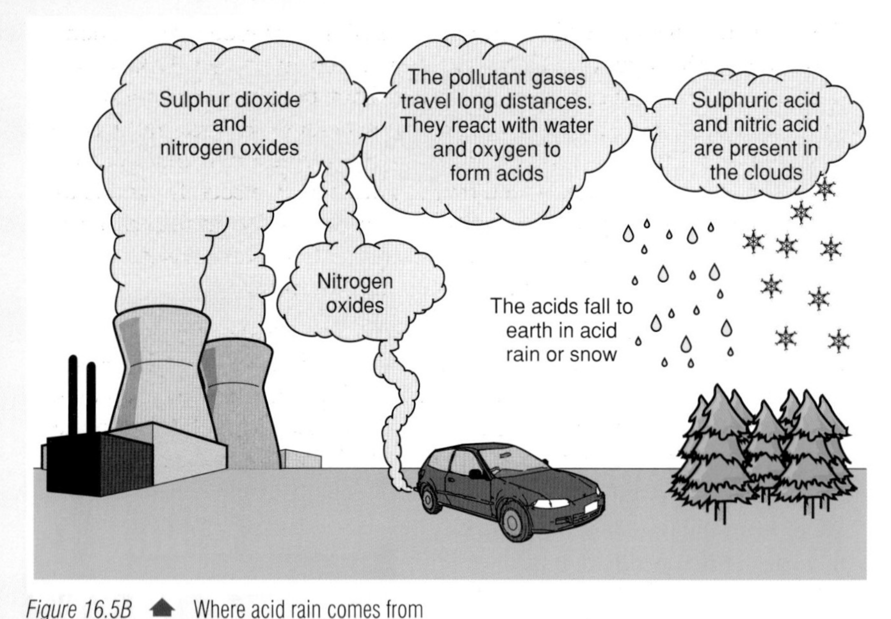

Figure 16.5B ▲ Where acid rain comes from

How do sulphuric acid and nitric acid get into rain water? Tall chimneys emit sulphur dioxide and other pollutant gases, such as oxides of nitrogen. Air currents carry the gases away. Before long, the gases react with water vapour and oxygen in the air. Sulphuric acid, H_2SO_4, and nitric acid, HNO_3, are formed. The water vapour with its acid content becomes part of a cloud. Eventually it falls to earth as acid rain or acid snow which may turn up hundreds of miles away from the source of pollution (see Figure 16.5C).

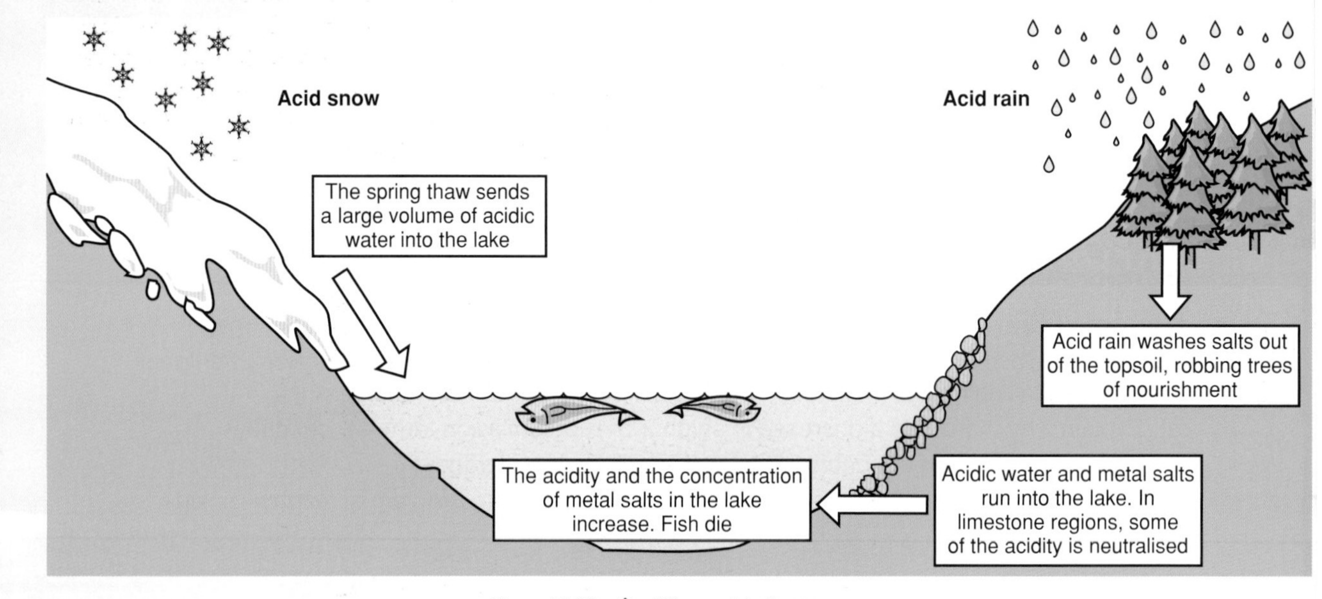

Figure 16.5C ▲ Where acid rain goes

Acid rain which falls on land is absorbed by the soil. At first, the nitrates in the acid rain fertilise the soil and encourage the growth of plants. But acid rain reacts with minerals, converting the metals in them into soluble salts. The rain water containing these soluble salts of calcium, potassium, aluminium and other metals trickles down through the soil into the subsoil where plant roots cannot reach. In this way, salts are leached out of the topsoil, and crops are robbed of nutrients. One of the salts formed by acid rain is aluminium sulphate. This salt damages the roots of trees. The damaged roots are easily attacked by viruses and bacteria, and the trees die of a combination of malnutrition and disease. The Black Forest is a famous beauty spot in Germany which makes money from tourism. About half the trees there are now damaged or dead. Pollution is an important issue in Germany. The Green Party is a major political party. It is compaigning for the reduction of pollution. There are dead forests also in the Czech Republic and Poland. In 1987, the UK Forestry Commission reported that damage to spruce and pine trees is as widespread in the UK as in Germany. One third of British trees are damaged, but not everyone agrees that the cause of the damage is acid rain.

It's a fact!

The most acidic rain ever recorded fell on Pitlochry in Scotland; it had a pH of 2.4.

Science at work

It is possible to monitor the emission from the chimneys of factories and power stations from a distance. A van can carry equipment using infra-red radiation to detect sulphur dioxide. The readings are automatically recorded by a computer.

The acidic rain water trickles through the soil until it meets rock. Then it travels along the layer of rock to emerge in lakes and rivers. Lakes are more affected by acid rain than rivers are. They become more and more acidic, and the concentrations of metal salts increase. Fish cannot live in acidic water. Aluminium compounds, e.g. aluminium hydroxide, come out of solution and are deposited on the gills. The fish secrete mucus to try to get rid of the deposit. The gills become clogged with mucus, and the fish die. An acid lake is perfectly transparent because plants, plankton, insects and other living things have perished.

Thousands of lakes in Norway, Sweden and Canada are now 'dead' lakes. One reason why these countries suffer badly is that acidic snow piles up during the winter months. In the spring thaw, the accumulated snow melts suddenly, and a large volume of acidic water flows into the lakes. Acid rain is partially neutralised as it trickles slowly through soil and over rock. Limestone, in particular, keeps damage to a minimum by neutralising some of the acidity. There is not time for this partial neutralisation to occur when acid snow melts and tonnes of water flow rapidly down the hills and into the lakes.

The UK is affected too. In 1982, lakes and rivers in south-west Scotland had become so acidic that the water companies started treating the lakes with calcium hydroxide (lime). The aim is to neutralise the acidic water and revive stocks of fish. In Wales, the water company has for some years poured tonnes of powdered limestone into acidic lakes. A number of lakes are 'dead' and the fish in many others are threatened.

It's a fact!

In 1984, the UK joined the **30 per cent club**. All the nations in the 'club' agreed to reduce their emissions of sulphur dioxide by 30 per cent. In 1990, the UK increased its target to a cut of 60%.

Figure 16.5D ▲ The effects of acid rain

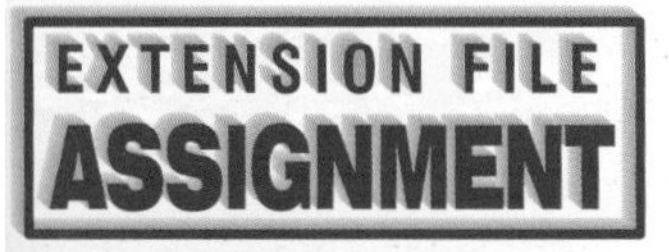

What can be done about acid rain?

There are three main methods of attacking the problem of acid rain. They all cost money, but then the damage done by acid rain costs money too.

1 Low-sulphur fuels can be used. Crushing coal and washing it with a suitable solvent reduces the sulphur content by 10 to 40 per cent. The dirty solvent must be disposed of without creating pollution on land or in rivers. Oil refineries could refine the oil which they sell to power stations. The cost of the purified oil would be higher, and the price of electricity would increase.

2 Flue gas desulphurisation, FGD, is the removal of sulphur from power station chimneys after the coal has been burnt and before the waste gases leave the chimneys. As the combustion products pass up the chimney, they are bombarded by jets of wet powdered limestone. The acid gases are neutralised to form a sludge. The method will remove 95 per cent of the acid combustion products. FGD can be fitted to existing power stations. One of the products is calcium sulphate, which can be sold to the plaster board industry and to cement manufacturers.

It's a fact!

Together, the USA, Canada and Europe send 100 million tonnes of sulphur dioxide into the air each year. Nine tenths of this comes from burning coal and oil.

Science at work

The CEGB fitted FGD to Drax power station in Yorkshire and Fidlers' Ferry in Wales. At a cost of £40 million, the FGD plant at Drax power station started operating in 1993.

3 Pulverised fluidised bed combustion, PFBC, uses a new type of furnace. The furnace burns pulverised coal (small particles) in a bed of powdered limestone. An upward flow of air keeps the whole bed in motion. The sulphur is removed during burning. The PFBC uses much more limestone than the FGD method: one power station needs 1 million tonnes of limestone a year (4 times as much as the FGD method). The PFBC method also produces a lot more waste material, which has to be dumped.

An international issue

Pollutants cross national boundaries. Some countries receive much more pollution than they produce. International agreement on reducing pollution is necessary.

Acid rain is a political issue. Some countries, such as Norway, receive much more sulphur dioxide pollution than they produce themselves. Other countries export more pollution than they receive. Sulphur dioxide produced in the UK is blown across the North Sea to Norway. The USA produces sulphur dioxide which crosses the border to Canada.

Some regions are more at risk than others from acid rain. Factors which produce acidic lakes and rivers are:

- high rainfall
- hills and mountains down which rain water cascades too fast for it to be neutralised by carbonates in the ground
- rocks which contain too little basic material to neutralise acidic rain
- acidic snow piling up in winter and thawing rapidly in the spring.

This is why Norway, Sweden and the Scottish highlands suffer more from acid rain than lowland areas. The regions which are most at risk from acid rain are not the regions which produce most of the world's sulphur dioxide.

In 1984 the UK joined the 'Thirty Percent Club', a group of nations who promised to reduce sulphur dioxide emission by 30% by the year 2000. In 1988 a group of European nations, including Germany, France, the Netherlands and Belgium, increased their target to a cut in sulphur dioxide emission of 70% by 2003. The UK is the largest producer of sulphur dioxide in Europe because the domestic coal has a high sulphur content and because other nations derive more of their energy from nuclear power stations. However in 1990 the UK promised to cut sulphur dioxide emission by 60% by 2003. In a few years this substantial reduction in sulphur dioxide emission across Europe will begin to show results across the continent.

CHECKPOINT

1 What is the advantage of building a power station close to a densely populated area? What is the disadvantage?

2 Why do power stations and factories have tall chimneys? Are tall chimneys a solution to the problem of pollution? Explain your answer.

3 Why does acid rain attack (a) iron railings (b) marble statues and (c) stone buildings?

4 (a) Why does Sweden suffer badly from acid rain?
(b) Why do lakes suffer more than rivers from the effects of acid rain?

5 A country decides to increase the price of electricity so that the power stations can afford to use refined low-sulphur fuel oil. In what ways will the country actually save money by reducing the emission of sulphur dioxide?

16.6 Carbon monoxide

Where does it come from?

Worldwide, the emission of carbon monoxide is 350 million tonnes a year. Most of it comes from the exhaust gases of motor vehicles. Vehicle engines are designed to give maximum power. This is achieved by arranging for the mixture in the cylinders to have a high fuel to air ratio. This design leads to incomplete combustion. The result is the discharge of carbon monoxide, carbon and unburnt **hydrocarbons**.

What harm does carbon monoxide do?

The root of the carbon monoxide problem is incomplete combustion of hydrocarbon fuels. The poisonous effect of carbon monoxide is due to combination with haemoglobin.

Oxygen combines with **haemoglobin**, a substance in red blood cells. Carbon monoxide is 200 times better at combining with haemoglobin than oxygen is. Carbon monoxide is therefore able to tie up haemoglobin and prevent it combining with oxygen. A shortage of oxygen causes headache and dizziness, and makes a person feel sluggish. If the level of carbon monoxide reaches 0.1% of the air, it will kill. Carbon monoxide is especially dangerous in that, being colourless and odourless, it gives no warning of its presence. Since carbon monoxide is produced by motor vehicles, it is likely to affect people when they are driving in heavy traffic. This is when people need to feel alert and to have quick reflexes.

What can be done?

Soil contains organisms which can convert carbon monoxide into carbon dioxide or methane. This natural mechanism for dealing with carbon monoxide cannot cope in cities, where the concentration of carbon monoxide is high and there is little soil to remove it. People are trying out a number of solutions to the problem.

- Vehicle engines can be tuned to take in more air and produce only carbon dioxide and water. Unfortunately, this increases the formation of oxides of nitrogen (see Topic 16.7).
- Catalytic converters are fitted to the exhausts of many cars. The catalyst helps to oxidise carbon monoxide in the exhaust gases to carbon dioxide (see Topic 16.7).
- New fuels may be used in the future. Some fuels, e.g. alcohol, burn more cleanly than hydrocarbons (see Topic 23.1).

SUMMARY

Carbon monoxide is emitted by vehicle engines. It is poisonous. Catalytic converters fitted in the exhaust pipes of cars reduce the emission of carbon monoxide.

CHECKPOINT

1. (a) What are the products of complete combustion of petrol?
   (b) What harm do these products do?
   (c) What conditions lead to the formation of carbon monoxide?
   (d) What harm does it do?
2. How does carbon monoxide act on the body?
3. Which types of people are likely to breathe in carbon monoxide? Is there anything they can do to avoid it?
4. A family was spending the weekend in their caravan. At night, the weather turned cold, so they shut the windows and turned up the paraffin heater. In the morning, they were all dead. What had gone wrong? Why did they have no warning that something was wrong?

16.7 Oxides of nitrogen

When fuels are burned in air, nitrogen is present. Combustion temperatures are high enough to make some nitrogen combine with oxygen. As a result, the gases nitrogen monoxide, NO, and nitrogen dioxide, NO_2, are formed. These gases enter the air from the chimneys of power stations and factories and from the exhausts of motor vehicles. This mixture of gases is sometimes shown as NO_x or even NOX.

What harm do they do?

Nitrogen monoxide, NO, is not a very dangerous gas. However, it quickly reacts with air to form nitrogen dioxide, NO_2. Nitrogen dioxide is highly toxic and irritates the breathing passages. It reacts with oxygen and water to form nitric acid, an ingredient of acid rain.

What can be done?

A reaction which can be used to reduce the quantity of nitrogen monoxide in exhaust gases is

Nitrogen monoxide + Carbon monoxide → Nitrogen + Carbon dioxide

$$2NO(g) + 2CO(g) \rightarrow N_2(g) + 2CO_2(g)$$

This reaction takes place in the presence of a catalyst. A metal cylinder containing the catalyst is fitted in the vehicle exhaust (see Figure 16.8B). All new UK cars are now fitted with these **catalytic converters**.

SUMMARY

Oxides of nitrogen, nitrogen monoxide, NO, and nitrogen dioxide, NO_2 (NO_x), are formed when oxygen and nitrogen combine at high temperature

- in vehicle engines
- in power stations
- in factory furnaces.

They react with air and water to form nitric acid. This strong acid is very toxic and corrosive. Catalytic converters can be fitted in cars to convert nitrogen oxides into nitrogen.

16.8 Hydrocarbons

Where do they come from?

Hydrocarbons are present naturally in air. Methane, CH_4, is one of the products of decay of plant material. Only 15 per cent of the hydrocarbons in the air come from human activities. They affect our health because they are concentrated in city air.

What harm do they do?

Hydrocarbons by themselves cause little damage. In intense sunlight, **photochemical reactions** occur ('photo' means light). Hydrocarbons react with oxygen and oxides of nitrogen to form irritating and toxic compounds.

What can be done?

The hydrocarbons in the exhausts of petrol engines can be reduced by increasing the oxygen supply so as to burn the petrol completely. This also decreases the formation of carbon monoxide. There is a snag, however. Increasing the oxygen supply increases the formation of oxides of nitrogen in the engine. The problem may have a solution. Research workers are trying the idea of running the engine at a lower temperature (to reduce the combination of oxygen and nitrogen) with a catalyst (to assist complete combustion of hydrocarbons at the lower temperature).

SUMMARY

Hydrocarbons from vehicle exhausts take part in photochemical reactions to form irritating and toxic compounds. The emission of hydrocarbons may be reduced by running the engine at a lower temperature with a catalyst.

Science at work

A microcomputer in a vehicle engine can reduce the emission of pollutants by adjusting the fuel to air ratio to the speed of the vehicle.

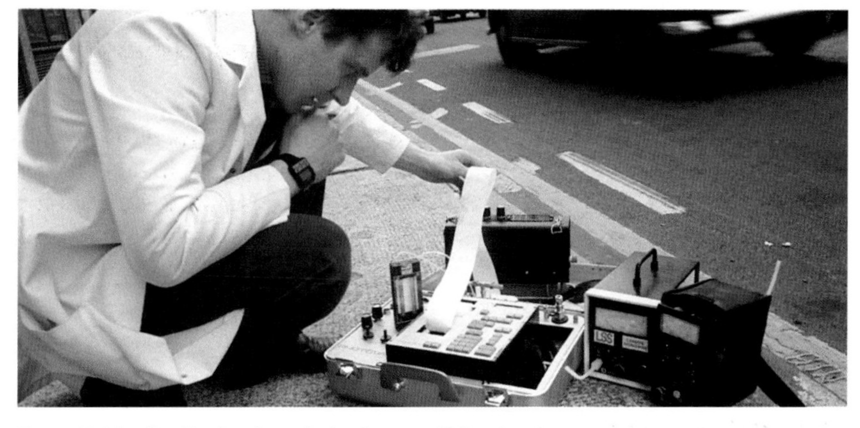

Figure 16.8A ▲ Testing the emission from a vehicle exhaust

SUMMARY

Catalytic converters reduce carbon monoxide, oxides of nitrogen and hydrocarbons in vehicle exhausts.

Catalytic converters

Pollution by carbon monoxide, oxides of nitrogen and hydrocarbons is reduced when catalytic converters are fitted to vehicle exhausts. Figure 16.8B shows the structure of a catalytic converter.

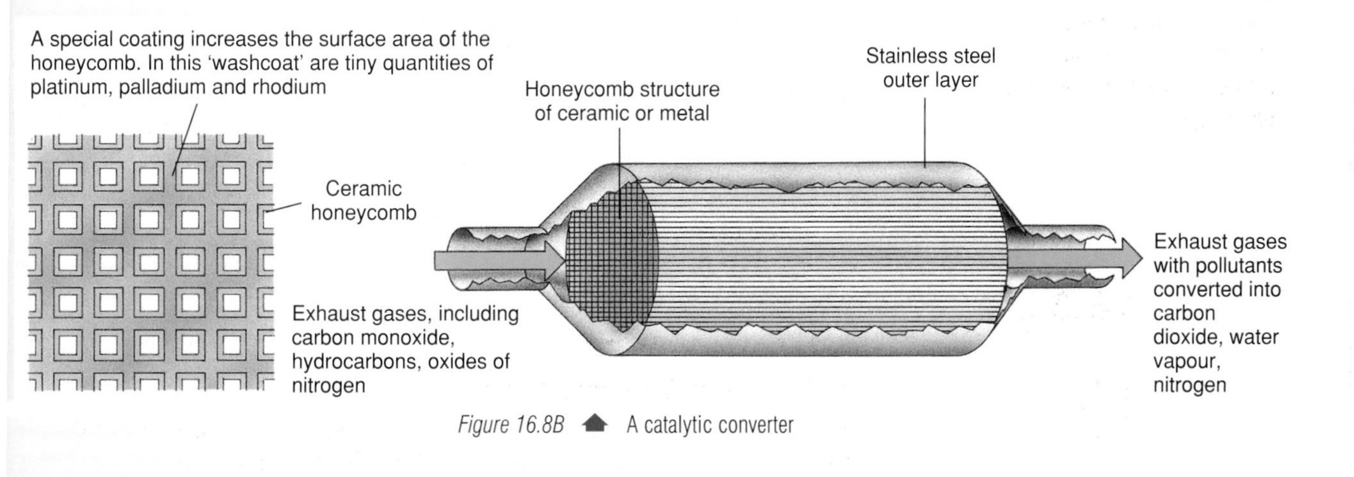

Figure 16.8B ▲ A catalytic converter

CHECKPOINT

1. What products are formed by the combustion of hydrocarbons in petrol engines?
2. How does the supply of air affect the course of combustion?
3. What is the advantage of increasing the air to fuel ratio in the combustion chamber?
4. What are the pollutants that form when the air to fuel ratio is high? What can be done about them?
5. Copy and complete this summary.
   In internal combustion engines, a high air to fuel ratio:
   decreases the emission of unburnt ____ A
   decreases the emission of ____ B
   increases the emission of ____ C
   A way of reducing the emission of C would be to run the engine at a lower temperature.
   A ____ would be needed to promote ____ combustion and reduce the emission of A and B.

6 The figures below show approximately how the emissions of carbon monoxide, oxides of nitrogen and hydrocarbons change with the speed of a vehicle. (Note that the scale for carbon monoxide goes up to 30 g/l, while that of the other pollutants goes up to 3 g/l.)

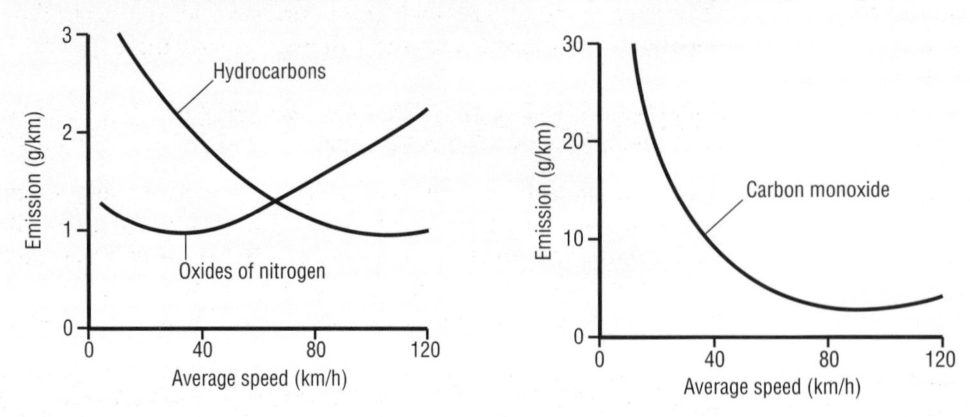

(a) Say what speed is best for reducing the emission of
  (i) carbon monoxide
  (ii) oxides of nitrogen
  (iii) hydrocarbons.

(b) (i) What speed would you recommend as the best to reduce overall pollution?
  (ii) What is this speed in miles per hour (5 mile = 8 km)?

16.9 Smoke, dust and grit

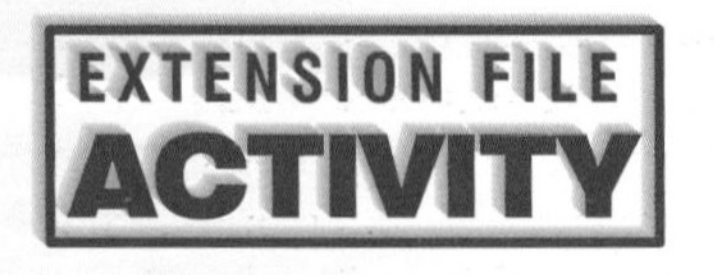

Millions of tonnes of smoke, dust and grit are present in the atmosphere. Dust storms, forest fires and volcanic eruptions send matter into the air. Human activities such as mining, land-clearing and burning coal and oil add to the solid matter in the air.

What harm do particles do?

Particles darken city air by scattering light. Smoke increases the danger of smog. Solid particles fall as grime on people, clothing, buildings and plants.

Sunlight which meets dust particles is reflected back into space and prevented from reaching the Earth. Some scientists believe that the increasing amount of dust in the atmosphere is serious. A fourfold increase in the amount of dust would make the Earth's temperature fall by about 3 °C. This would affect food production.

How can particles be removed?

Industries use a number of methods. These include:

- using sprays of water to wash out particles from their waste gases
- passing waste gases through filters,
- electrostatic precipitators, which remove dust particles from waste gases by **electrostatic attraction**

SUMMARY

Particles of smoke and dust and grit are sent into the air by factories, power stations and motor vehicles. Dirt damages buildings and plants. It pollutes the air we breathe; mixed with fog, it forms smog.

Dust is removed by washing, filtration, electrostatic precipitation.

16.10 Metals

Mercury and lead enter the air from the smelting of metal ores and the combination of fuels.

Much of the lead in the air is present because lead compounds are added to petrol. The amount is decreasing as sales of unleaded petrol increase.

Many heavy metals and their compounds are serious air pollutants. 'Heavy' metals are metals with a density greater than 5 g/cm^3.

Mercury

Earth-moving activities, such as mining and road-making, disturb soil and rock and allow the mercury which they contain to escape into the air. Mercury vapour is also released into the air during the smelting of many metal ores and the combustion of coal and oil. Both mercury and its compounds cause kidney damage, nerve damage and death.

Figure 16.10A ▲ City dwellers breathe exhaust gases

Lead

Where does it come from?

The lead compounds in the air all come from human activity. Vehicle engines, the combustion of coal and the roasting of metal ores send lead and its compounds into the air. For many years lead compounds have been purposely added to petrol to improve the performance of the engine.

What harm does it do?

Lead compounds settle out of the air on to plant crops, and contaminate our food. The level of lead in our environment is high: some areas still have lead plumbing; old houses may have peeling lead-based paint. Symptoms of mild lead poisoning are headache, irritability, tiredness and depression. Higher levels of lead cause damage to the brain, liver and kidneys.

What can be done?

This type of pollution is being remedied. Cars can run on petrol without added lead compounds. Research chemists have found other compounds which can be used to improve engine performance. Vehicles made in the UK after 1990 are adjusted to run on unleaded petrol. Petrol stations now sell mainly unleaded petrol. This must be used in vehicles fitted with catalytic converters because they are 'poisoned' by lead compounds.

SUMMARY

Heavy metals are serious air pollutants. Levels of mercury and lead and their compounds in the air are increasing.

CHECKPOINT

1. Name the pollutants which come from motor vehicles.
2. Name the pollutants which can be reduced by fitting catalytic converters into vehicle exhausts. What effect has this modification had on the price of cars?
3. What effect does the use of lead compounds have on the air, apart from the effect on catalytic converters?
4. In which ways does the control of pollution from vehicles cost money? In which ways does a reduction in the level of pollutants in the air save money? (Consider the effects of pollution on people and materials.) Is the expense worthwhile?

16.11 Chlorofluorohydrocarbons

Ozone, O_3, is an allotrope of oxygen, O_2. The ozone layer in the upper atmosphere protects us from receiving too much ultraviolet radiation from the Sun.

The ozone layer

Ozone is an allotrope of oxygen, O_2. There is a layer of ozone, O_3, surrounding the Earth. It is 5 km thick at a distance of 25–30 km from the Earth's surface. The ozone layer cuts out some of the ultraviolet light coming from the Sun. Ultraviolet light is bad for us and for crops. Long exposure to ultraviolet light can cause skin cancer. This complaint is common in Australia among people who spend a lot of time out of doors. If anything happens to decrease the ozone layer, the incidence of skin cancer from exposure to ultraviolet light will increase. An excess of ultraviolet light kills **phytoplankton**, the minute plant life of the oceans which are the primary food on which the life of an ocean depends.

The very unreactive compounds chlorofluorocarbons, CFCs, react with nothing in the lower atmosphere. When they reach the upper atmosphere they react with ozone. Oxides of nitrogen also react with ozone. CFCs and NO_x decrease the thickness of the ozone layer.

Figure 16.11A ▲ An aerosol can

The problem

Ozone is a very reactive element. If the upper atmosphere becomes polluted, ozone will oxidise the pollutants. Two pollutants are accumulating in the upper atmosphere. One is chlorofluorocarbons, CFCs. These are compounds of carbon, fluorine and chlorine, e.g. $CFCl_3$ and CF_2Cl_2. They are very unreactive compounds. They spread through the atmosphere without meeting any substances they can react with and drift into the upper atmosphere. There, the ultraviolet radiation is sufficiently intense to break covalent bonds in the CFC molecules. The carbon–chlorine bond breaks so that one of the bonding pair of electrons remains with the carbon atom and one remains with the chlorine atom. An atom or group with an unpaired electron , e.g. Cl· is called a **free radical.**

$$\text{CFC} + \text{Energy} \rightarrow \text{Very reactive free radicals}$$
$$CFCl_3 + UV \rightarrow CFCl_2\cdot + Cl\cdot$$

Free radicals are very reactive. They react with ozone to form oxygen and oxidation products of CFCs.

$$\text{Ozone} + \text{Free radicals} \rightarrow \text{Oxygen} + \text{Oxidation products}$$

CFCs were used as the propellants in aerosol cans (Figure 16.11A) and are used as refrigerant liquids in fridges, freezers and air conditioners. They seemed at first to be the ideal replacement for liquid ammonia and liquid sulphur dioxide which were used formerly as refrigerator coolants. Much later they were identified as the culprits damaging the ozone layer.

Another pollutant found at this height is nitrogen monoxide, NO. It comes from the exhausts of high-altitude aircraft, such as Concorde. Ozone oxidises nitrogen monoxide to nitrogen dioxide:

$$\text{Ozone} + \text{Nitrogen monoxide} \rightarrow \text{Oxygen} + \text{Nitrogen dioxide}$$
$$O_3(g) + NO(g) \rightarrow O_2(g) + NO_2(g)$$

Science at work

Microcomputers and sensors are taken up in aircraft to keep watch on the ozone layer. The sensor detects the thickness of the ozone layer and the microcomputer records the measurements.

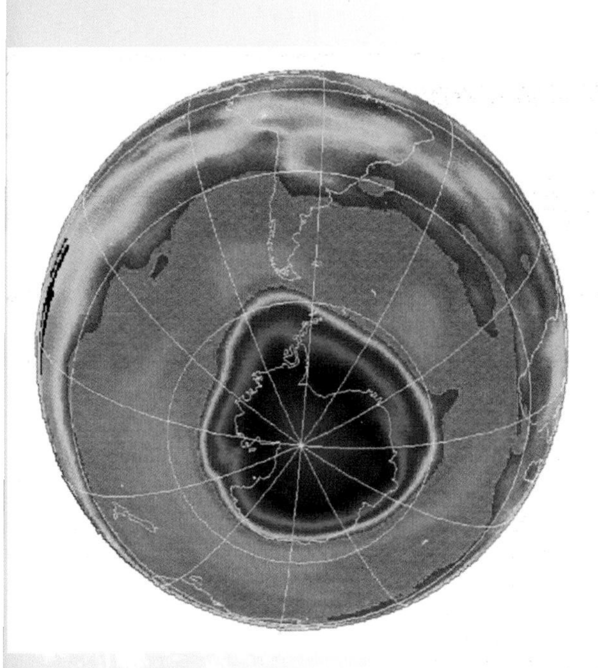

Figure 16.11B ▲ The ozone 'hole'

What should be done?

Is it happening? Is the ozone layer becoming thinner? In June 1980 the British Antarctic Expedition discovered that there was a gap in the ozone layer over Antarctica during certain months. In 1987, research workers in the US confirmed that there was a thinning of the ozone layer which was 'large, sudden and unexpected ... far worse than we thought'. In 1988 a team of scientists working in the Arctic Ocean discovered that the ozone layer over Northern Europe was thinner than it had been.

In 1990 a US plane flying at high altitude took samples which showed that levels of ozone-destroying gases are 50 times higher than expected over the Arctic. An ozone 'hole' over the Arctic would be even more serious than the 'hole' over the Antarctic because more people live in the northern hemisphere.

Knowing that the danger had appeared over more populated regions of the globe spurred many countries to take action. As a result of international meetings in Vienna in 1985, Montreal in 1987, Helsinki in 1989, London in 1990, Copenhagen in 1992 and Vienna in 1995, many countries agreed to phase out CFCs. Developed countries have ceased production of CFCs since 1996. Research chemists have come up with substitutes for CFCs. Hydrocarbons are now used as propellants in aerosol cans, which manufacturers label 'ozone friendly'. Hydrochlorofluorocarbons, **HCFCs**, are compounds containing hydrogen, chlorine, fluorine and carbon, e.g. CH_2FCCl_3. Hydrofluorocarbons, **HFCs** are compounds containing hydrogen, fluorine and carbon, e.g. CH_2F_2 and CH_2FCF_3. These HCFCs and HFCs are very stable compounds but are less stable than CFCs and do much less harm to the ozone layer. They are now used instead of CFCs in the production of polystyrene foams and as refrigerator coolants.

The replacements are more costly than CFCs. Developing countries are increasing their use of refrigeration at a fast rate. For examples the huge population of China is now buying refrigerators. These countries regard it as unfair that, while the rich nations created the CFC problem, the poor nations should be forced to use more expensive HCFs and HCFCs to help solve the problem. As a compromise, the agreement is that developing countries are allowed to go on producing and importing CFCs until 2010.

In spite of the progress which has been made in limiting production of CFCs, the problem of the ozone layer will continue because CFCs remain in the atmosphere for 40–150 years, depending on conditions. Their effect on the ozone layer will continue for decades after all production of CFCs has stopped.

SUMMARY

The ozone layer protects animals and plants from ultraviolet radiation. As it reacts with pollutants in the upper atmosphere, the ozone layer is becoming thinner. CFCs and nitrogen monoxide from high altitude planes are the culprits. The use of CFCs is being reduced.

CHECKPOINT

1. What does CFC stand for?
2. Why do bonds in CFC molecules break in the upper atmosphere?
3. Why were liquid ammonia and liquid sulphur dioxide replaced by CFCs as refrigerator fluids?
4. Many aerosol cans are labelled 'ozone friendly'. What does this mean?
5. Why is it taking so long to solve the problem of the ozone 'hole'?

17.1 Pollution by industry

FIRST THOUGHTS

What's the problem? We need clean drinking water – nothing could be more important. We need industry, and industry needs to dispose of waste products; in the process our water is polluted.

Controls

You will notice that many industrial firms are on river banks. These firms can get rid of waste products by discharging them into rivers. Until 1989, the quantities of waste which industries were allowed to discharge into rivers were controlled by the water authority of each region. Under the 1974 Control of Pollution Act, the water authorities had power to control pollution in inland rivers but not in tidal rivers, estuaries and the sea. In spite of the Act in 1990, more than 2800 km of Britain's largest rivers were too dirty and lacking in oxygen to keep fish alive.

In 1989 the UK Water Privatisation Bill became law. The water authorities were sold to private companies, and are now run for profit as other industries are. The Government set up a National Rivers Authority to watch over the quality of water and prosecute polluters. This has resulted in big improvements in UK rivers (see Topic 17.7).

SUMMARY

The National Rivers Authority controls pollution of inland rivers. It does not regulate the discharge of pollutants into tidal rivers, estuaries and the sea. The estuaries in the UK are heavily polluted by industry and by sewage.

Estuaries

Many of the worst polluters discharge into coastal waters and estuaries. The oil refineries, chemical works, steel plants and paper mills on coasts and estuaries can pour all the waste they want into estuaries and the sea. In the 1930s, fishermen could make a living in the Mersey. Now, it is too foul to keep fish alive. One reason is the discharge of raw sewage into the Mersey. The other is that too many firms pour waste into the estuary. There is unemployment in Merseyside, and the Government does not want to make life difficult for industry in the area. The industries on the banks of the Mersey have been given permission to fall below the standards of the Control of Pollution Act.

Other estuaries, such as the Humber, the Tees, the Tyne and the Clyde, are also polluted by industry.

www.keyscience.co.uk

Figure 17.1A ▲ The Mersey

SUMMARY

Many industrial firms do not keep their discharges of wastes within the limits set by law.

Thermal pollution

FIRST THOUGHTS

What's wrong with warming up the water?

SUMMARY

Thermal pollution means warming rivers and lakes. It reduces the concentration of oxygen dissolved in the water.

Industries use water as a coolant. A large nuclear power station uses 4000 tonnes of water a minute for cooling. River water is circulated round the power station, where its temperature increases by 10 °C, and is returned to the river. If the temperature of the river rises by many degrees, the river is **thermally polluted**. As the temperature rises, the solubility of oxygen decreases. At the same time, the **biochemical oxygen demand** increases. Fish become more active at the higher temperature, and need more oxygen. The bacteria which feed on decaying organic matter become more active and use more oxygen.

CHECKPOINT

1. When a car engine has an oil change, the waste oil is sometimes poured down the drain. What is wrong with doing this?
2. Does it matter whether rivers are clean and stocked with fish or foul and devoid of life? Explain your answer.
3. Why is warming the water in a river regarded as pollution?

Pollution by sewage

FIRST THOUGHTS

The population of the UK is increasing. One result is the need for more sewage works. What happens when a country does not keep up with this need?

In Topic 15.4, you read how sewage is treated before it is discharged into rivers or the sea. Unfortunately, as some water companies do not have enough plants to treat all their area's sewage, they discharge some raw sewage into rivers and estuaries. The Mersey receives raw sewage from Liverpool and other towns. Some sewage treatment works are inadequate and sewage is discharged into the sea.

The quality of the water at many of Britain's bathing beaches fails to meet standards set by the European Community (EC). Many British beaches have more coliform bacteria and faecal bacteria in the water than the EC standard.

The Third World

Of the four billion people in the world, two billion have no toilets, and one billion have unsafe drinking water. In Third World countries (the developing countries) three out of five people have difficulty in obtaining clean water. Some Third World communities have to use a river as a source of drinking water as well as for disposal of their sewage. Bacteria are present in faeces, and they infect the water. Many diseases are spread by contaminated water. They include cholera, typhoid, river blindness, diarrhoea and schistosomiasis. Four-fifths of the diseases in the Third World are linked to dirty water and lack of sanitation. Five million people each year are killed by water-borne disease.

Estuaries and bathing beaches suffer when untreated sewage is discharged into estuaries and the sea.

SUMMARY

During the 1980s, the United Nations set a target of safe water and sanitation for all by 1990. The aim was to provide wells and pumps, kits for disinfecting water and hygienic toilets. The sum needed was £25 billion, slightly more than the world spends on its armies in one month. The target was not reached by 1990, but the work is continuing.

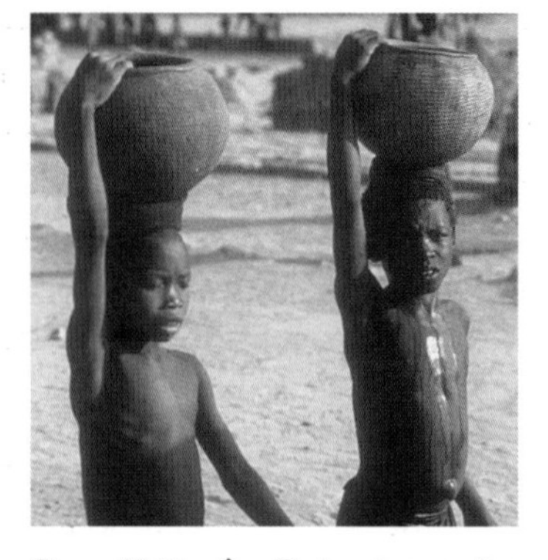

Figure 17.3A ▲ Their water supplies

17.4 Pollution by agriculture

FIRST THOUGHTS

Farmers need to use fertilisers. What happens when a crop does not use all the fertiliser applied to it? There can be pollution, as this section explains.

Science at work

Research chemists are working on a microbiological method for stripping nitrates from drinking water. A bacterium converts nitrates and hydrogen into nitrogen and water. The bacterium is bound to beads of calcium alginate. Hydrogen and water containing nitrates are passed through a tube containing the beads. Nitrogen is formed, and nitrate-free water flows out of the tube.

Fertilisers

A lake has a natural cycle. In summer, algae grow on the surface, fed by nutrients which are washed into the lake. In autumn the algae die and sink to the bottom. Bacteria break down the algae into nutrients. Plants need the elements carbon, hydrogen, oxygen, nitrogen and phosphorus. Water always provides enough carbon, hydrogen and oxygen; plant growth is limited by the supply of nitrogen and phosphorus. Sometimes farm land surrounding a lake receives more fertiliser than the crops can absorb. Then the unabsorbed nitrates and phosphates in the fertiliser wash out of the soil into the lake water. When fertilisers wash into a lake, they upset the natural cycle. The algae multiply rapidly to produce an **algal bloom**. The lake water comes to resemble a cloudy greenish soup. When the algae die, bacteria feed on the dead material and multiply. The increased bacterial activity consumes much of the dissolved oxygen. There is little oxygen left in the water, and fish die from lack of oxygen. The lake becomes difficult for boating because masses of algae snag the propellers. The name given to this accidental fertilisation of lakes and rivers is **eutrophication**.

Many parts of the Norfolk Broads are now covered with algal bloom. The tourist industry centred on the Broads would like to see them restored to their former condition.

Fertiliser which is not absorbed by crops can be carried into the ground water (the water in porous underground rock). Ground water provides one third of Britain's drinking water. The EC has set a maximum level of nitrates in drinking water at 50 mg/l. Four out of the ten water companies in England and Wales have drinking water which exceeds this nitrate level. The figure of 50 mg/l is well below the level at which the nitrate concentration would be harmful. In regions where river water contains more than 50 mg/l of nitrate, for example some farming areas, the water must be treated before it can be supplied for drinking. There are two methods in use.

Figure 17.4A ▲ Algal bloom

1 Denitrifying bacteria can be added to convert nitrate ions into nitrogen gas which is released into the air.
2 The water can be passed through an ion exchange resin which replaces nitrate ions with other ions, e.g. chloride ions.

There are two health worries over nitrates. Nitrates are converted into nitrites (salts containing the NO_2^- ion). Some chemists think that nitrites are converted in the body into nitrosoamines. These compounds cause cancer. The other worry is that nitrites oxidise the iron in haemoglobin. The oxidised form of haemoglobin can no longer combine with oxygen. The extreme form of nitrite poisoning is 'blue baby' syndrome, in which the baby turns blue from lack of oxygen. The last fatality in the UK was in the 1950s, and the last case, in which the baby survived, was in 1972. Against this risk we have to weigh the importance of nitrate fertilisers in feeding the Earth's population. The Worldwatch Institute warned in 1990 that the Earth's food output would fall by 40 % if the use of nitrate fertilisers were to stop. This would mean that millions of people could no longer be fed.

The level of nitrites in drinking water permitted by the EC is 0.1 mg/l. Some parts of London have nitrite levels which are higher than this. The UK Government has agreed to bring the UK into line with the rest of Europe. To install nitrate-stripping equipment would cost £200 million. *Should the Government reduce the use of fertilisers? How? Should the Government introduce a tax on fertilisers or ration fertilisers?*

SUMMARY

When a crop receives more fertiliser than it can use, nitrates and phosphates wash into lakes and rivers. There, they stimulate the growth of weeds and algae. When the plants die, bacterial decay of the dead material uses oxygen. The resulting shortage of dissolved oxygen kills fish. The level of nitrates in ground water, from which we obtain much of our drinking water, is rising. Pollution can be reduced by reducing the application of fertilisers and by omitting phosphates from detergents.

Pesticides

FIRST THOUGHTS

What are the 'drins'? Why is the EC worried about the level of drins in UK water?

Other pollutants which must worry us are the pesticides dieldrin, endrin and aldrin (sometimes called the 'drins'). They cause liver cancer and affect the central nervous system. The EC sets a maximum level of 5×10^{-9} g/l for 'drins'. Half the water in the UK exceeds this level. The danger with the 'drins' is that fish take them in and do not excrete them. The level of 'drins' in fish may build up to 6000 times the level in water. Figure 17.4B shows what happened when DDT, another powerful insecticide, was used to spray Clear Lake in California to get rid of mosquitoes. It is an example of pollutants being concentrated by a food chain. Although the level of DDT in the water was low, the level of DDT in acquatic birds was high enough to kill them.

SUMMARY

Pesticides are serious pollutants, especially when they can be concentrated through a food chain.

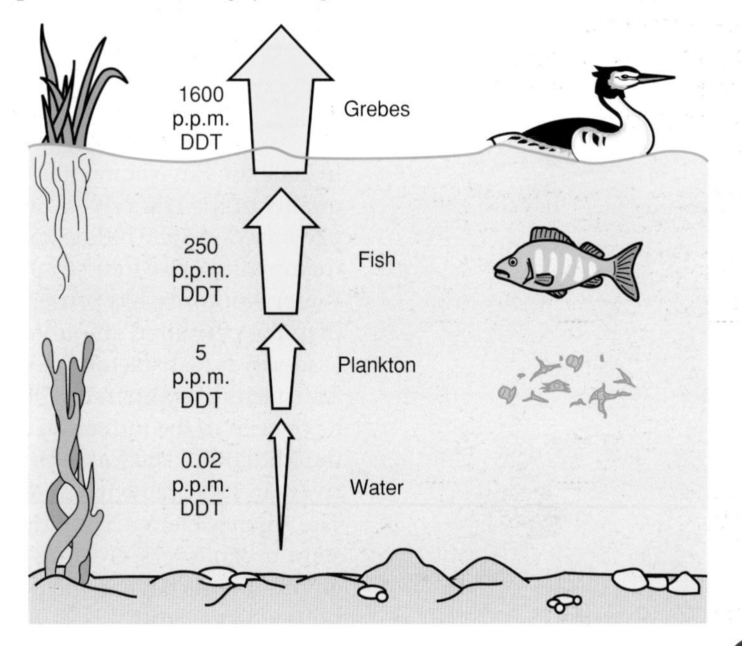

Figure 17.4B ▸ A food chain in Clear Lake, California

CHECKPOINT

1. Groups of settlers in North America always built their villages on river banks, and discharged their sewage into the river. How did the river dispose of the sewage? Why can this method of sewage disposal not be used for larger settlements?
2. British Tissues make toilet paper, paper towels, paper handkerchiefs, etc. They use a lot of bleach on the paper, and this bleach is one of the chemicals which the firm has to dispose of. Can you suggest how the firm could reduce the problem of bleach disposal?
3. (a) Why do some lakes develop a thick layer of algal bloom?
   (b) Why is algal bloom less likely to occur in a river?
   (c) What harm does algal bloom do to a lake that is used as
      (i) a reservoir
      (ii) a fishing lake
      (iii) a boating lake?
4. The concentration of nitrates in ground water is rising. Explain:
   (a) why this is happening,
   (b) why some people are worried about the increase.
5. Water companies can tackle the problem of high nitrate levels by:
   - blending water from high-nitrate sources with water from low-nitrate sources
   - closing some sources of water
   - treating the water with chemicals
   - ion exchange
   - microbiological methods

   (a) Say what you think are the advantages and disadvantages of each of these methods.
   (b) Which do you think would be the most expensive treatments? How will water companies be able to pay for the treatment?
   (c) Suggest a different method of reducing the level of nitrates in ground water.
   (d) Say who would pay for the method which you mention in (c) and how they would find the money.

17.5 Improvements in river water

In 2000 the Environmental Agency reported big improvements in the quality of the UK's rivers and canals. Since the water industry was privatised in 1989 billions of pounds have been spent on sewage treatment works. Industrial discharges have been reduced, and the Water Authority has prosecuted industries which discharge more waste than the permitted amounts. In Scotland 97% of river lengths are assessed as 'satisfactory'. In England and Wales 92% of rivers are 'satisfactory' compared with 85% in 1990. The biggest improvements have been in the industrial regions of northwest England, Yorkshire and the Midlands. The Calder in Yorkshire was one of the country's dirtiest rivers in 1980. Now it is well stocked with fish and popular with anglers. The Agency claims 'real improvements in the quality of many urban and semi-urban rivers, creating real amenity for people in areas where river environments have in the past been poor or in some cases lifeless, polluted eyesores'.

17.6 Pollution by oil

FIRST THOUGHTS

Huge oil tankers sail the sea. Sometimes one has an accident – not often, but often enough to cause a pollution disaster. The first big oil spill off the UK coast was when the oil tanker, the Torrey Canyon, sank off Cornwall in 1967. The pollution fouled beaches in Cornwall and killed thousands of sea birds. Now, tankers of up to 550 000 tonnes are in use, and accidents can happen on a much larger scale.

Prince William Sound was a beautiful unspoiled bay in Alaska. It was home to a huge variety of sea animals. Death struck over an area of 1300 square kilometres in 1989. The supertanker *Exxon Valdez* left Valdez with a cargo of oil from Alaska. Only 40 km out of port, the tanker hit a submerged reef and 60 million litres of crude oil leaked from her tanks. Fish, sea mammals and migrating birds from all parts of the American continent perished in the giant oil slick.

Sea birds dive to obtain food. If the sea is polluted by a discharge of oil, sea birds may find themselves in an oil slick when they surface. Then oil sticks to their feathers and they cannot fly. They drift on the surface, becoming more and more waterlogged until they die of hunger and exhaustion. Thirty five thousand sea birds died in the *Exxon Valdez* disaster.

The *Exxon Valdez* disaster was by no means the biggest ever. The increase in the size of tankers since 1945 has been spectacular. A tanker built in 1980 is 35 times as heavy as a tanker built in 1945. The very large crude carriers are difficult to stop and to change direction, and when accidents happen the results are more serious. The *Exxon Valdez* accident led to a law to prevent such disasters. In 1990 the Oil Pollution Act was passed in the USA, stating that in future tankers that want to operate in US waters must have a double hull, with a 3 m gap between the outer hull and the oil tanks inside.

A double hull is not the answer to all accidents, but it would have saved the *Sea Empress* when she grounded off the UK in 1996. In February 1996, the oil tanker *Sea Empress* ran on to rocks outside Milford Haven refinery. A stretch of Welsh coastline 190 km long was contaminated by oil. Oil slicks hit the marine nature reserves of Skomer and Lundy Islands and six sites of special scientific interest (see the map). These include major bird colonies and the country's only coastal national park. Within a month, over 1000 oiled birds had been counted, and 300 dead birds washed ashore. The leakage of oil was 70 000 tonnes.

More recent accidents involved the *Exxon Valdez* 1988 (60 000 tonnes of oil lost) the *Sea Empress* in 1996 (70 000 tonnes of oil lost).

A double hull is a protection against accidents

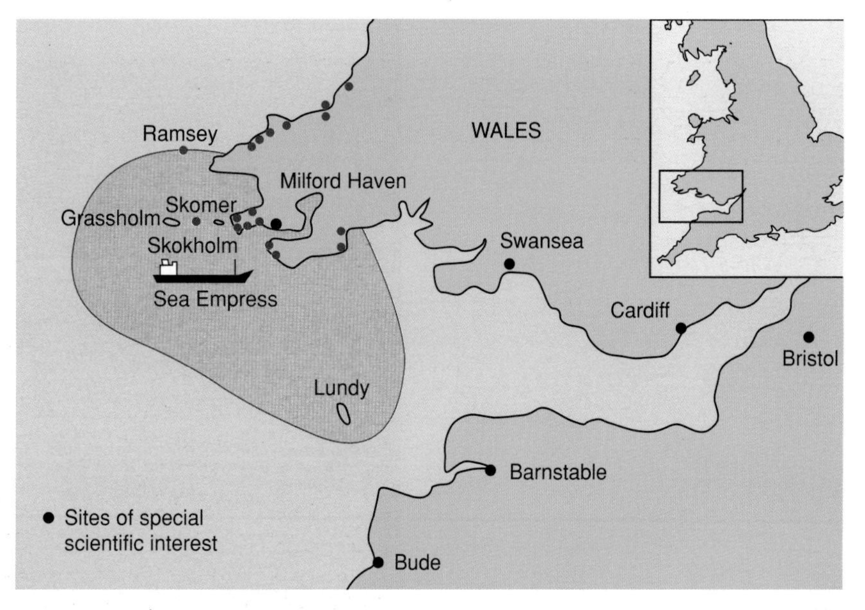

Figure 17.6A ▲ Milford Haven in South Wales

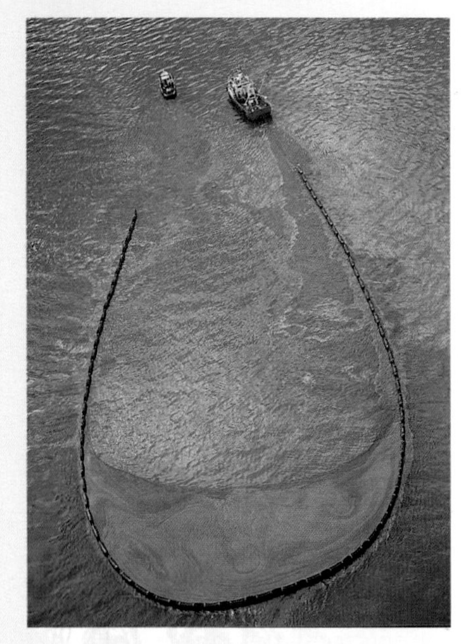

Figure 17.6B ▲ Using a boom to clean up after the *Sea Empress* disaster

Figure 17.6C ▲ The *MV Braer* after running aground on the Shetland Islands 1993

Figure 17.6D ▲ Ten thousand sea otters perished in the *Exxon Valdez* spill. Some ate poisoned fish. Others drowned when their fur became clogged with oil

Figure 17.6E ▲ Oil spill in Alaska

Figure 17.6F ▶ More victims

Oil spills at sea are the results of capsizings, collisions and accidental spills during loading and unloading at oil terminals. There is another source. After a large tanker has unloaded, it may have 200 tonnes of oil left in its tanks. While in port, the tanker is flushed out with water sprays, and the cleaning water is collected in a special tank, where the oil separates. Some captains save time by flushing out their tanks at sea and pumping the wash water overboard. This is illegal. Maritime nations have tried to set up standards to stop pollution of the seas, but several nations have not signed the agreements. Enforcing agreements is very difficult as it is impossible to detect everything that happens at sea.

Various methods have been tried for the removal of oil from the surface of the sea.

- **Dispersal** Chemicals are added to emulsify the oil. The danger is that they may be toxic to marine life.
- **Sinking** Oil may be treated with sand and other fine materials to make it sink. A danger is that the sunken oil may cover and destroy the feeding areas of marine creatures.
- **Burning** Burning oil is dangerous as a fire can spread rapidly over the sea. Research has been done on safe methods of burning oil, but they leave 15 per cent of the oil behind as lumps of tar.
- **Absorbing** Absorbents do not work well in the open sea. They provide the best way of cleaning a beach or preventing an oil spill from reaching the shore.
- **Skimming off** The method of surrounding an oil spill with a line of booms to prevent it spreading and then pumping oil off the surface has been used with some success.
- **Solidifying** Scientists at British Petroleum have discovered chemicals which will solidify oil spills. The chemicals must be sprayed on to the oil slick from the air. They convert the oil into a rubber-like solid which can be skimmed off the surface in nets.
- **Bacteria** There are bacteria which will decompose petroleum. A mixture of bacteria (of the correct strain) and nutrients is sprinkled on to the spill from the air.

Science at work

Bacteria can be used to clean out a tanker's storage compartment. The empty tank is filled with sea water, nutrient, air and bacteria. When the tanker reaches its destination, the tank contains clean water, a small amount of recoverable oil and an increased number of bacteria. The bacteria can be used as animal feed.

SUMMARY

Spillage of oil from large tankers is a source of pollution at sea. It kills marine animals and washes ashore to pollute beaches.

CHECKPOINT

1 (a) What are the causes of oil spills at sea?
(b) What damage do they do?
(c) Who pays to clean up the mess?
(d) Suggest what can be done to stop pollution of the sea by oil.

18.1 The structure of the Earth

FIRST THOUGHTS

The Earth's crust is a layer of solid rock. It is 50 km thick, a small fraction of the Earth's diameter of 12 700 km. This layer of rock provides the raw materials we need for our industries.

www.keyscience.co.uk

The Earth's crust is a layer of solid rock. It is 50 km thick, a small fraction of the Earth's diameter of 12 700 km. This layer of rock provides the raw materials we need for our industries.

Earth was formed as a molten mass cooled down over millions of years. Dense materials sank into the centre to form a core of dense molten rock. Less dense materials remained on the surface to form a **crust** of solid rock (50 km thick). As Earth cooled, water vapour condensed to form rivers, lakes and oceans on the surface of the Earth. Figure 18.1A shows the structure of the Earth today. It is composed of the **core, mantle, crust and atmosphere**.

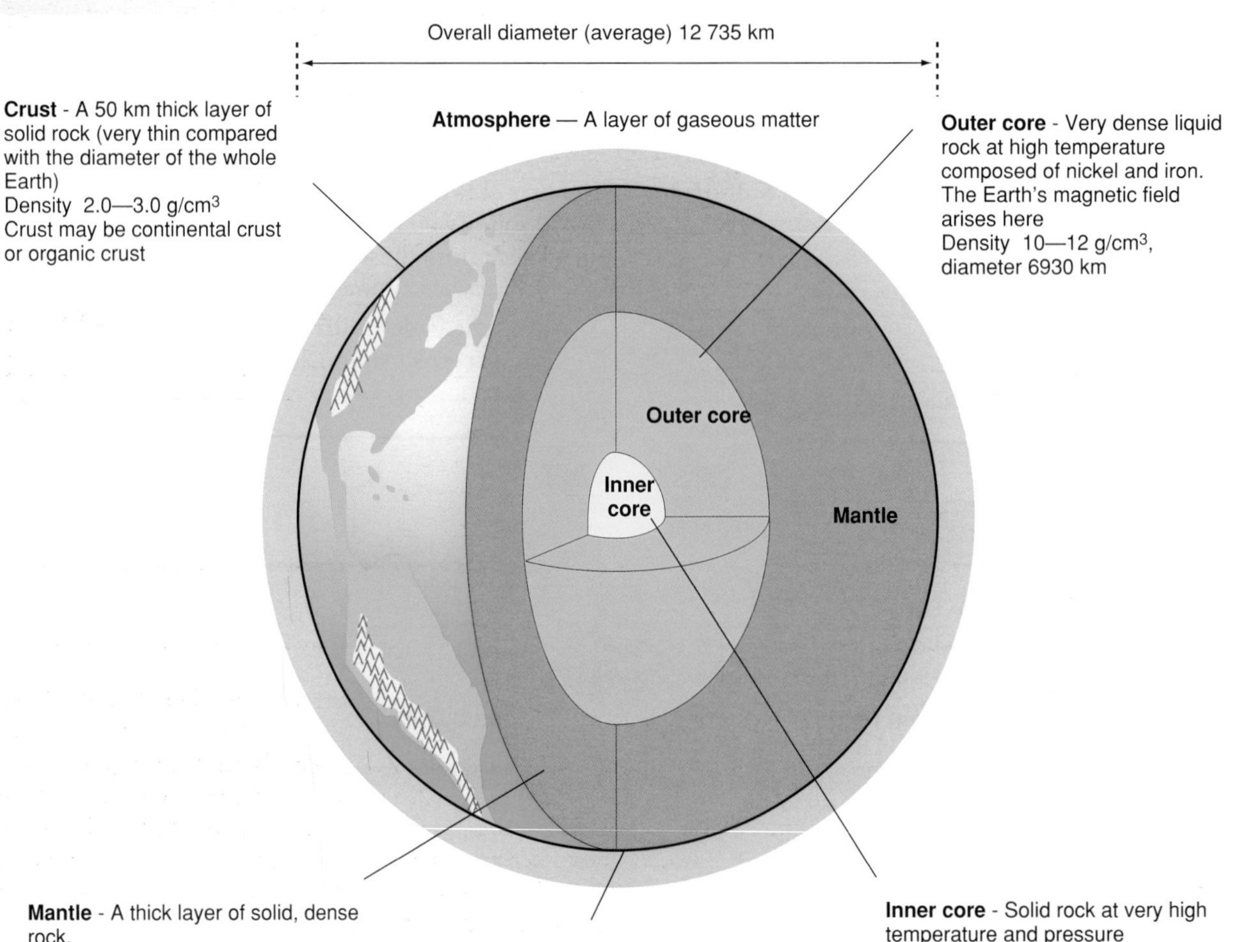

Figure 18.1A ▲ The structure of the Earth

18.2 Earth's crust

Check that you know the meaning of:

- crust
- continental crust
- oceanic crust
- mantle
- outer core
- inner core
- Moho

Earth's crust is composed of **rocks** and **soils**. Rocks are composed of compounds, e.g. carbonates, oxides and silicates, and some free elements, e.g. sulphur and copper. The elements and compounds which occur naturally in Earth's crust are called **minerals**. Soils have been formed by the breakdown of rocks and vegetation.

The crust is divided into

- **continental crust**, forming continents, which has the same composition as granite rock
- **oceanic crust**, forming the deep sea floor, which has the same composition as basalt rock.

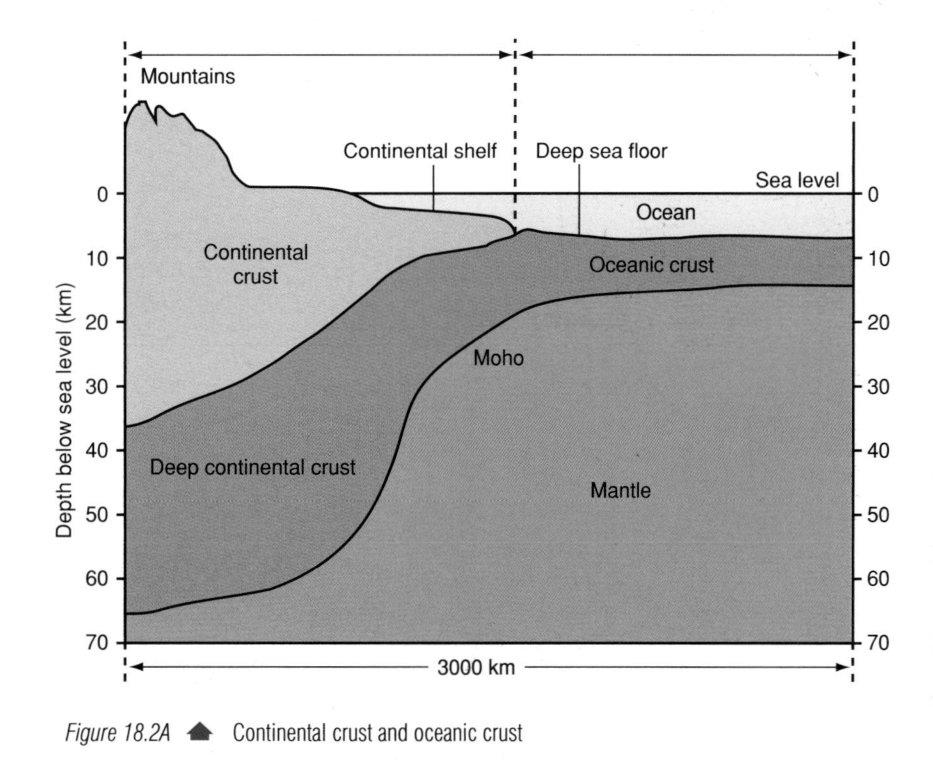

Figure 18.2A ▲ Continental crust and oceanic crust

SUMMARY

The Earth has a layered structure: inner core, outer core, mantle, crust and atmosphere. The crust is composed of oceanic crust beneath the ocean floors and continental crust, which forms the Earth's land masses.

18.3 Plate tectonics

FIRST THOUGHTS

Is Earth's crust really composed of separate moving pieces? It sounds amazing, but this is one of the newest scientific theories.

People who study the structure of the Earth are called **geologists**. They believe that the outer layer of Earth is made up of separate pieces called **tectonic plates**. Each plate consists of a piece of **lithosphere** (crust and uppermost layer of mantle) 80–120 km thick. The plates rest on the lower part of the mantle. Parts of the mantle move very slowly, and the plates which it carries move at a rate of a few centimetres a year. This model of the Earth, in which rigid slabs of the crust are moving very slowly on the surface of a sphere, is called **plate tectonics**.

Boundaries

As a result of the movement of the mantle, plates sometimes rub against each other. If the stress builds up to a large extent, the plates may bend.

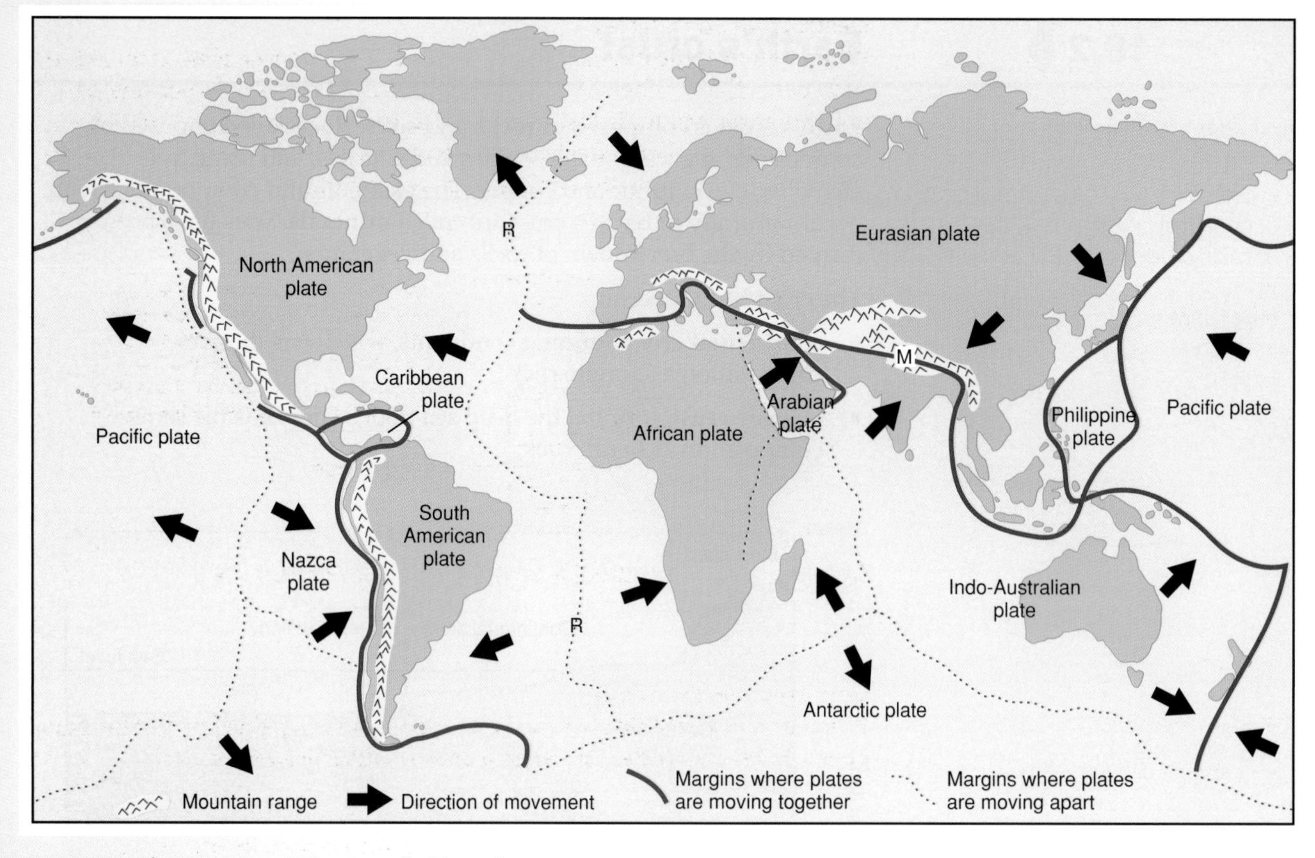

Figure 18.3A ▲ The plates which make up Earth's crust

Boundaries between plates may be

- constructive – where new crust is formed between plates moving apart
- destructive – where crust is forced downwards as plates collide
- conservative – where plates slide past one another with no gain or loss of crust

When they spring back into shape, the ground shakes violently: there is an earthquake. Other results of the movement of tectonic plates are shown in Figures 18.3B, C, D.

Constructive boundaries

In some regions tectonic plates are moving apart. At the boundary between the plates a gap forms, and this allows molten mantle (magma) to escape between the plates. It erupts as a volcano, shooting out as **lava** (mantle that has reached the surface). Lava solidifies to form new rocks along the edges of the plates. Since matter is added to the plates, a boundary where plates are moving apart is called a **constructive boundary.** As this process takes place time and time again over millions of years mountain ranges form.

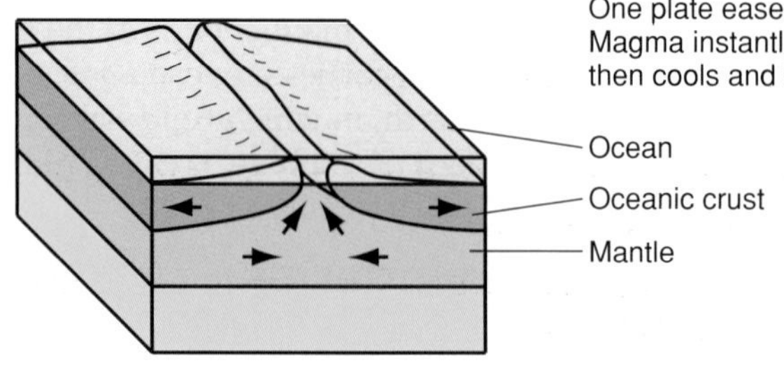

Figure 18.3B ▲ A constructive plate boundary: rocks are formed

Destructive boundaries

When two tectonic plates collide, the edge of the denser plate is forced to slide beneath the other. The descending plate edge melts to become part of the mantle. Matter is lost from the plates to the mantle, and this type of boundary is called a **destructive boundary**. At some time in the future the mantle may reach the surface and cool to form new rock.

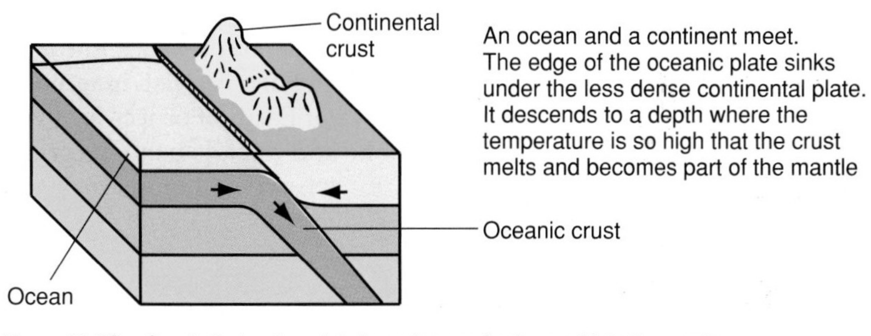

Figure 18.3C ▲ A destructive plate boundary: rocks descend into the mantle

Conservative boundaries

When two plates slide past each other, no matter is gained or lost by the plates. Such a boundary is called a **conservative boundary**.

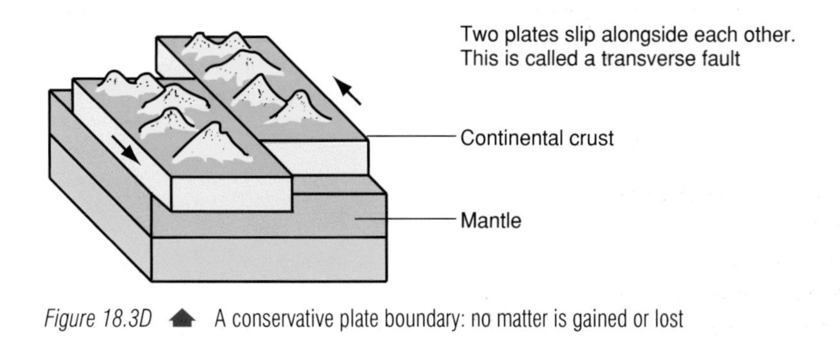

Figure 18.3D ▲ A conservative plate boundary: no matter is gained or lost

CHECKPOINT

▶ **1** Take a piece of string 4 m long. Imagine that this length represents the 4000 million years that have passed since the first living things appeared on Earth. Mark on the string the length that represents the 2.5 million years since the early human beings appeared.

▶ **2** Name the parts of the Earth which are
- (a) gaseous
- (b) magnetic
- (c) dense rock, parts of which move slowly
- (d) a 50 km thick layer
- (e) the material of which Earth's land masses are composed
- (f) the material of which the deep sea floor is composed.

▶ **3** Explain what is meant by a tectonic plate. Name the three different types of boundary between tectonic plates, and state what happens at each.

18.4 Types of rock

FIRST THOUGHTS

There are three main types of rock: igneous, sedimentary and metamorphic. In this section, you can find out which is which.

Igneous rocks

Sometimes enough heat is generated in the crust and upper mantle to melt rocks. The molten rock is called **magma**. Once formed, magma tends to rise. If it reaches Earth's surface, it is called **lava**. When cracks appear in Earth's crust, magma is forced out from the mantle on to the surface of Earth. It erupts as a **volcano**, a shower of burning liquid, smoke and dust. When lava crystallises above the surface of the Earth, **extrusive igneous rock** is formed. When magma crystallises below Earth's surface, **intrusive igneous rock** is formed.

The rocks that are formed when volcanic lava cools on the surface of the Earth are called **extrusive igneous rocks**.

The rocks that are formed when magma crystallises inside the Earth are called **intrusive igneous rocks**.

Extrusive igneous rocks formed when the lava erupted from a volcano cools are: **basalt**, from free-flowing mobile lava; **rhyolite**, from slow-moving lava; **pumice**, from a foam of lava and volcanic gases.

Types of rock erupted by a volcano are: **agglomerate**, the largest rock fragments which settle close to the vent; **volcanic ash**, finer fragments of rock; **tuff**, compacted volcanic ash; **dust**.

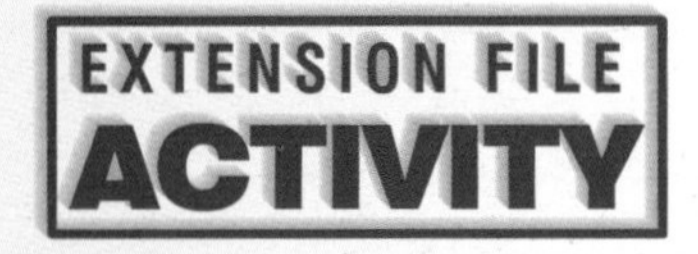

In appearance, igneous rocks can be seen to be composed of randomly arranged, interlocking crystals of different minerals. If the crystals are small, this indicates that the rock was formed by magma cooling rapidly, as when magma is erupted from a volcano. If the crystals are large, this indicates that the rock was formed by magma cooling more slowly, within Earth's crust.

Liquid lava. When lava cools crystals form in it. The crystals grow and interlock to form hard rock. Rocks formed from molten material in this way are called igneous rocks

Rising magma

Alternate layers of solidified lava and erupted solid rock

Vent The opening in the volcano. Through it come volcanic gases: water vapour, carbon dioxide, sulphur dioxide, hydrogen sulphide, etc. at about 1000 °C

If the lava solidifies in the vent, gas pressure builds up and there is likely to be a violent eruption. If this happens, lava and rock are forced out of the vent in a jet of volcanic gas. The mixture can travel rapidly down the side of the volcano causing death and destruction in its path.

Figure 18.4A ▲ A volcano

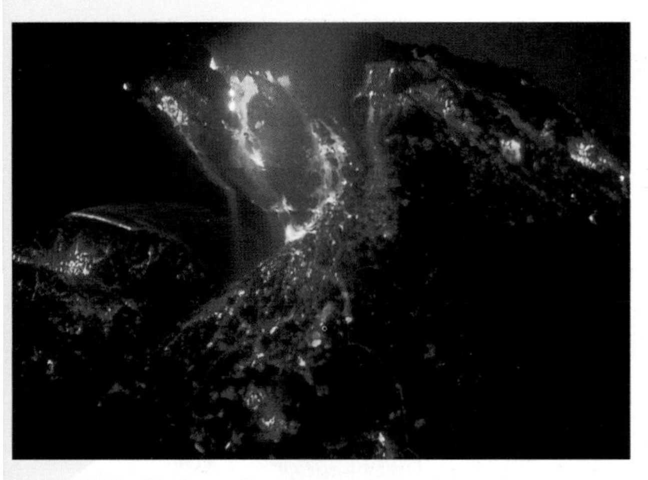

Figure 18.4B ▲ Lava rolling down the slopes of Mount Etna, one of the most active volcanoes

Figure 18.4C ▲ Mount St. Helens during its second eruption. Dust and ash were spread for many miles around and stayed in the atmosphere for months

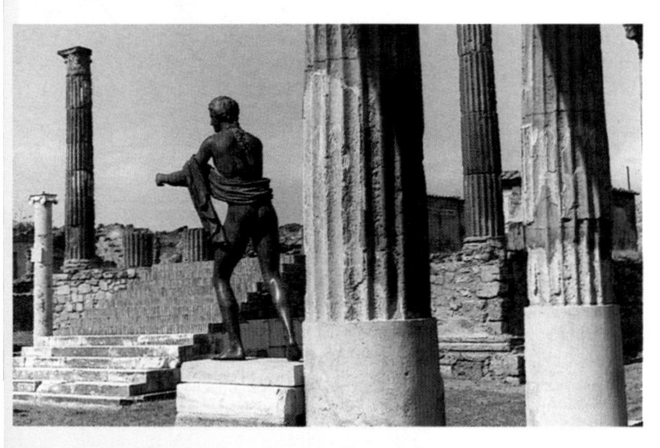

Figure 18.4D ▲ In 79AD, the city of Pompeii was destroyed when a mixture of lava, rock and ash travelled quickly down the side of the volcano Vesuvius

Figure 18.4E ▲ The Giant's Causeway in Northern Ireland is made from basalt, solidified lava

Sedimentary rocks

The rocks formed by compressing solid particles are called **sedimentary rocks**.

The formation of sedimentary rocks begins when solid particles settle out of a liquid or an air stream to form a **sediment**. The solid material comes from older rocks or from living organisms. All rocks exposed on Earth's surface are worn away by **weathering** and by **erosion**. The material that is worn away is transported by gravity, wind, ice, rivers and seas. The transported material may be fragments of rock, pebbles and grains of sand, or it may be dissolved in water. Eventually the transported material is deposited as a **bed** (layer) of sediment. It may be deposited on a sea bed, on a sea shore or in a desert. The beds of sediment are slowly compacted (pressed together) as other material is deposited above. Eventually, after millions of years, the pieces of sediment become joined together into a sedimentary rock. This process is called **lithification**. As the layers of sediment are laid down, remains of plants and animals may be trapped and preserved as fossils. The rocks may contain ripple marks, caused by currents or waves as they were laid down.

Examples of sedimentary rocks are: **limestone**, formed from the shells of dead animals; **coal**, formed from the remains of dead plants; **sandstone**, compacted grains of sand.

It's a fact!

There are 540 active volcanoes, including 80 on the sea bed. In addition to **active** volcanoes (which have erupted in the past 80 years) there are **dormant** (resting) and **extinct** (dead) volcanoes. Rocks found in the Lake District and in North Wales prove that volcanoes were erupting there 450–500 million years ago, showing that Britain must have been in contact with plate margins in the past.

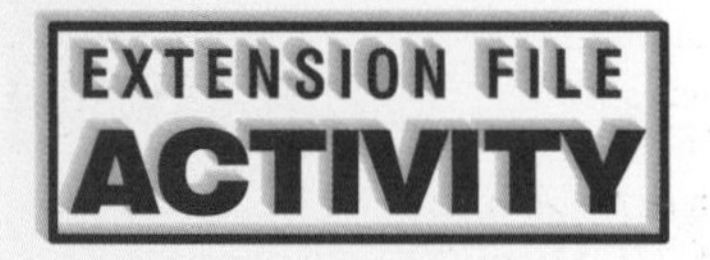

Rocks formed by the action of high pressure and high temperature on other types of rock are called **metamorphic rocks.**

Metamorphic rocks

The movement of tectonic plates can cause igneous and sedimentary rocks to become buried deep underground. There they are compressed and heated. This treatment may change the structure of a rock without melting it. The changed rock is called a **metamorphic rock** (from the Greek word for 'change of shape'). Rocks composed of bands of interlocking crystals are likely to be metamorphic. Examples of metamorphic rocks are:

- **marble**, formed when limestone is close to hot igneous rocks,
- **slate**, formed from clay, mud and shale at high pressure,
- **metaquartzite**, from metamorphism of sandstone.

The composition of the Earth's crust is: igneous rocks 65%, sedimentary rocks 8% and metamorphic rocks 27%. The differences are summarised in Table 18.1. Rocks are composed of compounds, e.g. carbonates, oxides and silicates, and some free elements, e.g. sulphur and copper. The elements and compounds which occur naturally in Earth's crust are called **minerals**.

Table 18.1 Types of rock

	Igneous	*Sedimentary*	*Metamorphic*
Type of grain	Crystalline Large cystals – extrusive Small crystals – intrusive	Fragmental: grains do not usually interlock (They do in some limestones)	Crystalline
Direction of grain	Grains usually not lined up	Grains usually not lined up	Grains usually lined up
Mode of formation	Crystallisation of magma	Deposition of particles	Recrystallisation of other rocks
Fossil remains	Absent	May be present	Absent
Appearance when broken	Shiny	Usually dull	Shiny
Ease of breaking	Hard, not easily split, may crumble if weathered	May be soft and crumble, but some are hard to break	Hard, but may split in layers, may crumble if weathered
Examples	Basalt, granite, rhyolite, pumice	Limestone, clay, sandstone, mudstone	Marble, hornfels, slate, schist

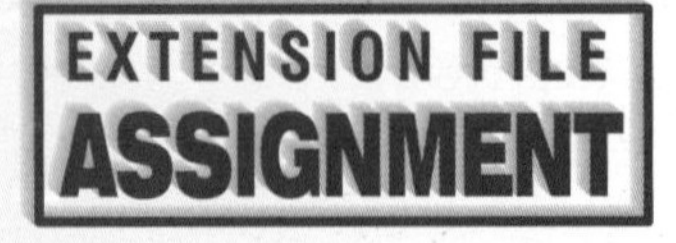

The rock cycle

Only igneous rocks are formed from new material brought into the crust. The original crust of Earth must have been made entirely from igneous rocks. The slow processes by which metamorphic and sedimentary rocks are formed from igneous rock and also converted back into igneous rock is called the rock cycle (see Figure 18.4G).

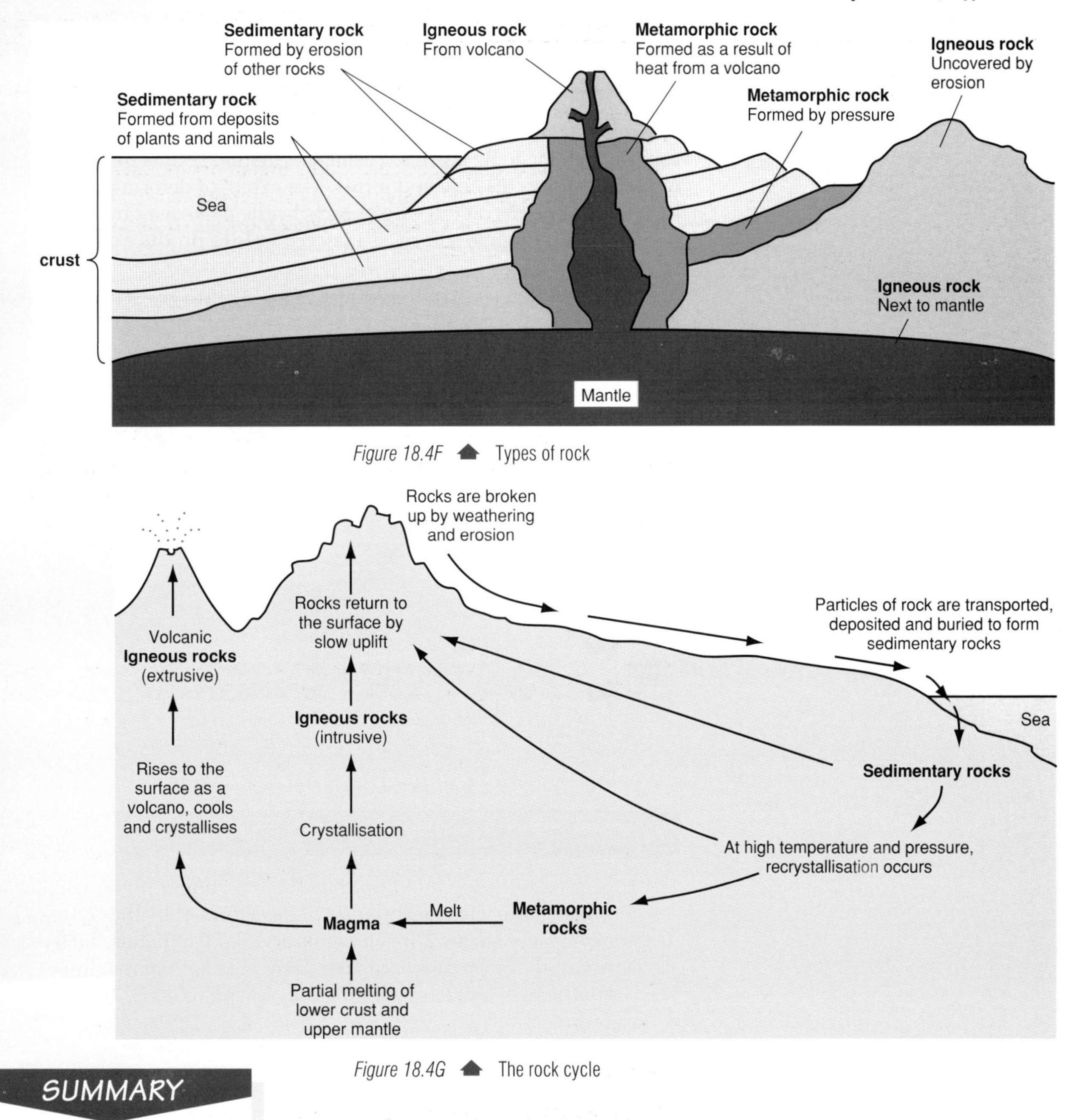

Figure 18.4F ▲ Types of rock

Figure 18.4G ▲ The rock cycle

SUMMARY

When volcanoes erupt, they emit volcanic gases, lava, ash and pieces of solid rock. Lava cools and solidifies to form igneous rock. The solid rock settles as agglomerate. Ash is compacted to form tuff. Rocks are weathered into smaller particles. These particles are deposited as a sediment, which becomes compacted to form sedimentary rock. Igneous and sedimentary rocks may be changed by high temperature and pressure into metamorphic rock. The slow processes by which rock material is recycled are called the rock cycle.

CHECKPOINT

1 Copy and complete this passage.

When volcanoes erupt, molten rock called _____ streams out of the Earth. It solidifies to form _____ rocks, e.g. _____. When deposits of solid materials are compressed to form rocks, _____ rocks are formed, e.g. _____. The action of heat and pressure can turn _____ rocks and _____ rocks into _____ rocks.

2 Which type of rock is

(a) limestone

(b) granite

(c) marble?

18.5 Deformation of rocks

The movement of tectonic plates (Topic 18.3) causes deformation of the rocks which form the Earth's crust. Deformation results in the formation of **folds, faults, cleavage** and **joints**. The extent of deformation produced by a force depends on the type of rock: brittle rocks may fracture to produce a fault, while soft rocks may crumple to produce a fold.

Folding

When rocks are compressed (squeezed) they may become folded. Sedimentary rocks are often folded. The beds of rock which have been laid down are no longer horizontal; they are folded to bulge upwards (an **anticline**) or downwards (a **syncline**).

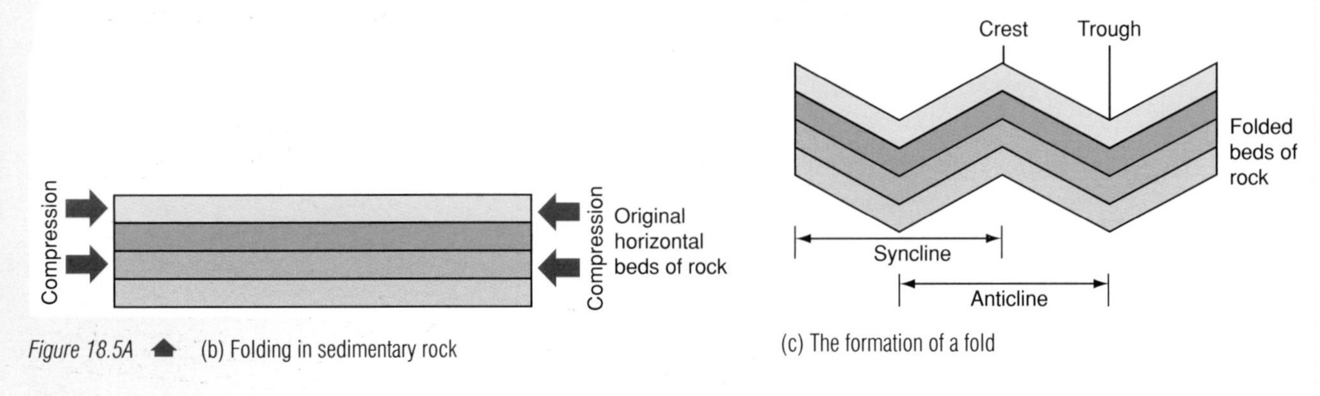

Figure 18.5A (b) Folding in sedimentary rock

(c) The formation of a fold

Cleavage

Cleavage is the splitting into thin sheets of rock under pressure. Slate is easily broken along **cleavage planes**. Slate is formed by the metamorphism of shales. Clay minerals and flaky minerals, such as mica, are recrystallised to lie perpendicular to the direction of maximum stress. The slate which results has a weakness in one plane along which it can be easily broken.

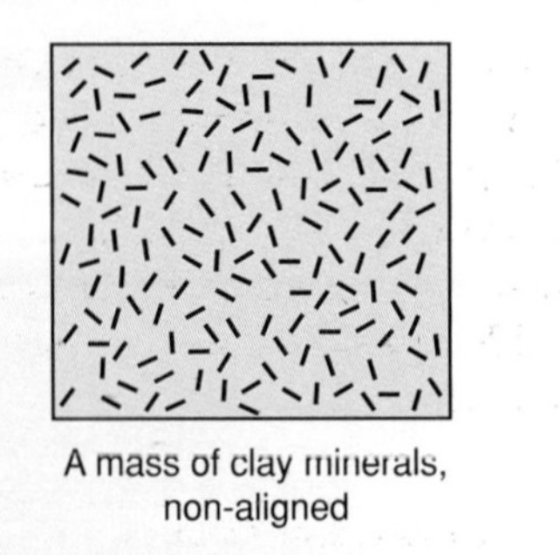

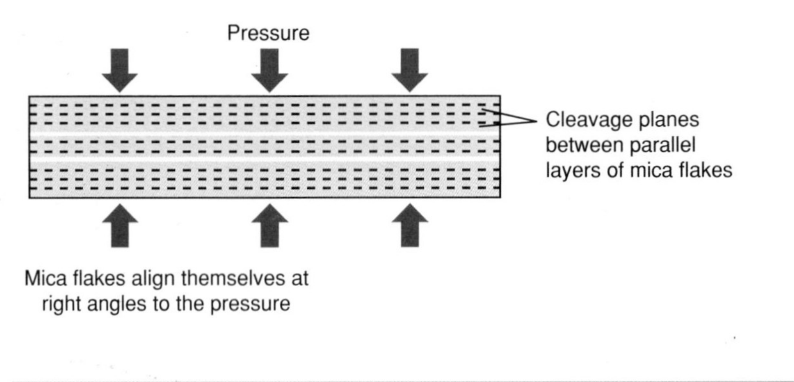

Figure 18.5B The formation of slate

Faults

A powerful force can break rocks and cause faults (breaks) in the rock. A break allows rocks on one side of a fault to move against rocks on the other side of the fault. As long as forces keep acting, rocks will continue to keep moving against each other along the line of the fault. Since there is friction between the edges of rocks, the movement takes place in jerks. Each of these jerks may cause an earthquake.

Figure 18.5C ▲ (a) A vertical fault

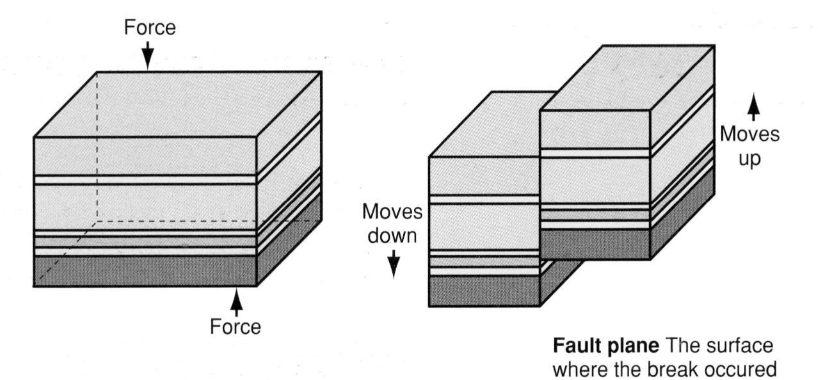

(b) A fault developing

In Figure 18.5C the faults are vertical displacements. Horizontal displacements occur at faults where two of the Earth's plates slide past each other. The San Andreas fault which lies beneath California in the USA is of this type. There were severe earthquakes in California in 1838, 1906 and 1989.

Forces acting within the crust deform rocks.

The result may be

- a fold (an anticline or a syncline)
- a fault
- a cleavage
- a joint.

Check that you know what each of these terms means.

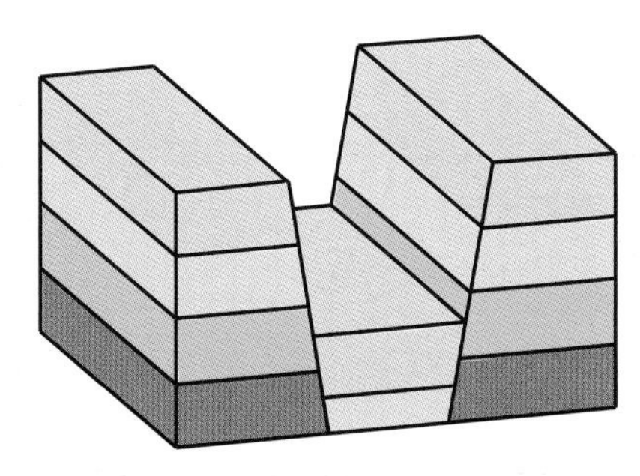

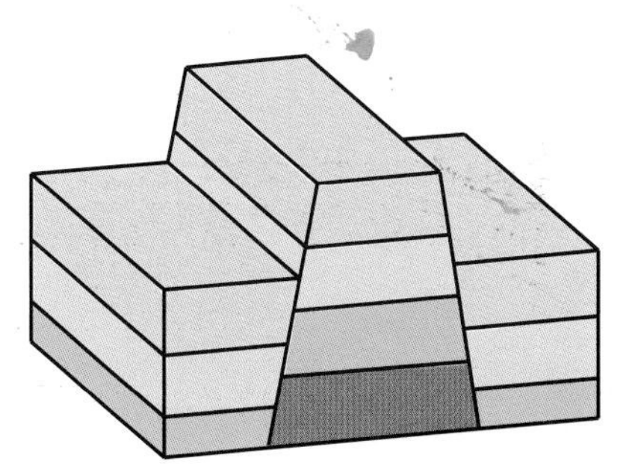

Figure 18.5D ▲ (a) A rift valley. The valley has subsided between faults. It may become the course of a river or the site of a lake. (b) A horst left standing when rocks on both sides subside

It's a fact!

Ever since 'the trembler' of 1906, California has been waiting for 'the big one'. The earthquake of 1906 registered 8.3 on the Richter scale and flattened three-quarters of San Francisco leaving 700 dead. Scientists predicted that there would be another earthquake. They predicted the site and the size of the future earthquake but could not predict the date. In 1989 an earthquake, measuring 6.9 on the Richter scale hit San Francisco killing 300 people. This time the damage was limited because engineers had designed buildings to withstand an earthquake. Scientists predict that another earthquake, of magnitude 8, is still to come.

Joints

A fracture may occur without the rocks on either side moving relative to one another. Such a break cannot be called a fault because there is no displacement so it is called a **joint**. Joints in igneous rock may be caused by cooling and shrinking. Figure 18.5F shows joints in the limestone pavement at Malham Cove. Joints in sedimentary rocks may be caused by loss of water. Joints make a rock permeable to water. They provide weaknesses which may be affected by weathering.

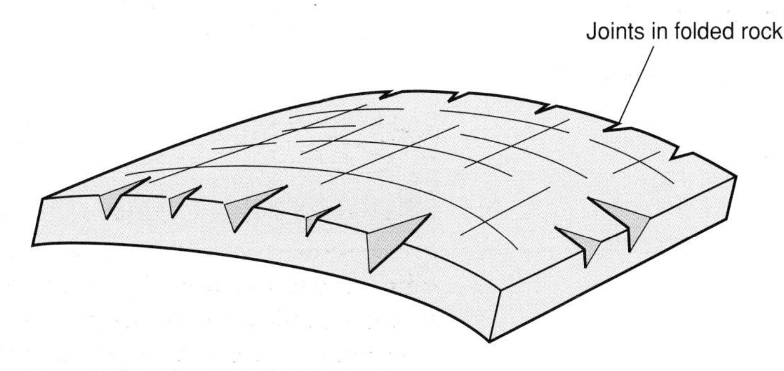

Figure 18.5E ▲ Joints in folded rock

SUMMARY

Sedimentary rock is laid down in horizontal layers. These may be deformed by:

- folding, with the formation of synclines and anticlines
- cleavage, splitting into thin sheets
- faulting, developing breaks which allow slabs of rock to slide against one another
- jointing, developing breaks without any movement of rock.

Figure 18.5F ▲ Limestone pavement at Malham Cove

CHECKPOINT

1. The figure shows a section through some layers of rock.

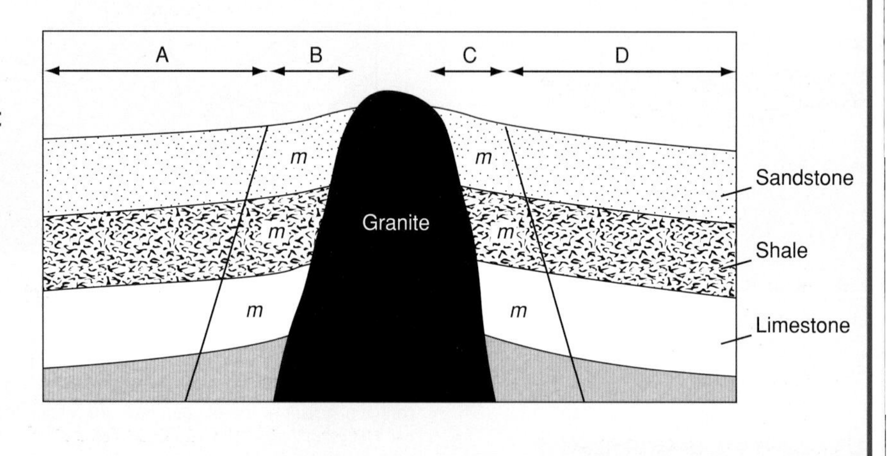

(a) Explain the statement: *Limestone, shale and sandstone are **sedimentary** rocks.*

(b) What type of rock is granite?

(c) Explain how the granite could have pushed through the layers of sedimentary rock.

(d) Explain the statement: *The layers of sedimentary rock in regions B and C have been **metamorphosed**.*

(e) Why have the sedimentary rocks at *A* and *D* not been metamorphosed?

18.6 The forces which shape landscapes

FIRST THOUGHTS

What forces shaped the varied landscape that we see around us? What forces pushed up the mountain ranges, smoothed the plains and carved out the valleys?

Rocks are continually being broken down into smaller particles by forces in the environment. These processes are called **weathering.** Weathering may be brought about by:

- **physical forces**, especially in deserts and high mountains,
- **chemical reactions**, especially in warm, wet climates.

Rain

Water is an important weathering agent. Water expands when it freezes. If water enters a crack in a rock and then freezes, it will force the crack to open wider. When the ice thaws, water will penetrate further into the rock. After cycles of freeze and thaw, pieces of rock will break off.

Water reacts with some minerals, like mica. The reaction produces tiny particles which are easily transported away and deposited as a sediment of mud or clay.

EXTENSION FILE ASSIGNMENT

Rivers and streams

Rivers and streams carry water back to the oceans as part of the **water cycle**. A fast-flowing stream can carry a lot of particles in suspension (see Figure 18.6A). A very fast stream can push sand and pebbles along with it.

Running water causes erosion. The bed load and the suspension load rub against the bed and sides of the river channel. In addition, there are chemical reactions between water and rocks.

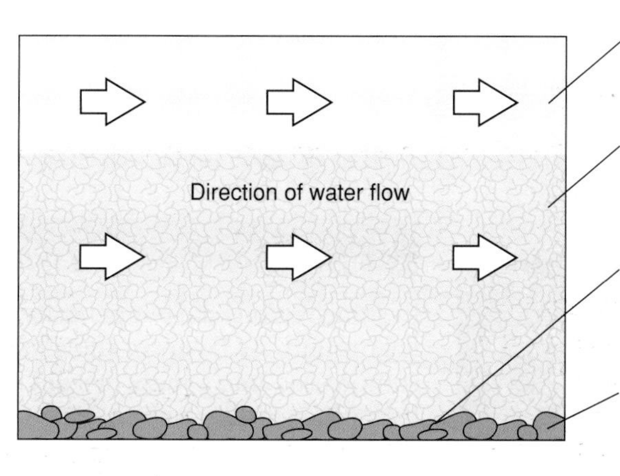

Figure 18.6A ▲ The load carried by a stream

SUMMARY

Some of the weathering processes are:

- expanding ice breaks up rock
- water breaks up some minerals, e.g. mica, and dissolves others
- running water brings about erosion.

Sediments are deposited when rivers lose speed: on the inside curve of river bends and where rivers flow into lakes and seas.

■ Deposition

Rivers deposit the sand and gravel which they carry. The deposits form when rivers lose speed:

- on the inside curves of river bends,
- when a river flows into the sea or a lake.

Underground water

Water can pass through certain kinds of rocks. It can seep through joints in beds of sedimentary rock. Such rocks, like limestone, are described as **permeable**. Other rocks, like sandstone, have tiny spaces between their mineral grains which allow water to enter. These rocks are **porous**.

Water held in rocks below the surface is called **ground water**. Some of it finds a way back to the surface to become a **spring**. Some of it remains as an underground reserve. About one third of the UK water supply is drawn from ground water.

Rain water causes weathering of rocks. This happens also below ground. Rain water permeates the ground and dissolves some minerals; gypsum is dissolved by rain. The acid in rain water reacts with limestone to form soluble compounds. Small openings in the rock become wider and in time form large underground passages which carry underground rivers and streams. When you visit **underground caves** you walk along dried-up river beds.

SUMMARY

Underground water reacts with limestone and other rocks to form soluble compounds. In time, underground caverns are formed.

The sea

Erosion by the sea is illustrated in Figure 18.6B The cave has been gouged out by waves at a weak point in the cliff, e.g. a joint. The arch has been formed by two caves meeting back to back. The sea stack has been left where the top of an arch has fallen away.

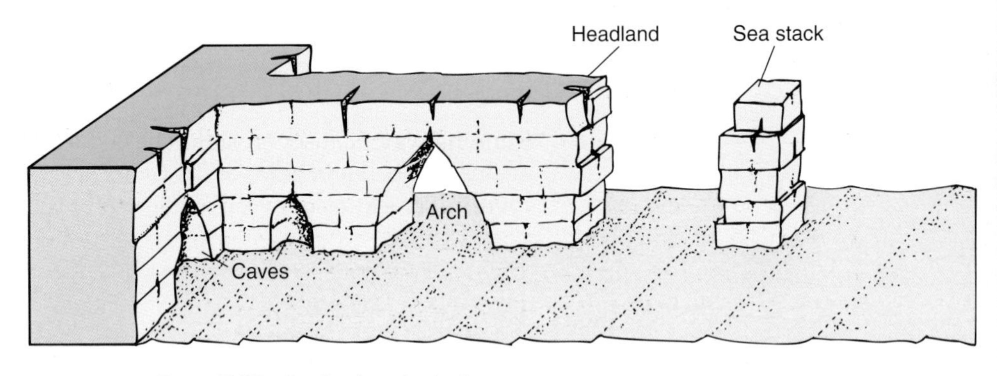

Figure 18.6B ▲ Landscaping by the sea

Glaciers

Figure 18.6C ▲ Glacier

In some regions the temperature is low enough for snow to exist all the year round. As layers of snow build up, the lower levels become compressed into a mass of solid ice. When the ice begins to move under the influence of gravity, a **glacier** develops. Glaciers move slowly downhill, usually at less than 1 m/day. As glacial ice moves over a land surface, it wears down rocks by:

- **plucking**, freezing round pieces of rock and carrying them along,
- **grinding**, wearing down the rocks over which it is moving by means of the sharp rocks which become attached to the bottom of the glacier.

When a glacier melts, all the material which it carries is deposited.

Wind

In dry desert regions, wind is a landscaping agent, for example in shaping sand dunes. Moisture holds particles of sand and soil together and makes it much more difficult for the wind to remove them. In moist regions, plants grow, and their roots bind the soil, making it much more difficult to erode.

Figure 18.6D ▲ Sand dunes

SUMMARY

Forces which mould the landscape are:

- Rain, causing the formation of cracks in rock when it freezes
- Rivers and streams, transporting and depositing rock fragments
- Underground water, dissolving soluble minerals, e.g. gypsum, and reacting with limestone and other rocks to form soluble substances
- The sea, eroding rocks
- Glaciers, plucking rocks from the landscape and grinding land surfaces
- Wind, eroding sand and soil, especially in dry regions.

CHECKPOINT

1. Which of the following are needed for the formation of a glacier?
   (a) mountains (b) steep-sided valleys (c) heavy snowfall (d) heavy rainfall (e) low temperatures
2. Which of the following are examples of (a) erosion and (b) weathering?
   (i) Waves breaking against a cliff.
   (ii) Rocks splitting after a cold winter.
   (iii) Soil carried by the wind.
   (iv) Sand carried along the bed of a river.
   (v) The surface of a rock cracking after repeated heating and cooling.
3. What kind of rock (sedimentary, igneous or metamorphic) is formed under each of the following conditions?
   (a) Fragments of rock are formed by the action of frost and fall to the foot of a mountain.
   (b) Particles of clay come out of suspension in still water.
   (c) Dead plants sink to the bottom of a swamp.
   (d) Shells and shell fragments are rolled along a sea floor.

18.7 The geological time scale

FIRST THOUGHTS

Geologists can date the different layers of rocks which they unearth. Inspecting the fossils which they find helps, and the radioactivity of the rock tells a story.

Figure 18.7A shows the **geological column**. It divides Earth's history (4600 million years) into **eras**. Each era is divided into **periods**. Human life evolved in the Quaternary period. When geologists describe a rock as being of the Silurian age, they mean that the rock was formed between 435 and 395 million years ago. It was during the Pre-Cambrian period that the Earth's crust solidified, oceans and atmospheres developed, and the first living organisms appeared.

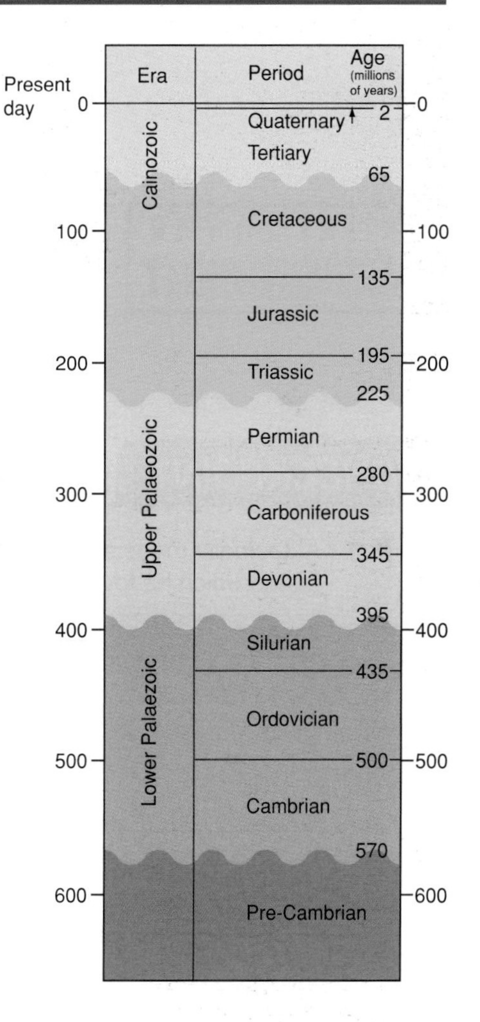

Figure 18.7A ▲ The geological column

Fossils

Geologists are able to say what period a rock dates from by examining the fossils that the rock contains. Fossils are the preserved remains of, or marks made by, dead plants and animals (see Figure 19.5B). If a rock contains the imprints of the shells of creatures known to have been living 300 million years ago, the rock may be Carboniferous.

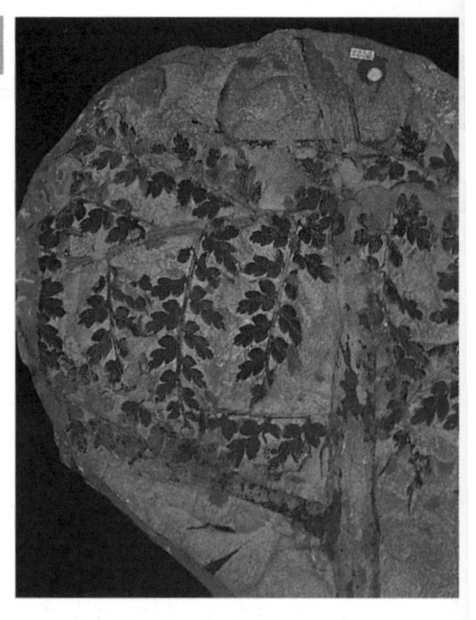

Figure 18.7B ▲ Carboniferous fossils

Relative dating

Relative dating does not give the age of rocks but enables you to classify them (arrange them) in order of age. If one sedimentary rock lies above another, it is very likely that the upper rock is younger than the lower one (although folding of the rocks can reverse the order). Fragments of rock included in another rock must be older than the rock that surrounds them.

Dating from radioactivity

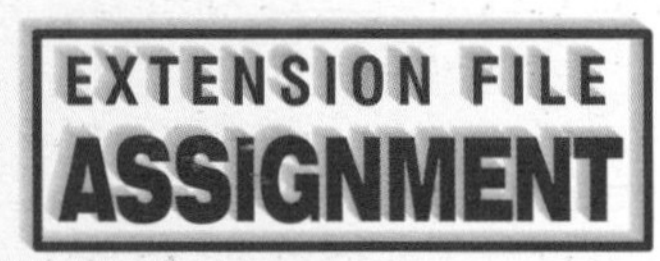

Some elements are **radioactive**. They have unstable atoms which split up (decay) to form atoms of stable elements. Imagine that a rock contains the radioactive element A, which decays very slowly to form element B. Then the ratio of B to A in the rock increases as the years go by. A measurement of the ratio of B to A will give the length of time for which A has been decaying, that is, the age of the rock. It is by radioactive dating that the age of Earth has been established as 4600 million years.

CHECKPOINT

➧ **1** A geologist made a sketch of the beds of rock in a quarry (see the figure below). He tabulated the fossils which he found in the four layers of rock.

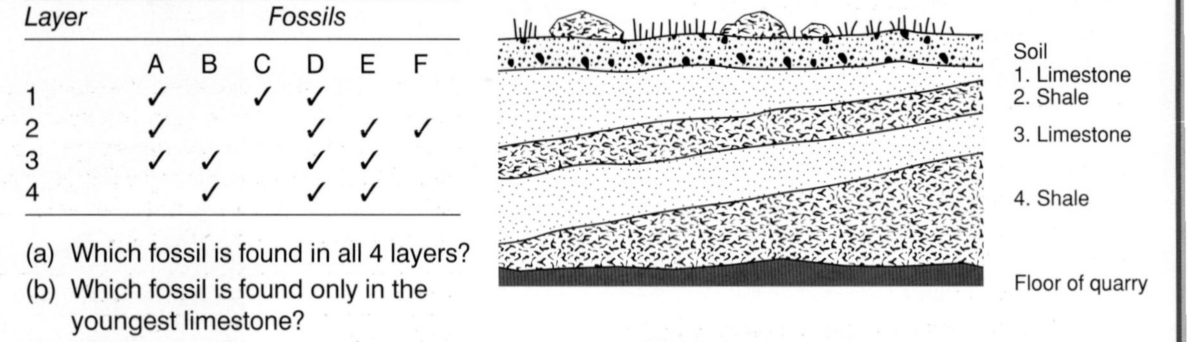

Layer	*Fossils*					
	A	B	C	D	E	F
1	✓		✓	✓		
2	✓			✓	✓	✓
3	✓	✓		✓	✓	
4		✓		✓	✓	

(a) Which fossil is found in all 4 layers?
(b) Which fossil is found only in the youngest limestone?
(c) Which fossil can be used to give the age of layer 2? Explain your answer.

1 (a) What happens to the temperature of a gas if the gas is compressed suddenly?
(b) What happens to the temperature of a gas if the gas is allowed to expand suddenly? How is this effect used to liquefy air? Why is this method chosen for the liquefaction of air?
(c) Boiling points are nitrogen, −196 °C; oxygen, −183 °C.
Explain how the difference in boiling points makes it possible to separate oxygen and nitrogen from liquid air.
(d) Give one-large scale industrial use for (i) oxygen (ii) nitrogen and (iii) another gas which is obtained from liquid air. Say why each gas is chosen for that particular use.

2 Three of the gases in air dissolve in water.
(a) Which of them dissolves to give an acidic solution? What use is made of this solution?
(b) Which of the three gases is a nuisance to deep-sea divers? Explain why, and say how the problem has been solved.
(c) The life processes of plants and animals depend on the solubility of two of these gases. Explain why.

3 The table shows the composition of inhaled air and exhaled air, excluding water vapour, and a comment on the content of water vapour.

	Percentage by volume	
	Inhaled air	*Exhaled air*
Oxygen	21	17
Nitrogen	78	78
Carbon dioxide	0.03	4
Noble gases	1	1
Water vapour	Variable	Saturated

(a) Describe the differences between inhaled air and exhaled air.
(b) Briefly give the cause of each of these differences.

4 A classroom contains 36 pupils. The doors and windows are closed for half an hour. Answer these questions about the air at the end of the half hour.
(a) Will the air temperature be higher or lower? Explain your answer.
(b) Will the air be more or less humid (moist)? Explain your answer.
(c) Will the percentage of carbon dioxide in the air be higher or lower? Explain your answer. Say how the change in carbon dioxide content will affect the class.

5 You are given four gas jars. One contains oxygen, one nitrogen, one carbon dioxide and one hydrogen. Describe how you would find out which is which.

6 The diagram shows an apparatus which is being used to pass a sample of air slowly over heated copper.

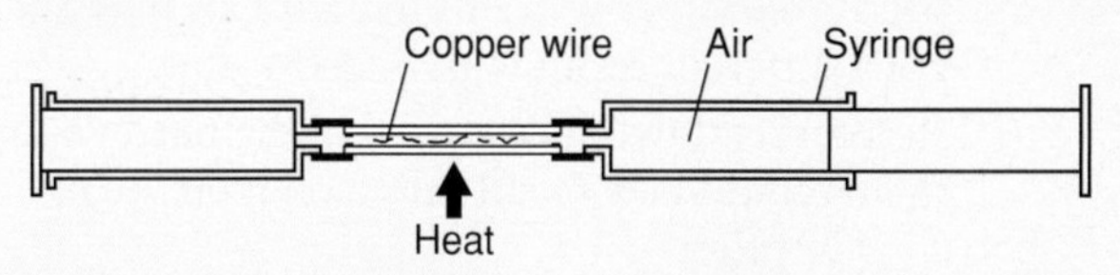

(a) Describe how the appearance of the copper changes.
(b) Which gas is removed from the air by copper? Name the solid product formed.
(c) If 250 cm^3 of air are treated in this way, what volume of gas will remain?
(d) Name the chief component of the gas that remains.
(e) Name two other gases that are present in air.

7 When petrol burns in a car engine, carbon dioxide and carbon monoxide are two of the products.
(a) Write the formula of (i) carbon dioxide (ii) carbon monoxide.
(b) Explain the statement 'Carbon monoxide is a product of incomplete combustion.'
(c) Red blood cells contain haemoglobin. What vital job does haemoglobin do in the body?
(d) If people breathe in too much carbon monoxide, it may kill them. How does carbon monoxide cause death?
(e) Explain how people can be poisoned accidentally by carbon monoxide.
(f) What precautions can people take to make sure that carbon monoxide is not formed in their homes?
(g) The blood of people who smoke contains more carbon monoxide than the blood of non-smokers. Can you explain why?

8 Ruth carried out an experiment to compare the hardness of the water from three towns. She measured 50 cm^3 of each water sample into separate conical flasks. She added soap solution gradually to each flask, shaking them until a lather was formed. Her results are shown in the table.

Water sample	*Volume of soap solution* (cm^3)
Distilled water	2.0
Johnstown water	7.5
Mansville water	10.0
Rumchester water	4.0

(a) Say what piece of apparatus Ruth could use for measuring (i) the 50 cm^3 samples of water and (ii) the volume of soap solution added.
(b) Explain why she did a test on distilled water.
(c) Why does distilled water require the smallest volume of soap solution to form a lather?
(d) Which town has the hardest water? Explain your answer.
(e) When Ruth boiled a 50 cm^3 sample of Rumchester water before testing it, she found that the volume of soap solution needed to produce a lather was 2.0 cm^3. Explain why she got a different result with boiled water.
(f) Recommend two measures that the hard water towns could take to cut down on their soap consumption.

9 The diagram shows rain falling on the ground and trickling over underground rocks.

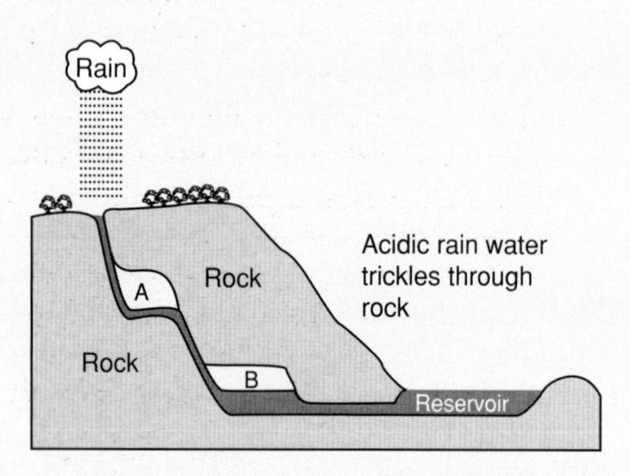

(a) Explain why natural rain water is weakly acidic.
(b) Name a type of rock which will be attacked by acidic rain water.
(c) Name the type of cavity, A, that is formed as a result of the action of rain water on the rock.
(d) Name the type of water that accumulates at B.
(e) Water flows from B into a reservoir. What treatment does the water need before it is fit to drink? Explain why this water receives a different treatment from river water.
(f) Will the water in the reservoir be hard water or soft water? Explain your answer.
(g) State one advantage and one disadvantage of hard water compared with soft water.

10 The diagram shows acid rain falling on the shores of three lakes.

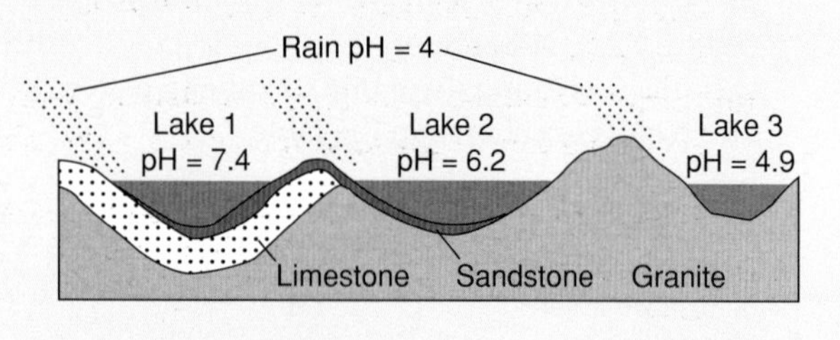

(a) Unpolluted rain water has a pH of 6.8. What gives it this weak acidity?
(b) By acid rain, we mean rain with a pH below 5.6. Name two substances that react with rain to make it strongly acidic.
(c) Explain why Lake 3 is more acidic than Lake 2.
(d) Explain why Lake 2 is more acidic than Lake 1.
(e) Lakes in Sweden become more acidic in the spring. Suggest an explanation.
(f) Acidic lakes in Sweden are treated with crushed limestone. Explain how this reduces the acidity. Give two disadvantages of this solution to the problem.

11 Many industrial plants take water from a river and then return it at a higher temperature. What harm can this do? What name is given to this practice?

12 The diagram shows how rocks are broken down and new rocks formed in the rock cycle.

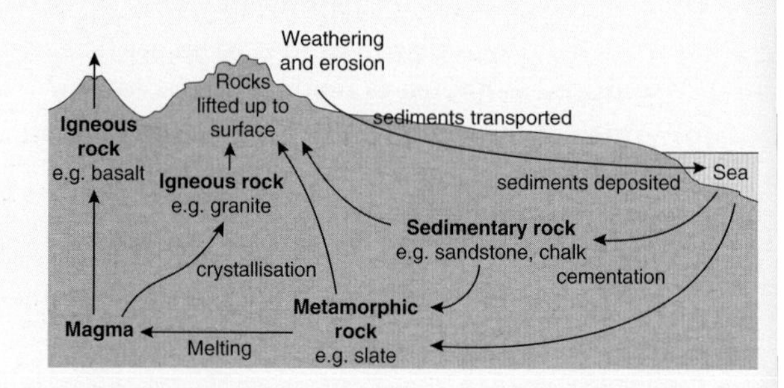

(a) These are four processes involved in forming *sedimentary* rocks. They are in the wrong order. *A* depositing sediments; *B* cementation; *C* transporting sediments; *D* weathering and erosion. List them in the correct order. [3]
(b) Write down *two* processes taking place when *metamorphic* rocks turn into *igneous* rocks. [2]
(c) What conditions of temperature and pressure are needed to turn *sedimentary* rocks into *metamorphic* rocks? [2]
(d) This table shows some information about slate, chalk and granite.

Rock	*Is the rock crystalline?*	*Can the rock contain fossils?*
Slate	no	yes
Chalk		
Granite		

Finish a copy of the table by putting 'yes' or 'no' in each of the four spaces. The diagram may help you. [2]
(e) Igneous rocks can be described as *extrusive* or *intrusive*.
Basalt is an extrusive rock. Granite is an intrusive rock.
Describe how extrusive rocks are formed from magma. [2] (OCR)

13 Analysis of a water supply produced the following data.

Ion		*Concentration in mg per dm³*
Calcium	(Ca^{2+})	104.0
Magnesium	(Mg^{2+})	1.4
Sodium	(Na^{+})	8.0
Potassium	(K^{+})	1.0
Iron	(Fe^{3+})	0.02
Hydrogencarbonate	(HCO_3^-)	293.0
Chloride	(Cl^-)	15.0
Sulphate	(SO_4^{2-})	12.0
Nitrate	(NO_3^-)	5.0
Fluoride	(F^-)	0.1

(a) (i) Which *two* elements in this water supply are in Group 2 of the Periodic Table? [1]

(ii) Write down the name of a transition metal which is present in this water supply. [1]

(iii) Write down the names of *two* ions in this water supply which can combine together to form a compound of formula XY_2, where X is a metal. [2]

(b) Sana tests for ions found in this water supply. She concentrates the ions by evaporating some of the water before doing the tests.
Copy and finish the table. There are *four* gaps.

Test	*What is seen*	*Ion present*
(i) Add dilute hydrochloric acid. Test the gas given given off with limewater.	It fizzes, and the gas given off turns the limewater _____.	HCO_3^- [1]
(ii) Add dilute nitric acid and silver nitrate solution.	A white precipitate is formed.	_____ [1]
(iii) Add dilute nitric acid and barium chloride solution.	A _____ precipitate is . formed.	SO_4^{2-} [1]
(iv) Add sodium hydroxide solution.	A brown precipitate is formed.	_____[2]

(c) A hospital process has to let 30 dm³ of water flow through pipes in 24 hours.
If more than 120 mg per hour of calcium flows through the pipes, the process becomes damaged.
Decide whether this water supply is suitable for use in the process.
You *must* show how you work out your answer. [2]

(d) *All* the hardness in this water has to be removed before the water is used in a steam-iron.

Which *two* of these methods could be used successfully?
Boil the water, cool it and then filter it; distil the water; trickle the water down a zeolite ion exchange column; treat the water with calcium hydroxide, then filter it. [2]

(e) Glyn always makes tea using this water supply.
He boils the water in an electric kettle.
Describe and explain the problem caused by the frequent use of this water supply in the kettle. [3] (OCR)

14 The diagram shows three calcium compounds.

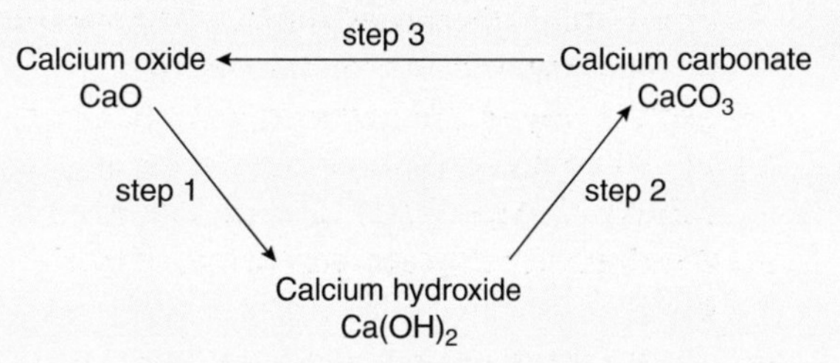

(a) (i) Which compound occurs widely in nature?

(ii) Which compound is present in solution in limewater?

(iii) What state symbol is used to show limewater is a solution?

(iv) Which compound is formed on boiling temporary hard water?

(v) What is added to calcium oxide to carry out step 1?

(vi) How can step 3 be carried out?

(vii) Which step occurs during the test for carbon dioxide? [7]

(b) Two different white powders are thought to be calcium carbonate and calcium hydroxide.

(i) Describe a test to prove that both powders are calcium compounds. [2]

(ii) Describe a test to find out which one of the powders is calcium carbonate. [2] (Edexcel)

15 Rainwater falling on limestone rocks can form caves.

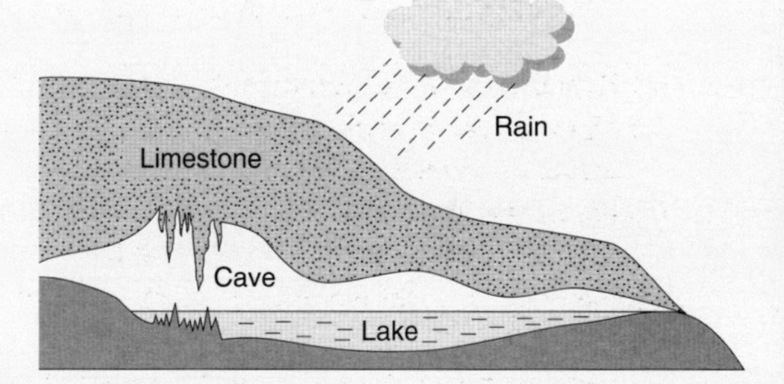

(a) Complete the sentences by choosing the correct words from the box.

acidic	alkaline	dissolves	
hard	reacts	soft	tastes

You may use each word once or not at all.
Rainwater is an ______ solution which ______ with limestone. The solution formed in the lake is known as ______ water. One advantage of drinking the water from the lake is that it ______ better than rainwater. [4]

(b) Samples of water were tested by shaking with soap solution. The results are shown in the table.

Water sample (50 cm³)	*Volume of a soap solution to form a lather in cm³*
Lake	15
Boiled lake	3
Rain	1

(i) What is seen when only 10 cm^3 of soap solution is shaken with 50 cm^3 of water from the lake? [1]

(ii) Why did the rainwater need only 1 cm^3 of soap solution to form a lather? [1]

(iii) Why did the water from the lake need 15 cm^3 of soap solution to form a lather? [1]

(iv) Explain why boiled water from the lake needed only 3 cm^3 of soap solution to form a lather. [2] AQA(SEG) 1999

16 (a) The sketch below was made on a field trip in Wales.

A, *B* and *C* are three different types of rock.

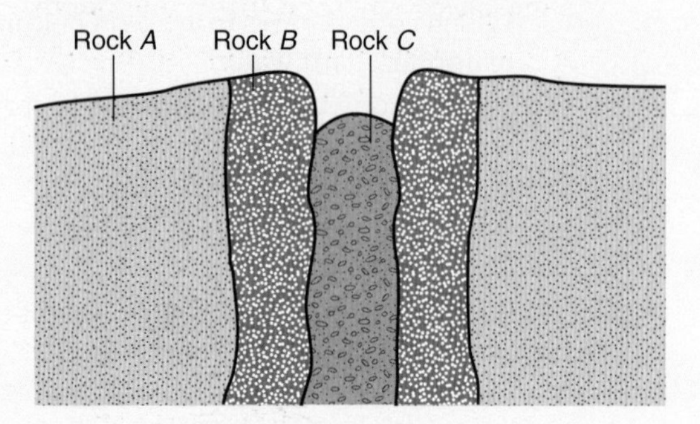

The diagrams below show what the three rock types look like under a microscope.

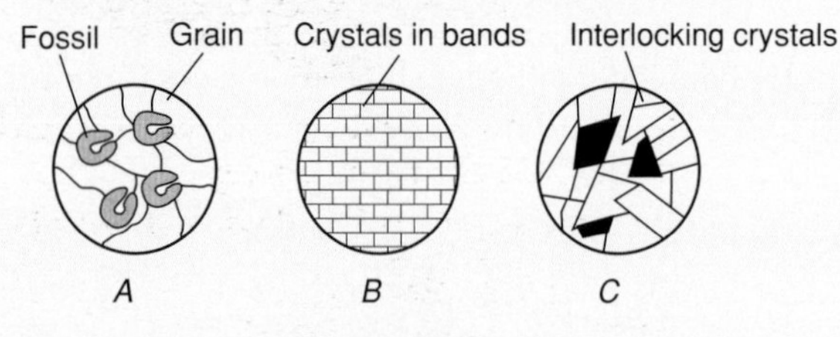

(i) Using the words in the box below complete a copy of the table that follows:

granite igneous limestone marble metamorphic sedimentary

Rock	*Type of rock*	*Name of rock*
A		
B		
C		

[3]

(ii) Place the rocks, *A*, *B* and *C* in order of their age. [1]

(b) Describe how the following rock types are formed.

(i) Sedimentary. [2]

(ii) Metamorphic. [2] (WJEC)

17 (a) The air is a mixture of many gases. Some of these gases are shown in the table.

Name	*Chemical formula*
Nitrogen	N_2
Oxygen	O_2
Argon	Ar
Carbon dioxide	CO_2

(i) Which of these gases are: (i) elements (ii) compounds [2]

(ii) Give *one* important use of nitrogen. [1]

(iii) State the correct percentage of carbon dioxide in the air.

0.03% 0.9% 78% [1]

(iv) The amount of carbon dioxide in the air varies from place to place.

The amount of carbon dioxide in the countryside is often lower than in towns and cities. Explain why. [2]

(b) An experiment to find how much oxygen is in the air is shown.

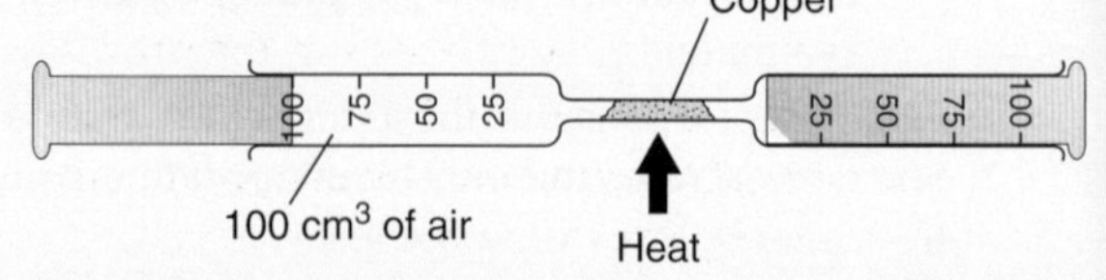

The 100 cm^3 of air was pushed from one syringe to the other over hot copper. This was repeated many times.

(i) Complete the word equation to show how oxygen was removed from the air. [1]

copper + oxygen →

(ii) The surface of the copper changed colour when it was heated in air.

The colour before heating was ______. [2]

The colour after heating was ______.

(iii) The experiment showed that there was 21% oxygen in the air.

What volume of the air was left in the syringe at the end of the experiment? [1] AQA(SEG) 1999

18 A sample of tap water contains the following dissolved salts:

calcium hydrogencarbonate, calcium sulphate, magnesium sulphate, potassium chloride, sodium chloride

(a) (i) Hard water can be formed when water is in contact with the rock, gypsum. Which salt in the list is present in gypsum? [1]

(ii) Name TWO other salts in the list which make water hard. [2]

(b) Some methods of treating water are given below. *A* adding chlorine; *B* adding a fluoride (fluoridation); *C* adding sodium carbonate; *D* boiling the water.

Which ONE of the methods (*A*, *B*, *C* or *D*) removes:

(i) temporary hardness but NOT permanent hardness; [1]

(ii) both temporary hardness *and* permanent hardness? [1]

(c) Ten drops of soap solution are shaken with a sample of hard tap water. The mixture turns cloudy but it does not form a lather. On shaking with ten more drops of soap solution, a lather is formed.

(i) What causes the cloudiness when soap solution is first mixed with this tap water? [1]

(ii) Predict what you would SEE when distilled water is shaken with ten drops of soap solution. Explain your answer. [2]

(d) Calcium hydrogencarbonate is formed when water and carbon dioxide are in contact with limestone. This reaction removes carbon dioxide gas from the atmosphere.

(i) Write the word equation for this reaction. [1]

(ii) Plants are also able to remove carbon dioxide from the atmosphere, forming glucose and oxygen gas in the process.
Give the name of this process and state ONE essential condition needed for it to take place. [2]

(e) If the percentage of carbon dioxide in the Earth's atmosphere increases, the average temperature of the atmosphere may also increase.
What is the name given to this effect?
[1] (Edexcel)

19 (a) The main components of the **original** Earth's atmosphere were carbon dioxide and water. The pie chart below shows the approximate composition of dry air in the atmosphere **today**.

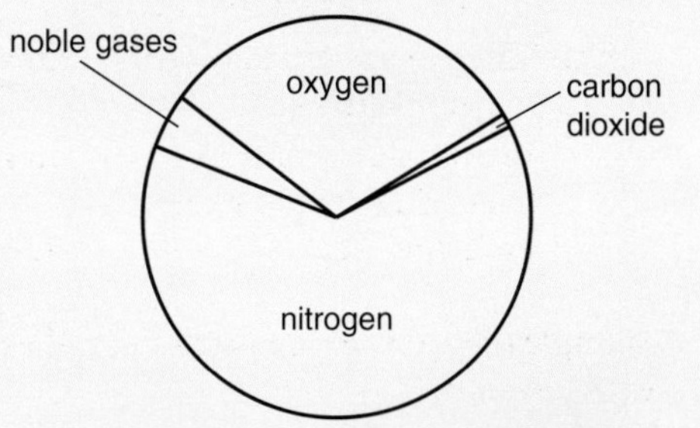

(i) State the source of the original Earth's atmosphere. [1]

(ii) Name the gas in the pie chart which is entirely biological in origin. [1]

(iii) There has been a drastic reduction in the amount of water vapour in the air over geological time. Explain how this decrease occurred. [2]

(b) The Earth's atmosphere is surrounded by a layer of ozone. State the importance of the ozone layer to our health. [1]

(c) Give *one* use to which oxygen is put. [1]

(d) The Carbon Cycle helps to maintain atmospheric composition. Name *two* processes which have the opposite effect to photosynthesis in the Carbon Cycle. [2] (WJEC)

20 The Earth formed as a mass of molten rock.

(a) The Earth cooled forming a layered structure. Label the mantle, crust and core on a copy of the diagram.

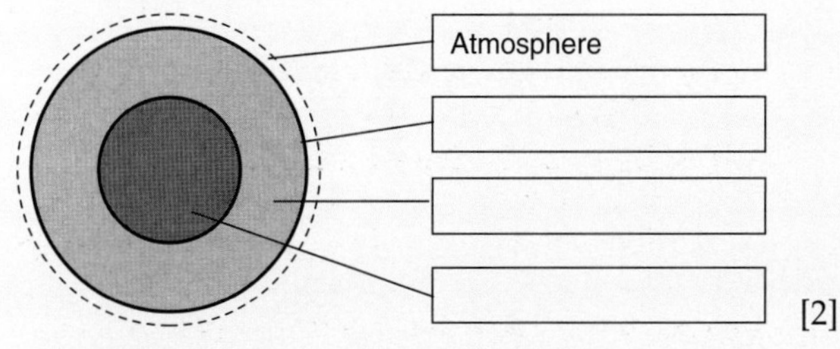

[2]

(b) As the Earth cooled volcanic activity on the surface formed the first atmosphere. This atmosphere contained these compounds: ammonia (NH_3); carbon dioxide (CO_2); methane (CH_4); water vapour (H_2O).
Which *two* elements in these compounds make up over 98% of our atmosphere? [2]

(c) There are still many active volcanoes.

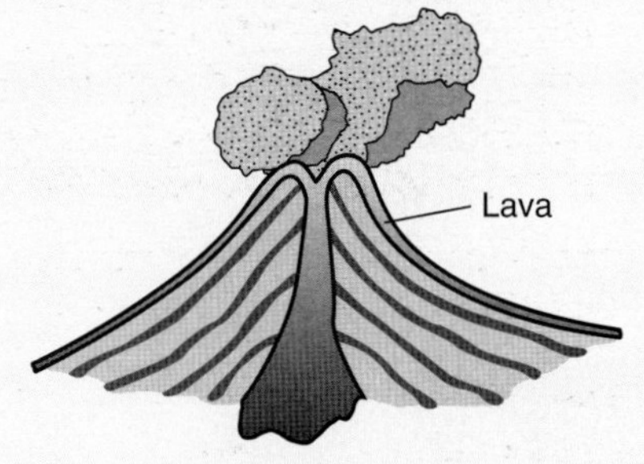

In the following sentences, choose in each box the *two* words that are wrong.
Volcanic activity is found where some of the

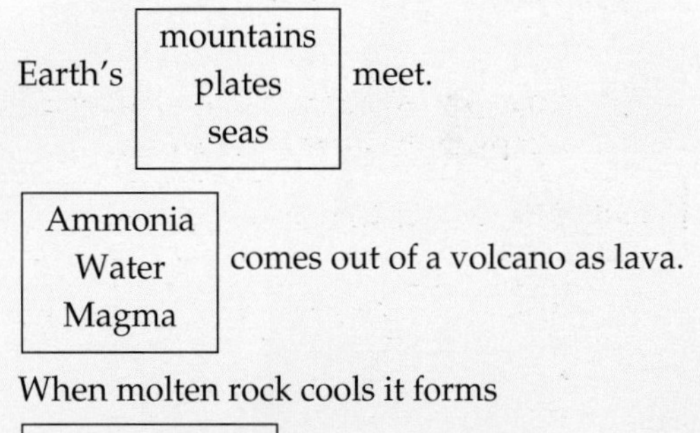

Earth's [mountains / plates / seas] meet.

[Ammonia / Water / Magma] comes out of a volcano as lava.

When molten rock cools it forms

[igneous / metamorphic / sedimentary] rock.

[3]

(d) The rocks, limestone and marble, are made up of calcium carbonate.

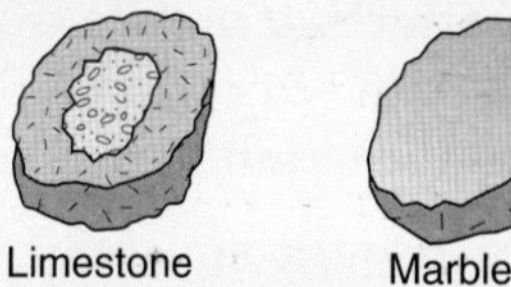

(i) Limestone is a sedimentary rock. How are sedimentary rocks formed? [1]

(ii) Marble can be made from limestone. Explain why the marble contains no fossils.
[2] AQA (SEG) 2000

Theme E

Using
th's
rces

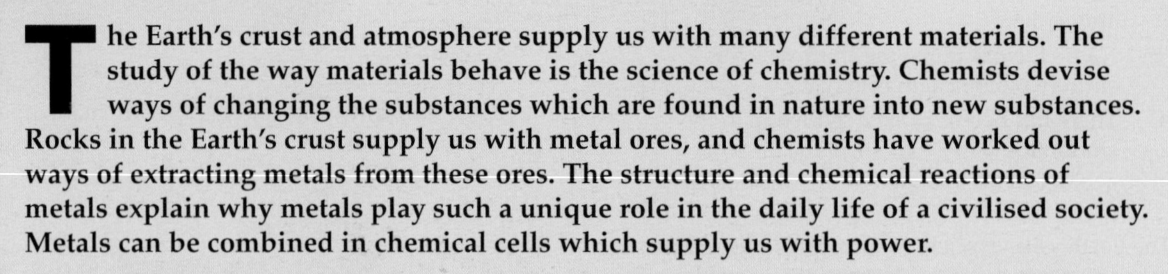

The Earth's crust and atmosphere supply us with many different materials. The study of the way materials behave is the science of chemistry. Chemists devise ways of changing the substances which are found in nature into new substances. Rocks in the Earth's crust supply us with metal ores, and chemists have worked out ways of extracting metals from these ores. The structure and chemical reactions of metals explain why metals play such a unique role in the daily life of a civilised society. Metals can be combined in chemical cells which supply us with power.

Rocks are also the source of the essential building materials limestone, concrete and glass. Rock salt is the starting point in the manufacture of many important chemicals.

Nitrogen from the air and natural gas are the starting materials in the manufacture of fertilisers.

Topic 19 Metals

19.1 Metals and alloys

FIRST THOUGHTS

Metals and alloys have played an important part in history. The discovery of bronze made it possible for the human race to advance out of the Stone Age into the Bronze Age. Centuries later, smiths found out how to extract iron from rocks, and the Iron Age was born. In the nineteenth century, the invention of steel made the Industrial Revolution possible.

www.keyscience.co.uk

Metals are strong materials. They are used for purposes where strength is required. Metals can be worked into complicated shapes. They can be ground to take a cutting edge. They conduct heat and electricity. As science and technology advance, metals are put to work for more and more purposes.

Alloys

An alloy is a mixture of metallic elements and in some cases non-metallic elements also. Many metallic elements are not strong enough to be used for the manufacture of machines and vehicles which will have to withstand stress. Alloying a metallic element with another element is a way of increasing its strength. Steel is an alloy of iron with carbon and often other metallic elements. Duralumin is an alloy of aluminium with copper and magnesium. This alloy is much stronger than pure aluminium and is used for aircraft manufacture (Figure 19.1A).

Alloys have different properties from the elements of which they are composed. Brass is made from copper and zinc. It has a lower melting point than either of these metals. This makes it easier to cast, that is, to melt and pour into moulds. Brass musical instruments have a more pleasant and sonorous sound than instruments made from either of the two elements copper or zinc.

SUMMARY

An alloy is a mixture of metals, with, in some cases, non-metallic elements also. Alloys have different properties from the elements of which they are composed.

The differences between the physical properties of metals and non-metallic elements are summarised in Table 4.1.

Figure 19.1A ▲ Concorde

19.2 The metallic bond

FIRST THOUGHTS

Metals possess their remarkable and useful properties because of the type of chemical bond between the atoms: the metallic bond.

A piece of metal consists of positive metal ions and free electrons (see Figure 19.2A). The free electrons are the outermost electrons which break free when the metal atoms form ions. Free electrons move about between the metal ions. This is what prevents the metal ions from being driven apart by repulsion between their positive charges.

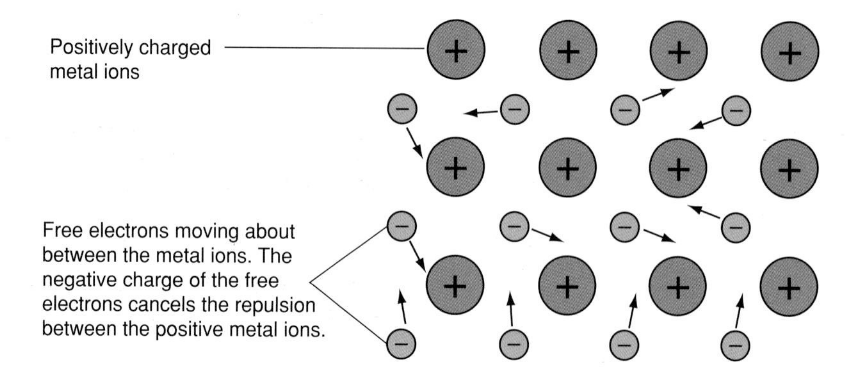

Figure 19.2A The metallic bond

The metallic bond explains how metals **conduct electricity**. It also explains how metals can change their shape without breaking. When a metal is bent, the shape changes but the free electrons continue to hold the metal ions together.

SUMMARY

The metallic bond is a strong bond. It gives metals their strength and allows them to conduct electricity.

19.3 The chemical reactions of metals

Reactions of metals with air

FIRST THOUGHTS

You have already met the reaction between metals and acids in Topic 9.1 and the reaction between metals and air in Topic 14.4. In this section, you will find out how to classify metals on the basis of the vigour of their chemical reactions.

Many metals react with the oxygen in the air (see Table 19.1).

Table 19.1 The reactions between metals and oxygen

Metal	Symbol	Reaction when heated in air	Reaction with cold air
Potassium	K	Burn in air to form oxides	React slowly with air to form a surface film of the metal oxide. This reaction is called **tarnishing**
Sodium	Na		
Calcium	Ca		
Magnesium	Mg		
Aluminium	Al		
Zinc	Zn		
Iron	Fe		
Tin	Sn	When heated in air, these metals form oxides without burning	
Lead	Pb		
Copper	Cu		
Silver	Ag	Do not react	Silver tarnishes in air
Gold	Au		Do not react
Platinum	Pt		

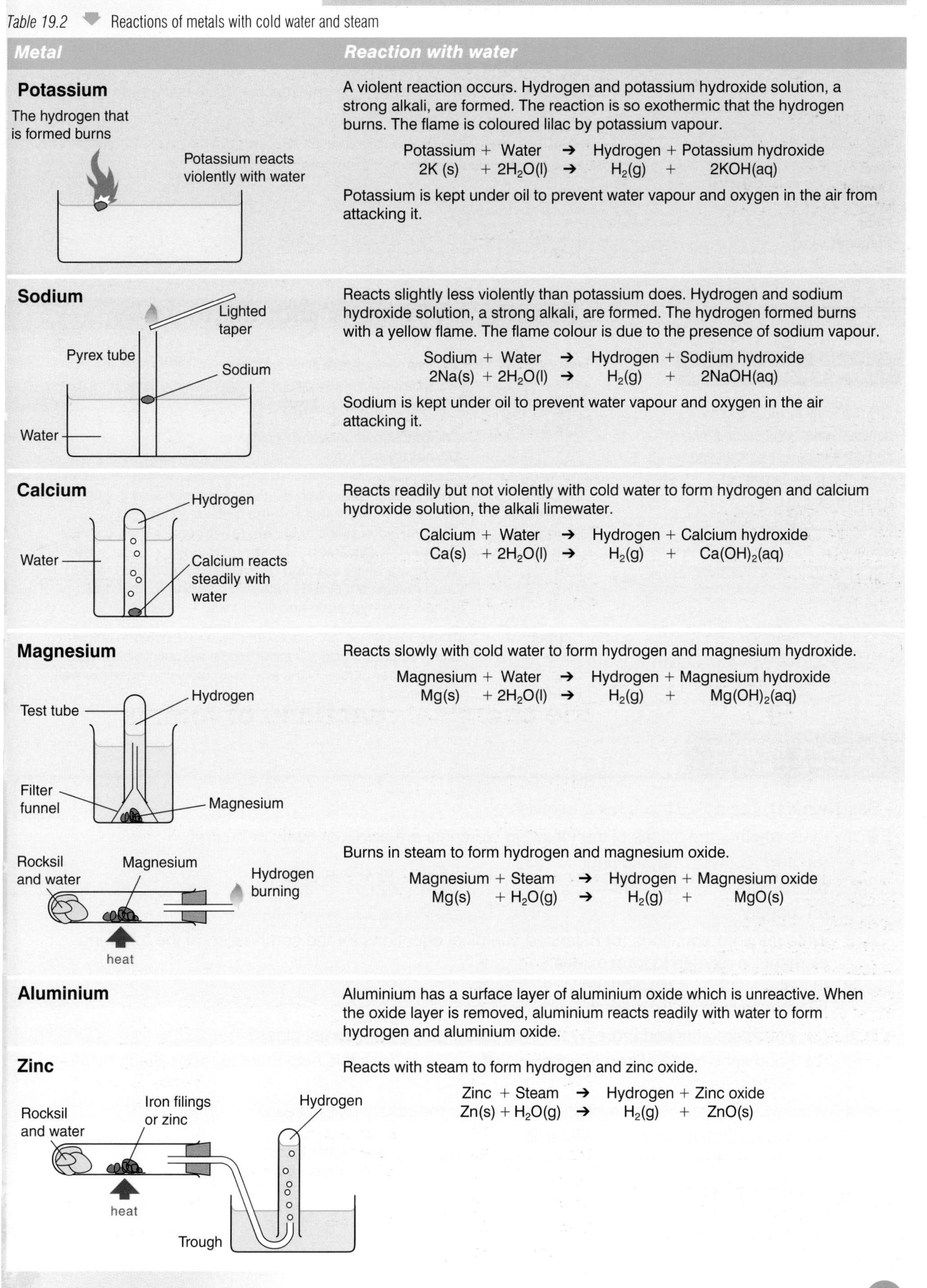

Reactions of metals with water

Table 19.2 Reactions of metals with cold water and steam

Metal	Reaction with water
Potassium	A violent reaction occurs. Hydrogen and potassium hydroxide solution, a strong alkali, are formed. The reaction is so exothermic that the hydrogen burns. The flame is coloured lilac by potassium vapour. Potassium + Water → Hydrogen + Potassium hydroxide $2K(s) + 2H_2O(l) \rightarrow H_2(g) + 2KOH(aq)$ Potassium is kept under oil to prevent water vapour and oxygen in the air from attacking it.
Sodium	Reacts slightly less violently than potassium does. Hydrogen and sodium hydroxide solution, a strong alkali, are formed. The hydrogen formed burns with a yellow flame. The flame colour is due to the presence of sodium vapour. Sodium + Water → Hydrogen + Sodium hydroxide $2Na(s) + 2H_2O(l) \rightarrow H_2(g) + 2NaOH(aq)$ Sodium is kept under oil to prevent water vapour and oxygen in the air attacking it.
Calcium	Reacts readily but not violently with cold water to form hydrogen and calcium hydroxide solution, the alkali limewater. Calcium + Water → Hydrogen + Calcium hydroxide $Ca(s) + 2H_2O(l) \rightarrow H_2(g) + Ca(OH)_2(aq)$
Magnesium	Reacts slowly with cold water to form hydrogen and magnesium hydroxide. Magnesium + Water → Hydrogen + Magnesium hydroxide $Mg(s) + 2H_2O(l) \rightarrow H_2(g) + Mg(OH)_2(aq)$ Burns in steam to form hydrogen and magnesium oxide. Magnesium + Steam → Hydrogen + Magnesium oxide $Mg(s) + H_2O(g) \rightarrow H_2(g) + MgO(s)$
Aluminium	Aluminium has a surface layer of aluminium oxide which is unreactive. When the oxide layer is removed, aluminium reacts readily with water to form hydrogen and aluminium oxide.
Zinc	Reacts with steam to form hydrogen and zinc oxide. Zinc + Steam → Hydrogen + Zinc oxide $Zn(s) + H_2O(g) \rightarrow H_2(g) + ZnO(s)$

Table 19.2 Reactions of metals with cold water and steam (continued)

Metal	Reaction with water
Iron	Reacts with steam to form hydrogen and the oxide, Fe_3O_4, tri-iron tetraoxide, which is blue-black in colour. Iron + Steam → Hydrogen + Tri-iron tetraoxide $3Fe(s) + 4H_2O(g) \rightarrow 4H_2(g) + Fe_3O_4(s)$
Tin **Lead** **Copper** **Silver** **Gold** **Platinum**	Do not react

Reactions of metals with dilute acids

Safety matters

The reactions in Table 19.2 between water and potassium or sodium and between steam and magnesium are carried out behind a safety screen. Observers must wear safety glasses.

Table 19.3 The reactions of metals with dilute acids

Metal	Reaction with dilute acid
Potassium Sodium Lithium	The reaction is dangerously violent **Do not try it**
Calcium Magnesium Aluminium Zinc Iron Tin Lead	These metals react with dilute hydrochloric acid to give hydrogen and a solution of the metal chloride, e.g. Zinc + Hydrochloric acid → Hydrogen + Zinc chloride $Zn(s) + 2HCl(aq) \rightarrow H_2(g) + ZnCl_2(aq)$ The vigour of the reaction decreases from calcium to lead. Lead reacts very slowly. With dilute sulphuric acid, the metals give hydrogen and sulphates.
Copper Silver Gold Platinum	These metals do not react with dilute hydrochloric acid and dilute sulphuric acid. (Copper reacts with dilute nitric acid to form copper nitrate. Nitric acid is an oxidising agent as well as an acid.)

CHECKPOINT

See Theme D, Topic 14.5 if you need to revise.

1 State whether the oxides of the following elements are acidic or basic or neutral.
(a) iron (b) carbon (c) copper (d) sulphur (e) zinc

2 Write (a) word equations (b) balanced chemical equations for the combustion of the following elements in oxygen to form oxides.
(i) zinc (ii) magnesium (iii) sodium (iv) carbon

3 (a) Which are attacked more by acid rain: lead gutters or iron fall pipes?
(b) Food cans made of iron are coated with tin. How does this help them to resist attack by the acids in foods?

4 Write word equations and symbol equations for the reactions between:
(a) magnesium and hydrochloric acid,
(b) iron and hydrochloric acid to form iron(II) chloride,
(c) zinc and sulphuric acid,
(d) iron and sulphuric acid to form iron(II) sulphate.

19.4 The reactivity series

FIRST THOUGHTS

There are over seventy metallic elements. One way of classifying them is to arrange them in a sort of league table, with the most reactive metals at the top of the league and the least reactive metals at the bottom. This section shows how it is done.

There are over 70 metals in the Earth's crust. Table 19.4 summarises the reactions of metals with oxygen, water and acids. You will see that the same metals are the most reactive in the different reactions.

Table 19.4 Reactions of metals

Metal	Reaction when heated in oxygen	Reaction with cold water	Reaction with dilute hydrochloric acid
Potassium	Burn to form oxides	Displace hydrogen; form alkaline hydroxides	React dangerously fast
Sodium			
Lithium			
Calcium			Displace hydrogen; form metal chlorides
Magnesium		Slow reaction	
Aluminium		No reaction, except for slow rusting of iron; all react with steam	
Zinc			
Iron			
Tin	Oxides form slowly without burning	Do not react, even with steam	React very slowly
Lead			
Copper			Do not react
Silver	Do not react		
Gold			
Platinum			

The metals can be listed in order of **reactivity**, in order of their readiness to take part in chemical reactions. This list is called the **reactivity series** of the metals. Table 19.5 shows part of the reactivity series.

Table 19.5 Part of the reactivity series

Metal	Symbol
Potassium	K
Sodium	Na
Calcium	Ca
Magnesium	Mg
Aluminium	Al
Zinc	Zn
Iron	Fe
Tin	Sn
Lead	Pb
Copper	Cu
Silver	Ag
Gold	Au
Platinum	Pt

Reactivity decreases from top to bottom

SUMMARY

- Many metals react with oxygen; some metals burn in oxygen. The oxides of metals are bases.
- A few metals (for example sodium) react with cold water to give a metal hydroxide and hydrogen. Some metals react with steam to give a metal oxide and hydrogen.
- Many metals react with dilute acids to give hydrogen and the salt of the metal.

Metals fall into the same order of reactivity in all these reactions. This order is called the reactivity series of the metals.

If you have used aluminium saucepans, you may be surprised to see aluminium placed with the reactive metals high in the reactivity series. In fact, aluminium is so reactive that, as soon as aluminium is exposed to the air, the surface immediately reacts with oxygen to form aluminium oxide. The surface layer of aluminium oxide is unreactive and prevents the metal from showing its true reactivity.

CHECKPOINT

1. In some parts of the world, copper is found 'native'. Why is zinc never found native?
2. The ancient Egyptians put gold and silver objects into tombs. Explain why people opening the tombs thousands of years later find the objects still in good condition. Why are no iron objects found in the tombs?
3. The following metals are listed in order of reactivity:

   sodium > magnesium > zinc > copper

   Describe the reactions of these metals with (a) water, (b) dilute hydrochloric acid. Point out how the reactions illustrate the change in reactivity.

19.5 Predictions from the reactivity series

FIRST THOUGHTS

The classifications you have been learning about are useful: they enable you to make predictions about chemical reactions.

Competition between metals for oxygen

Aluminium is higher in the reactivity series than iron. When aluminium is heated with iron(III) oxide, a very vigorous, exothermic reaction occurs

$$\text{Aluminium} + \text{Iron(III) oxide} \rightarrow \text{Iron} + \text{Aluminium oxide}$$
$$2Al(s) + Fe_2O_3(s) \rightarrow 2Fe(s) + Al_2O_3(s)$$

This reaction is called the **thermit reaction** (therm = heat). It is used to mend railway lines because the iron formed is molten and can weld the broken lines together (see Figure 19.5A).

Where in the Periodic Table do metals occur?

Revise Topic 11.2 if you are in doubt!

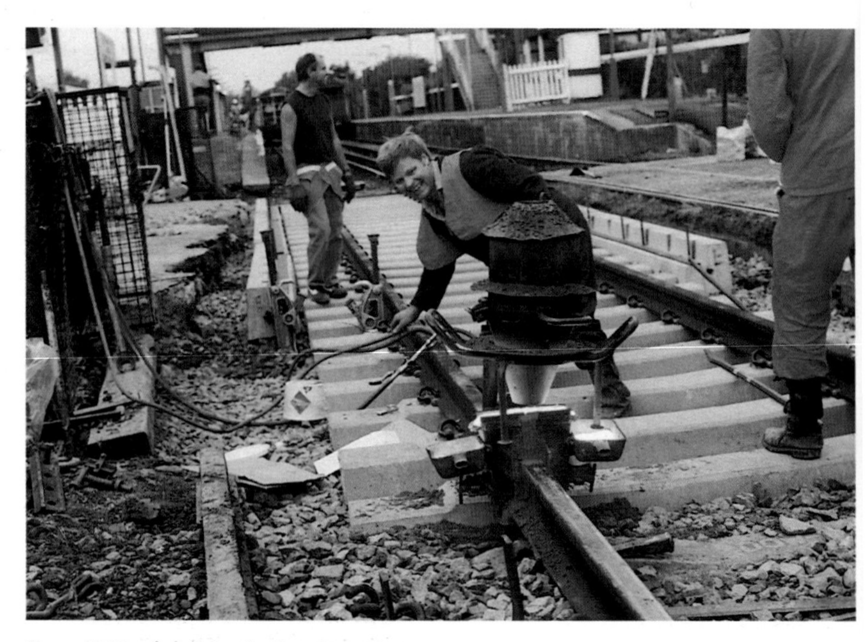

Figure 19.5A ▲ Using the thermit reaction

SUMMARY

Reactive metals displace metals lower down the reactivity series from their compounds.

Displacement reactions

Metals can displace other metals from their salts. A metal which is higher in the reactivity series will displace a metal which is lower in the reactivity series from a salt. Zinc will displace lead from a solution of lead nitrate.

Extraction of metals

The method which could be used to extract a metal from its ore can be predicted from the position of the metal in the reactivity series (see Topic 19.8).

CHECKPOINT

1 Which metal is best for making saucepans: zinc, iron or copper? Explain your choice.

2 Gold is used for making electrical contacts in space capsules. Explain (a) why gold is a good choice and (b) why it is not more widely used.

3 A metal, romin, is displaced from a solution of one of its salts by a metal, sarin, A metal, tonin, displaces sarin from a solution of one of its salts. Place the metals in order of reactivity.

4 The following metals are listed in order of reactivity, with the most reactive first:
Mg, Zn, Fe, Pb, Cu, Hg, Au
List the metals which will:
(a) occur 'native',
(b) react with cold water,
(c) react with steam,
(d) react with dilute acids,
(e) displace copper from copper(II) nitrate solution.

19.6 Uses of metals and alloys

FIRST THOUGHTS

As you read through this section, think about the reasons which lead to the choice of a metal or alloy for a particular purpose.

Metals and alloys have thousands of uses. The purposes for which a metal is used are determined by its physical and chemical properties. Sometimes a manufacturer wants a material for a particular purpose and there is no metal or alloy which fits the bill. Then metallurgists have to invent a new alloy with the right characteristics. Table 19.6 gives some examples.

Table 19.6 What are metals and alloys used for?

Metal/Alloy	Characteristics	Uses
Aluminium (Duralumin is an important alloy)	Low density Never corroded Good electrical conductor Good thermal conductor Reflector of light	Aircraft manufacture (Duralumin) Food wrapping Electrical cable Saucepans Car headlamps
Brass (alloy of copper and zinc)	Not corroded Easy to work with Sonorous Yellow colour	Ships' propellers Taps, screws Trumpets Ornaments
Bronze (alloy of copper and tin)	Harder than copper Not corroded Sonorous	Coins, medals Statues, springs Church bells

It's a fact!

One cm^3 of gold can be hammered into enough thin gold leaf to cover a football pitch. Gold foil is used for decorating books, china, etc.

Table 19.6 What are metals and alloys used for? (continued)

Metal/Alloy	Characteristics	Uses
Copper	Good electrical conductor Not corroded readily	Electrical circuits Water pipes and tanks
Gold	Beautiful colour Never tarnishes Easily worked	Jewellery Electrical contacts Filling teeth
Iron	Hard, strong, inexpensive, rusts	Motor vehicles, trains, ships, buildings
Lead	Dense Unreactive	Protection from radioactivity Was used for all plumbing (Lead is no longer used for water pipes as it reacts very slowly with water)
Magnesium	Bright flame	Distress flares, flash bulbs
Mercury	Liquid at room temperature	Thermometers Electrical contacts Dental amalgam for filling teeth
Silver	Good electrical conductor Good reflector of light Beautiful colour and shine (tarnishes in city air)	Electrical contacts Mirrors Jewellery
Sodium	High thermal capacity	Coolant in nuclear reactors
Solder (alloy of tin and lead)	Low melting point	Joining metals, e.g. in an electrical circuit
Steel (alloy of iron)	Strong	Construction, tools, ball bearings, magnets, cutlery, etc.
Tin	Low in reactivity series	Coating 'tin cans'
Titanium	Low in density Stays strong at high and low temperatures	Supersonic aircraft
Zinc	High in reactivity series	Protection of iron and steel; see Table 19.11.

It's a fact!

At the launch of the 'Platinum 1990' exhibition, the Johnson Matthey precious metals group presented a £300,000 wedding dress made of platinum. They had made it by lining super-thin platinum foil with paper, shredding the platinum-paper combination into strands and weaving the strands into a fabric. The Japanese designer who made the dress sent instructions to the exhibition, *Ironing the dress is strictly forbidden.*

SUMMARY

Metals and alloys are essential for many different purposes. Metals and alloys are chosen for particular uses because of their physical properties and their chemical reactions.

CHECKPOINT

1. Name the metal which is used for each of these purposes. Explain why that metal is chosen.
   (a) Thermometers
   (b) Window frames
   (c) Sinks and draining boards
   (d) Radiators
   (e) Water pipes
   (f) Household electrical wiring
   (g) Scissor blades
2. Explain the following:
   (a) Some mirrors have aluminium sprayed on to the back of the glass instead of silver, which was used previously.
   (b) Although brass is a colourful, shiny material, it is not used for jewellery.
   (c) Titanium oxide has replaced lead carbonate as the pigment in white paint.

19.7 Compounds and the reactivity series

The stability of metal oxides

- Reactive metals form compounds readily.
- The compounds of reactive metals are difficult to split up.

Magnesium is a reactive metal, and magnesium oxide is difficult to reduce to magnesium.

Copper is an unreactive metal, and copper(II) oxide is easily reduced to copper. Hydrogen will reduce hot copper(II) oxide to copper.

$$\text{Copper(II) oxide} + \text{Hydrogen} \xrightarrow{\text{Heat}} \text{Copper} + \text{Water}$$
$$CuO(s) + H_2(g) \rightarrow Cu(s) + H_2O(l)$$

Carbon is another reducing agent. When heated, it will reduce the oxides of metals which are fairly low in the reactivity series. Reduction by carbon is often the method employed to obtain metals from their ores.

$$\text{Zinc oxide} + \text{Carbon} \xrightarrow{\text{Heat}} \text{Zinc} + \text{Carbon monoxide}$$
$$ZnO(s) + C(s) \rightarrow Zn(s) + CO(g)$$

Carbon and hydrogen are reducing agents. They can reduce many metal oxides.

Neither hydrogen nor carbon will reduce the oxides of metals which are high in the reactivity series. Aluminium is high in the reactivity series; its oxide is difficult to reduce. The method used to obtain aluminium from aluminium oxide is electrolysis.

If a metal is very low in the reactivity series, its oxide will decompose when heated. The oxides of silver and mercury decompose when heated.

SUMMARY

The oxides of metals low in the reactivity series (e.g. Cu) are easily reduced by carbon and hydrogen. The oxides of metals high in the reactivity series are difficult to reduce.

Compounds of metals low in the reactivity series are decomposed by heat. Compounds of metals high in the reactivity series are stable to heat.

The stability of other compounds

When compounds are described as stable it means that they are difficult to decompose (split up) by heat. Table 19.7 shows how the stability of the compounds of a metal is related to the position of the metal in the reactivity series.

The sulphates, carbonates and hydroxides of the most reactive metals are not decomposed by heat. Those of other metals decompose to give oxides.

Table 19.7 Action of heat on compounds

<table>
<tr><th rowspan="2">Cation</th><th colspan="5">Anion</th></tr>
<tr><th>Oxide</th><th>Chloride</th><th>Sulphate</th><th>Carbonate</th><th>Hydroxide</th></tr>
<tr><td>Potassium</td><td rowspan="9">No decomposition</td><td rowspan="11">No decomposition</td><td colspan="3" rowspan="2">No decomposition</td></tr>
<tr><td>Sodium</td></tr>
<tr><td>Calcium</td><td rowspan="7">Oxide and sulphur trioxide $MO + SO_3$ Some also give SO_2</td><td rowspan="7">Oxide and carbon dioxide $MO + CO_2$</td><td rowspan="7">Oxide and water $MO + H_2O$</td></tr>
<tr><td>Magnesium</td></tr>
<tr><td>Aluminium</td></tr>
<tr><td>Zinc</td></tr>
<tr><td>Iron</td></tr>
<tr><td>Lead</td></tr>
<tr><td>Copper</td></tr>
<tr><td>Silver</td><td>Metal + oxygen</td><td rowspan="2">Metal + $O_2 + SO_3$</td><td rowspan="2">Metal + $O_2 + CO_2$</td><td rowspan="2">Do not form hydroxides</td></tr>
<tr><td>Gold</td><td>Not formed</td></tr>
</table>

19.8 Extraction of metals from their ores

FIRST THOUGHTS

One of the important jobs which chemists do is to find ways of extracting metals from the rocks of the Earth's crust.

This section tells you
- what is meant by a metal occurring 'native',
- when a rock is called an ore,
- how chemists decide on the method to use for extracting a metal from its ore.

Metals are found in rocks in the Earth's crust. A few metals, such as gold and copper, occur as the free metal, uncombined. They are said to occur 'native'. Only metals which are very unreactive can withstand the action of air and water for thousands of years without being converted into compounds. Most metals occur as compounds.

Rock containing the metal compound is mined. Then machines are used to crush and grind the rock. Next a chemical method must be found for extracting the metal. All these stages cost money. If the rock contains enough of the metal compound to make it profitable to extract the metal, the rock is called an **ore**.

Figure 19.8A 'Native' copper

The method used to extract a metal from its ore depends on the position of the metal in the reactivity series (see Table 19.8).

Table 19.8 Methods used for the extraction of metals

Metal	Method
Potassium Sodium Calcium Magnesium	The anhydrous chloride is melted and electrolysed
Aluminium	The anhydrous oxide is melted and electrolysed
Zinc Iron Lead Copper	Found as sulphides and oxides. The sulphides are roasted to give oxides; the oxides are reduced with carbon
Silver Gold	Found 'native' (as the free metal)

Sodium

Sodium is obtained by the electrolysis of molten dry sodium chloride. The same method is used for potassium, calcium and magnesium. This is an expensive method of obtaining metals because of the cost of the electricity consumed.

Aluminium – the 'newcomer' among metals

Aluminium is mined as **bauxite**, an ore which contains aluminium oxide, $Al_2O_3.2H_2O$. This ore is very plentiful, yet aluminium was not extracted from it until 1825. Another 60 years passed before a commercial method of extracting the metal was invented. *What was the problem?*

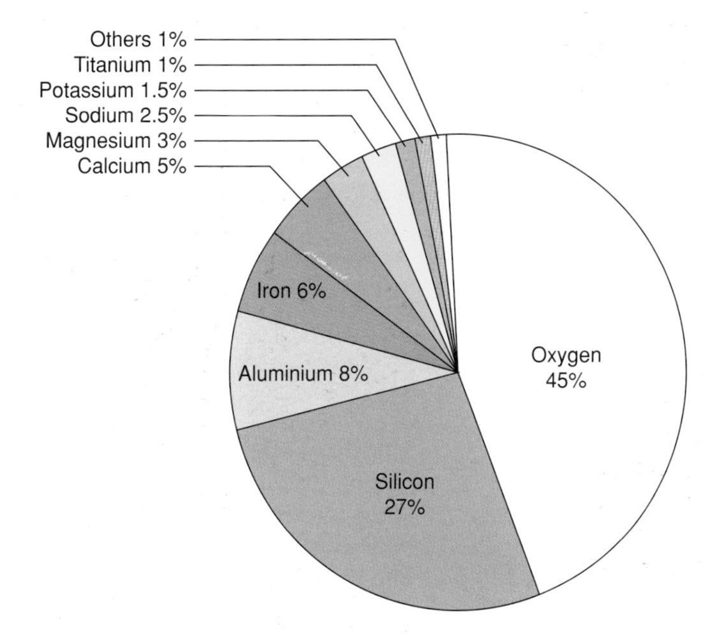

Figure 19.8B ▲ Elements in the Earth's crust

KEY SCIENTISTS

The story of how aluminium came into use tells of contributions from a number of scientists of different ages and different nationalities over a period of 60 years. They were: Hans Christian Oersted, the Danish academic; Henri Sainte-Claire Déville, the French industrial chemist; Charles Martin Hall, an American student; Paul Héroult, a French student.

A Danish scientist called Hans Christian Oersted succeeded in obtaining aluminium from aluminium oxide in 1825. First he made aluminium chloride from aluminium oxide. Then he used potassium amalgam (an alloy of potassium and mercury) to displace aluminium from aluminium chloride.

The German chemist Friedrich Wöhler altered the method somewhat. He used potassium (instead of potassium amalgam).

A French chemist, Henri Sainte-Claire Déville, tackled the problem of scaling up the reaction. He used sodium in the displacement reaction. In 1860 he succeeded in making the reaction yield aluminium on a large scale. The price of aluminium tumbled. Instead of being an expensive curiosity, it became a useful commodity. With its exceptional properties, aluminium soon found new uses, and the demand for the new metal increased.

Aluminium was expensive because of the cost of the sodium used in its extraction. Chemists were keen to find a less costly method of extracting the new metal. Many thought it should be possible to obtain aluminium by electrolysing molten aluminium oxide. The difficulty was the high melting point, 2050 °C, which made it impossible to keep the compound molten while a current was passed through it.

The big breakthrough came in 1886. Two young chemists working thousands of miles apart made the discovery at the same time. An American called Charles Martin Hall, aged 21, and a Frenchman called Paul Héroult, aged 23, discovered that they could obtain a solution of aluminium oxide by dissolving it in molten cryolite, Na_3AlF_6, at 700 °C. On passing electricity through the melt, the two men succeeded in obtaining aluminium. Their method is still used. The Hall-Héroult cell is shown in Figure 19.8C. Anhydrous pure aluminium oxide must be obtained from the ore and then electrolysed in the cell.

Science at work

A firm in Northampton had the idea of buying empty aluminium drink cans for recycling. They installed machines which pay for aluminium cans fed in but not for steel cans. *How can the machines tell the difference between aluminium cans and steel cans?*

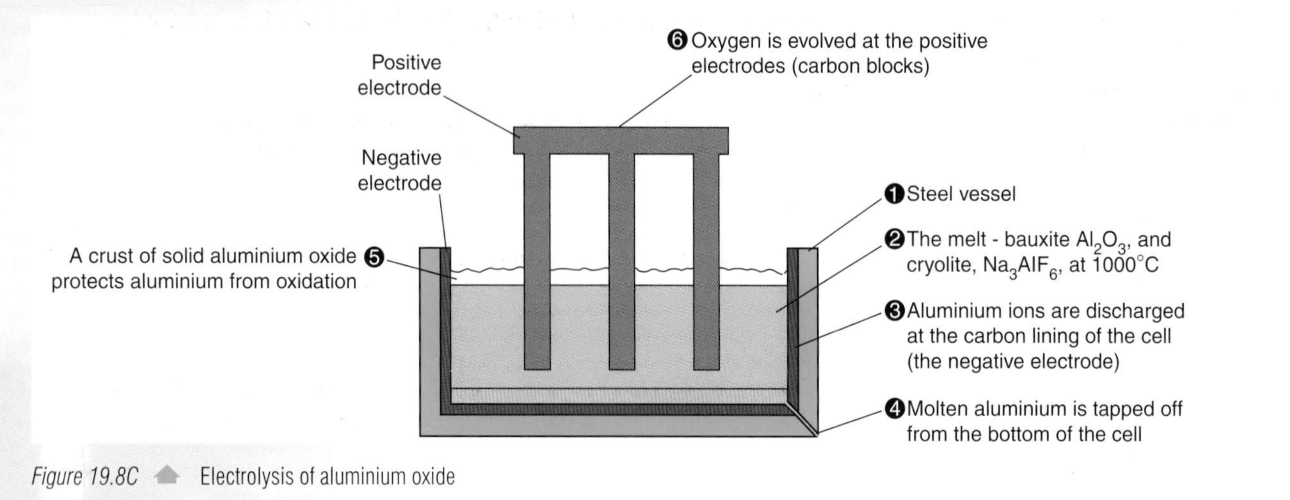

Figure 19.8C Electrolysis of aluminium oxide

Iron

Many countries have plentiful resources of iron ores, **haematite**, Fe_2O_3, **magnetite**, Fe_3O_4 and **iron pyrites**, FeS_2. The sulphide ore is roasted in air to convert it into an oxide. The oxide ores are reduced to iron in a **blast furnace** (Figure 19.8D). Iron ore and coke and limestone are fed into the furnace. Iron ores and limestone are plentiful resources. Coke is made by heating coal.

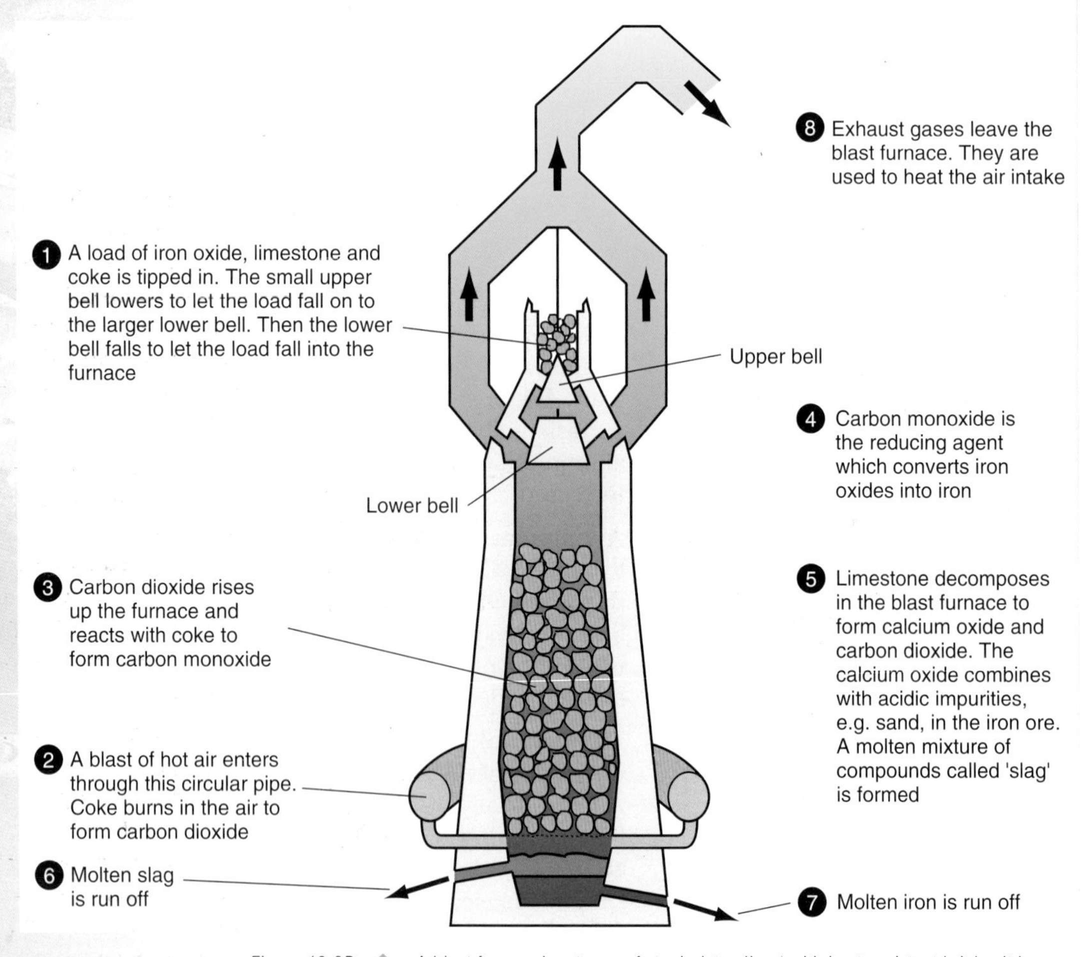

Figure 19.8D A blast furnace is a tower of steel plates lined with heat-resistant bricks. It is about 50 m high

Science at work

A few years ago, explorers discovered that there was a rich source of minerals on the sea bed. Lumps or 'nodules' of metal ore were found. They consist chiefly of manganese ores but contain other metals such as zinc and nickel also. Methods of scooping the nodules off the sea bed are being developed. There are political problems as well as technical problems. The industrialised countries can afford to develop the technology for recovering the nodules. Sometimes nodules are discovered on the floor of the ocean within the territorial limits of a poor country, which cannot afford to develop the technology. A fair arrangement must be worked out to split the profits from recovering the nodules between the country which owns the nodules and the country which provides the technology.

The chemical reactions which take place in the blast furnace are:

Step 2

Carbon (coke) + Oxygen → Carbon dioxide

$$C(s) + O_2(g) \rightarrow CO_2(g)$$

Step 3

Carbon dioxide + Carbon (coke) → Carbon monoxide

$$CO_2(g) + C(s) \rightarrow 2CO(g)$$

Step 4

Iron(III) oxide + Carbon monoxide → Iron + Carbon dioxide

$$Fe_2O_3(s) + 3CO(g) \rightarrow 2Fe(s) + 3CO_2(g)$$

Step 5

(a) Calcium carbonate (limestone) → Calcium oxide + Carbon dioxide

$$CaCO_3(s) \rightarrow CaO(s) + CO_2(g)$$

(b) Calcium oxide + Silicon(IV) oxide (sand) → Calcium silicate(slag)

$$CaO(s) + SiO_2(s) \rightarrow CaSiO_3(l)$$

The blast furnace runs continuously. The raw materials are fed in at the top, and molten iron and molten slag are run off separately at the bottom. The slag is used in foundations by builders and road-makers. The process is much cheaper to run than an electrolytic method. With the raw materials readily available and the cost of extraction low, iron is cheaper than other metals.

Copper

Copper is low in the reactivity series. It is found 'native' (uncombined) in some parts of the world. More often, it is mined as the sulphide. This is roasted in air to give impure copper. Pure copper is obtained from this by the **electrolytic** method shown in Figure 19.8E.

After the cell has been running for a week, the negative electrode becomes very thick. It is lifted out of the cell, and replaced by a new thin sheet of copper. When all the copper has dissolved out of the positive electrode, a new piece of impure copper is substituted.

Other metals are present as impurities in copper ores. Iron and zinc are more reactive than copper, and their ions therefore stay in solution while copper ions are discharged. Silver and gold are less reactive than copper. They do not dissolve and are therefore present in the anode sludge. They can be extracted from this residue.

1. The electrolyte is copper(II) sulphate solution
2. The negative electrode is a strip of pure copper. Copper ions are discharged and copper atoms are deposited on the electrode. The strip of pure copper becomes thicker $Cu^{2+}(aq) + 2e^- \rightarrow Cu(s)$
3. The positive electrode is a lump of impure copper. Copper atoms supply electrons to this electrode and become copper ions which enter the solution. The lump of impure copper becomes smaller $Cu(s) \rightarrow Cu^{2+}(aq) + 2e^-$
4. Anode sludge, this is the undissolved remains of the lump of impure copper

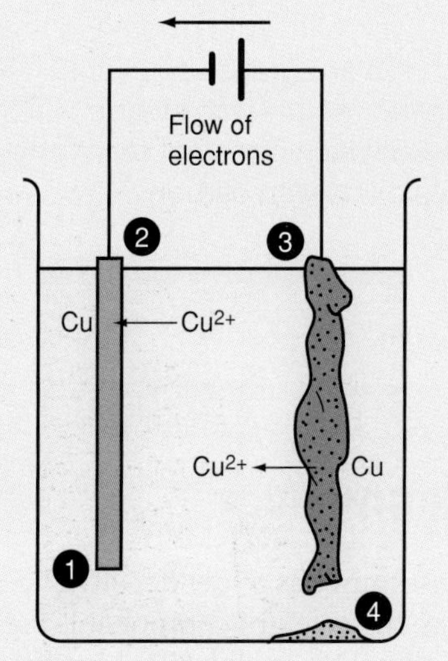

Figure 19.8E Purification of copper

SUMMARY

The method used for extracting a metal from its ores depends on the position of the metal in the reactivity series. Very reactive metals, such as sodium, are obtained by electrolysis of a molten compound. Less reactive metals, such as iron, are obtained by reducing the oxide with elements such as carbon or hydrogen. The metals at the bottom of the reactivity series occur 'native'.

Silver and gold

Silver and gold are found 'native'. A new deposit of gold was discovered in the Sperrin mountains of Northern Ireland in 1982.

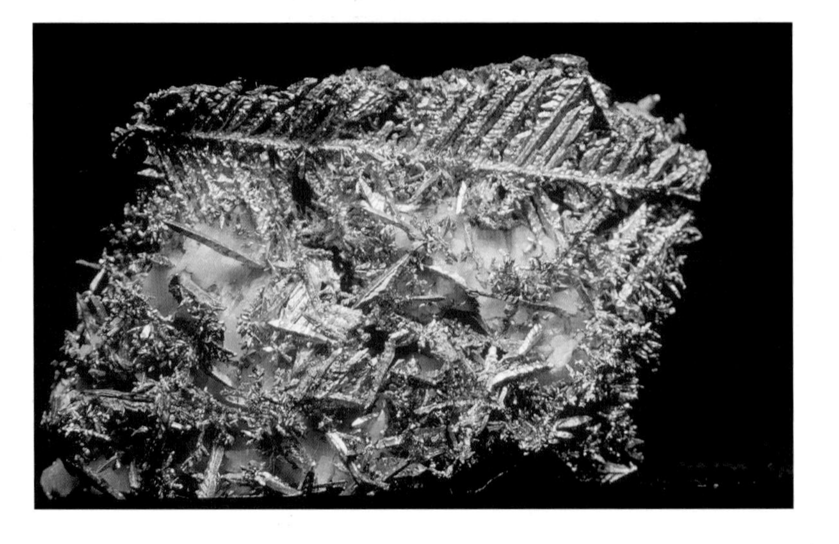

Figure 19.8F ➧ Natural silver crystals

CHECKPOINT

1 The Emperor Napoleon invested in the French research on new methods of obtaining aluminium. He was interested in the possibility of aluminium suits of armour for his soldiers. What advantage would aluminium armour have had over iron or steel?

2 (a) Arrange in order of the reactivity series the metals copper, iron, zinc, gold and silver.
(b) Arrange the same metals in order of their readiness to form ions, that is, the ease with which the reaction $M(s) \rightarrow M^{n+}(aq) + ne^-$ takes place.
(c) Arrange the same metals in order of the ease of discharging their ions, that is, the ease with which the reaction $M^{n+}(aq) + ne^- \rightarrow M(s)$ takes place.
(d) Refer to Figure 19.8E showing the electrolytic purification of copper. Use your answers to (b) and (c) to explain why copper is deposited on the negative electrode but iron, zinc, gold and silver are not.

3 Explain why (a) the human race started using copper, silver and gold long before iron was known and (b) iron tools were a big improvement on tools made from other metals.

4 Give the names of metals which fit the descriptions A, B, C, D and E.
A Reacts immediately with air to form a layer of oxide and then reacts no further.
B Reacts violently with water to form an alkaline solution.
C Reacts slowly with cold water and rapidly with steam.
D Is a reddish-gold coloured metal which does not react with dilute hydrochloric acid.
E Can be obtained by heating its oxide with carbon.

It's a fact!

All the magnesium that has been produced so far could have been extracted from 4 cubic kilometres of sea water. The sea is an almost inexhaustible reserve of magnesium.

Magnesium

Magnesium has a low density (1.7 g/cm^3) and its alloys are widely used where lightweight components are required, e.g. the aircraft industry. The alloy Dowmetal, 89% Mg, 9% Al, 2% Zn, has a tensile strength which is close to steel. Magnesium is present to the extent of 0.14% in sea water. Figure 19.8G illustrates the extraction of magnesium by the Dow process.

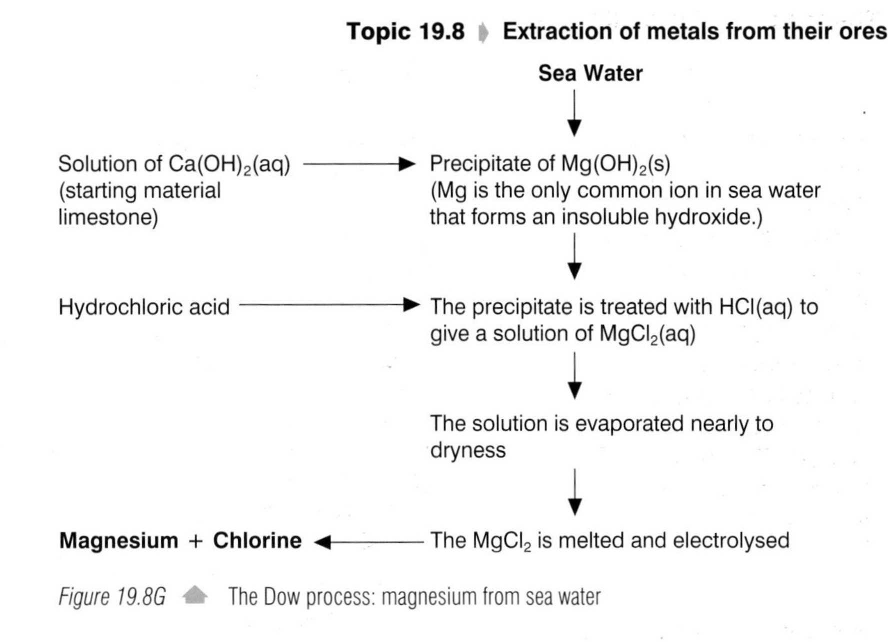

Figure 19.8G The Dow process: magnesium from sea water

Titanium

Titanium is sometimes described as the **aerospace metal**. It has a unique set of properties: low density, high strength, high resistance to corrosion and high melting point. The low density and the ability to retain its strength at high temperatures make titanium and its alloys valuable in the manufacture of spacecraft and aircraft designed to fly at high speeds.

Many transition metals are obtained by reducing the metal oxides. Carbon and carbon monoxide are often used as reducing agents. In the blast furnace for the reduction of iron oxides, carbon monoxide is the reducing agent.

Titanium is mined as rutile, titanium(IV) oxide, TiO_2. Titanium(IV) oxide is more difficult to reduce than iron oxides, and carbon monoxide cannot be used. The Kroll process is used. Titanium(IV) chloride, $TiCl_4$, is reduced with magnesium. Magnesium, being higher than titanium in the reactivity series, can displace titanium from its compound.

The steps in the extraction are:

1 Rutile is mixed with coke and ground. It is heated in a furnace, through which chlorine passes.

titanium(IV) oxide + chlorine + carbon → titanium(IV) choride + carbon monoxide

$$TiO_2(s) + 2Cl_2(g) + 2C(s) \rightarrow TiCl_4(g) + 2CO(g)$$

2 The titanium(IV) chloride formed is purified by fractional distillation.

3 The Kroll process. Gaseous titanium(IV) chloride is passed into a reactor containing molten magnesium. Argon is passed through the reactor. Reduction takes place.

titanium(IV) chloride + magnesium → titanium + magnesium chloride

$$TiCl_4(g) + 2Mg(l) \rightarrow Ti(s) + 2MgCl_2(s)$$

4 A spongy deposit of titanium forms on the walls of the reactor. A batch process is employed because the reactor must be allowed to cool so that the deposit of titanium can be scraped off. The metal is distilled and then melted to form ingots.

5 Magnesium chloride is run off from the reactor and electrolysed to give magnesium and chlorine for recycling.

SUMMARY

Titanium and its alloys are used in spacecraft and high-speed aircraft. Titanium is mined as the oxide TiO_2. This is converted into the chloride $TiCl_4$. Magnesium displaces titanium from its chloride.

19.9 Iron and steel

FIRST THOUGHTS

The machines that manufacture our possessions, our means of transport and the frameworks of our buildings: all these depend on the strength of steel.

The iron that comes out of the blast furnace is called **cast iron** (or pig iron). It contains three to four per cent carbon. The carbon content makes it brittle, and cast iron cannot be bent without snapping. The impurity makes the melting point lower than that of pure iron so that cast iron is easier to cast – to melt and mould – than pure iron. Cast iron expands slightly as it solidifies. This helps it to flow into all the corners of a mould and reproduce the shape exactly. By casting, objects with complicated shapes can be made, for example the cylinder block of a car engine (which contains the cylinders in which the combustion of petrol vapour in air takes place).

Iron which contains less than 0.25% carbon is called wrought iron. Wrought iron is strong and easily worked (Figure 19.9A). Nowadays, mild steel has replaced wrought iron.

Figure 19.9A Wrought iron

Steel

Steel is made by reducing the carbon content of cast iron, which makes it brittle, to less than one per cent. Carbon is burnt off as its oxides, the gases carbon monoxide, CO, and carbon dioxide, CO_2. Iron is less easily oxidised. The sulphur, phosphorus and silicon in the iron are also converted into acidic oxides. These are not gases, and a base, such as calcium oxide, must be added to remove them. The base and the acidic oxides combine to form a slag (a mixture of compounds of low melting point).

Figure 19.9B shows a basic oxygen furnace. In it, cast iron is converted into steel. One converter can produce 150–300 tonnes of steel in an hour.

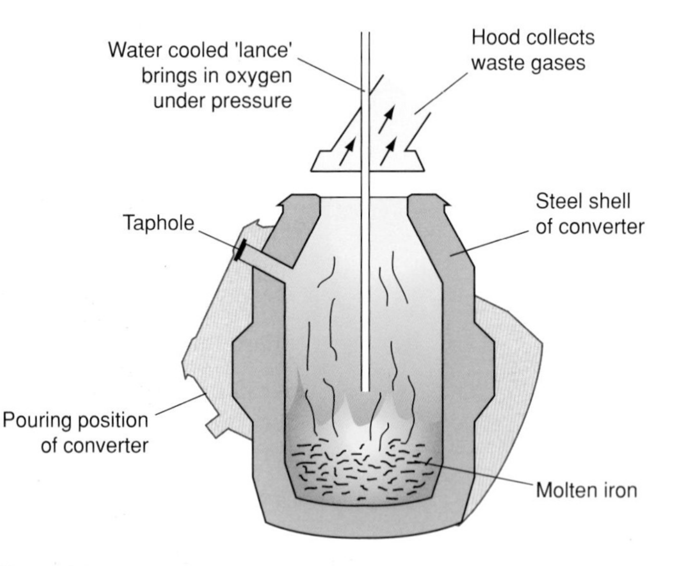

Figure 19.9B A basic oxygen furnace holding 150–300 tonnes of steel

There are various types of steel. They differ in their carbon content and are used for different purposes. Low-carbon steel (mild steel) is pliable; high-carbon steel is hard.

SUMMARY

Cast iron (from the blast furnace) is easy to mould, but is brittle. Wrought iron (the purest form of iron) is easily worked without breaking. Alloy steels contain other elements in addition to iron and carbon. Different steels are suited to different uses.

Alloy steels

Many elements are alloyed with iron and carbon to give alloy steels with different properties. They all have different uses, for example nickel and chromium give stainless steel, manganese and molybdenum increase strength, and vanadium increases springiness.

CHECKPOINT

1 What is the difference between cast iron and wrought iron in (a) composition (b) strength and (c) ease of moulding? Name two objects made from cast iron and two objects made from wrought iron.

The uses of alloy steels

Some steels are alloys of iron and carbon. They have more uses than cast iron (see Table 19.9).

Table 19.9 Cast iron and steel and their uses

Type of steel	Description	Uses
Mild steel (<0.25%) carbon)	Pliable (can be bent without breaking)	Chains and pylons
Medium steel (0.25–0.45% carbon)	Tougher than mild steel, more springy than high carbon steel	Nuts and bolts Car springs and axles Bridges
High carbon steel (0.45–1.5% carbon)	The carbon content makes it both hard and brittle	Chisels, files, razor blades, saws, cutting tools
Cast iron (2.5–4.5% carbon)	Cheaper than steel, easily moulded into complicated shapes	Fire grates, gear boxes, brake discs, engine blocks. These articles will break if they are hammered or dropped

SUMMARY

Steels contain carbon. Low-carbon steel is pliable; high-carbon steel is hard. Cast iron contains up to 4% carbon.

Many elements are used for alloying with steel. The different alloy steels have different properties which fit them for different uses (see Table 19.10).

Table 19.10 Some alloy steels

Element	Properties of alloy	Uses
Chromium	Prevents rusting if >10%	Stainless steels, acid-resisting steels, cutlery, car accessories, tools
Cobalt	Takes a sharp cutting edge Can be strongly magnetised	High-speed cutting tools Permanent magnets
Manganese	Increases strength and toughness	Some is used in all steels; steel in railway points and safes contains a high percentage of manganese
Molybdenum	Strong even at high temperatures	Rifle barrels, propeller shafts
Nickel	Resists heat and acids	Stainless steel cutlery, industrial plants which must withstand acidic conditions
Tungsten	Stays hard and tough at high temperatures	High-speed cutting tools
Vanadium	Increases springiness	Springs, machinery

SUMMARY

Alloy steels contain added metals. The different elements give a range of steels with different characteristics suited to different uses.

CHECKPOINT

1. Think what is the most important characteristic of the steel that must be used for making the articles listed below. Should the steel be pliable or springy or hard? Then say whether you would use mild steel or medium steel or high-carbon steel or cast iron for making the following articles. (a) car axles (b) axes (c) car springs (d) ornamental gates (e) drain pipes (f) chisels (g) sewing needles (h) picks (i) saws (j) food cans.
2. Suggest which type of alloy steel could be chosen for: (a) an electric saw (b) a car radiator (c) the suspension of a car (d) a gun barrel (e) a set of steak knives.

19.10 Rusting of iron and steel

(a) Paint protects the car

(b) Chromium plating protects the bicycle handlebars

(c) Galvanised steel girders

(d) Zinc bars protect the ship's hull

Figure 19.10A Rust prevention

Iron and most kinds of steel rust. Rust is hydrated iron(III) oxide $Fe_2O_3.nH_2O$ (n, the number of water molecules in the formula, varies).

Some of the methods which are used to protect iron and steel against rusting are listed in Table 19.11. Some methods use a coating of some substance which excludes water and air. Other methods work by sacrificing a metal which is more reactive than iron.

Table 19.11 Rust prevention

Method	Where it is used	Comment
A coat of paint	Ships, bridges, cars, other large objects (see Figure 19.10A(a))	If the paint is scratched, the exposed iron starts to rust. Corrosion can spread to the iron underneath the paintwork which is still sound.
A film of oil or grease	Moving parts of machinery, e.g. car engines	The film of oil or grease must be renewed frequently.
A coat of plastic	Kitchenware, e.g. draining rack	If the plastic is torn, the iron starts to rust.
Chromium plating	Kettles, cycle handlebars (see Figure 19.10A(b))	The layer of chromium protects the iron beneath it and also gives a decorative finish. It is applied by electroplating.
Galvanising (zinc plating)	Galvanised steel girders are used in the construction of buildings and bridges (see Figure 19.10A(c))	Zinc is above iron in the reactivity series: zinc will corrode in preference to iron. Even if the layer of zinc is scratched, as long as some zinc remains, the iron underneath does not rust. Zinc cannot be used for food cans because zinc and its compounds are poisonous.
Tin plating	Food cans	Tin is below iron in the reactivity series. If the layer of tin is scratched, the iron beneath it starts to rust.
Stainless steel	Cutlery, car accessories, e.g. radiator grille	Steel containing chromium (10–25%) or nickel (10–20%) does not rust.
Sacrificial protection	Ships (see Figure 19.10A(d))	Blocks of zinc are attached to the hulls of ships below the waterline. Being above iron in the reactivity series, zinc corrodes in preference to iron. The zinc blocks are sacrificed to protect the iron. As long as there is some zinc left, it protects the hull from rusting. The zinc blocks must be replaced.

The Earth contains huge deposits of iron ores. We need iron for our machinery and for our means of transport. During the twentieth century, we have used more metal than in all the previous centuries put together. If we keep on mining iron ores at the present rate, the Earth's resources may one day be exhausted. We allow tonnes of iron and steel to rust every year. We throw tonnes of used iron and steel objects on the scrap heap. The Earth's iron deposits will last longer if we take the trouble to collect scrap iron and steel and recycle it, that is, melt it down and reuse it.

Figure 19.10B ▲ Scrap iron dump

SUMMARY

Exposure to air and water in slightly acidic conditions makes iron and steel rust. Some of the treatments for protecting iron and steel from rusting are:

- oil or grease,
- a protective coat of another metal, e.g. chromium, zinc, tin,
- alloying with nickel and chromium,
- attaching a more reactive metal, e.g. magnesium or zinc, to be sacrificed.

It's a fact!

The rusting of iron is an expensive nuisance. Replacing rusted iron and steel structures costs the UK £500 million a year.

CHECKPOINT

1. Say how the rusting of iron is prevented:
   (a) in a bicycle chain,
   (b) in a food can,
   (c) in parts of a ship above the water line,
   (d) in parts of a ship below the water line,
   (e) in a galvanised iron roof.
2. 'The Industrial Revolution would not have been possible without steel.' Say whether or not you agree with this statement. Give your reasons.
3. The map opposite shows possible sites, A, B, C and D, for a steelworks. Say which site you think is the best, and explain your choice. You will have to consider the need for:
   - iron ore, coke (from coal), limestone,
   - a work-force,
   - transporting iron and steel to customers,
   - removing slag.

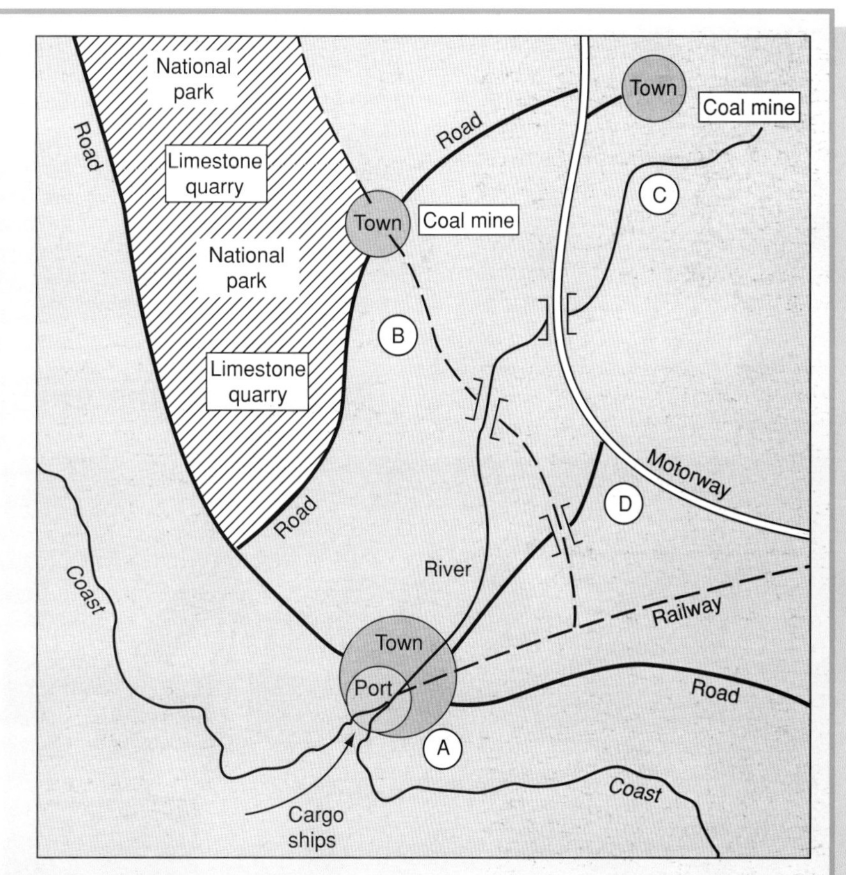

19.11 Aluminium

FIRST THOUGHTS

Aluminium is the most plentiful metal in the Earth's crust, yet aluminium was not manufactured until the nineteenth century. The twentieth century has seen aluminium finding more and more vitally important uses.

Uses of aluminium

Aluminium is a metal with thousands of applications. Some of these are listed in Table 19.12. Pure aluminium is not a very strong metal. Its alloys, such as duralumin (which contains copper and magnesium), are used when strength is needed.

For anodising aluminium see Figure 7.7E.

Note how the different uses of aluminium and its alloys depend on their various properties.

Table 19.12 Some uses of aluminium and its alloys

Property	Uses for aluminium which depend on this property
Never corroded (except by bases)	Door frames and window frames are often made of aluminium. 'Anodised aluminium' is used. The thickness of the protective layer of aluminium oxide has been increased by anodising (making it the positive electrode in an electrolytic cell).
Low density	Packaging food: milk bottle tops, food containers, baking foil. The low density and resistance to corrosion make aluminium ideal for aircraft manufacture. Alloys such as duralumin are used because they are stronger than aluminium.
Good electrical conductor	Used for overhead cables. The advantage over copper is that aluminium cables are lighter and need less massive pylons to carry them.
Reflects light when polished	Car headlamp reflectors
Good thermal conductor	Saucepans, etc.
Reflects heat when highly polished	Highly polished aluminium reflects heat and can be used as a thermal blanket. Aluminium blankets are used to wrap premature babies. They keep the baby warm by reflecting heat back to the body. Firefighters wear aluminium fabric suits to reflect heat away from their bodies.

The cost to the environment

Bauxite is found near the surface in Australia, Jamaica, Brazil and other countries. The ore is obtained by open cast mining (Figure 19.11A). A layer of earth 1 m to 60 m thick is excavated, and the landscape is devastated. In some places, mining companies have spent money on restoring the landscape after they have finished working a deposit.

Figure 19.11A An open cast bauxite mine

Pure aluminium oxide must be extracted from bauxite before it can be electrolysed. Iron(III) oxide, which is red, is one of the impurities in bauxite. The waste produced in the extraction process is an alkaline liquid containing a suspension of iron(III) oxide. It is pumped into vast red mud ponds.

Jamaica has a land area of 11 000 km^2. Every year, 12 km^2 of red mud are created. The Jamaicans are worried about the loss of land. Even when it dries out, red mud is not firm enough to build on. They also worry about the danger of alkali seeping into the water supply. The waste cannot be pumped into the sea because it would harm the fish.

Purified aluminium oxide is shipped to an aluminium plant. The electrolytic method of extracting aluminium is expensive to run because of the electricity it consumes. Aluminium plants are often built in areas which have hydroelectric power (electricity from water-driven generators). This is relatively cheap electricity. The waterfalls and fast-flowing rivers which provide hydroelectric power are found in areas of natural beauty. Local residents often object to the siting of aluminium plants in such beauty spots.

We want to use aluminium – thousands of tonnes a year – and we also want to retain the beauty of the countryside. Sometimes there is a conflict.

There may be other difficulties over hydroelectric power. Purified aluminium oxide has to be transported to a remote area and aluminium has to be transported away. If there are not enough local workers, a workforce may have to be brought into the area and provided with housing. Often it pays to build an aluminium plant in an area which has a big population and good transport, even if the cost of electricity is higher.

CHECKPOINT

1. The percentages of some metals recycled in the UK are shown below.
   (a) Explain what 'recycled' means. (b) Plot the figures as a bar chart or as a pie chart.
   (c) Why is lead easy to recycle? (d) How can iron be separated from other metals for recycling?
   (e) What resources are saved by recycling aluminium?

Metal	*Aluminium*	*Zinc*	*Iron*	*Tin*	*Lead*	*Copper*
Percentage recycled	28	30	50	30	56	19

SUMMARY

Aluminium has brought us many benefits. There is, however, a cost to the environment. The open cast mining of bauxite spoils the landscape. The purification creates unsightly red mud ponds. The extraction can cause pollution through fluoride emission.

Factors which decide the siting of aluminium plants include
- the cost of electricity,
- the cost of transporting the raw material and the product,
- the availability of a workforce.

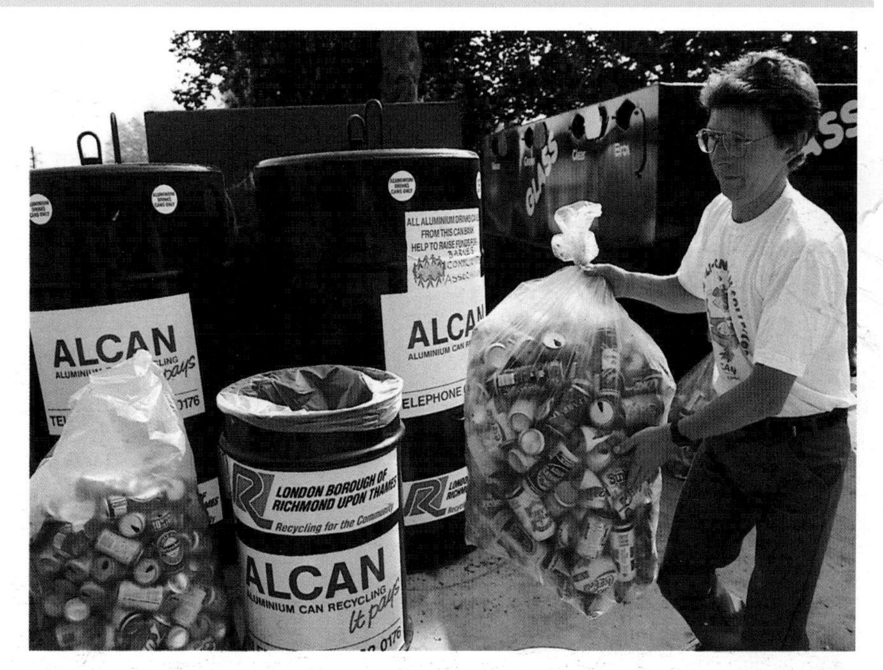

Figure 19.11B ▲ Recycling aluminium

19.12 Recycling

Few materials are destroyed during their use. Recycling is a possible method of saving resources and energy. Metals and alloys are prime candidates for recycling (see Table 19.13). Other materials are wood, paper, plastics and textiles. Some objects are easier to recycle than others. An iron machine has a high scrap value, but iron bars embedded in concrete are more difficult to reuse. In many cases a number of different metals are used in the manufacture of an item, e.g. a motor vehicle. A motor vehicle contains about 1% by mass of copper and this must be reduced to 0.1% before the scrap can go to the steelworks.

Table 19.13 Recovery of metals

Metal	Recovery
Aluminium	About 50% is recycled. Recycling uses only 5% of the energy need to make aluminium from its ore
Iron	About 50% of the feedstock for steel furnaces is scrap
Copper	Pipes, vehicle radiators, etc are recycled
Lead	Batteries, pipes, sheet metal, type metal are recycled
Zinc	Recovery is from alloys
Tin	Recovery is from alloys and from tin-plate
Mercury	A large percentage is recovered from mercury cells, instruments and apparatus
Gold, silver and platinum	Recovery is high from jewellery, watches and chemical plants

SUMMARY

Recycling metals saves Earth's resources of metal ores and saves energy. Do you play your part in recycling?

CHECKPOINT

1. Why is the recovery of gold and silver high?
2. Why do people find it more convenient to recycle mercury than to dump it?
3. Why is copper an easy metal to recycle?
4. How can other metals be separated from scrap iron?

Topic 20 Chemical cells

20.1 Simple chemical cells

FIRST THOUGHTS

A car needs a battery to start the motor and to power the instruments. A battery is a series of chemical cells. A chemical cell is a system for converting the energy of a chemical reaction into electricity. This topic will tell you more about how chemical cells work.

www.keyscience.co.uk

EXTENSION FILE ACTIVITY

You have studied metals in this theme. You have learned that metals are made up of positive ions and a cloud of moving electrons (Figure 19.2A). What happens when a strip of metal is put into water or a solution of an electrolyte? Figure 20.1A(a) shows what happens when the metal is a reactive metal such as zinc. Some zinc ions pass into solution, leaving their electrons behind on the metal:

$$Zn(s) \rightarrow Zn^{2+}(aq) + 2e^-$$

As a result, the strip of zinc becomes negatively charged. After a while the negative charge on the zinc builds up to a level which prevents any more Zn^{2+} ions from leaving the metal.

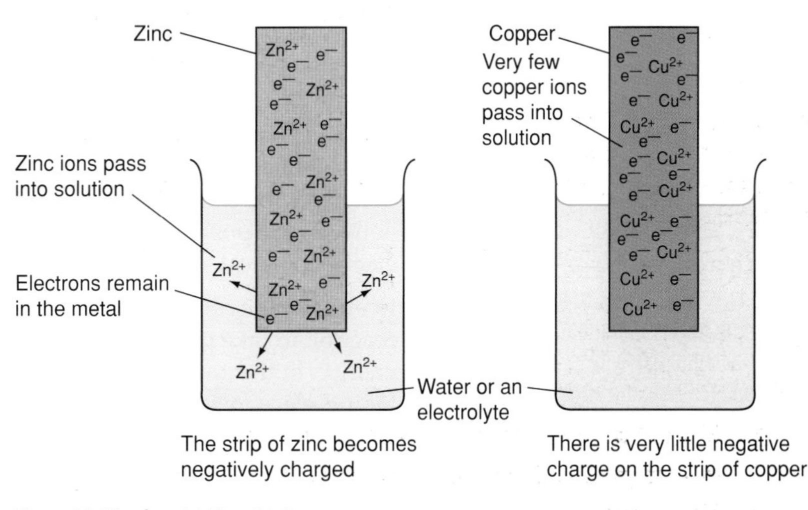

Figure 20.1A (a) Zinc (b) Copper

The reactivity series of metals (see Topic 19.4) is based on the observation that some metals are more ready to form ions (and are therefore more reactive) than others. Copper is one of the less reactive metals. A strip of copper placed in water or a solution of an electrolyte has very little tendency to form ions and to acquire a negative charge (see Figure 20.1A(b)).

Perhaps you can predict what will happen if you immerse a strip of zinc and a strip of copper in a solution and then connect the two metals. Electrons flow through the external circuit from zinc, which is negatively charged, to copper, which has very little charge (see Figure 20.1B).

Electrons flow from the negative electrode to the positive electrode. Conventional electricity flows from the positive electrode to the negative electrode

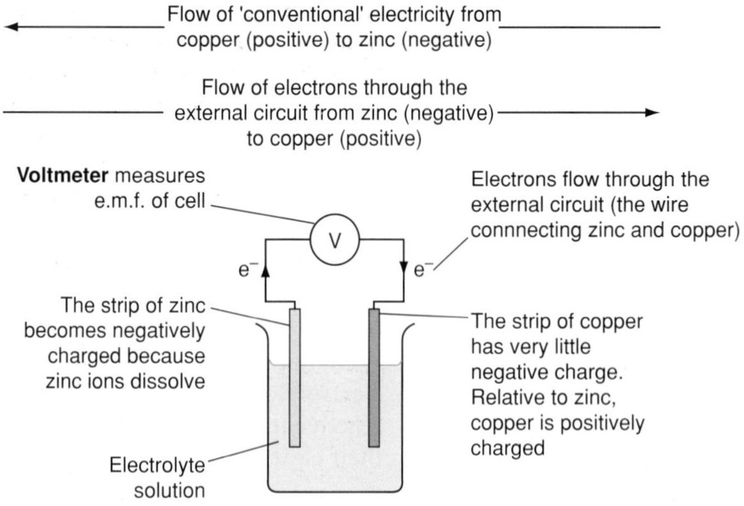

Figure 20.1B ▲ A zinc–copper chemical cell

SUMMARY

When a metal is placed in a solution of an electrolyte, it acquires an electric charge. When two metals which dip into an electrolyte are connected, there is a difference in electric potential between them, called the e.m.f. of the cell. An electric current flows from one metal to the other. The difference in electric potential between the two metals is called the e.m.f. of the cell. The further apart the metals are in the reactivity series, the greater is the e.m.f. of the cell.

In Figure 20.1B, electrons flow from zinc to copper through the external circuit. Before the nature of electricity was understood, the flow of electricity was regarded as taking place from the positive electrode to the negative electrode, in this case from copper to zinc. The flow of 'conventional' electricity is from copper to zinc (right to left in Figure 20.1B). The direction of flow of electrons is from a metal higher in the reactivity series to a metal lower in the series. The difference in electric potential between the two electrodes is called the electromotive force (e.m.f.) of the cell. The e.m.f. of the cell depends on the difference between the positions of the two metals in the reactivity series (and the electrochemical series; see Table 20.1).

In electrolytic cells (see Topic 7), an electric current causes a chemical reaction to take place. Electrical energy is converted into chemical energy. In the zinc–copper cell described above, the chemical reaction taking place inside the cell causes a current to flow through the external circuit. Chemical energy is converted into electrical energy. This kind of cell is called a **chemical cell** (or a **galvanic cell** or a **voltaic cell**).

Arranging metals in order of their tendency to lose electrons

In the zinc–copper cell (Figure 20.1A), zinc is the negative electrode and copper is the positive electrode. Copper is not always the positive electrode. If copper is paired with silver in a chemical cell, the reaction:

$$Cu(s) \rightarrow Cu^{2+}(aq) + 2e^-$$

happens to a greater extent than the reaction:

$$Ag(s) \rightarrow Ag^{+}(aq) + e^-$$

The build-up of electrons on copper is greater than the build-up of electrons on silver. Copper is the negative electrode of the cell. Paired with a more reactive metal, e.g. zinc, copper is the positive electrode; paired with a less reactive metal, e.g. silver, copper is the negative electrode.

It is possible to construct a 'league table' of metals, arranging them in order of their tendency to lose electrons. The method is to choose one metal as a reference electrode, say copper, and measure the e.m.f. of a number of different metal–copper cells. The values of e.m.f. place the metals in order of their readiness to give electrons and form cations in aqueous solution. The order is called the **electrochemical series**. Table 20.1 shows a section of the electrochemical series including hydrogen. It also puts cations in the reverse order of the ease with which they accept electrons and are discharged in electrolysis (see Topic 7.5).

SUMMARY

When you place metals in order of their tendency to lose electrons and gain a positive charge, you obtain the electrochemical series. This is similar to the reactivity series of metals. Some metals occupy different positions in the two series. This happens when an oxide coating or a film of insoluble salt decreases the true reactivity of the metal.

Table 20.1 Part of the electrochemical series

Metal	
Potassium, K	↑
Calcium, Ca	
Sodium, Na	Increasing ease
Magnesium, Mg	of losing electrons
Aluminium, Al	to form
Zinc, Zn	positive ions
Iron, Fe	(cations)
Tin, Sn	Decreasing ease
Lead, Pb	of discharging ions
Hydrogen, H	in electrolysis
Copper, Cu	
Silver, Ag	
Gold, Au	

The reactivity series

The electrochemical series will remind you of the reactivity series of metals (Topic 19.4). The reactivity series does not always show the true ability of a metal to form ions. Some metals, e.g. aluminium, acquire a coating of the metal oxide which prevents the metal from showing its true reactivity. Other metals, e.g. calcium, form insoluble salts and this slows down their reactions with solutions, e.g. with water and acids. The electrochemical series, on the other hand, shows the true ability of a metal to form ions.

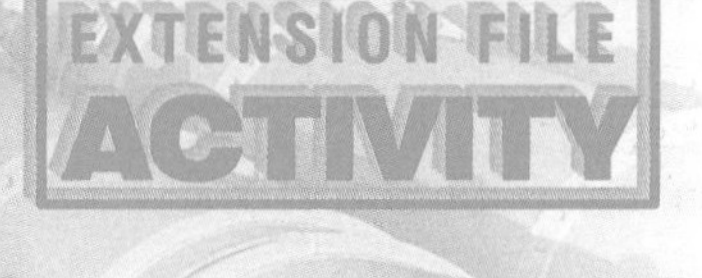

Two metals far apart in the electrochemical series give a cell with a bigger e.m.f. than two metals close together.

EXTENSION FILE ACTIVITY

Many metals can be paired up in chemical cells. The further apart the metals are in the electrochemical series, the bigger is the e.m.f. produced by the cell. Cells can be combined in series to form a battery. Three cells each with an e.m.f. of 1.5 V combine to make a battery with an e.m.f. of 4.5 V. The word 'battery' is sometimes used incorrectly for a single cell. For example, what we call a torch 'battery' is often a single cell.

CHECKPOINT

1. Name the particles that conduct electricity through (a) a chemical cell and (b) an electrical circuit outside the cell.
2. The following pairs of metals are joined to form chemical cells. Place the pairs in list (a) and the pairs in list (b) in order of the e.m.f. of the cells:
   (a) zinc–copper, magnesium–copper, iron–copper
   (b) magnesium–zinc, iron–magnesium, magnesium–tin
3. Zinc and iron are connected to form a cell.
   (a) Which of the two metals is the more able to form ions?
   (b) Which metal will become the negative electrode of the cell and which the positive?
   (c) Sketch a zinc–iron cell showing the direction of flow of electrons in the external circuit.
4. Copper and lead are both low in the electrochemical series. When they are paired up to form a chemical cell, which of the two metals will become the negative electrode?

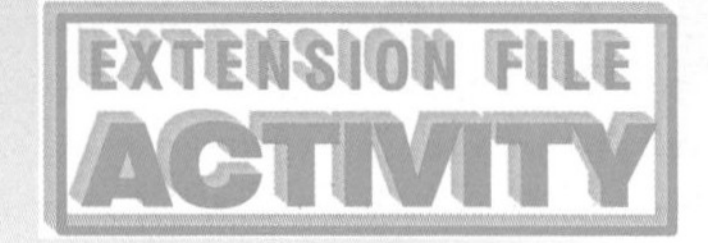

The lead–acid accumulator

The type of battery commonly used in motor vehicles is the lead–acid accumulator. The battery delivers 6 V or 12 V, depending on the number of 2 V cells which it contains (see Figure 20.1A). In each cell, one pole is a lead plate and the other is a lead grid packed with lead(IV) oxide, PbO_2. The electrolyte is sulphuric acid. When the battery supplies a current, the reactions which occur convert lead and lead(IV) oxide into lead(II) ions. When the reactants have been used up, the battery can no longer supply a current; it is 'flat'. It can, however, be recharged. When the vehicle engine is running, the dynamo (the alternator) rotates and generates electricity which recharges the battery. If the dynamo is faulty, the battery will become flat. It can be recharged by connecting it to a battery charger, which is a transformer connected to the mains. This reverses the sign of each electrode and reverses the chemical reactions that have occurred at the electrodes. The battery is again able to supply a current.

Figure 20.1C ▲ A car battery

SUMMARY

A vehicle battery consists of a number of chemical cells. Chemical reactions in the cells supply an electric current. When the chemicals have been used up they can be reformed by recharging the battery.

- During discharge, chemical energy is converted into electrical energy.
- During charge, electrical energy is converted into chemical energy.

CHECKPOINT

1. What is the advantage of the lead–acid accumulator over the zinc–copper cell (Figure 20.1B)?
2. What does the alternator in a vehicle engine do?
3. (a) Why do car batteries sometimes become flat?
   (b) How can they be given new life?

20.2 Dry cells

Do you use a portable radio or cassette player, a pocket calculator, a wrist watch or a mobile phone? All these depend on dry cells.

Dry cells are chemical cells which are used in everyday life. For use in batteries for torches, radios etc., a dry cell is used, one which has a damp paste instead of a liquid electrolyte.

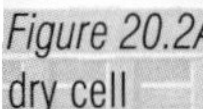

Figure 20.2A ▲ A pocket calculator uses a dry cell

The zinc–carbon dry cell

You will already be familiar with the zinc–carbon type of dry cell (see Figure 20.2B).

Figure 20.2B ▲ Dry cells

The zinc–carbon cell has an e.m.f. of 1.5 V.

The chemical reactions that take place are complex. A summary is:
Anode: Zinc atoms are oxidised to zinc ions.
Cathode: Manganese(IV) oxide is reduced to manganese(III) oxide.

When the chemicals have been used up, the cell can operate no longer: it is flat.

Cells of this type are 1–2 cm in diameter and 3–10 cm in length. They are used in flashlights, portable radios, toys etc. A disadvantage is that with heavy use the cell quickly becomes flat. Zinc–carbon dry cells cannot be recharged and have a short shelf life.

The alkaline manganese cell

The alkaline manganese cell also uses zinc as anode and manganese(IV) oxide as cathode. The electrolyte contains potassium hydroxide. The zinc anode is slightly porous, giving it a larger surface area. The alkaline manganese cell can therefore deliver more current than the zinc–carbon cell; it has an e.m.f. of 1.5 V. The alkaline manganese cells do not become flat as quickly in use and have a longer shelf life than zinc–carbon cells.

The silver oxide cell

Another type of dry cell is the silver oxide cell. Silver oxide cells measure 0.5–1 cm in diameter and 0.25–0.5 cm in height. These tiny 'batteries' are used in electronic wrist watches, in cameras and in electronic calculators. They have a long life because little current is drawn from them. They are rather expensive.

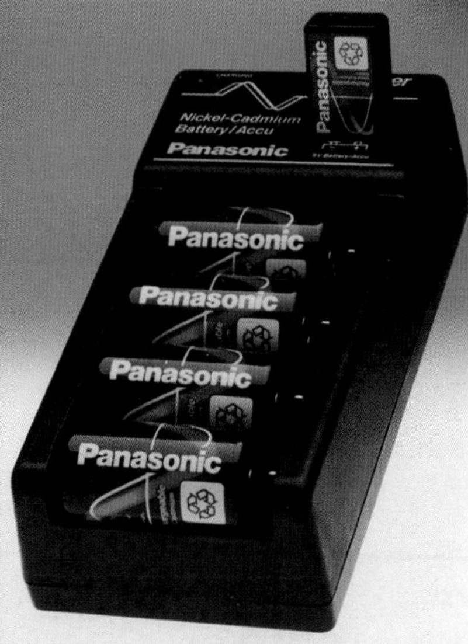

Figure 20.2C ▲ Nickel–cadmium cells and chargers

The nickel–cadmium cell

The dry cells described above have a limited life. Once the chemicals have reacted, there is no further source of energy. The nickel–cadmium cell (see Figure 20.2C) need not be thrown away when the chemical reaction that gives rise to the e.m.f. is finished. The cell can be recharged. This is done by connecting the cell to a source of direct current. The chemical reaction is reversed, and the cell has a new source of e.m.f. As the cell operates, cadmium is oxidised to cadmium(II) ions, and nickel(IV) oxide is reduced to nickel(II) hydroxide. During recharging, these reactions are reversed.

The nickel–cadmium cell has an e.m.f. of 1.4 V. It has a longer life than the lead storage battery. It can be packaged in a sealed unit ready for use in rechargeable calculators and photographic flash units etc.

Table 20.2 Dry cells

Type	Electrodes	Electrolyte	Voltage	Use	Characteristics
Zinc–carbon cell	Carbon (+) Zinc (−)	Ammonium chloride	1.5 V	Radios, cassette players, electrical toys, torches	Cheap. Relatively short life. Not rechargeable. Leakage of electrolyte occurs in the unsealed types of cell. Operates on low current; current drops gradually during discharge.
Alkaline manganese cell	Manganese(IV) oxide (+) Zinc (−)	Potassium hydroxide	1.5 V	Cassette players, electrical toys, appliances which are in use for long periods	Costs about twice as much as the zinc–carbon cell and lasts twice as long. Not rechargeable. Long shelf-life. Gives a large current; voltage remains steady during discharge.
Silver oxide cell	Silver oxide (+) Zinc (−)	Potassium hydroxide	1.5 V	Watches, calculators, cameras, hearing aids	High cost. Not rechargeable. Small in size; light-weight. Voltage remains constant in operation.
Nickel–cadmium cell	Nickel oxide (+) Cadmium (−)	Potassium hydroxide	1.5 V	Cassette players, electrical toys, radios	Expensive. Rechargeable. Gives a large current.

SUMMARY

Chemical cells transform the energy of chemical reactions into electrical energy. Examples of dry cells are: the carbon–zinc dry cell, the alkaline manganese cell, the silver oxide dry cell, the nickel–cadmium dry cell.

Fuel cells operate continuously. One kind uses the reaction between hydrogen and oxygen to make water.

Fuel cells

Space craft use **fuel cells**. A fuel cell is another means of converting chemical energy into electrical energy; the efficiency converting chemical energy into electrical energy is about 80%, compared with the efficiency of a power station which is about 40%. The fuel cell shown in Figure 20.2D uses the reaction:

Hydrogen + Oxygen → Water

The big advantage of a fuel cell over other cells is that it operates continuously with no need for recharging. As long as reactants are fed in, the cell can supply energy.

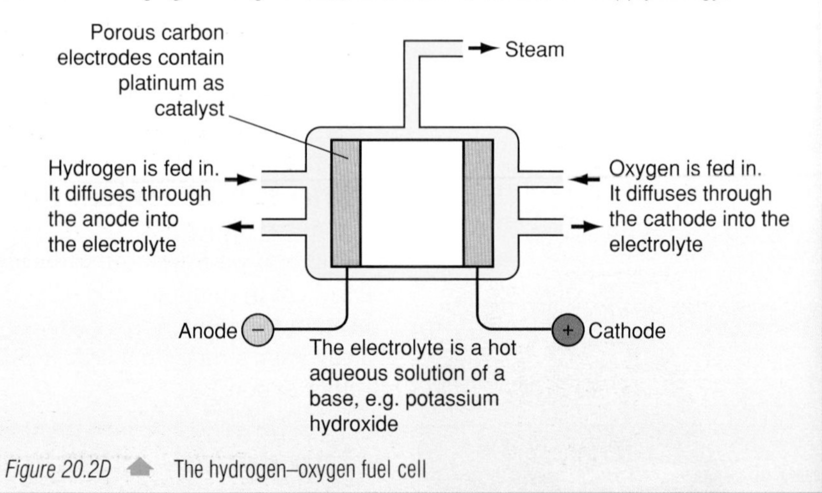

Figure 20.2D The hydrogen–oxygen fuel cell

CHECKPOINT

1. Why is the silver oxide 'battery' suitable for use in a camera? Why is it expensive?
2. Both the lead–acid accumulator and the nickel–cadmium cell are rechargeable.
   (a) What does 'rechargeable' mean?
   (b) What is the advantage of rechargeable cells over others?
   (c) For what uses is the nickel–cadmium cell more suited than the lead–acid accumulator?
   (d) What is the most widespread use of the lead–acid accumulator?
3. Why is the alkaline manganese cell preferred to the silver oxide cell for use in radios?

Topic 21 Building materials

21.1 Concrete

FIRST THOUGHTS

The UK quarries 90 million tonnes of limestone and chalk each year. Most of it is used in the building industry.

www.keyscience.co.uk

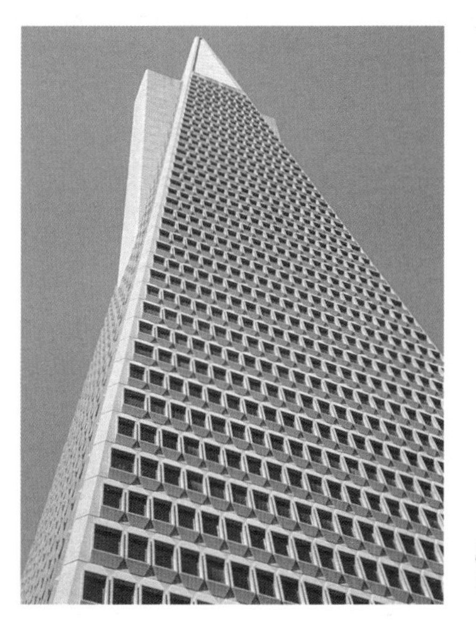

Figure 21.1A Concrete and glass

All around you, you can see buildings made of concrete, steel and glass. Concrete is used in the construction of all kinds of homes, schools, factories, skyscrapers, power stations and reservoirs. All our forms of transport depend on the concrete which is used in roads and bridges, docks and airport runways. For extra strength, for example in bridges, reinforced concrete is strengthened by steel supports.

The starting point in the manufacture of concrete is limestone or chalk. Both these minerals are forms of calcium carbonate. First, limestone or chalk is used with shale or clay to make cement. A small percentage of calcium sulphate (gypsum) is also needed (see Figure 21.1B). Figure 21.1C shows how concrete is made from cement.

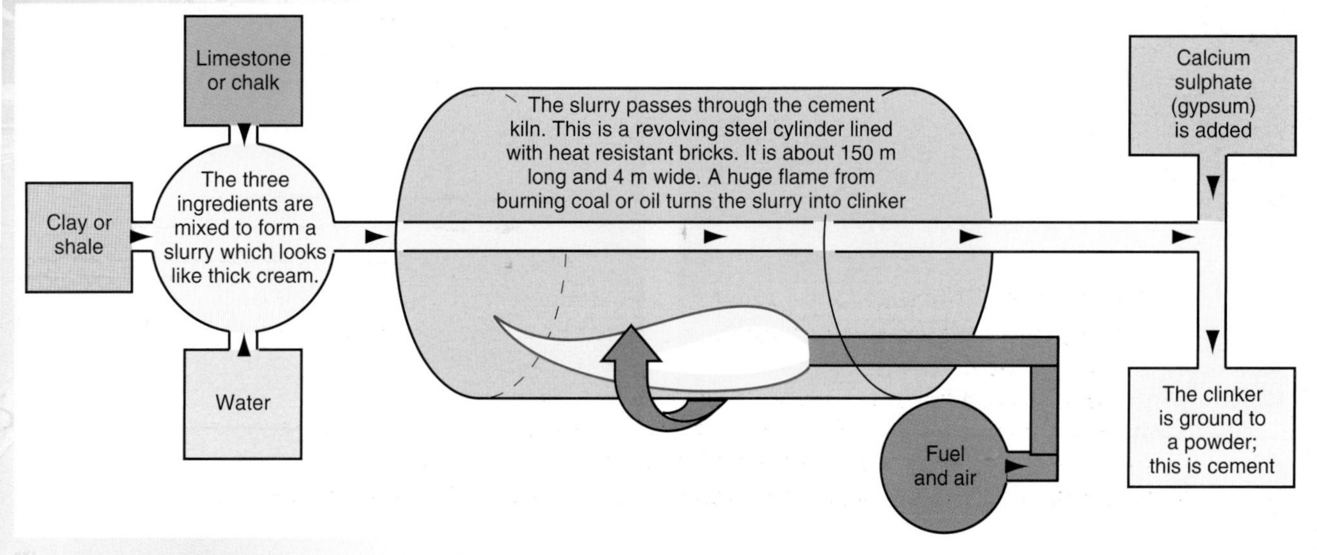

Figure 21.1B The manufacture of cement (after Blue Circle)

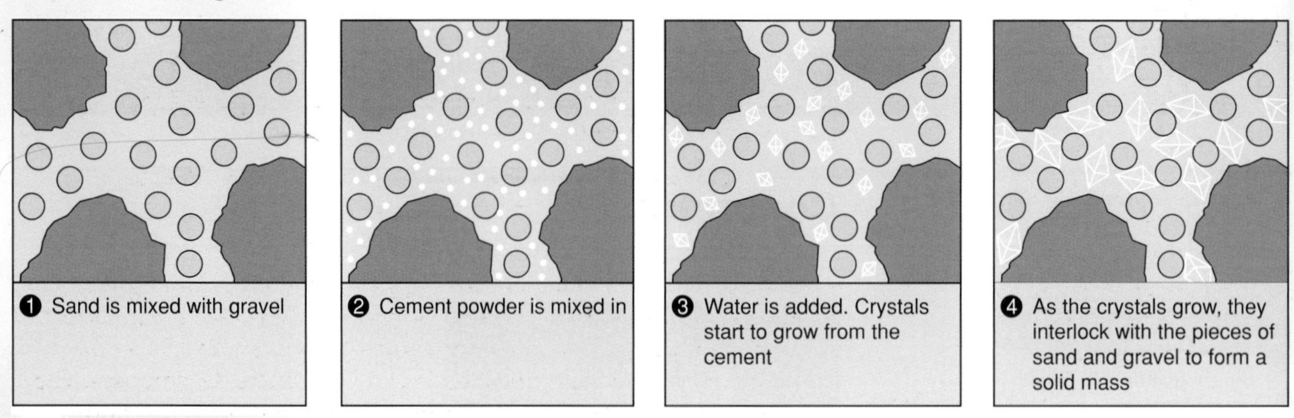

Figure 21.1C Making concrete (after Blue Circle)

SUMMARY

Concrete is a strong and versatile construction material. It is made from cement, sand, gravel and water. Cement is made from chalk or limestone and clay or shale. The quarrying of vast quantities of these raw materials creates environmental problems.

Chalk and limestone are calcium carbonate, $CaCO_3$. Clay and shale consist largely of silicon(IV) oxide, SiO_2, and aluminium oxide, Al_2O_3, with some other compounds. The chemical reactions which occur between them produce cement, which consists chiefly of calcium silicate, $CaSiO_3$, and calcium aluminate, $CaAl_2O_4$. A little calcium sulphate (gypsum) is added to slow down the rate at which concrete sets.

The UK has large deposits of both limestone and chalk. Limestone is quarried by blasting a hillside with an explosive. Chalk is dug out by mechanical excavators. Both methods devastate the landscape. It happens that these minerals often occur in regions of great natural beauty (Figure 21.1D). We have to balance the damage to our countryside against the useful materials which industry can obtain from limestone and chalk. Mining companies are required to restore the countryside after they exhaust a deposit of limestone or chalk (Figure 21.1E). Limestone is used in the manufacture of iron in blast furnaces, in the manufacture of glass and in the manufacture of lime.

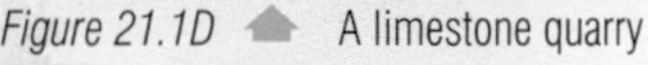

Figure 21.1D A limestone quarry

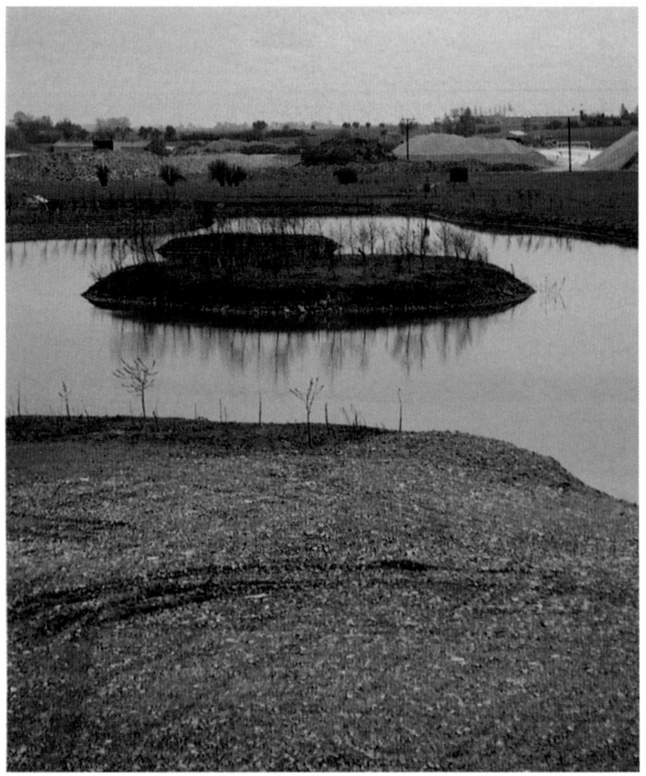

Figure 21.1E Restoring the scenery

Figure 21.1G ▲ Enjoying a 'carbonated' drink

EXTENSION FILE ACTIVITY

Lime

When calcium carbonate is heated strongly, it decomposes (splits up) to give calcium oxide and carbon dioxide.

$$\text{Calcium carbonate} \rightleftharpoons \text{Calcium oxide} + \text{Carbon dioxide}$$
$$CaCO_3(s) \rightleftharpoons CaO(s) + CO_2(g)$$

This is an example of **thermal decomposition**. The reaction is a **reversible** reaction: it can go from left to right and also from right to left, depending on the temperature and the pressure. The reaction is carried out industrially in towers called **lime kilns** (see Figure 21.1F). At the temperature of a lime kiln, 1000 °C, calcium carbonate decomposes. The through draft of air carries away carbon dioxide as it is formed and prevents it from recombining with calcium oxide. Otherwise there would be an acid-base reaction between the acid gas, carbon dioxide, and the base, calcium oxide.

Calcium oxide is called **lime** or **quicklime**. It reacts with water in an exothermic reaction to form calcium hydroxide, which is called **slaked lime**.

$$\text{Calcium oxide} + \text{Water} \rightarrow \text{Calcium hydroxide}$$
$$CaO(s) + H_2O(l) \rightarrow Ca(OH)_2(s)$$

On building sites, calcium hydroxide, slaked lime, is mixed with sand to give **mortar**. Mortar is used to hold bricks together. As mortar is exposed to the air, it becomes gradually harder as it reacts with carbon dioxide in the air to form calcium carbonate.

Calcium oxide, lime, is used in agriculture. Farmers spread it on fields to neutralise excess acid in soils.

The carbon dioxide produced in lime kilns is also useful. Carbon dioxide dissolves to a slight extent in water to give a solution of the weak acid, carbonic acid, H_2CO_3. Under pressure, the solubility of carbon dioxide increases. The basis of the soft drinks industry is dissolving carbon dioxide in water under pressure, and adding flavourings. When the cap is taken off a bottle, the pressure is decreased, and carbon dioxide comes out of solution.

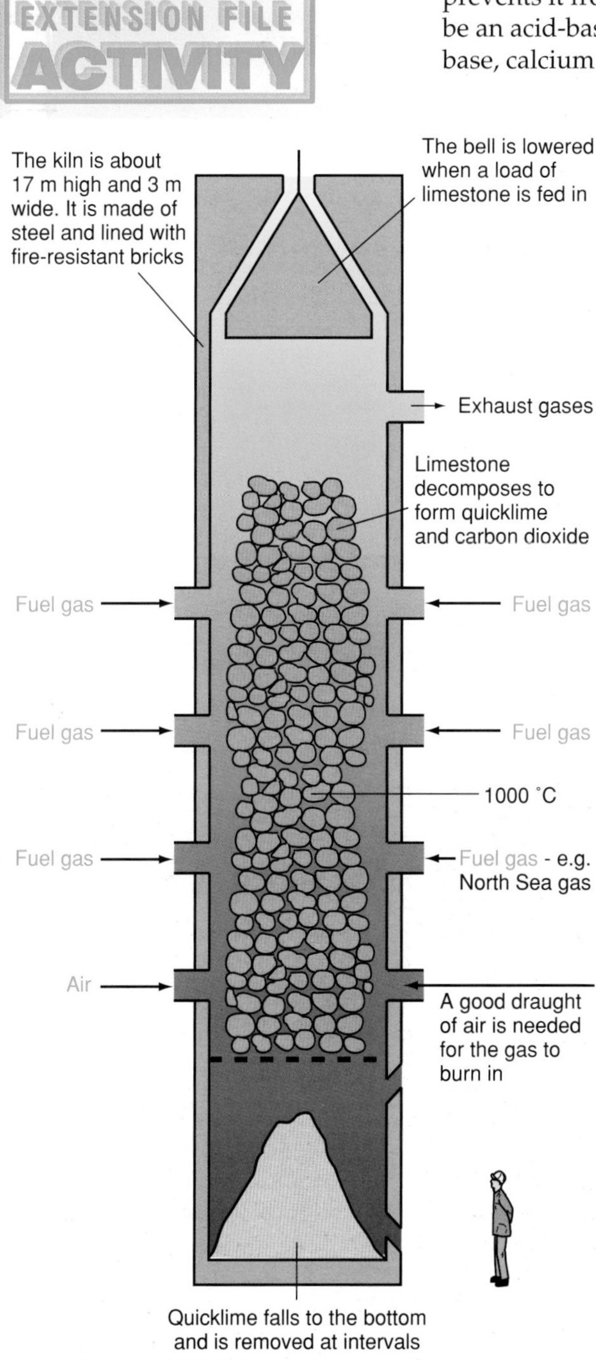

Figure 21.1F ▲ A lime kiln

SUMMARY

When calcium carbonate (limestone) is heated in lime kilns, calcium oxide (quicklime) and carbon dioxide are produced. On building sites, calcium hydroxide is used to make mortar. On farms, it is used to neutralise excessive acidity of soils. Carbon dioxide is used in the soft drinks industry and as a fire extinguisher.

21.2 Glass

FIRST THOUGHTS

The recipe for glass is 4500 years old. Egyptians discovered it when they melted sand with limestone and sodium carbonate. To their surprise, they obtained a transparent material, glass. Egypt is one of the few countries where sodium carbonate occurs naturally.

Glass is a mixture of calcium silicate, $CaSiO_3$, and sodium silicate, Na_2SiO_3. In structure, it resembles silicon(IV) oxide, SiO_2 (sand), which is a crystalline substance with a macromolecular structure (see Figure 21.2A). In glass, many of the Si—O bonds have been broken, and the structure is less regular. X-rays show that glass does not have the orderly packing of atoms found in other solids. Glass is neither a liquid nor a crystalline solid. It is a **supercooled liquid**: it appears to be solid, but has no sharp melting point.

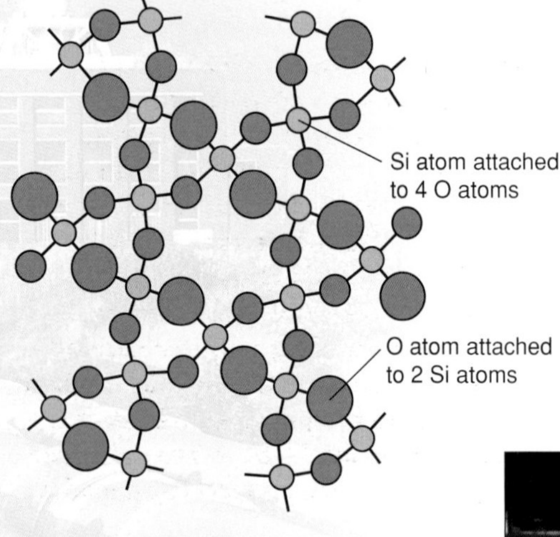

Figure 21.2A Bonding in crystalline silica, SiO_2

Some new uses for glass

Craftsmen worked with glass for thousands of years without knowing anything about its structure. When chemists began to study glass, they made many advances in glass technology. New types of glass were invented. Some of these are described in this section.

Pyrex® glass will stand sudden changes in temperature without cracking. It is made by adding boron oxide during the manufacture of glass.

Window glass consists of plates of glass which are the same thickness all over. Formerly, plates were made by grinding a sheet of glass to the required thickness and smoothness. In the process, 30% of the sheet was wasted. The Pilkington Glass Company invented the **float glass** process. Molten glass flows on to a bath of molten tin (Figure 21.2B). As the glass cools and solidifies, the top and bottom surfaces are both perfectly smooth and planar. While it is still soft, the glass is rolled to the required thickness and then cut into sections.

Figure 21.2B Float glass

Photosensitive glass is used in sunglasses which darken in bright light and lighten again when the light fades. The glass includes silver chloride.

Glass ceramic is almost unbreakable. It is made by heating photosensitive glass in a furnace. Glass ceramic is used in ovenware, electrical insulation, the nose cones of space rockets and the tiles of space shuttles.

Soluble glass has important uses. In many tropical countries, the disease, **bilharzia** (or schistosomiasis) is a blight. The snails which carry the disease can be killed by copper salts in quite low concentrations. The copper compounds can be incorporated into a soluble glass. Pellets of the copper-containing soluble glass can be put into the water. As they dissolve, they release copper compounds gradually into the snail-infested water. To make a soluble glass, phosphorus(V) oxide, P_2O_5, is used instead of silicon(IV) oxide.

SUMMARY

The glass industry makes:

- Pyrex® glass,
- plate glass,
- light-sensitive glass,
- glass ceramic,
- soluble glass.

CHECKPOINT

1 (a) Why must a bottle of a carbonated soft drink have a well-fitting cap?
(b) Why can you not see bubbles inside the closed bottle?
(c) Why can you see bubbles when the bottle is opened?

2 'Spreading calcium hydroxide on soil reduces the acidity of the soil.' Describe an experiment which you could do to check whether this statement is true.

3 'Mortar reacts with the air to form calcium carbonate.' Describe experiments which you could do on new mortar and old mortar to find out whether this statement is true.

21.3 Ceramics

Figure 21.3A Traditional ceramics

Ceramics are crystalline compounds which consist of an ordered arrangement of atoms. Many are compounds of metallic and non-metallic elements, e.g. silicon or aluminium combined with oxygen or nitrogen. Pottery is a ceramic. Brick is a ceramic. Traditional ceramics such as these are derived from the raw materials clay and silica. They are formed on, for example, a potter's wheel and hardened in a fire kiln. Examples of modern ceramics are silica glass, soda glass, aluminium oxide, Al_2O_3, silicon carbide, SiC, silicon nitride, SiN, zirconium oxide, ZrO_2, titanium carbide, TiC, zirconium carbide, ZrC, tungsten carbide, WC. They are made by grinding the components to fine powders, mixing them and heating to high temperatures under high pressure in electric kilns. New uses are being found for ceramics all the time. They are used in spacecraft, artificial bones, cutting tools, engine parts, turbine blades and electronic components (see next page).

The bonding in ceramics may be ionic or covalent. The arrangement of particles is more complicated than in metals. The bonds are directed in space and the structure is therefore more rigid than a metallic structure.

This structure makes ceramics

- crystalline
- harder than metals (able to cut steel and glass)
- of higher melting point than metals
- poor conductors of heat and electricity because they lack free electrons
- not ductile, malleable or plastic (except at high temperature)
- rather brittle because the bonds are rigid, so that it is difficult to change the shape without it breaking

Applications of ceramics

Vehicle components A ceramic coating, e.g. ZrO_2, on cast iron improves wear and tear and heat-resistance. It extends the life of diesel engine parts.

Machine tools made of ceramics can rotate twice as fast as metal tools without deforming or wearing out. *Sialon* is a ceramic made of silicon, aluminium, oxygen and nitrogen. It is almost as hard as diamond, and is as strong as steel and as low in density as aluminium. It can be used at up to 1300 °C and needs no lubricants.

How do ceramics compare with metals and plastics? Different properties mean different uses for these materials; see Topic 28.3.

Bioceramics Recently ceramics, e.g. aluminium oxide and silicon nitride, have been considered as implant materials. Alloys and polymers are used for replacement devices, e.g. hips, knees, teeth etc. Ceramics have a high strength to weight ratio, would wear better than alloys or polymers and would never corrode. For replacing damaged bone, a porous ceramic implant could be used, so as to allow bone to grow into and bond with the implant.

Space craft Ceramic tiles protect a space shuttle during re-entry into the Earth's atmosphere. The outside temperature can reach 1500 °C. Each tile consists of a cellular structure of very fine fibres coated with silica. The fibres are loosely packed: 95% of the tiles is air, which gives them their insulating property.

SUMMARY

Ceramics are crystalline compounds of oxygen or nitrogen with other elements, e.g. silicon, aluminium and some transition elements. Their structure makes them hard-wearing, resistant to corrosion, and of high melting point. Their resistance to heat and corrosion finds them many uses.

Figure 21.3B Space-age ceramics

Topic 22 Agricultural chemicals

22.1 Fertilisers

FIRST THOUGHTS

The world's population keeps on growing. Many people in the world are short of food. A major contribution to increasing the size of food crops is the use of manufactured fertilisers to supplement natural fertilisers

www.keyscience.co.uk

As the world population has increased, the demand for food has increased. In 1965 it was 3.5 billion, and in 2000 it reached 6.5 billion. In many parts of the world, people are either hungry or undernourished. Farming has to be intensive, that is, able to grow large crops on the available land. Crops take nutrients from the soil, and these must be replaced before the next crop is sown (see the nitrogen cycle, Topic 13.6).

Before the nineteenth century, farmers relied on the natural fertilisers, manure and compost, to replace nutrients in the soil. When the world population began to rise sharply in the nineteenth century, farmers needed more fertilisers. They asked the chemists to make some artificial fertilisers. Before the chemists could do this, they had to find out which chemicals in manure fertilised the soil. Then they would be able to make chemical fertilisers to do the same job.

The chemists' work showed that plants need these three groups of chemicals.

1 Large amounts of carbon, hydrogen and oxygen. There is no shortage of these elements. Carbon comes from carbon dioxide in the air; hydrogen and oxygen from water.
2 Small amounts of 'trace elements' (see Table 22.1). So little is needed that there is no need to add these elements to the soil.
3 Larger quantities of the other elements shown in Table 22.1.

The elements which plants obtain from soil are listed in Table 22.1.

Table 22.1 Elements obtained by plants from soil

Major elements	Importance	Trace elements
Nitrogen	Needed for protein synthesis	Manganese
Phosphorus	Needed for development of roots, energy-transfer reactions, nucleic acids	Copper Iron Zinc
Potassium	Needed in photosynthesis	Chlorine
Magnesium	Present in chlorophyll	Boron
Sulphur	Needed for synthesis of some proteins	Molybdenum
Calcium	Needed for transport	

It's a fact!

With 2–3 million tonnes of nitrogen (in the form of nitrates) going on to the soil of the UK each year, 600 000 tonnes of nitrate ions leach out every year.

How much fertiliser?

Most fertilisers concentrate on supplying the necessary nitrogen (N), phosphorus (P) and potassium (K). **NPK fertilisers** are consumed in huge quantities: in 1990, the world consumption was over 20 million tonnes. In the UK, the consumption of NPK fertilisers is about seven million tonnes a year. The fertilisers cost about £60 a tonne. Farmers and market gardeners need to invest a lot of money in fertilisers. They want good crops, but they do not want to spend more than they need on fertilisers. They can obtain expert advice from the Ministry of

Agriculture, Fisheries and Food. Agricultural chemists at the Ministry will advise them on the type and quantity of fertiliser to apply and the best season of the year for applying it. Every farmer and grower has a different problem. The agricultural chemists must weigh up the type of crop and the type of soil before they can recommend the most suitable treatment.

Nitrogen

Nitrogen can be absorbed by plants when it is combined as nitrates. Ammonia and ammonium salts can also be used as fertilisers because ammonium salts are converted into nitrates by organisms which live in the soil. All nitrates and all ammonium salts are soluble so they are easily absorbed. Sometimes concentrated ammonia is used as a fertiliser. Solid fertilisers, ammonium nitrate, ammonium sulphate and urea, (CON_2H_4) are applied in pellet form (see Figure 22.1A). The manufacture of nitrogenous fertilisers is described below. The use of more fertiliser than plants can absorb causes eutrophication of lakes and rivers and pollution of ground water (see Topic 17.4).

It's a fact!

There are islands off the coast of South America which are inhabited by large flocks of sea birds. Mounds of their droppings, called guano, accumulate. Since sea birds eat a fish diet, which is high in protein, guano is rich in nitrogen compounds. For a century, European farmers imported guano from South America to use as a fertiliser.

Manufactured fertilisers contain nitrogen, phosphorus and potassium. Nitrogenous fertilisers include ammonia, urea and ammonium salts, which you see being applied in the photograph.

Figure 22.1A ▲ Applying pellets of ammonium salts to a field

22.2 Manufacture of ammonia

Ammonium salts and nitrates are used as fertilisers. The first step in making them is to make ammonia.

There is plenty of nitrogen in the air. Making it combine with hydrogen to make ammonia was a difficult problem. Fritz Haber solved it.

The first nitrogen-containing fertiliser which the farmers of Europe used was sodium nitrate. They had to import it across the Atlantic from Chile. There were disadvantages to this practice. European chemists tackled the problem of making nitrogenous fertilisers. Nitrogen was the obvious starting material because every country has plenty of it in the air. Making nitrogen combine with other elements proved to be a problem. The problem was solved by a German chemist called Fritz Haber. In 1908, he succeeded in combining nitrogen with hydrogen to form ammonia

$$\text{Nitrogen} + \text{Hydrogen} \rightleftharpoons \text{Ammonia}$$
$$N_2(g) + 3H_2(g) \rightleftharpoons 2NH_3(g)$$

This reaction is reversible: it takes place from right to left as well as from left to right. Some of the ammonia formed dissociates (splits up) into nitrogen and hydrogen. A mixture of nitrogen, hydrogen and ammonia is formed. To increase the percentage of ammonia formed in the reaction, a high pressure and a low temperature are used. A low temperature reduces the dissociation of ammonia, but it also makes the reaction very slow. Modern industrial plants use a compromise temperature and speed up the reaction by means of a catalyst. (The buildings and equipment in which a manufacturing process is carried out are called a **plant**.) The yield of ammonia is about 10%. Ammonia is condensed out of the mixture. With a boiling point of −33 °C, which is higher than that of most gases, ammonia is easily liquefied. The nitrogen and hydrogen which have not reacted are recycled through the plant (see Figure 22.2A).

Conditions which the Haber Process uses to favour the formation of ammonia are high pressure, moderate temperature and a catalyst.

The hydrogen for the Haber process is obtained from natural gas. The process takes place in a number of stages. The overall reaction is:

Natural gas (methane) + Steam → Hydrogen + Carbon monoxide

$CH_4(g) + H_2O(g) \rightarrow 3H_2(g) + CO(g)$

Carbon monoxide is oxidised to carbon dioxide and removed to leave hydrogen.

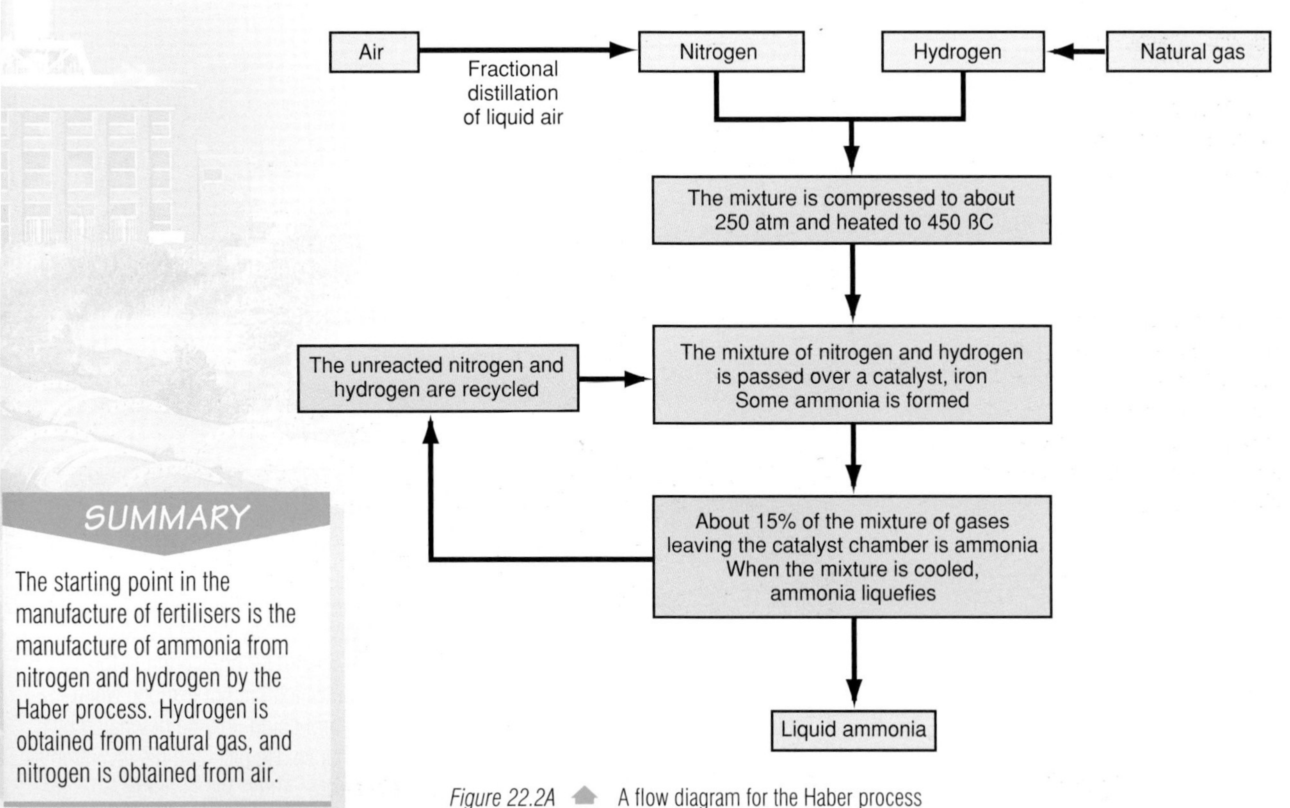

Figure 22.2A ▲ A flow diagram for the Haber process

SUMMARY

The starting point in the manufacture of fertilisers is the manufacture of ammonia from nitrogen and hydrogen by the Haber process. Hydrogen is obtained from natural gas, and nitrogen is obtained from air.

Making fertilisers from ammonia

A concentrated solution of ammonia can be used as a fertiliser. It is easier to store solid fertilisers, such as ammonium salts.

Being a base, ammonia is neutralised by acids to yield ammonium salts

Ammonia + Nitric acid → Ammonium nitrate

$NH_3(aq) + HNO_3(aq) \rightarrow NH_4NO_3(aq)$

Ammonia + Sulphuric acid → Ammonium sulphate

$2NH_3(aq) + H_2SO_4(aq) \rightarrow (NH_4)_2SO_4(aq)$

Ammonia + Phosphoric acid → Ammonium phosphate

$3NH_3(aq) + H_3PO_4(aq) \rightarrow (NH_4)_3PO_4(aq)$

SUMMARY

Ammonia is used as a fertiliser, but ammonium nitrate, sulphate and phosphate are preferred.

22.3 Manufacture of nitric acid

Nitric acid is made by the oxidation of ammonia (see Figure 22.3A).

SUMMARY

Nitric acid is made by the oxidation of ammonia.

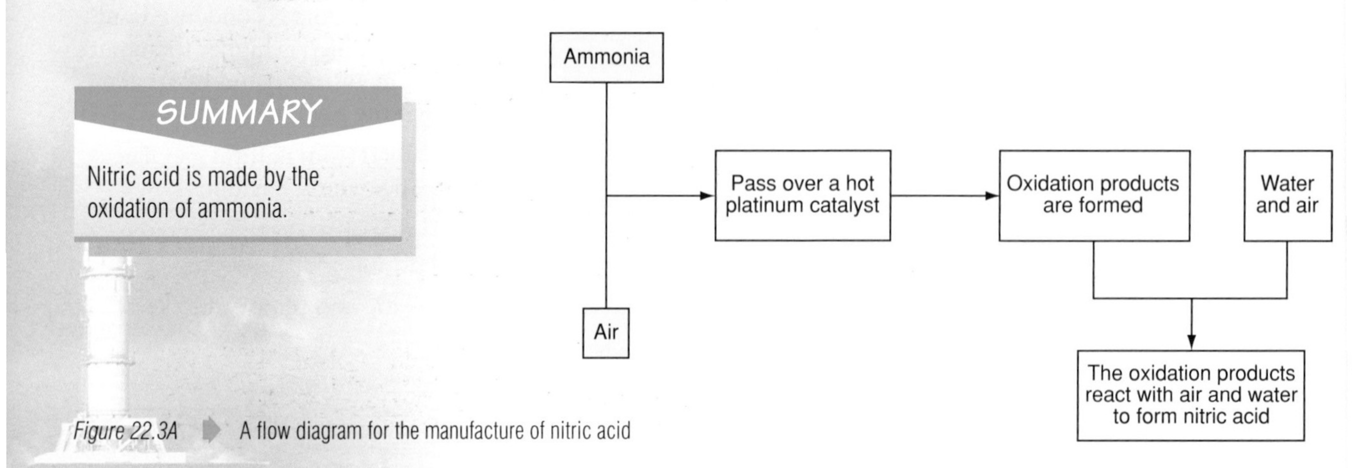

Figure 22.3A A flow diagram for the manufacture of nitric acid

In addition to its importance in the fertiliser industry, nitric acid is used in the manufacture of explosives, such as TNT (trinitrotoluene), and dyes.

KEY SCIENTIST

In 1846, Professor Sobrero, an Italian chemist, had the idea of reacting concentrated nitric acid with glycerol. He obtained a liquid which explodes when ignited by a lighted fuse. He called it **nitroglycerine**. The new compound was welcomed by miners who started to use it for blasting away rock. Unfortunately, there were accidents because nitroglycerine can explode if it receives a sudden blow, and this sometimes happened in transport.

This drawback did not stop Immanuel Nobel from beginning to manufacture nitroglycerine in 1863 in Sweden. A year later, an explosion killed four people in the factory, including his son, Emil. The other son, Alfred, continued manufacturing, but he looked for a method of stablising nitroglycerine to withstand mechanical shock. He found that a type of clay called kieselguhr would absorb nitroglycerine to form a mixture which did not explode when struck and could only be detonated by a lighted fuse. He patented his invention as **dynamite**. Alfred Nobel made a fortune out of his new, safer explosive. As well as being used to make tunnels, roads and railways and in mining, dynamite was sometimes used in warfare. Alfred Nobel was distressed to see his invention being misused in this way, and he became interested in promoting peace between nations. He financed the Nobel Foundation in Stockholm. This awards prizes each year to the people who have done the most effective work for peace and the most outstanding work in medicine, physics, chemistry and literature.

22.4 Manufacture of sulphuric acid

Sulphuric acid is made by the **contact process**. Sulphur dioxide and oxygen combine in contact with a catalyst, vanadium(V) oxide. Air is used as a source of oxygen. Sulphur dioxide is obtained by:

- Burning sulphur (deposits of sulphur occur in many countries).
- As a by-product of the extraction of metals from sulphide ores.
- As a by-product of petroleum oil and natural gas.

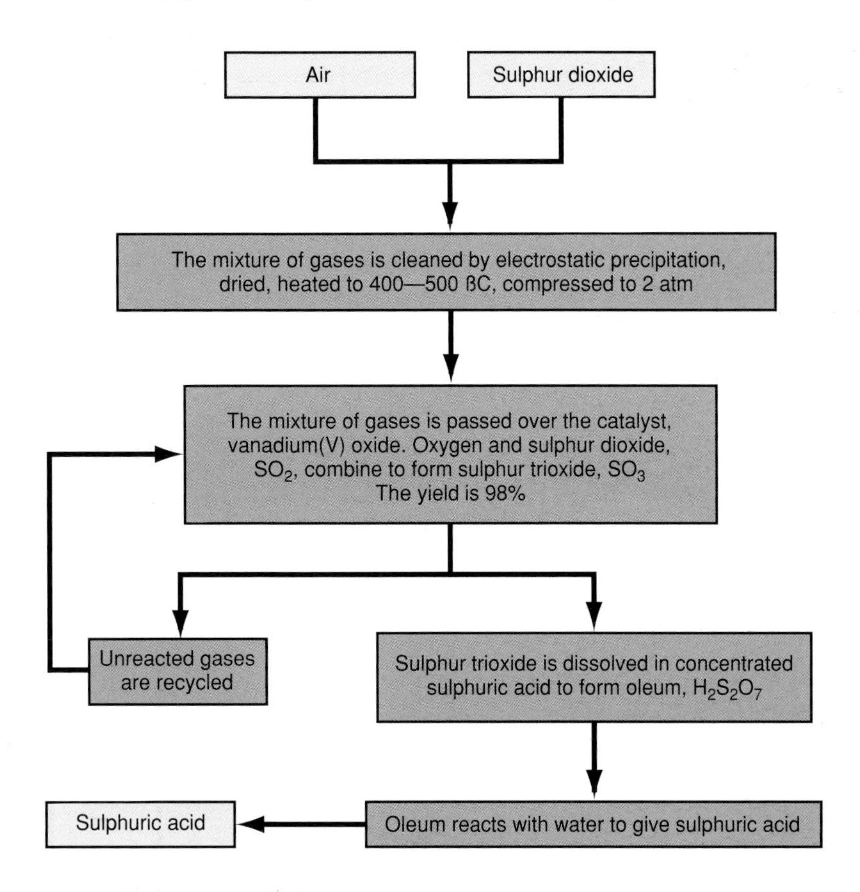

Figure 22.4A ▶ The contact process

Sulphur trioxide reacts with water to give sulphuric acid

$$\text{Sulphur trioxide} + \text{Water} \rightarrow \text{Sulphuric acid}$$
$$SO_3(s) + H_2O(l) \rightarrow H_2SO_4(l)$$

It is dangerous for industry to carry out the reaction in this way. As soon as sulphur trioxide meets water vapour, it reacts to form a mist of sulphuric acid. It is safer to absorb the sulphur trioxide in sulphuric acid to form oleum, $H_2S_2O_7$, and then convert the oleum into sulphuric acid.

Some of the uses of sulphuric acid are shown in Figure 22.4B.

SUMMARY

Sulphuric acid is made by the contact process. Sulphur or a metal sulphide is oxidised to sulphur dioxide. Some sulphur dioxide is obtained from natural gas. Sulphur dioxide is oxidised to sulphur trioxide. This reacts with concentrated sulphuric acid to form oleum, which reacts with water to form sulphuric acid. The flow diagram shows the steps involved.

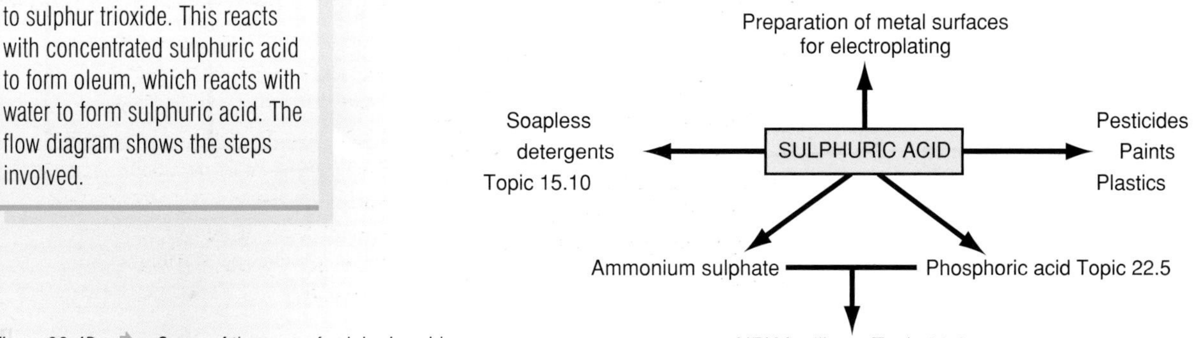

Figure 22.4B ▶ Some of the uses of sulphuric acid

22.5 Manufacture of phosphoric acid

SUMMARY

Phosphoric acid is made from calcium phosphate.

Phosphoric acid is made from the widespread ore calcium phosphate by a reaction with sulphuric acid.

Calcium phosphate + Sulphuric acid → Phosphoric acid + Calcium sulphate

22.6 Making NPK fertilisers

SUMMARY

The most popular fertilisers are those which contain compounds of nitrogen, phosphorus and potassium: the NPK fertilisers.

Ammonium nitrate and ammonium phosphate are manufactured. Potassium chloride is mined. The three salts are mixed to make NPK fertilisers.

Figure 22.6A shows a flow diagram for the manufacture of NPK fertilisers. It shows how ammonium nitrate and ammonium phosphate are manufactured. There is no need to make potassium chloride because there are plentiful deposits of it in the UK.

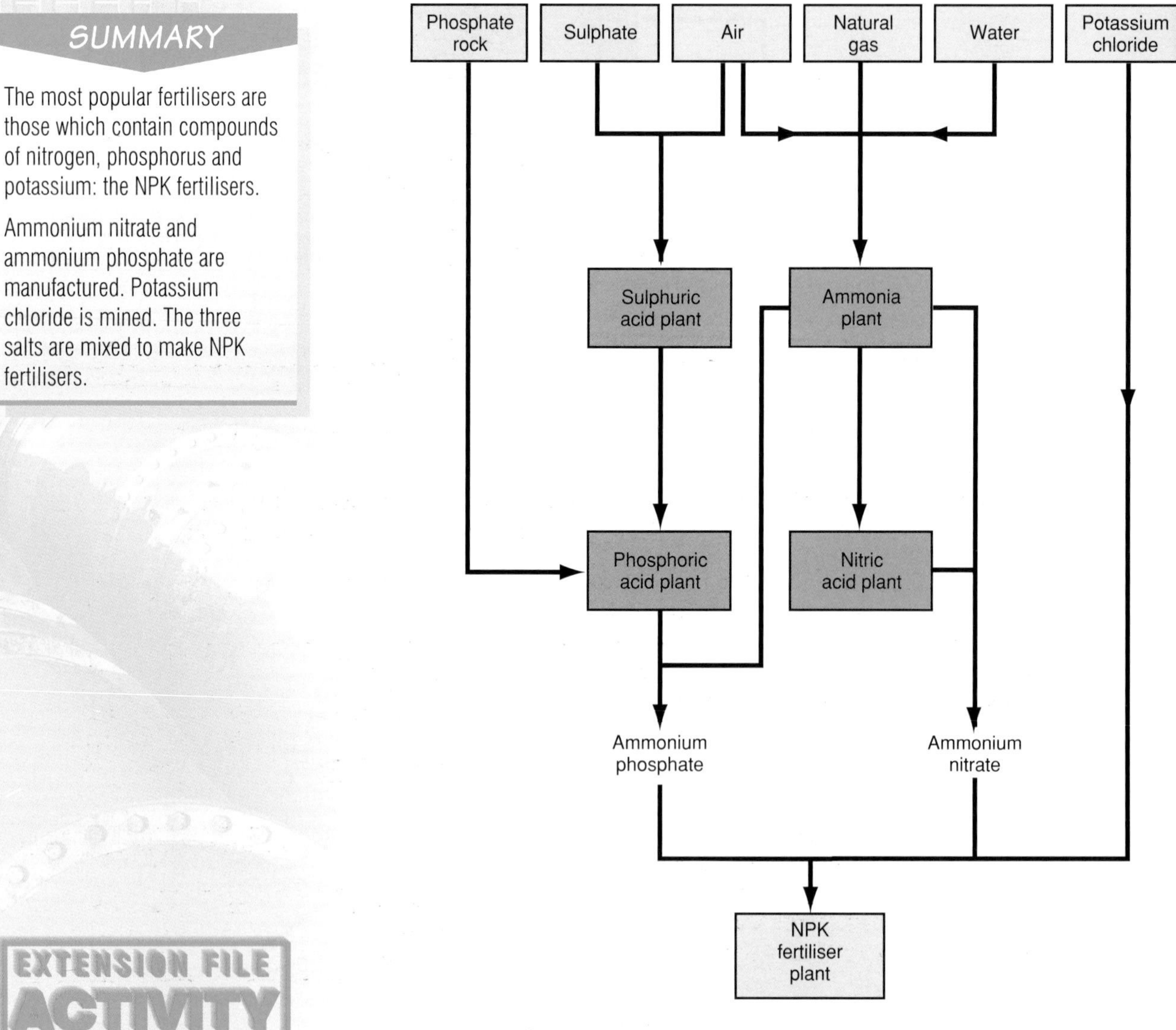

Figure 22.6A Manufacture of NPK fertilisers

EXTENSION FILE ACTIVITY

CHECKPOINT

1. State two advantages of the fertiliser ammonium phosphate over the insoluble salt calcium phosphate.
2. What methods do farmers use for applying (a) ammonia and (b) ammonium sulphate?
3. Why are farmers prepared to pay a lot for fertilisers? Why do all fertilisers contain nitrogen?
4. What is the advantage of using nitrogen as a starting material in the manufacture of fertilisers? What difficulty had to be overcome before manufacture started? Who solved the difficulty and how?
5. NPK fertilisers are popular. What do the letters NPK stand for? Explain why it is important that fertilisers contain (a) N, (b) P and (c) K.
6. State three routes by which nitrogen from the air finds its way into the soil. Why is the nitrogen content of the air not used up?
7. The figure below shows how the yields of winter wheat and grass increase as more nitrogenous fertiliser is used. After studying the graphs, say what mass of nitrogen you would apply to a 100 hectare field to avoid waste and to give a maximum yield of (a) winter wheat and (b) grass.

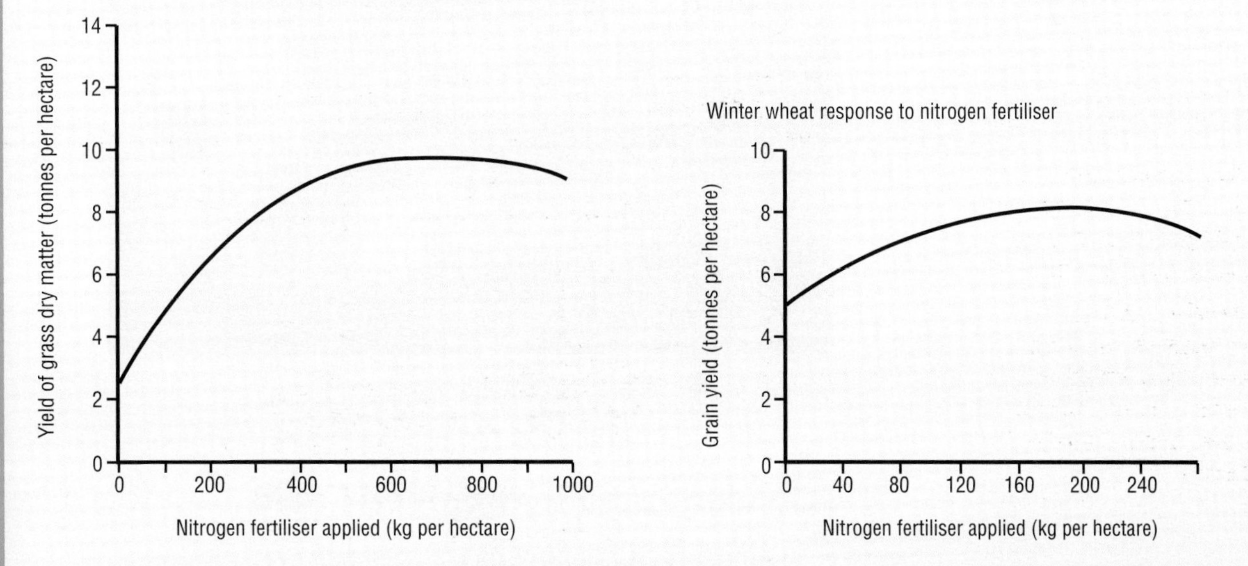

8. A dairy farmer spreads 240 tonnes of nitrochalk on his pasture. It costs £50 per tonne. If his milk cheque is £6000 a month, how long will it take him to recoup the cost of the fertiliser?
9. (a) The table opposite shows the mass of nitrogen, phosphorus and potassium (in kg/hectare) removed from the soil when different crops are grown. Name the crops which take out a large quantity of potassium. How do these crops differ from the rest?

Crop	*Mass* (kg/hectare)		
	N	*P*	*K*
Wheat grain	115	22	26
Oat grain	72	13	18
Sugar beet root	86	14	302
Potatoes	109	14	133
Pasture grass	128	14	100

(b) The table opposite shows the composition of three NPK fertilisers that a farmer uses. Which fertiliser should she use (i) on her pasture (ii) on her wheat and (iii) on her sugar beet? Explain your choices.

Fertiliser	*Composition* (percentage mass)		
	N	*P*	*K*
Fertiliser A	62	13	25
Fertiliser B	22	27	51
Fertiliser C	44	19	37

22.7 Concentrated acids

Figure 22.7A ▲ A Hazard sign for a corrosive substance

Check that you understand the difference between a concentrated acid, a strong acid (see 9.1) and a corrosive acid.

The photograph shows a spillage of corrosive acid being diluted and washed away.

A concentrated solution is one that contains a large amount of solute per litre of solution. Concentrated acids take part in the same chemical reactions as dilute acids but react more vigorously. In addition, concentrated sulphuric acid and concentrated nitric acid take part in some reactions which are not shared by the dilute acids (see below). Concentrated solutions of acids are extremely corrosive. They attack skin and clothing. Great care must be taken in handling concentrated acids and in transporting them. Occasionally, a lorry carrying a corrosive acid is involved in an accident (see Figure 22.7B). Two methods are employed to deal with a spill. The acid may be diluted with plenty of water and then washed away or it may be diluted and then neutralised with a base.

Figure 22.7B ▲ Tanker in simulated road accident

Concentrated hydrochloric acid

Concentrated hydrochloric acid has the same reactions as dilute hydrochloric acid. The concentrated acid reacts more vigorously with metals, bases and carbonates.

Concentrated sulphuric acid

Concentrated sulphuric acid takes part in the same reactions as dilute sulphuric acid and has two additional properties. It is an **oxidising agent** and a **dehydrating agent**.

Oxidising agent

1 Copper and other metals low in the reactivity series do not react with dilute sulphuric acid. Concentrated sulphuric acid reacts with these metals. With copper it forms copper(II) sulphate, sulphur dioxide and water. Sulphuric acid has acted as an oxidising agent; it has taken electrons from copper atoms to form copper ions.

Copper + Conc. sulphuric acid → Copper(II) sulphate + Sulphur dioxide + Water

$Cu(s) + 2H_2SO_4(aq) \rightarrow CuSO_4(aq) + SO_2(g) + 2H_2O(l)$

2 Hot, dilute sulphuric acid reacts with iron to form iron(II) sulphate, $FeSO_4$. With concentrated sulphuric acid, the product is iron(III) sulphate, $Fe_2(SO_4)_3$, showing that concentrated sulphuric acid has acted as an oxidising agent, taking electrons from Fe^{2+} ions to form Fe^{3+} ions.

Dehydrating agent

Ask your teacher to let a drop of concentrated sulphuric acid fall on to a piece of cloth. You will see a brown patch develop, which looks exactly as though the cloth has been burnt. **Don't try it** – but people who have accidentally spilt concentrated sulphuric acid on their skins will tell you that the same thing happens: it feels very much like being burnt by fire. The reason is that concentrated sulphuric acid removes water from the cloth, from skin and from other substances. A reagent which removes water from a compound is called a **dehydrating agent**.

Concentrated sulphuric acid has acidic properties. It is also an oxidising agent, as in its reactions with e.g. copper and iron(II) salts.

Concentrated sulphuric acid is a dehydrating agent removing the elements of water from e.g. copper(II) sulphate-5-water and sucrose.

1 Copper(II) sulphate crystallises with water of crystallisation as copper(II) sulphate-5-water, $CuSO_4.5H_2O$. Concentrated sulphuric acid can remove the water of crystallisation.

Copper(II) sulphate-5-water (blue crystals) $\xrightarrow{\text{Conc. sulphuric acid}}$ Copper(II) sulphate (white powder) + water; Heat is given out

The water of crystallisation combines with the sulphuric acid to leave anhydrous (without water) copper(II) sulphate, which is a white powder. Evidently, the water of crystallisation gives copper(II) sulphate-5-water its colour and its crystalline form.

2 Cane sugar, sucrose, is a carbohydrate (see Topic 30.2) of formula $C_{12}H_{22}O_{11}$. Concentrated sulphuric acid dehydrates sucrose (removes the elements of water from it) to leave a form of carbon known as 'sugar charcoal'.

Sucrose $\xrightarrow{\text{Conc. sulphuric acid}}$ Carbon + Water; Heat is given out

$$C_{12}H_{22}O_{11}(s) \rightarrow 12C(s) + 11H_2O(l)$$

3 The reaction between concentrated sulphuric acid and water itself is very exothermic. This makes it difficult to prepare a dilute solution of sulphuric acid from concentrated sulphuric acid. Never attempt to make a dilute solution by adding water to concentrated sulphuric acid. Heat will be generated in a small volume of acid and will make the added water boil and splash out of the container, bringing with it a shower of concentrated sulphuric acid. Instead, add concentrated sulphuric acid slowly, with stirring, to a large volume of water. Then the heat generated will be spread through a large volume of water.

The figure shows the care that must be taken in diluting concentrated sulphuric acid.

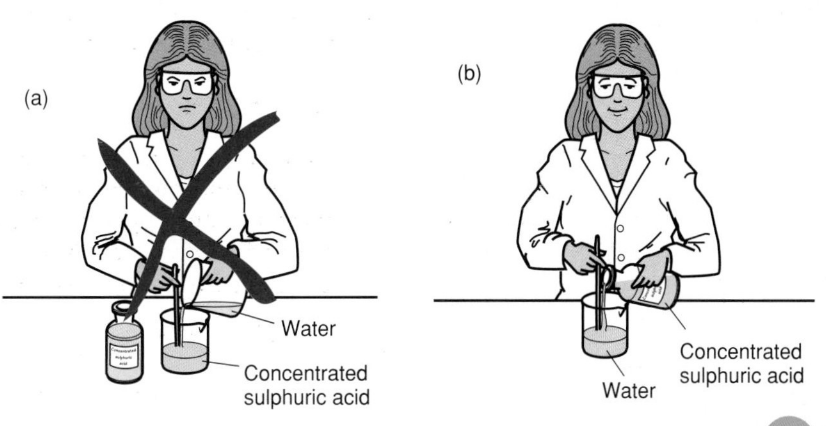

Figure 22.7C ▸
(a) Don't try it this way
(b) This is how to dilute a concentrated acid safely – but wear safety goggles.

Nitric acid – concentrated and dilute

Concentrated nitric acid takes part in the same reactions as dilute nitric acid and also acts as an oxidising agent. It is an even more powerful oxidising agent than concentrated sulphuric acid.

Dilute nitric acid is an oxidising agent. When heated, it reacts with copper to form copper(II) nitrate. Hydrogen is not formed. The colourless gas nitrogen oxide, NO, and the pungent brown gas nitrogen dioxide, NO_2, are formed. This gas is toxic, and the reaction must therefore be carried out in a fume cupboard.

When you need to work with a concentrated acid or alkali it is essential to wear safety goggles (see Figure 22.7C). Spilling a concentrated acid or alkali on your skin is painful and dangerous, and you should immediately wash off the spill with plenty of cold water. The danger to your eyes is very much greater. Firstly, the tissues of the eye are very delicate and easily damaged. Secondly, if you get something in your eye, you cannot see your way to the cold water tap to wash it out. Concentrated alkalis are also extremely dangerous.

SUMMARY

Concentrated sulphuric acid is an oxidising agent and a dehydrating agent. Nitric acid, both concentrated and dilute, is an oxidising agent. In addition, the concentrated acids take part in the same chemical reactions as the dilute acids.

CHECKPOINT

1. Name two types of reaction which concentrated sulphuric acid takes part in but dilute sulphuric acid cannot bring about. Give one example of each.
2. 'Here lies Susan still and placid.
   She added water to the acid.'

   Suggest which acid Susan could have been using to result in an accident. Explain why her method was dangerous. Describe how she should have diluted the acid.
3. 'Dilute nitric acid is an oxidising agent.' Give two reactions which support this statement.
4. Explain why the following reactions are oxidation reactions (see Topic 12).
   (a) Copper is converted into copper(II) sulphate
   (b) Iron is converted into iron(II) sulphate
   (c) Iron is converted into iron(III) sulphate
   (d) Iron(II) sulphate is converted into iron(III) sulphate.

Topic 23 Fuels

23.1 Methane

FIRST THOUGHTS

Developing countries need more fuel. Industrial countries have difficulty in disposing of all their waste. One solution to both problems is biogas. You can find out about biogas in this section.

Father Conlon and Father Williams are monks at Bethlehem Abbey in Northern Ireland. In 1987 they won a 'pollution abatement technology award'. They feed manure from their farm into a **biomass digester**. Biomass is material of plant or animal origin. Bacteria feed on biomass and make it decay. Under anaerobic conditions (in the absence of air), a gas called **biogas** forms. Biogas is a fuel, and the monks burn it to provide the monastery with heating. The solid remains of the biomass form an odourless compost which they bag and sell as a fertiliser for use on gardens, potato farms and golf courses.

Many farms in the UK now have biomass digesters. There are millions of such digesters in India and China. India's huge cattle population supplies plenty of dung for digestion. The biogas produced is used as a fuel for cooking and heating. It burns with a hotter, cleaner flame than cattle dung. The residual sludge which gathers in the digester is a better fertiliser than raw dung. Both the gas and the sludge are clean and odourless.

Science at work

Every year, 25 million tonnes of organic waste goes into landfill sites in the UK. Biogas forms as the rubbish decays. Most landfill operators burn biogas to get rid of it. Others sink pipes into the landfill and pump out biogas for sale. In Bedfordshire, biogas from a landfill is used to heat the kilns in a brickworks. On Merseyside, biogas is used to heat the ovens in a Cadburys' biscuit factory.

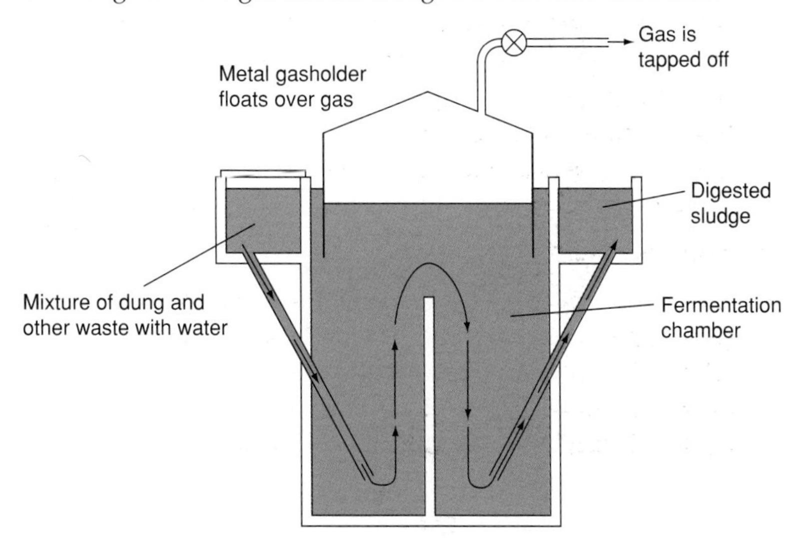

Figure 23.1A A biomass digester

www.keyscience.co.uk

Figure 23.1B A gas rig in the North Sea

The chief component of biogas is **methane**. Methane is the gas we burn in Bunsen burners and gas cookers. It is the valuable fuel we call **natural gas** or **North Sea gas**. Reserves of North Sea gas are usually found together with North Sea oil. Methane is also the gas called **marsh gas** which bubbles up through stagnant water. It collects in coalmines, and methane explosions have been the cause of many pit disasters.

SUMMARY

Biogas is formed when biomass (plant and animal matter) decays in the absence of air. The chief component of biogas is methane. Biogas can be collected from landfill rubbish sites, from sewage works and on farms. It can be burned as a fuel. Biogas generators are common in India and China. Methane accumulates in coal mines, where it can cause explosions.

CHECKPOINT

1. (a) What are the advantages of biogas generators for developing countries?
   (b) Why does India have plenty of cattle dung?
   (c) In what ways is biogas a more convenient fuel than dung?
   (d) What advantages do biogas generators have for industrial countries?
2. (a) What is biomass? Briefly describe how biomass can be fermented to produce biogas. What other product is formed?
   (b) What are the problems in adapting the process to produce fuel gas for domestic use?
   (c) Can you point out any situations in which a biogas generator might both save money and also benefit the environment?

23.2 Alkanes

FIRST THOUGHTS

Methane is one member of a family of compounds, the alkanes. Composed of hydrogen and carbon only, the alkanes are hydrocarbons. You will meet another family of hydrocarbons, the alkenes, in Topic 29.1.

Methane is a **hydrocarbon**, a compound of hydrogen and carbon. With formula CH_4, it is the simplest of the hydrocarbons. Figure 23.2A shows how, in a molecule of methane, four covalent bonds join hydrogen atoms to a carbon atom.

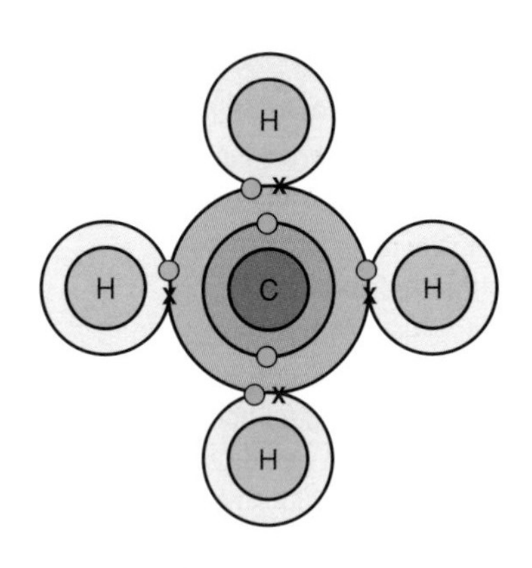

Figure 23.2A ▲ Bonding in methane

Hydrocarbons are **organic compounds**. Originally, the term 'organic compound' was applied to compounds which were found in plant and animal material, for example sugars, fats and proteins. All these compounds contain carbon, and the term 'organic compound' is now used for all carbon compounds, whether they have been obtained from plants and animals or made in a laboratory. However, simple compounds like carbon dioxide and carbonates are not usually described as organic compounds. Most organic compounds are covalent. The salts of organic acids contain ionic bonds.

Methane is one of a **series** of hydrocarbons called the **alkanes**. The next members of the series are ethane, C_2H_6, propane, C_3H_8, and butane, C_4H_{10} (see Figure 23.2B). Many hydrocarbons have much larger molecules.

You often meet the term 'organic compound'. It means a carbon compound, with the exception of simple compounds such as carbonates.

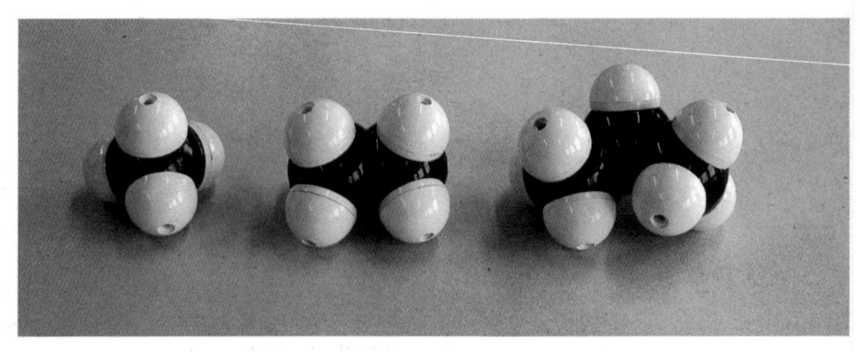

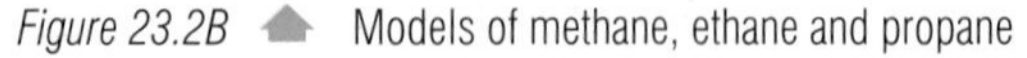
Figure 23.2B ▲ Models of methane, ethane and propane

As well as molecular formulas, CH_4, C_2H_6 and C_3H_8, the compounds have **structural formulas**. A structural formula shows the bonds between atoms.

The srtructural formula of a compound shows the bonds between atoms.

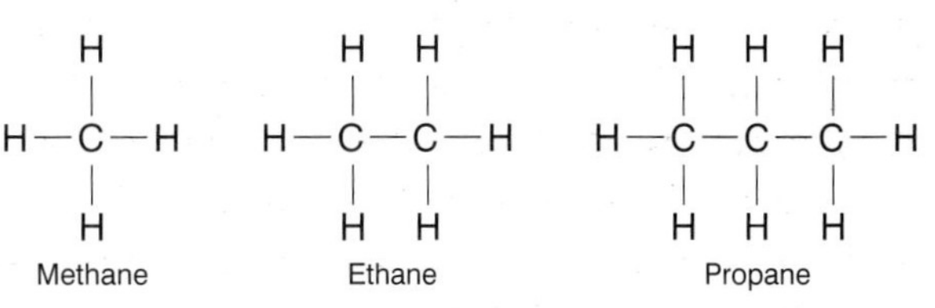

Each compound differs from the next in the series by the group

```
  H
  |
—C—
  |
  H
```

Table 23.1 ▾ The alkanes

Alkanes	Formula C_nH_{2n+2}
Methane	CH_4
Ethane	C_2H_6
Propane	C_3H_8
Butane	C_4H_{10}
Pentane	C_5H_{12}
Hexane	C_6H_{14}
Heptane	C_7H_{16}
Octane	C_8H_{18}

SUMMARY

Carbon compounds are called organic compounds.

Hydrocarbons are compounds of carbon and hydrogen. The alkanes are hydrocarbons with the general formula C_nH_{2n+2}. They are a homologous series: each member differs from the next by a CH_2 group.

A set of chemically similar compounds in which each member differs from the next in the series by a CH_2 group is called a **homologous series**. Table 23.1 lists the first members of the **alkane series**. Their formulas all fit the general formula C_nH_{2n+2}, for example for pentane $n = 5$, giving the formula C_5H_{12}. The alkanes are described as **saturated** hydrocarbons. This means that they contain only single bonds between carbon atoms. This is in contrast to the **alkenes**. The alkanes are unreactive towards acids, bases, metals and many other chemicals. Their important reaction is combustion.

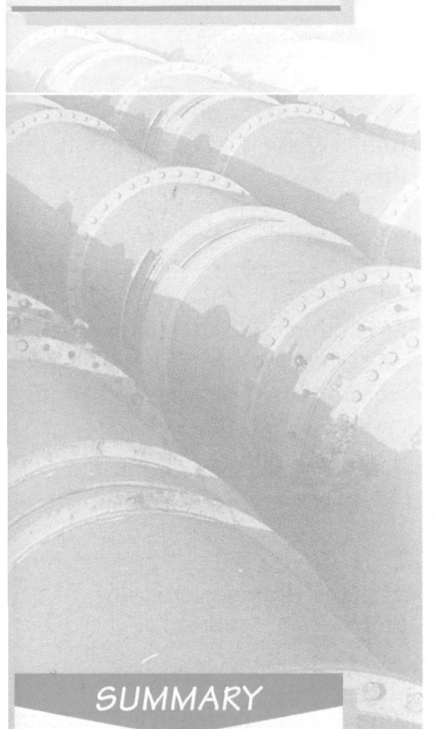

Isomerism

Sometimes it is possible to write more than one structural formula for a molecular formula. For the molecular formula C_4H_{10}, there are two possible structures

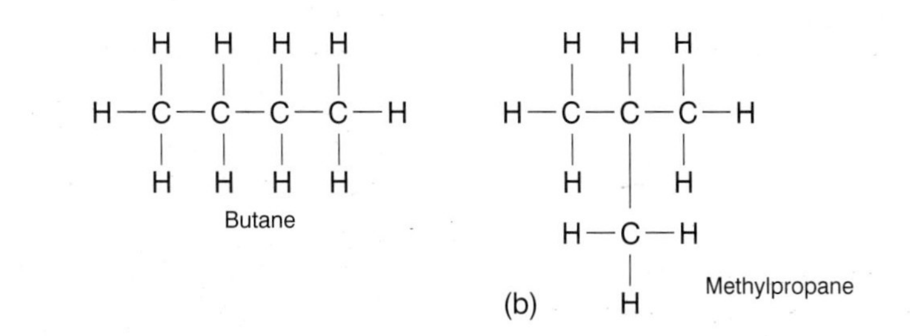

The difference is that (a) has an unbranched chain of carbon atoms and (b) has a branched chain. The formulas belong to different compounds, which differ in boiling point and other physical properties. The compound with formula (a) is called butane; the compound with formula (b) is called methylpropane. These compounds are **isomers**. Isomers are compounds with the same molecular formula and different structural formulas.

SUMMARY

Isomers have the same molecular formula and different structural formulas.

CHECKPOINT

1. Explain what is meant by a 'homologous series'.
2. (a) Explain what is meant by the term 'alkane'.
   (b) Why are alkanes very important in our way of life?
   (c) Where do we obtain alkanes?
   (d) The molecular formula for propane is C_3H_8. Write the structural formula for propane.
   (e) What information does the structural formula of a compound give that the molecular formula does not give?
   (f) Explain what is meant by the term 'isomerism'. Illustrate your answer by referring to pentane, C_5H_{12}.

Substitution reactions

Alkanes react with chlorine in sunlight. The products are a **chloroalkane** and hydrogen chloride. For example,

Methane	+	Chlorine	→	Chloromethane	+	Hydrogen chloride
$CH_4(g)$	+	$Cl_2(g)$	→	$CH_3Cl(g)$	+	$HCl(g)$

```
     H                      H
     |                      |
 H — C — H + Cl — Cl → H — C — Cl + H — Cl
     |                      |
     H                      H
```

The reaction is called a **substitution reaction** because one atom in the molecule of alkane is replaced by another atom. It is called **chlorination** because it is a chlorine atom that is substituted into the alkane molecule. It is described as a **photochemical reaction** because it takes place only in sunlight. The reaction can continue further.

Chloromethane	+	Chlorine	→	Dichloromethane	+	Hydrogen chloride
$CH_3Cl(g)$	+	$Cl_2(g)$	→	$CH_2Cl_2(g)$	+	$HCl(g)$

A mixture of products, including trichloromethane, $CHCl_3$ (chloroform) and tetrachloromethane, CCl_4 (also called carbon tetrachloride), is formed. Trichloromethane, $CHCl_3$, is chloroform, which was used as an anaesthetic.

Bromine also reacts with alkanes in sunlight. The substitution product which is formed is called a **bromoalkane**. Hydrogen bromide is the other product, e.g.

Hexane	+	Bromine	→	Bromohexane	+	Hydrogen bromide
$C_6H_{14}(l)$	+	$Br_2(l)$	→	$C_6H_{13}Br(l)$	+	$HBr(g)$

Iodination is very slow and is a reversible reaction. Fluorination is a dangerously violent reaction. The substitution of a halogen in a compound is called **halogenation**. Chlorination and bromination are both halogenation reactions. Compounds in which halogen atoms have been substituted for one or more of the hydrogen atoms in an alkane are called **halogenoalkanes**. Chloromethane, tetrachloromethane and bromohexane are all halogenoalkanes.

In a substitution reaction, an atom or atoms in a molecule are replaced by another atom or atoms. An example is the halogenation of alkanes. This includes chlorination to form chloroalkanes and bromination to form bromoalkanes.

SUMMARY

Alkanes take part in substitution reactions, e.g. halogenation.

CHECKPOINT

1. (a) Write the equation for the reaction between ethane and chlorine to form chloroethane.
   (b) Write the structural formulas for all the substances.
   (c) Construct molecular models of all the substances.
2. (a) Write the structural formulas for iodomethane, dibromomethane, fluoroethane, tetrachloromethane.
   (b) Make molecular models of the molecules.
3. Is the chlorination of methane exothermic or endothermic? Explain your answer.

23.3 Combustion

SUMMARY

The combustion of hydrocarbons is exothermic. The products of complete combustion are carbon dioxide and water. Incomplete combustion produces carbon and poisonous carbon monoxide.

Hydrocarbons burn in a plentiful supply of air to form carbon dioxide and water. The reaction is **exothermic**: energy is released.

Methane + Oxygen → Carbon dioxide + Water; Energy is released

$CH_4(g) + 2O_2(g) \rightarrow CO_2(g) + 2H_2O(l)$

Butane + Oxygen (in Camping Gaz) → Carbon dioxide + Water; Energy is released

Octane + Oxygen (in petrol) → Carbon dioxide + Water; Energy is released

CHECKPOINT

Revise what you learned about combustion in Topic 14.6.

1. (a) What harmless combustion products are formed when hydrocarbons burn in plenty of air?
   (b) When does incomplete combustion take place?
   (c) What are the products of incomplete combustion? What is dangerous about incomplete combustion? How can it be avoided?

23.4 Petroleum oil and natural gas

FIRST THOUGHTS

Prospecting for oil is an even more important business than prospecting for gold. *How do prospectors find oil? And how was oil formed in the first place?* You can find out in this section.

In some parts of the world, a black, treacle-like liquid seeps out of the ground. At one time, farmers in Texas, USA, used to burn this substance to get rid of it when it formed troublesome pools on their land. The black liquid was called **petroleum oil** or **crude oil**. It is now regarded as one of the most valuable resources a country can have. The petrochemicals industry obtains valuable fuels such as petrol, and thousands of useful materials such as plastics, from crude oil. Natural gas is found in the same deposits as crude oil.

We call oil and gas fossil fuels because they were formed by slow decay of the remains of dead sea creatures.

Oil and gas are **fossil fuels**. They were formed millions of years ago when a large area of the Earth was covered by sea. Tiny sea creatures called plankton died and sank to the sea bed, where they became mixed with mud. Bacteria in the mud began to bring about the decay of the creatures' bodies. Decay took place slowly because there is little oxygen dissolved in the depths of the sea. As the covering layer of mud and silt grew thicker over the years, the pressure on the decaying matter increased. Bacterial decay at high pressure with little oxygen turned the organic matter into crude oil and natural gas.

The sediment on top of the decaying matter became compressed to form rock. Rocks formed in this way are called **sedimentary rocks**. Some of these rocks are porous: they contain tiny passages through which liquid and gas can pass. Others are impermeable: they do not let any substances through. Ground water carries crude oil and natural gas upward through porous rocks. They may reach the surface, but more often they become trapped by a **cap rock** of impermeable rock (see Figure 23.4A). There the crude oil and natural gas remain unless an oil prospector drills down to them.

It's a fact!

Have you heard of the petrol tree? The gopher tree grows well in desert areas. Its sap contains about 30% hydrocarbons. Petrol could be made from the sap of this tree.

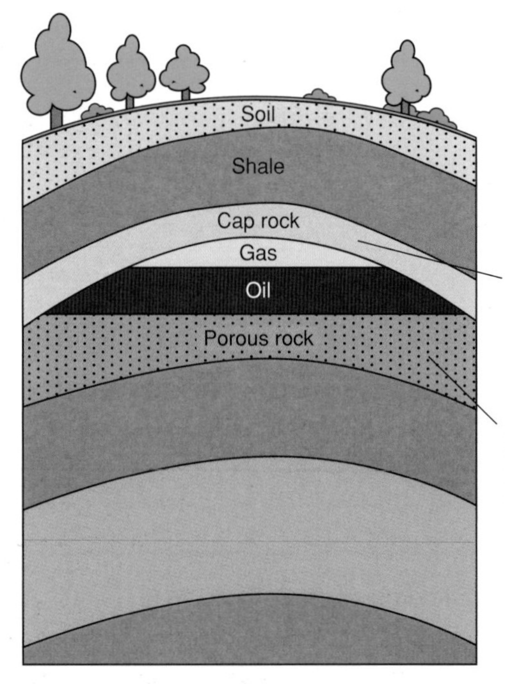

Figure 23.4A ▲ An oil trap

Many countries have deposits of oil and gas, for example the USA, the USSR, Iran, Nigeria and the countries in the Arabian Gulf. The UK and Norway have oil and gas beneath the North Sea. The UK has piped ashore oil and gas since 1972. There are two ways of getting oil ashore from an oil well in the sea. One way is to lay a pipeline along the sea bed and pump the oil through it. The other method is to use 'shuttle tankers' which pick up the oil from the oilfield and transport it to a terminal on land. Natural gas is almost always brought ashore by pipeline.

Crude oil is transported from oil wells to refineries. Pipelines carry oil overland. Tankers carry oil overseas. When a tanker has an accident at sea, oil can escape to form a huge oil slick. The damage which the oil may cause to wildlife and coastlines is described in Topic 17.6.

Crude oil does not burn very easily. **Fractional distillation** is used to separate crude oil into a number of important fuels. The fractions are separated on the basis of their boiling point range. They are not pure compounds: they are mixtures of alkanes with similar boiling points.

SUMMARY

Crude oil and natural gas were formed by the slow bacterial decay of animal and plant remains in the absence of oxygen. These fuels are found in many parts of the world.

Figure 23.4B ▼ An oil rig being towed out to sea

Table 23.2 ▼ Petroleum fractions and their uses

Fraction	Approximate boiling point range (°C)	Approximate number of carbon atoms per molecule	Use
Petroleum gases	below 0	1–4	Petroleum gases are liquefied and sold in cylinders as 'bottled gas' for use in gas cookers and camping stoves. They burn easily at low temperatures. Smelly sulphur compounds must be removed to make bottled gas pleasant to use and non-polluting.
Gasoline (petrol)	0–65	5–6	Petrol is liquid at room temperature, but vaporises easily at the temperature of vehicle engines.
Naphtha	65–170	6–10	Naphtha is used by the petrochemicals industry as the source of a huge number of useful chemicals, e.g. plastics, drugs, medicines, fabrics (see Figure 23.4C overleaf).
Kerosene	170–250	10–14	Kerosene is a liquid fuel which needs a higher temperature for combustion than petrol. The major use of kerosene is as aviation fuel. It is also used in 'paraffin' stoves.
Diesel oil	250–340	14–19	Diesel oil is more difficult to vaporise than petrol and kerosene. The diesel engine has a special fuel injection system which makes the fuel burn. It is used in buses, lorries and trains.
Lubricating oil	340–500	19–35	Lubricating oil is a viscous liquid. With its high boiling point range, it does not vaporise enough to be used as a fuel. Instead, it is used as a lubricant to reduce engine wear.
Fuel oil	340–500	Above 20	Fuel oil has a high ignition temperature. To help it to ignite, fuel oil is sprayed into the combustion chambers as a fine mist of small droplets. It is used in ships, heating plants, industrial machinery and power stations.
Bitumen	Above 500	Above 35	Bitumen has too high an ignition temperature to be used as a fuel. It is used to tar roads and to waterproof roofs and pipes.

Physical properties of alkanes which depend on the size of the molecules include

- viscosity
- flash point
- ignition temperature
- boiling point.

They decide the uses that are made of petroleum fractions

Alkanes with small molecules boil at lower temperatures than those with large molecules. Alkanes with large molecules are more viscous than alkanes with small molecules. The fractions also differ in the ease with which they burn: in their **flash points** and **ignition temperatures**.

- When a fuel is heated, some of it vaporises. When the fuel reaches a temperature called the **flash point**, there is enough vapour to be set alight by a flame. Once the vapour has burned, the flame goes out.
- The **ignition temperature** of a fuel is higher than its flash point. It is the temperature at which a mixture of the fuel with air will ignite and continue to burn steadily.

The use that is made of each fraction depends on its boiling point range, flash point, ignition temperature and viscosity; see Table 23.2.

It's a fact!

A petroleum fraction called white spirit is widely used as a solvent. In November, 1988, a fire at King's Cross underground station in London killed 30 people. Under the escalators, investigators found empty drums that had held 150 litres of white spirit. The staff use white spirit for cleaning the treads of the escalators. The presence of flammable white spirit could help to explain the speed with which the fire spread.

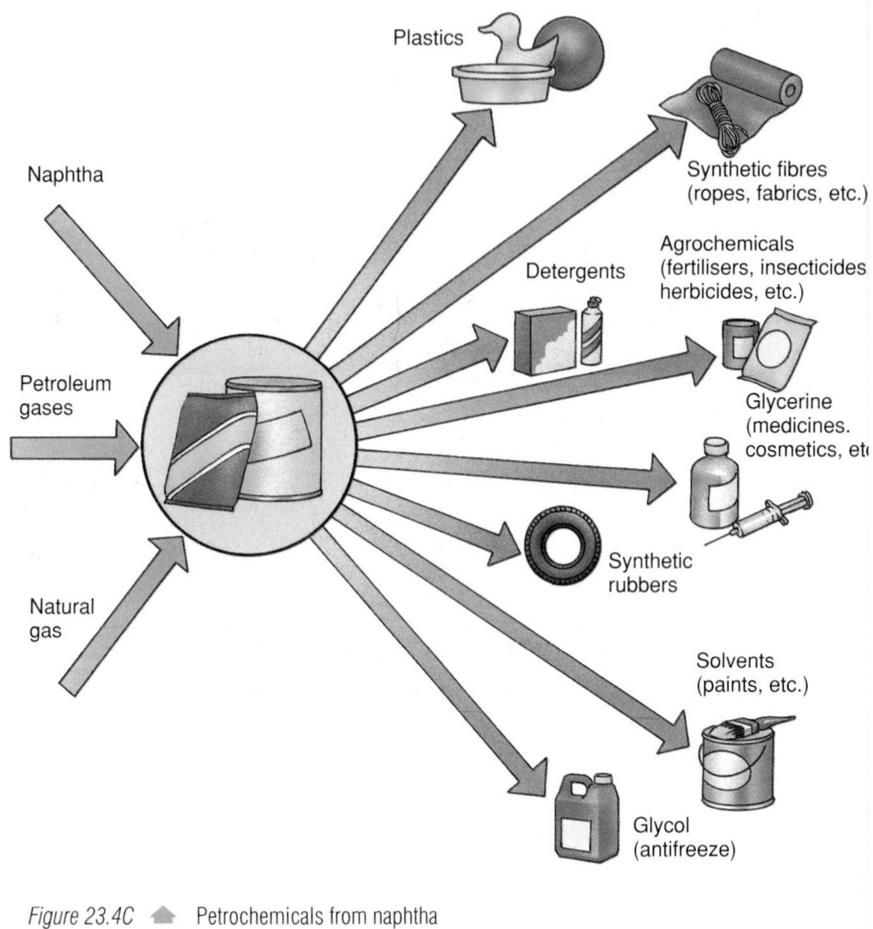

Figure 23.4C ▲ Petrochemicals from naphtha

It's nice to meet a term that means what it sounds like. 'Cracking' means cracking large molecules into smaller molecules.

Cracking

We use more petrol, naphtha and kerosene than heavy fuel oils. Fortunately, chemists have found a way of making petrol and kerosene from the higher-boiling fractions, of which we have more than enough. The technique used is called **cracking**. Large hydrocarbon molecules are cracked (split) into smaller hydrocarbon molecules. A heated catalyst (aluminium oxide or silicon(IV) oxide) is used.

Vapour of hydrocarbon with large molecules and high b.p. —CRACKING (passed over a heated catalyst)→ Mixture of hydrogen and hydrocarbons with smaller molecules and low b.p.

SUMMARY

Crude oil is separated into useful components by fractional distillation. The use that is made of each fraction depends on its boiling point range, ignition temperature and other properties. The fuels obtained from crude oil are listed in Table 23.2. Cracking is used to make petrol and kerosene from heavy fuel oils. The petrochemicals industry makes many useful chemicals from petroleum.

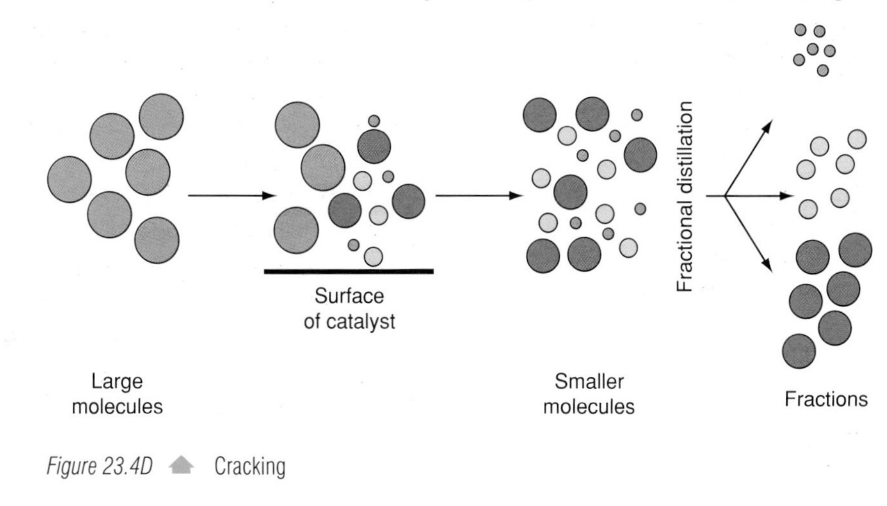

Figure 23.4D ▲ Cracking

CHECKPOINT

▸ **1** The use of crude oil fractions in the UK is as follows:

Road transport 37%
Heating and power for industry 21%
Heating and power for other consumers 19%
Making chemicals 11%
Power stations 8%
Heating houses 4%

Show this information in the form of a pie-chart or a bar graph.

▸ **2** The figure below shows the percentage of crude oil that distils at various temperatures.

(a) From the graph, read the percentage of crude oil that distils:

(i) below 70 °C,
(ii) between 120 and 170 °C,
(iii) between 170 and 220 °C,
(iv) between 220 and 270 °C,
(v) between 270 and 320 °C,
(vi) above 320 °C.

(b) Draw a pie-chart to show these figures.

(c) The percentages vary from one sample of oil to another. North Sea oil contains a larger percentage of lower boiling point compounds than Middle East oil does. Which oil should sell for a higher price? Explain your answer.

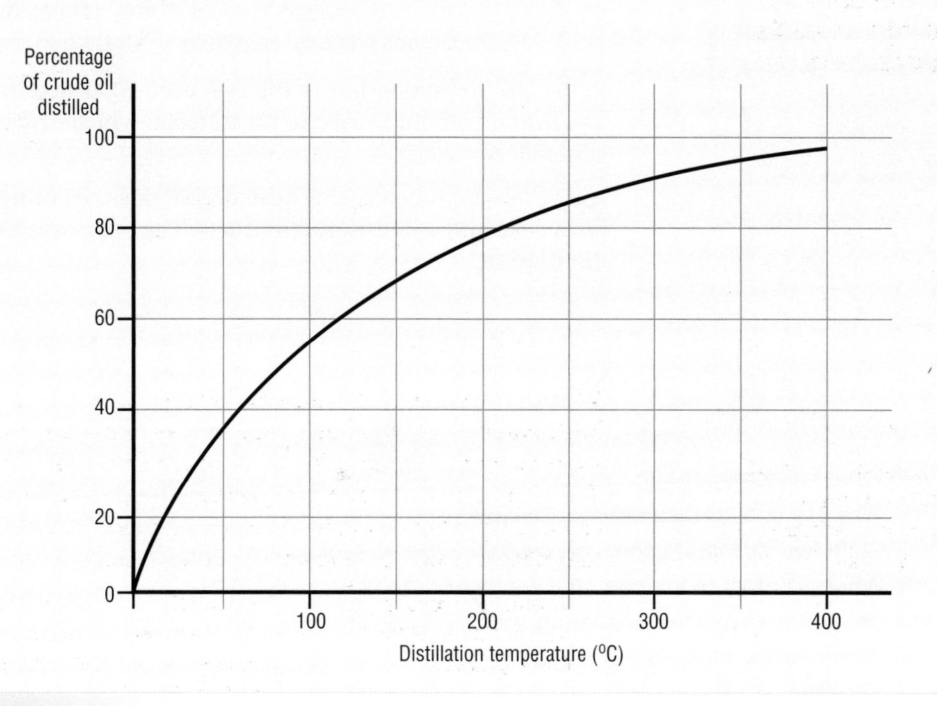

3 A barrel of Middle East oil contains 160 litres. From it are obtained:
natural gas (7 litres) kerosene (7 litres)
petrol (9 litres) gas oil (41 litres)
naphtha (23 litres) fuel oil (73 litres)
Show these figures as a pie-chart.

4 (a) How are petrol and kerosene obtained from crude oil?
(b) What is the name of the process for converting heavy fuel oil into petrol? What are the economic reasons for carrying out this reaction?

5 (a) What is crude oil?
(b) Why is it described as a fossil fuel?
(c) What useful substances are obtained from crude oil?
(d) Why is an increase in the price of crude oil such a serious matter?

23.5 Coal

FIRST THOUGHTS

Coal is an important source of energy. It fuelled the Industrial Revolution and is still a major source of energy in the UK today.

Between 200 and 300 million years ago, the Earth was covered with dense forests. When plants and trees died, they started to decay. In the swampy conditions of that era, they formed peat. Gradually, the peat became covered by layers of mud and sand. The pressure of these layers and high temperatures turned the peat into coal. The mud became shale and the sand became sandstone. Coal is a fossil fuel: it was formed from the remains of living things.

Many countries have coal deposits. The countries with the largest coal-mining industries are the countries of the former USSR, the USA, China, Poland and the UK. Coal is a mixture of carbon, hydrocarbons and other compounds. When it burns, the main products are carbon dioxide and water.

Carbon (in coal) + Oxygen → Carbon dioxide
Hydrocarbons (in coal) + Oxygen → Carbon dioxide + Water

In the UK, three-quarters of the coal used is burned in power stations. The heat given out is used to raise steam, which drives the turbines that generate electricity.

If air is absent when coal is heated, coal does not burn. It decomposes to form coke and other useful products. The process is called **destructive distillation**.

SUMMARY

Coal is a fossil fuel derived from plant remains. Most of the coal mined is burned in power stations. The destructive distillation of coal gives useful products.

CHECKPOINT

1 (a) What is a fuel?
(b) Why are coal and oil called fossil fuels?
(c) State two properties that make a fuel a 'good' fuel.
(d) Where is most of the coal that we use burned?
(e) What use is made of coal apart from burning it?

Theme Questions

1 Iron is extracted from its ores in a blast furnace. Most iron is converted into steel in a basic oxygen furnace.

(a) Name the two substances, in addition to iron ore, which are fed into the top of the blast furnace.
(b) Explain how the ore haematite, Fe_2O_3, is reduced to iron.
(c) Molten iron and slag form at the base of the blast furnace. Explain how slag is formed.
(d) What impurities are removed from iron in the basic oxygen furnace?
(e) Explain the chemical reaction that removes them.

2 Aluminium is mined as its oxide. It is extracted from its ore in a plant called an aluminium smelter.

(a) Name this aluminium oxide ore.
(b) State what process is used to extract aluminium from its oxide.
(c) Why can a blast furnace not be used for the extraction of aluminium?
(d) An aluminium ore called cryolite is used in the extraction of aluminium from aluminium oxide. What part does it play in the process?
(e) What environmental damage can be caused by an aluminium smelter?
(f) What two economies are made when aluminium is recycled?

3 The table shows part of the reactivity series of metals.

Aluminium
Zinc
Iron
Tin
Lead
Copper

Use the table to explain the following.

(a) Iron food cans are coated with tin.
(b) Zinc bars are attached to the hulls of ships below the waterline.
(c) When zinc powder is dropped into a solution of copper(II) sulphate, the colour of the solution fades.
(d) Galvanised (zinc-coated) steel does not rust.
(e) A mixture of aluminium and iron oxide is used to mend gaps in railway lines.

4 Explain why:

(a) electrical wiring is made of copper
(b) saucepans are made from aluminium
(c) aeroplanes are made from aluminium alloys
(d) bells are made from bronze
(e) trumpets are made from brass
(f) bridges are made from steel
(g) baking foil is made from aluminium
(h) solder is made from brass and tin
(i) lead was for many years used for water pipes
(j) lead is no longer used for water pipes
(k) dental amalgams are made from mercury
(l) teeth can be fitted with gold caps.

5 A geologist finds a green compound in a sample of rock. He does a number of experiments on the sample. The results are shown in the diagram.

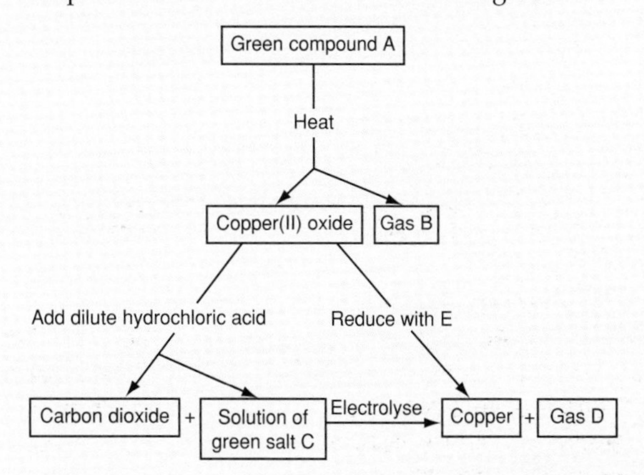

(a) Name the substances A, B, C and D.
(b) Name one substance which could be E. Describe (with a diagram if you wish) how this reaction could be carried out. Say, with an explanation, whether this reaction is an oxidation or a reduction.
(c) Is the formation of copper from C by electrolysis oxidation or reduction? Explain your answer.

6 (a) The table shows the prices and the lifetimes of two kinds of AA-size dry cell.

	Alkaline manganese dry cell	*Zinc-carbon dry cell*
Price of cell	£0.45	£0.20
Life when used in		
(i) electronic flash	180 flashes	20 flashes
(ii) walkman	14 hours	7 hours

Which of the two dry cells is the better buy when used in (i) an electronic flash (ii) a walkman? Explain your choices.

7 Read the following passage and answer the questions on it.

Mercury

In ancient times, the material used for filling cavities in teeth was a cement made of metal oxides. In medieval times, the method of clearing the cavity of decayed matter and then filling it with gold leaf was employed. Early in the nineteenth century the first dental amalgam was used. (An amalgam is an alloy of mercury with other metals.) The original dental amalgam of bismuth, lead, tin and mercury melted at about 100 °C and was poured into the cavity at this temperature. Later the composition was altered to make an amalgam which melted at 66 °C.

In the late nineteenth century, dentistry was revolutionised by G.V. Black. He measured the force required to chew different foods and the pressure which fillings could stand without cracking. Black's formula, 65% silver, 27% tin, 6% copper and 2% zinc is still used today, although many manufacturers use more copper and less tin. Dentists mix this powder with mercury, put it into the cavity, 'carve' it into shape, and allow it to set for 5–10 minutes. After 48 hours, the amalgam has hardened.

(a) What is the name for a mixture of metals?
(b) What is the name for a mixture of mercury with other metals?
(c) Why is gold a good metal to use for filling teeth?
(d) Why does a cavity have to be cleaned before it is filled?
(e) What is the advantage of gold over the metal oxides used to fill cavities?
(f) What is the drawback to gold?
(g) How did scientific instruments help dentistry?
(h) What was the disadvantage of the original dental amalgam?
(i) Give three reasons why mercury is a good basis for a dental amalgam.
(j) Your little brother finds out that his fillings contain mercury. He becomes very alarmed because he has heard of mercury poisoning. How would you explain to him why the mercury in his teeth does not poison him?

8 The following metals are listed in order of decreasing reactivity. X and Y are two unknown metals.

K X Ca Mg Al Zn Y Fe Cu

(a) Will X react with cold water?
(b) Will Y react with cold water?
(c) Will Y react with dilute hydrochloric acid?

Explain how you arrive at your answers.

(d) What reaction would you expect between zinc sulphate solution and (i) X (ii) Y?
(e) Which is more easily decomposed, XCO_3 or YCO_3?

9 Look at the following pairs of chemicals. If a reaction happens, copy the word equation, and complete the right hand side.

(a) Copper + Oxygen →
(b) Calcium + Hydrochloric acid →
(c) Copper + Sulphuric acid →
(d) Carbon + Lead(II) oxide →
(e) Hydrogen + Calcium oxide →
(f) Aluminium + Tin(II) oxide →
(g) Gold + Oxygen →
(h) Zinc + Copper(II) sulphate solution →
(i) Magnesium + Sulphuric acid →
(j) Hydrogen + Silver oxide →
(k) Carbon + Magnesium oxide →
(l) Lead + Copper(II) sulphate solution →
(m)Hydrogen + Potassium oxide →

10 Explain these statements about aluminium.

(a) Although aluminium is a reactive metal, it is used to make doorframes and windowframes.
(b) Although aluminium conducts heat, it is used to make blankets which are good thermal insulators.
(c) Although aluminium oxide is a common mineral, people did not succeed in extracting aluminium from it until seven thousand years after the discovery of copper.
(d) Recycling aluminium is easier than recycling scrap iron.

11 (a) Explain how steel is made from cast iron.
(b) Explain what advantages steel has over cast iron.
(c) Explain how the following methods protect iron against rusting: painting, galvanising, tin-plating, sacrificial protection.

12 Imagine that you live in a beautiful part of Northern Ireland. A firm called Alumco wants to build a new aluminium plant in your area so that they can use a river as a source of hydroelectric power.

(a) Write a letter from a local farmer to the Secretary of State for Northern Ireland. Say what you fear may happen as a result of pollution from the plant.
(b) Write a letter from a group of environmentalists to the Secretary of State, opposing the plan and giving your reasons.
(c) Write a letter from an unemployed couple to the Secretary of State saying that you welcome the coming of new industry to the area.
(d) Write a letter from the local Council to the Secretary of State. Tell him or her that there is very little unemployment in the area. Say that the new plant would have to bring in workers from outside the region. Explain that there is not enough housing in the area for newcomers.
(e) Write a letter from Alumco to the Secretary of State for Northern Ireland. Tell the Secretary of State of the importance of aluminium. Point out the many uses of aluminium. Explain that to keep up with increasing demand you have to build another plant to supply aluminium.

(If five letters are too many for you, divide the work among the class. Then get together to read out the letters. Have a discussion to decide what the Secretary of State ought to do.)

13 Concentrated sulphuric acid has some properties which are not shared by dilute sulphuric acid. It is a dehydrating agent and an oxidising agent.
(a) Give two examples of the action of concentrated sulphuric acid as a dehydrating agent.
(b) Give one example of concentrated sulphuric acid acting as an oxidising agent.
(c) Give two examples of the reactions of dilute sulphuric acid.
(d) Name three substances which are attacked by the sulphuric acid in acid rain.

14 Calcium carbonate is quarried as limestone. It is heated in lime kilns to form calcium oxide (quicklime) and a second product.
(a) What is the second product?
(b) State two uses of this product.
(c) Explain why quicklime is spread by some farmers on their fields.
(d) Explain the advantage of using powdered limestone in place of quicklime.
(e) Most of the limestone which is quarried is used in the manufacture of concrete. Briefly explain the process involved.
(f) What are the properties of concrete that make it a useful building material?
(g) When all the limestone has been extracted from a quarry, what should be done to the quarry?

15 (a) What type of plants can use atmospheric nitrogen as a nutrient?
(b) Why can't most plants use atmospheric nitrogen in this way?
(c) Name the nitrogen-containing compounds that are built by plants.
(d) Name one natural process that produces nitrogen compounds that enter the soil and are used by plants.
(e) Explain why there is not enough nitrogen from natural sources to support the growth of crop after crop on the same land.

16 (a) Explain why ammonium salts are used as fertilisers even though plants cannot absorb them (see Topic 13.6).
(b) Describe how you could make ammonium sulphate crystals (see Topic 10.5).

17 (a) Explain what is meant by the **nitrogen cycle**.
(b) Are we likely to run out of nitrogen?
(c) Say how the human race alters the natural nitrogen cycle (i) by taking nitrogen out of the cycle and (ii) by adding nitrogen.
(d) Explain how fertilisers have created a problem concerning nitrogen compounds.
(e) Suggest two ways in which this problem could be attacked.

18 Farmer Short had a field in which his crops did not grow well. One year, he added fertiliser to the soil and his crops grew better. However, a pond near the field became stagnant and full of algae, and the fish in the pond died. When the pond was tested, it was found to contain fertiliser.

Farmer Long did not use fertiliser. She rotated her crops between fields so that every few years each field grew peas, beans or clover.
(a) Explain how fertiliser got into the pond.
(b) Why did the algae in the pond increase?
(c) Why did the fish die?
(d) How do crops of peas, beans and clover make up for the lack of fertiliser?
(e) Give one advantage and one disadvantage of Farmer Long's system compared with Farmer Short's.

19 Explain each of the following statements:
(a) Prolonged use of artificial fertilisers is bad for the soil.
(b) Extensive use of fertilisers on arable land can lead to the pollution of waterways.
(c) Growing clover improves the fertility of the soil.

20 The diagram shows the nitrogen cycle. The labelled arrows represent different processes.

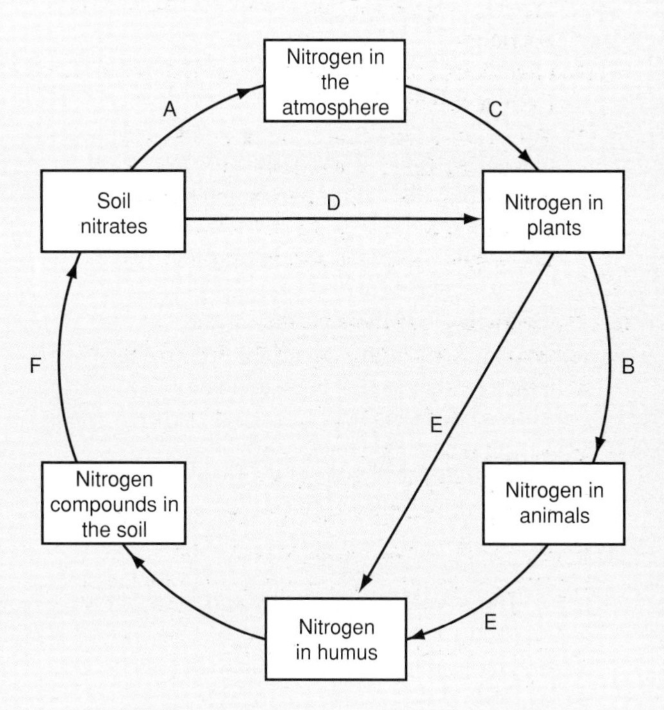

(a) Name the processes A–F, e.g. A = denitrification.
(b) (i) Why is process C useful to plants?
(ii) Process C takes place in clover but not in grass. Why is this so?
(c) What organisms are responsible for process E?
(d) In what form is nitrogen passed in process B?

21 (a) What is meant by 'recycling' glass?

(b) The value of used glass from a Bottle Bank just pays for the cost of its collection. Is it worth the trouble of collecting glass?

(c) Why do Bottle Banks collect brown, green and colourless glass separately?

(d) Why are people asked to remove metal bottle tops before placing bottles in the Bottle Banks?

(e) In which section of the Periodic Table would you be likely to find metal oxides to colour glass blue, green, brown and other colours?

22 (a) Iron corrodes in air forming rust.

(i) Give the chemical name for rust.

(ii) Name the chemical process for rusting. [2]

(b) The rusting of five identical nails was investigated by treating each nail as shown in the table below. All five nails were left exposed to the atmosphere for a few months.
One of the results in the table is incorrect.

Nail	*Treatment used*	*Cost of treatment*	*Mass of nail and coating before exposure to the atmosphere (grams)*	*Mass of nail and coating after exposure to the atmosphere (grams)*
A	oiled	cheap	2.0	2.3
B	chromium plated	expensive	2.0	2.0
C	painted	cheap	2.0	2.2
D	galvanised	fairly expensive	2.0	1.2
E	untreated	nil	1.9	2.7

Using only the information in the table above give the treatment

(i) which gives the best protection,

(ii) which is usually used to protect steel bridges from rusting,

(iii) which is usually used to protect steel garden tools from rusting,

(iv) which has the incorrect result. [4]

(c) Explain why, although aluminium is more reactive than iron, it appears to be corrosion-resistant. [2] (WJEC)

23 There are four main steps in the manufacture of the fertiliser ammonium nitrate.

1. The reaction of methane with steam to produce hydrogen.
2. The reaction of hydrogen and nitrogen to produce ammonia.
3. The oxidation of ammonia to produce nitric acid.
4. The reaction of ammonia with nitric acid to produce ammonium nitrate.

(a) Balance the equation for the reaction of methane with steam.

$\ldots CH_4 + \ldots H_2O \rightarrow \ldots H_2 + \ldots CO_2$ [1]

(b) (i) Suggest the source of nitrogen for step 2. [1]

(ii) Write the balanced equation, including state symbols, for the reaction in step 2. [3]

(c) (i) What type of reaction is taking place in step 4? [1]

(ii) What is the formula of ammonium nitrate? [1]

(d) The fertiliser ammonium nitrate is very soluble in water.
Describe the advantages and disadvantages of this. [4] (Edexcel)

24 (a) Magnesium is manufactured by electrolysis of magnesium chloride.

(i) Explain why this process is expensive to operate. [1]

(ii) Complete the word equation for the electrolysis of magnesium chloride. [1]

Magnesium chloride → Magnesium + . . .

(b) (i) Draw a diagram to show the electronic structure of an *atom* of magnesium. [2]

(ii) How does a magnesium atom (Mg) change when it forms a magnesium ion (Mg^{2+})? [2]

(c) Magnesium (Mg) burns in oxygen to form magnesium oxide (MgO).

(i) Write a balanced equation for this reaction. [2]

(ii) Explain why magnesium is said to be oxidised in this reaction. [1]

(d) The reaction between magnesium and dilute sulphuric acid is exothermic.

(i) State what is meant by the term *exothermic*. [1]

(ii) Describe how you could show that this reaction is exothermic. [2] (Edexcel)

25 (a) Most metals are found in rocks as compounds in their ores.

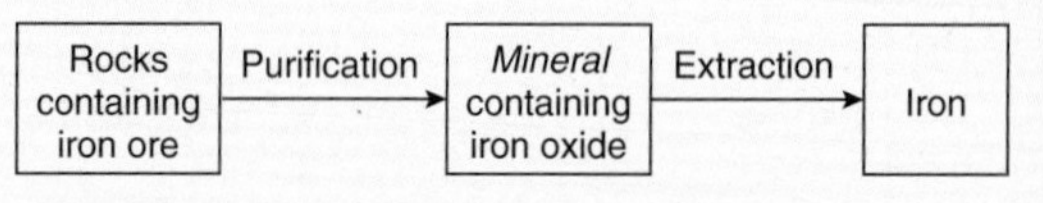

(i) Give *one* reason why purification of the ore causes an environmental problem.

(ii) What is meant by the word **mineral**? [2]

(b) Iron can be extracted from iron oxide in a blast furnace.

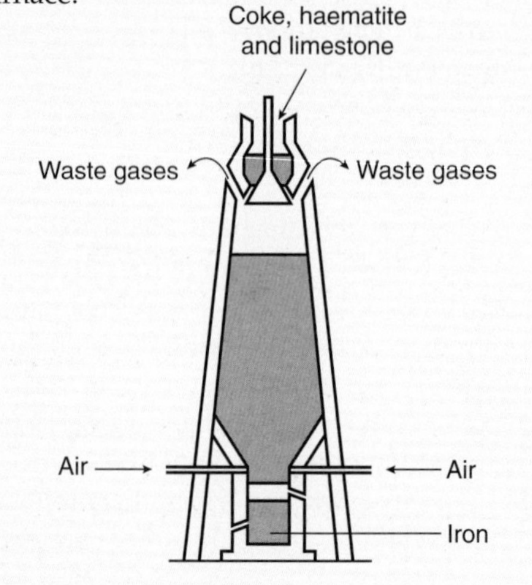

(i) Which one of the substances added to the furnace contains the iron oxide? [1]

(ii) Inside the furnace the coke (carbon) burns in air to release heat.

[A] Name the type of reaction that transfers heat to the surroundings. [1]

[B] Write a word equation to represent this chemical reaction. [2]

(c) Carbon monoxide is formed in the furnace. The carbon monoxide reacts with iron oxide to make iron. Balance the chemical equation for this reaction.

$\ldots Fe_2O_3 + \ldots CO \rightarrow \ldots Fe + \ldots CO_2$ [1]

(d) (i) The waste gases from a blast furnace contain carbon dioxide. Give *one* reason why releasing large amounts of carbon dioxide into the atmosphere causes an environmental problem. [1]

(ii) The extraction of iron causes air pollution. Give *two* other environmental problems caused by the extraction of iron. [2]

(e) Aluminium can be extracted from a mineral called bauxite.

(i) Bauxite contains aluminium oxide. Describe how aluminium is extracted from aluminium oxide. [3]

(ii) About five thousand years passed between the first extraction of iron and that of aluminium. Explain why.

2 AQA(SEG) 1988

26 The three main steps in the manufacture of sulphuric acid are:

Step 1

sulphur + oxygen → sulphur dioxide;

Step 2

sulphur dioxide + oxygen ⇌ sulphur trioxide;

Step 3

sulphur trioxide + water → sulphuric acid.

(i) Give the name of the raw material which supplies the oxygen in *Steps 1* and *2*? [1]

(ii) What is meant by ⇌ in *Step 2*? [1]

(iii) Written below are the conditions used in some industrial processes.

atmospheric pressure;
200–300 atmospheres pressure;
400–500 °C; room temperature;
iron; vanadium oxide.

Select the conditions of temperature, pressure and catalyst used to make sulphur trioxide in *Step 2*. [3]

(iv) The addition of sulphur trioxide to water is too dangerous to carry out in practice. State how sulphur trioxide is converted into sulphuric acid in the industrial process. [2]

(v) State *one* large scale use of sulphuric acid. [1] WJEC

27 This question is about fertilisers and the chemicals used to make them.

(a) Using too much fertiliser can cause pollution in rivers and kill fish.

(i) Here are five sentences describing how this happens. They are in the wrong order. Fill in a copy of the boxes to show the right order. The first one has been done for you.

A Algae grow well on the fertiliser and cover the river.

B Excess fertiliser dissolves in rain and drains into rivers.

C There is little oxygen left for the fish and they die.

D The algae die and bacteria decompose them.

E The bacteria use up most of the oxygen in the water.

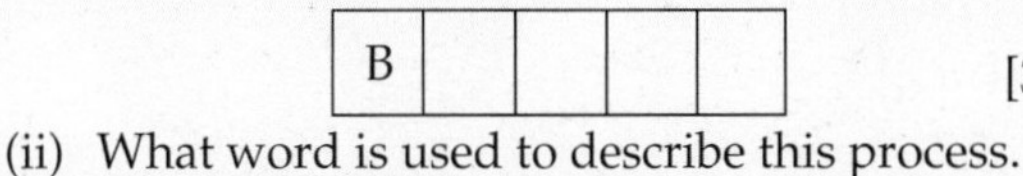

[3]

(ii) What word is used to describe this process.

distillation *eutrophication* *fertilisation* *neutralisation* [1]

(b) Ammonium nitrate, NH_4NO_3, can be used as a fertiliser. The flow chart shows how ammonium nitrate can be made.

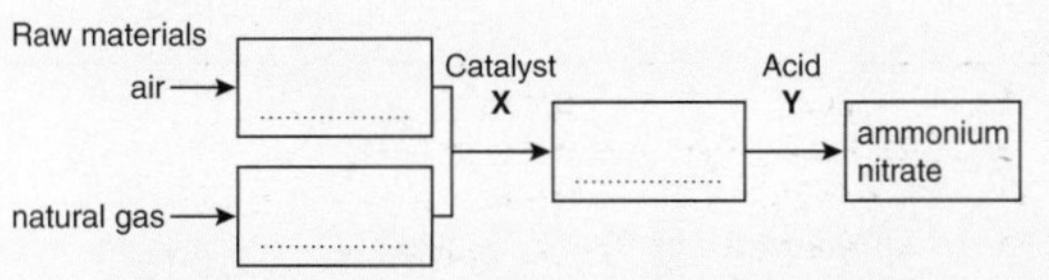

(i) Finish a copy of the flow chart by adding the three missing labels. [2]

(ii) Write down the name of catalyst *X*. [1]

(iii) Write down the name of acid *Y*. [1] (OCR)

28 Ammonia (NH_3) is manufactured from hydrogen in the Haber Process.

(a) (i) Write a balanced equation for the formation of ammonia in the Haber Process. [2]

(ii) Draw a dot and cross diagram to show the bonding in a molecule of ammonia. [2]

(iii) Explain, in terms of the bonds broken and formed, why the formation of ammonia from nitrogen and hydrogen is exothermic. [3]

(b) The manufacture of methanol from carbon monoxide and hydrogen requires similar conditions to those used in the Haber process. The equation for the manufacture of methanol is

$$CO\,(g) + 2H_2\,(g) \rightleftharpoons CH_3OH\,(g)$$

This reaction is exothermic. The reaction conditions are a pressure of 200 atm and a temperature of 400 °C.

(i) State ONE advantage of using a pressure higher than 200 atm. Explain your answer. [3]

(ii) State ONE disadvantage of using a pressure higher than 200 atm. [1]

(iii) State ONE advantage of using a temperature lower than 400 °C. Explain your answer. [3]

(iv) State ONE disadvantage of using a temperature lower than 400 °C. Explain your answer. [2] (Edexcel)

Theme F

Measurement

How fast will a chemical reaction take place?
Can we speed it up or slow it down?
Will the reaction go to completion or reach an equilibrium?

How much product will be formed?

Can we increase the yield of product?

Which fuel shall we use to heat the plant?

These are questions to which the manager of an industrial plant must find answers. In this theme, we show how the answers can be found.

Topic 24 Analytical chemistry

FIRST THOUGHTS

Does a sample of drinking water contain nitrates? Does a brand of paint contain lead compounds? Which metals are present in an alloy steel? To find out the answers to such questions, we use analytical chemistry.

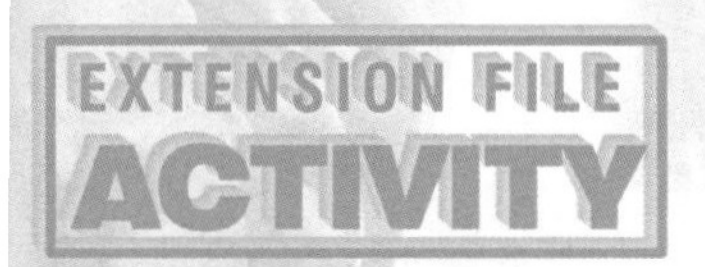

To **analyse** means to separate something into its component parts in order to learn more about the nature of these components. The branch of chemistry that deals with the analysis of chemical compounds and mixtures is called **analytical chemistry**. Sometimes we need to know only which substances are present in a sample; not the quantities of those substances. Which elements are present in this sample of soil? Which pigments are present in this food dye? Does this sample of zinc contain a trace of zinc sulphide? To find answers to such questions, we use **qualitative analysis**. If we need to know the precise quantity of one or more components of a sample, we use **quantitative analysis**. This might entail finding the percentage of nickel in a nickel ore or the number of parts per million of mercury in a fish. Quantitative analysis can be done by means of **volumetric analysis**, in which reactions take place in solution. An alternative is **gravimetric analysis**, in which the masses of reacting substances are found by the use of an accurate balance. In this topic, we shall be dealing with qualitative analysis.

Chemical industries need accurate methods for the analysis of their products. Concern in society over pollution of the environment has created demands for monitoring the state of the air, water and soil. These needs have been met by modern methods of analysis which use instruments. For example, spectrometers measure the concentration of a substance by measuring the amount of light of a certain wavelength which it absorbs. This is much faster than doing a titration. A pH meter connected to a computer will print out a graph showing the progress of an acid–base titration and read out the end point. These instrumental methods depend on electronics and computers.

24.1 Flame colour

EXTENSION FILE ACTIVITY

Some metals and their compounds give characteristic colours when heated in a Bunsen flame. The colour can be used to identify the metal ion present in a compound.

Table 24.1 Some flame colours

Metal	Colour of flame
Barium	Apple-green
Calcium	Brick-red
Copper	Green with blue streaks
Lithium	Crimson
Potassium	Lilac
Sodium	Yellow

24.2 Tests for ions in solution

The following tests can be done on a solution. A soluble solid can be analysed by dissolving it in distilled water and then applying the tests.

Table 24.2 ▼ Tests for anions in solution

Anion	Test and observation
Chloride, $Cl^-(aq)$	Add a few drops of dilute nitric acid followed by a few drops of silver nitrate solution. A white precipitate of silver chloride is formed. The precipitate is soluble in ammonia solution.
Bromide, $Br^-(aq)$	Add a few drops of dilute nitric acid followed by a few drops of silver nitrate solution. A pale yellow precipitate of silver bromide is formed. The precipitate is slightly soluble in ammonia solution.
Iodide, $I^-(aq)$	Add a few drops of dilute nitric acid followed by a few drops of silver nitrate solution. A yellow precipitate of silver iodide is formed. It is insoluble in ammonia solution.
Sulphate, $SO_4^{2-}(aq)$	Add a few drops of barium chloride solution followed by a few drops of dilute hydrochloric acid. A white precipitate of barium sulphate is formed.
Sulphite, $SO_3^{2-}(aq)$	Test as for sulphate. A white precipitate of barium sulphite appears and then reacts with acid to give sulphur dioxide.
Carbonate, $CO_3^{2-}(aq)$	Add dilute hydrochloric acid to the solution (or add it to the solid). Bubbles of carbon dioxide are given off.
Nitrate, $NO_3^-(aq)$	Make the solution strongly alkaline. Add Devarda's alloy and warm. Ammonia is given off. (If the solution contains ammonium ions, warm with alkali to drive off ammonia before testing for nitrate.)

Analytical chemists have traditionally made good use of these tests. Now they also have instrumental methods of analysis to help them.

Table 24.3 ▼ Tests for cations in solution

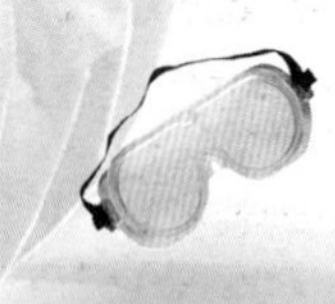

Cation	Add sodium hydroxide solution	Add ammonia solution
Ammonium, $NH_4^+(aq)$	Warm. Ammonia is given off	–
Copper, $Cu^{2+}(aq)$	Blue jelly-like precipitate, $Cu(OH)_2(s)$	Blue jelly-like ppt., dissolves in excess to form a deep blue solution
Iron(II), $Fe^{2+}(aq)$	Green gelatinous ppt., $Fe(OH)_2(s)$	Green gelatinous ppt., $Fe(OH)_2(s)$
Iron(III), $Fe^{3+}(aq)$	Rust-brown gelatinous ppt., $Fe(OH)_3(s)$	Rust-brown gelatinous ppt., $Fe(OH)_3(s)$
Lead(II), $Pb^{2+}(aq)$	White ppt., $Pb(OH)_2(s)$ dissolves in excess NaOH(aq)	White ppt., $Pb(OH)_2$
Zinc, $Zn^{2+}(aq)$	White ppt., $Zn(OH)_2(s)$ dissolves in excess NaOH(aq	White ppt., $Zn(OH)_2(s)$ dissolves in excess $NH_3(aq)$
Aluminium, $Al^{3+}(aq)$	Colourless ppt., $Al(OH)_3(s)$ dissolves in excess NaOH(aq)	Colourless ppt., $Al(OH)_3(s)$

Note: ppt. = precipitate

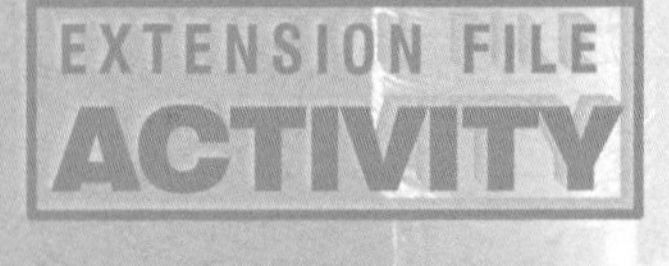

24.3 Identifying some common gases

If you are testing the smell of a gas, you should do so cautiously (see Figure 24.3A). Tests for common gases are given in Table 24.4.

Figure 24.3A ▶ How to smell a gas

Table 24.4 ▼ Testing gases

Gas	Colour and smell	Test and observation
Ammonia†	Colourless Pungent smell	Hold damp red litmus paper or universal indicator paper in the gas. The indicator turns blue.
Bromine	Reddish-brown Choking smell	Hold damp blue litmus paper in the gas. Litmus turns red and is then bleached.
Carbon dioxide	Colourless Odourless	Bubble the gas through limewater (calcium hydroxide solution). A white solid precipitate appears, making the solution appear cloudy.
Chlorine†	Poisonous Green gas Choking smell	Test a very small quantity in a fume cupboard. Hold damp blue litmus paper in the gas. Litmus turns red and is quickly bleached. Chlorine turns damp starch-iodide paper blue-black.
Hydrogen	Colourless Odourless when pure	Introduce a lighted splint. Hydrogen burns with a squeaky 'pop'.
Hydrogen chloride	Colourless Pungent smell	Hold damp blue litmus paper in the gas. Litmus turns red. With ammonia, a white smoke of NH_4Cl forms.
Iodine	Black solid or purple vapour	Dissolve in trichloroethane (or other organic solvent). Gives a violet solution.
Nitrogen dioxide†	Reddish-brown Pungent smell	Hold damp blue litmus paper in the gas. Litmus turns red but is not bleached.
Oxygen	Colourless Odourless	Hold a glowing wooden splint in the gas. The splint bursts into flame.
Sulphur dioxide†	Colourless Choking smell	Dip a filter paper in potassium dichromate solution, and hold it in the gas. The solution turns from orange through green to blue. Potassium manganate(VII) solution turns very pale pink.

†These gases are poisonous. Test with care.

EXTENSION FILE ACTIVITY

24.4 Solubility of some compounds

The solubilities of some compounds are tabulated in Table 24.5.

Table 24.5 ▼ Soluble and insoluble compounds

Soluble	Insoluble
All sodium, potassium and ammonium salts	
All nitrates	
Most chlorides, bromides and iodides	Chlorides, bromides and iodides of silver and lead
Most sulphates	Sulphates of lead, barium and calcium
Sodium, potassium and ammonium carbonates	Most other carbonates
Sodium, potassium and calcium oxides	Most other oxides
Sodium, potassium and calcium hydroxides	Most other hydroxides

CHECKPOINT

1 A body is found at the bottom of a clay pit. The dead man is known to have quarrelled violently with a neighbour, and the neighbour has clay on his shoes. The clay in the clay pit contains a high percentage of iron(II). How can you test the clay on the neighbour's shoes for Fe^{2+}?

2 After a visit to Somerset, some students bring home a sample of rock from an underground cavern. Luke says that the rock is calcium carbonate. Natalie believes that it may be magnesium carbonate. Rosalie suggests that it may be barium carbonate. How can the students investigate whether the rock is (a) a carbonate (b) a calcium compound (c) a barium compound (d) a magnesium compound?

3 A packet is labelled 'bicarb of soda'. How can you test to see whether it contains (a) a sodium compound (b) either a carbonate or a hydrogencarbonate?

4 A solution contains the metal ions, $Q^{2+}(aq)$ and $R^{2+}(aq)$. When a solution of sodium hydroxide is added, a precipitate is obtained. Addition of excess of sodium hydroxide to this precipitate gives a green precipitate containing Q and a colourless solution containing R. When dilute hydrochloric acid is added to this solution, a white precipitate is obtained. Deduce what Q and R may be, explaining your reasoning.

5 Old Sir Joshua Vellof often falls asleep after meals. He is a suspicious old gentleman, and he wonders whether his no-good nephew Jake is putting some of the tranquilliser potassium bromide into the salt cellar. Sir Joshua asks you to analyse the contents of the salt cellar to see whether it contains sodium chloride or potassium bromide. What do you do?

6 A factory orders calcium oxide, magnesium oxide and barium oxide. A lorry driver deposits three sacks at the factory and drives off without saying which is which. The works chemist has to sort it out. Suggest two tests which he could use to tell which sack is which.

7 A solution contains the sulphates of three metals, A, B and C. Explain the reactions shown in the flow chart and identify A, B and C.

Solution of sulphates of A, B and C

↓ $NaOH(aq)$

Blue-green precipitate

↓ Excess $NaOH(aq)$

Blue-green precipitate + Colourless solution

Blue-green precipitate ↓ $NH_3(aq)$ → Blue solution + Green precipitate

Colourless solution ↓ $HCl(aq)$ → White precipitate ↓ $NH_3(aq)$ → Colourless solution

24.5 Methods of collecting gases

Gases can be bought in cylinders from industrial manufacturers or made in the laboratory by a chemical reaction. You may need to obtain gas jars full of gas. The following methods can be used to collect gases from either source.

Collecting over water

A gas which is insoluble in water can be collected over water (s 24.5A). The gas displaces the water in the gas jar. When the gas , it must be replaced by another. This method can be used for hydrogen, oxygen, chlorine, carbon dioxide and other gases. A little of the gas will dissolve in the water.

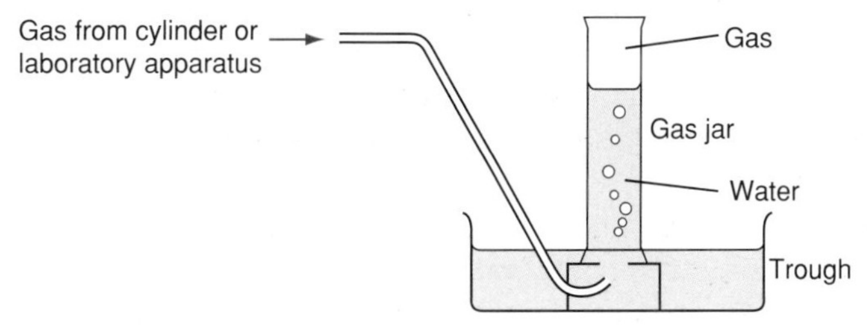

Figure 24.5A ▲ Collecting over water (for an insoluble gas)

Which method is best for collecting a gas? It depends on whether the gas is soluble or insoluble in water, denser or less dense than air. A gas syringe is a useful piece of equipment for this job.

Collecting downwards by displacement

A gas which is denser than air can be collected downwards by displacement of air (see Figure 24.5B). The gas displaces the air in the gas jar. This method can be used for chlorine, carbon dioxide, sulphur dioxide and other dense gases.

Collecting upwards by displacement

A gas which is less dense than air can be collected upwards by displacement of air (see Figure 24.5C). This method can be used for ammonia.

Gas from cylinder or laboratory apparatus → Lid, Gas jar

Figure 24.5B ▲ Collecting downwards by displacement (for a dense gas)

Gas jar, Gas from cylinder or laboratory apparatus →

Figure 24.5C ▲ Collecting upwards by displacement (for a gas of low density)

Safety matters

When collecting hydrogen, take care that there are no flames about. Hydrogen forms an explosive mixture with air. Chlorine, sulphur dioxide, ammonia and other toxic gases should be collected in a fume cupboard.

Collecting in a gas syringe

A gas supply can be connected to an empty gas syringe. As gas enters, it drives the plunger down the barrel (see Figure 24.5D). When the syringe is full, it must be replaced by another.

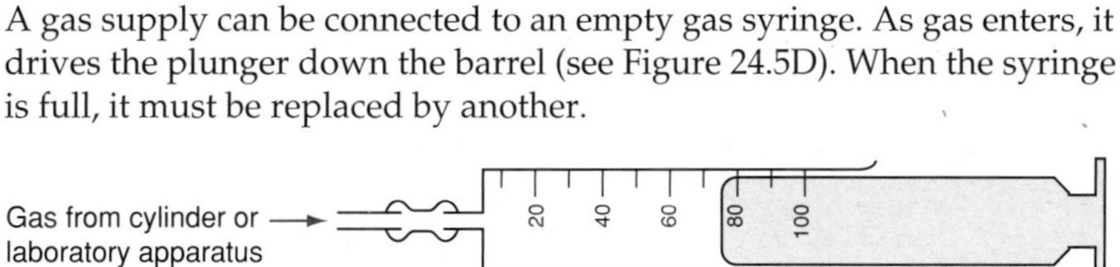

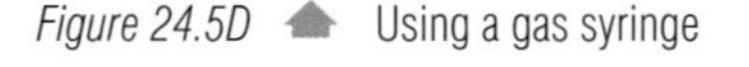

Figure 24.5D ▲ Using a gas syringe

CHECKPOINT

1. (a) Why is the method shown in Figure 24.5A not used for collecting ammonia?
   (b) Why is the method shown in Figure 24.5C not used for collecting sulphur dioxide?
2. Hydrogen can be collected by either of the methods shown in Figures 24.5A and C.
   (a) State an advantage of the method shown in Figure 24.5A over that in Figure 24.5C.
   (b) Sketch the apparatus which you would use to collect dry hydrogen.

Topic 25 Chemical reaction speeds

25.1 Why reaction speeds are important

FIRST THOUGHTS

Who is interested in the speeds of chemical reactions? If you were a cheese manufacturer, you would be interested in speeding up the chemical reactions which produce cheese. The more tonnes of cheese you could produce in a month, the more profit you would make. If you were a butter manufacturer, you would want to slow down the chemical reactions which make your product turn rancid.

In a chemical reaction, the starting materials are called the **reactants**, and the finishing materials are called the **products**. It takes time for a chemical reaction to happen. If the reactants take only a short time to change into the products, that reaction is a **fast reaction**. The **speed** or **rate** of that reaction is high. If a reaction takes a long time to change the reactants into the products, it is a **slow reaction**. The speed or rate of that reaction is low.

Many people are interested in knowing how to alter the speeds of chemical reactions. The factors which can be changed are:

- the size of the particles of a solid reactant,
- the concentrations of reactants in solution,
- the temperature,
- the presence of light,
- the addition of a catalyst.

A **catalyst** is a substance which can alter the rate of a chemical reaction without being used up in the reaction.

Examples of reactions which occur at different speeds are: chemical weathering of rocks (Topic 18), takes millions of years; rusting of iron (Topic 14.7) takes 1-50 years, depending on conditions; milk turning sour happens in a few days; reaction between iron and dilute acid (Topic 9.1, 19.3), takes 1–30 minutes, depending on conditions; reaction between sodium and water (Topic 19) takes about 1 minute; explosion of hydrogen and oxygen (Topic 14.1) takes a fraction of a second.

CHECKPOINT

Give an example of a reaction which would be complete in

(a) 1 day (b) 1 week (c) 1 month (d) 1 year.

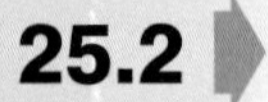

25.2 Particle size and reaction speed

Carbon dioxide can be prepared in the laboratory by the reaction

Calcium carbonate	+	Hydrochloric acid	→	Carbon dioxide	+	Calcium chloride	+	Water
$CaCO_3(s)$	+	$2HCl(aq)$	→	$CO_2(g)$	+	$CaCl_2(aq)$	+	$H_2O(l)$

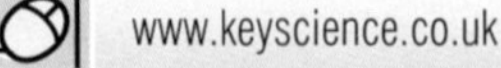

www.keyscience.co.uk

One of the reactants, calcium carbonate (marble) is a solid. You can use this reaction to find out whether large lumps of a solid react at the same speed as small lumps of the same solid. Figure 25.2A shows a method for finding the rate of the reaction. It can be used when one of the products is a gas. As carbon dioxide escapes from the flask, the mass of the flask and contents decreases.

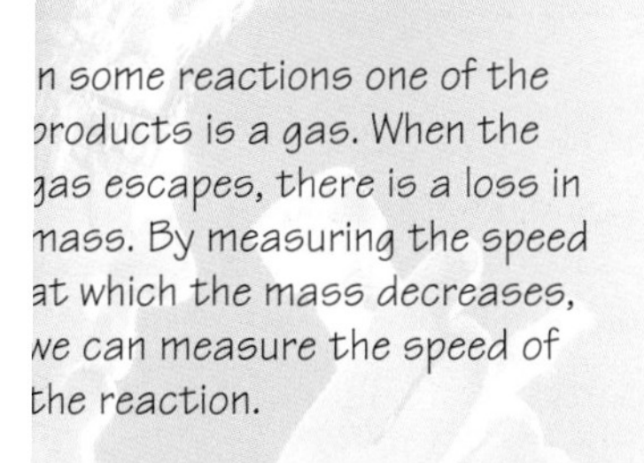

n some reactions one of the products is a gas. When the gas escapes, there is a loss in mass. By measuring the speed at which the mass decreases, we can measure the speed of the reaction.

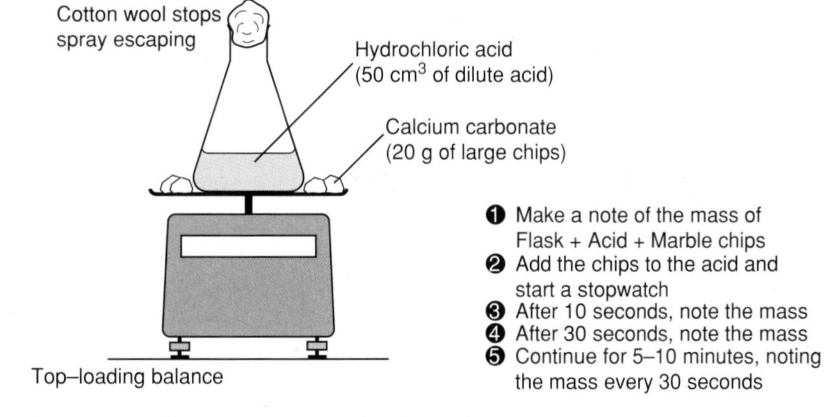

Figure 25.2A ▲ Apparatus for following the loss in mass when a gas is evolved

The reaction starts when the marble chips are dropped into the acid. The mass of the flask and contents is noted at various times after the start of the reaction. The mass can be plotted against time. Figure 25.2B shows typical results.

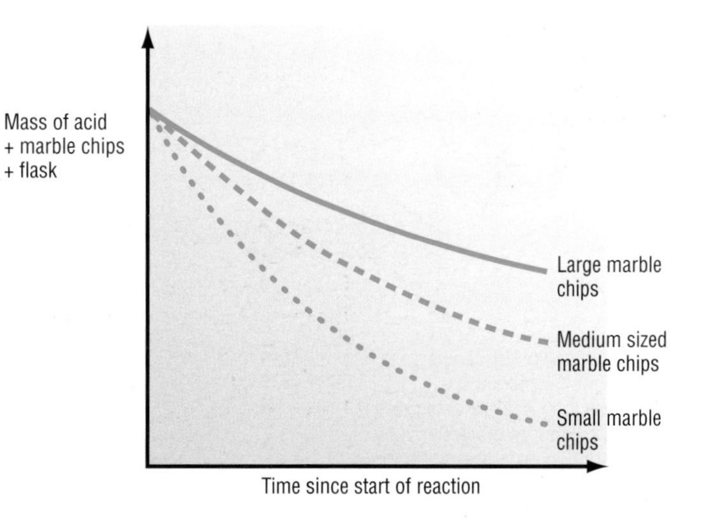

Figure 25.2B ▲ Results obtained with different sizes of marble chips

Science at work

Use data logging to analyse the results of a study of the progress of a reaction.

The results show that the smaller the size of the particles of calcium carbonate, the faster the reaction takes place. The difference is due to a difference in surface area. There is a larger surface area in 20 g of small chips than in 20 g of large chips. The acid attacks the surface of the marble. It can therefore react faster with small chips than with large chips.

SUMMARY

Reactions in which one reactant is a solid take place faster when the solid is divided into small pieces. The reason is that a certain mass of small particles has a larger surface area than the same mass of large particles.

CHECKPOINT

1. When potatoes are cooked, a chemical reaction occurs. What can you do to increase the speed at which potatoes cook?
2. There is a danger in coal mines that coal dust may catch fire and start an explosion. Explain why coal dust is more dangerous than coal.
3. 'Alko' indigestion tablets and 'Neutro' indigestion powder are both alkalis. Which do you think will act faster to cure acid indigestion? Describe how you could test the two remedies in the laboratory with a bench acid to see whether you are right.

25.3 Concentration and reaction speed

EXTENSION FILE ACTIVITY

Many chemical reactions take place in solution. One such reaction is

Sodium thiosulphate + Hydrochloric acid → Sulphur + Sodium chloride + Sulphur dioxide + Water

$$Na_2S_2O_3(aq) + 2HCl(aq) \rightarrow S(s) + 2NaCl(aq) + SO_2(g) + H_2O$$

Sulphur appears in the form of very small particles of solid. The particles do not settle: they remain in suspension. Figure 25.3A shows how you can follow the speed at which sulphur is formed.

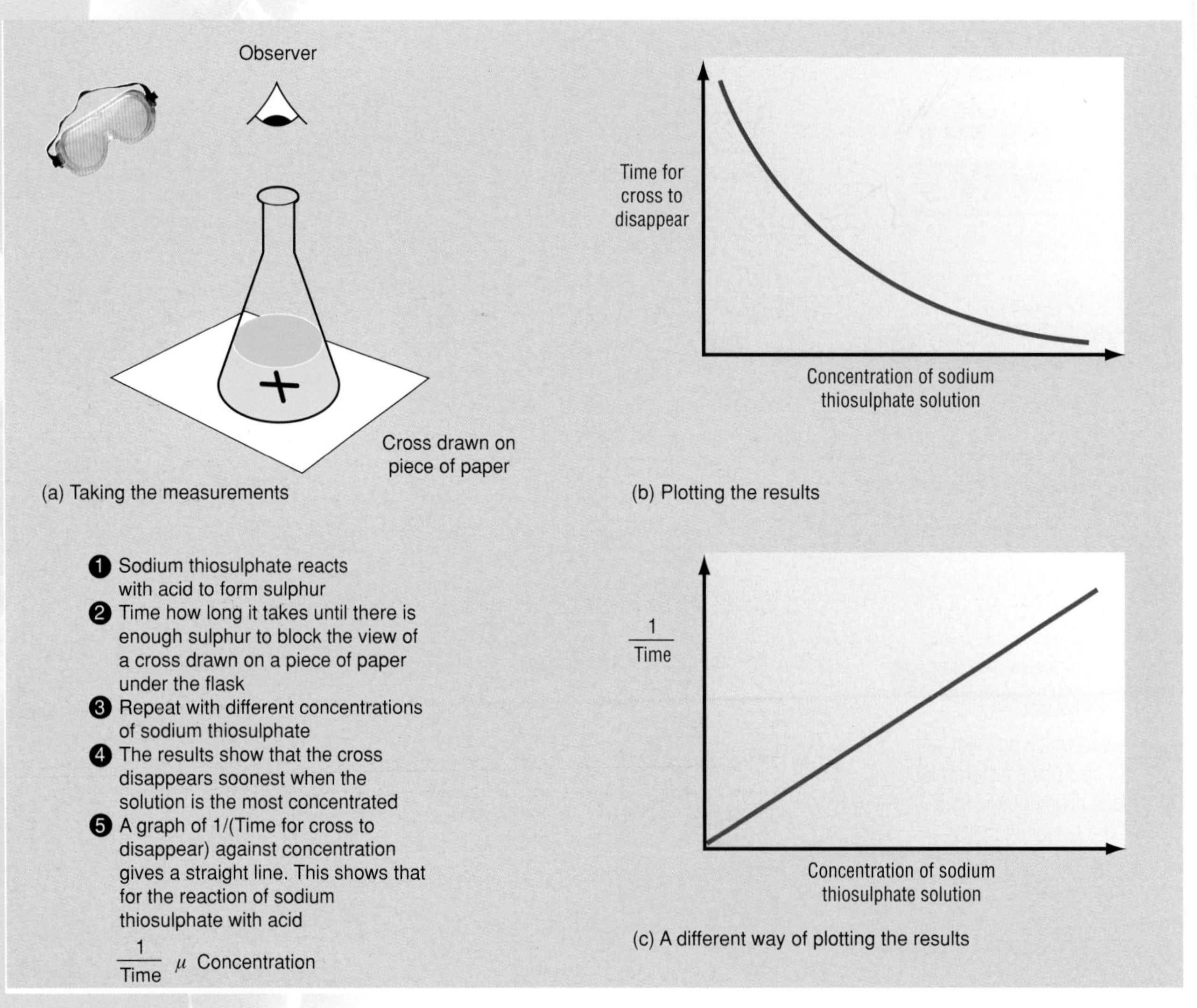

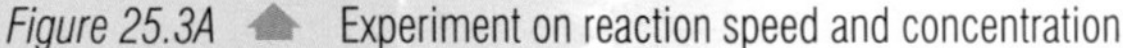

Figure 25.3A Experiment on reaction speed and concentration

The speed at which the precipitate of a solid product appears can be used to measure the speed of a reaction.

The faster a reaction takes place, the shorter is the time needed for the reaction to finish. To be more precise, the speed of the reaction is **inversely proportional** to the time taken for the reaction to finish

$$\text{Speed of reaction} \propto 1/\text{Time}$$

You can see from Figure 25.3A(c) above that

$$1/\text{Time} \propto \text{Concentration}$$

Therefore

$$\text{Speed of reaction} \propto \text{Concentration}$$

SUMMARY

For the reactions in solution mentioned here, the speed of the reaction is proportional to the concentration of the reactant (or reactants). That is, the speed doubles when the concentration is doubled.

In this experiment, only one concentration was altered. A variation is to keep the concentration of sodium thiosulphate constant and alter the concentration of acid. Then the speed of the reaction is found to be proportional to the concentration of the acid. If the acid concentration is doubled, the speed doubles. The reason for this is that the ions are closer together in a concentrated solution. The closer together they are, the more often the ions collide. The more often they collide, the more chance they have of reacting.

CHECKPOINT

1 Molly is asked to investigate the marble chips–acid reaction. She must find out what effect changing the concentration of the acid has on the speed of the reaction. Explain how Molly could adapt the experiment shown in Figure 25.2A to carry out her investigation.

25.4 Pressure and reaction speed

SUMMARY

The speed of a reaction between gases increases when the pressure is increased.

Pressure has an effect on reactions between gases. The speed of the reaction increases when the pressure is increased. The reason is that increasing the pressure pushes the gas molecules closer together. The molecules therefore collide more often, and the gases react more rapidly.

25.5 Temperature and reaction speed

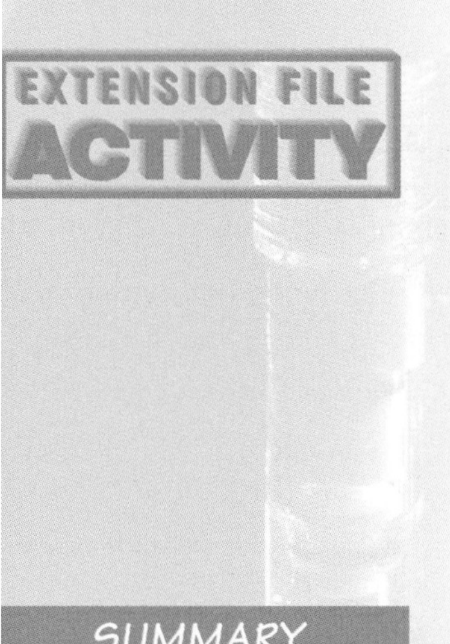

You met the reaction between sodium thiosulphate and acid in Topic 25.3. This reaction can also be used to study the effect of temperature on the speed of a chemical reaction. Warming the solutions makes sulphur form faster. There is a steep increase in the speed of the reaction as the temperature is increased.

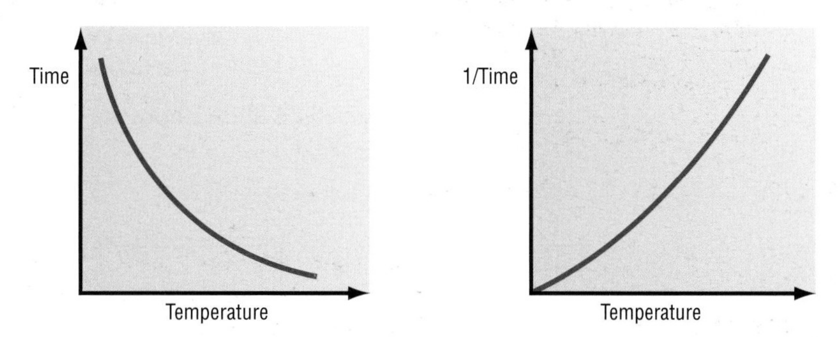

Figure 25.5A ▲ The effect of temperature on the speed of reaction

SUMMARY

The speed of a reaction increases when the temperature is raised.

This reaction goes approximately twice as fast at 30 °C as it does at 20 °C. It doubles in speed again between 30 °C and 40 °C and so on.

At the higher temperature, the ions have more kinetic energy. Moving through the solution more rapidly, they collide more often and more vigorously and so there is a greater chance that they will react.

25.6 Light and chemical reactions

SUMMARY

Some chemical reactions are speeded up by light.

Heat is not the only form of energy that can speed up reactions. Some chemical reactions take place faster when they absorb light. The formation of silver from silver salts takes place when a **photographic film** is exposed to light. In sunlight, green plants are able to carry on the process of **photosynthesis**.

CHECKPOINT

1. Polly's project is to find out what effect temperature has on the speed at which milk goes sour. Suggest what measurements she should make.
2. Ismail's project is to find out whether iron rusts more quickly at higher temperatures. Suggest a set of experiments which he could do to find out.
3. You are asked to study the reaction

   Magnesium + Sulphuric acid ➔ Hydrogen + Magnesium sulphate

   You are provided with magnesium ribbon, dilute sulphuric acid, a thermometer and any laboratory glassware you need. Describe how you would find out what effect a change in temperature has on the speed of the reaction. Say what apparatus you would use, what measurements you would make and what you would do with your results.
4. Magnesium reacts with cold water slowly

   Magnesium + Water ➔ Magnesium hydroxide + Hydrogen

   If there is phenolphthalein in the water, it turns pink, showing that an alkali has been formed. Describe experiments which you could do to find the effect of increasing the temperature on the speed of this reaction. Say what you would measure and what you would do with your results. With your teacher's approval, try out your ideas.

25.7 Catalysis

A reaction used to prepare oxygen is

$$\text{Hydrogen peroxide} \rightarrow \text{Oxygen} + \text{Water}$$
$$2H_2O_2(aq) \rightarrow O_2(g) + 2H_2O(l)$$

Figure 25.7A shows how the oxygen can be collected and measured in a gas syringe.

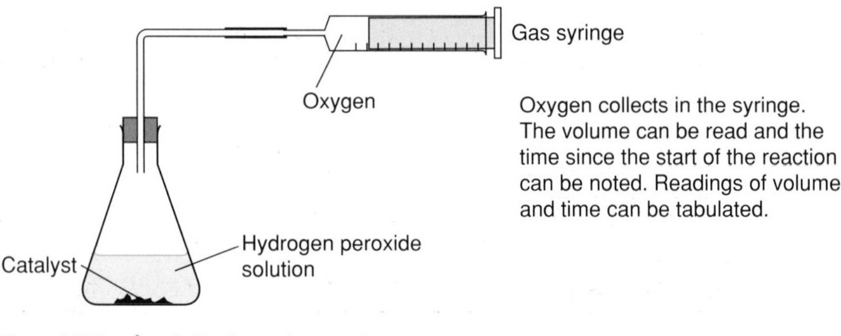

Figure 25.7A ⬆ Collecting and measuring a gas

The apparatus shown in Figure 25.7A can be used whenever you want to collect the gas that is formed in a reaction. The speed of the reaction can be measured by noting the volume of the gas formed at different time intervals.

A substance which speeds up a reaction without being used up in the reaction is called a catalyst.

The formation of oxygen is very slow at room temperature. The reaction can be speeded up by the addition of certain substances, for example manganese(IV) oxide. When manganese(IV) oxide is added to hydrogen peroxide, the evolution of oxygen takes place much more rapidly (see Figure 25.7B). Manganese(IV) oxide is not used up in the reaction. At the end of the reaction, the manganese(IV) oxide can be filtered out of the solution and used again. A substance which increases the speed of a chemical reaction without being used up in the reaction is called a **catalyst**. Different reactions need different catalysts.

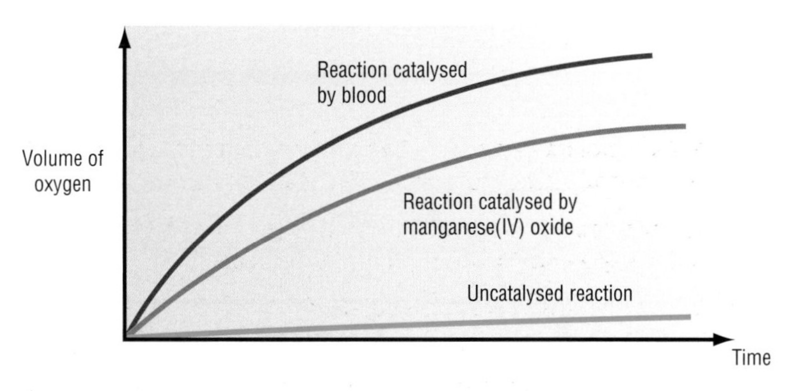

Figure 25.7B ▲ Catalysis of the decomposition of hydrogen peroxide

SUMMARY

Some chemical reactions can be speeded up by adding a substance which is not one of the reactants, which is not used up in the reaction. Such a substance is called a catalyst.

A certain catalyst will catalyse a limited number of reactions. Catalysts are of great importance to industry. They may diminish the need to carry out a reaction at high temperature or high pressure and so cut production costs.

Any individual catalyst will only catalyse a certain reaction or group of reactions. Platinum catalyses a number of oxidation reactions. Nickel catalyses hydrogenation reactions. Industries make good use of catalysts. If manufacturers can produce their product more rapidly, they make bigger profits. If they can produce their product at a lower temperature with the aid of a catalyst, they save on fuel. The reactions which make plastics take place under high pressure. The construction of industrial plastics plants which are strong enough to withstand high pressure is expensive. A plastics manufacturer therefore tries to find a catalyst which will enable the reaction to give a good yield of plastics at a lower pressure. Then the plant will not have to withstand high pressures. Less costly materials can be used in its construction. Industrial chemists are constantly looking for new catalysts.

Some reactions only give good yields of product at high temperatures. Such a reaction costs a manufacturer high fuel bills. Industrial chemists look for a catalyst which will make the reaction take place more readily at a lower temperature. This will cut running costs and increase profits.

CHECKPOINT

1 Catalysts A and B both catalyse the decomposition of hydrogen peroxide. The following figures were obtained at 20 °C for the volume of oxygen formed against the time since the start of the reaction.

Time (minutes)	0	5	10	15	20	25	30	35
Volume of oxygen with catalyst A (cm^3)	0	4	8	12	16	17	18	18
Volume of oxygen with catalyst B (cm^3)	0	5	10	15	16.5	18	18	18

(a) Plot a graph to show both sets of results.
(b) Say which is the better catalyst, A or B.
(c) Explain why both experiments were done at the same temperature.
(d) Explain why both sets of figures stop at 18 cm^3 of oxygen.
(e) Add a line to your graph to show the shape of the graph you would obtain for the uncatalysed reaction.

2 Someone tells you that nickel oxide will catalyse the decomposition of hydrogen peroxide to give oxygen. How could you find out whether this is true? Draw the apparatus you would use and state the measurements you would make.

25.8 Enzymes

SUMMARY

Enzymes are proteins. They catalyse reactions which take place in living organisms. An enzyme is specific for a certain reaction or type of reaction. Enzymes are powerful catalysts.

The reactions which take place inside the cells of living organisms are catalysed by **enzymes**. Enzymes are proteins. They are powerful catalysts which enable reactions to take place at relatively low temperatures, e.g. body temperature. An enzyme is **specific** for a certain reaction or type of reaction. For example the enzyme maltase catalyses the reaction of the sugar maltose with water to form glucose. The substance which the enzyme helps to react, e.g. maltose, is called the **substrate** (see Key Science Biology, Topic 10.3).

Like other proteins (Topic 30.6), enzymes have a complex three-dimensional structure. A protein can lose this structure – can be **denatured** – at high pH, low pH and high temperature (45 °C for many enzymes). When this happens an enzyme loses its catalytic power. Every enzyme has an optimum pH range over which it works best.

Many industrial processes are catalysed by enzymes. The advantage is that reactions will take place at a lower temperature and pressure than would otherwise be needed. This saves on the expense of fuel and on the construction of plants that will withstand high pressure.

Many industrial processes are **batch processes**. The enzyme and substrate are mixed in a container and left until a good yield of product has been formed. Then the container is emptied, and the batch of product is extracted. The container is then refilled with reactants and enzyme. The cost of the process can be reduced if the enzyme can be recovered and used again. Binding the enzyme to an inert support, e.g. a resin, solves the problem. The technique is called **immobilising** the enzyme. The resin, with the enzyme attached, is packed into a column and the substrate is passed through it. The reaction can run as a **continuous process**, with reactants fed in at the top of the column and products taken out at the bottom.

Figure 25.8A What has this box of chocolates got to do with enzymes?

SUMMARY

Protein molecules have a three-dimensional structure. If this structure is lost, an enzyme is 'denatured' and loses its catalytic activity.

Science at work

Starch is hydrolysed (split by water) to form sugars. All it takes is boiling for an hour with hydrochloric acid. Or do it the easy way: add the enzyme amylase. Then the reaction happens at body temperature in a few minutes.

SUMMARY

Enzymes have many uses in industry.

Examples of the use of enzymes

1 *Baking:* Enzymes catalyse the hydrolysis of the starch in flour to sugars. Then enzymes in yeast catalyse the fermentation of sugars with the production of carbon dioxide. This makes bread and cakes 'rise'.

2 *Dairy produce:* The enzyme rennin is present in the extract called rennet which is obtained from the digestive juices in calves' stomachs (or from bacteria by genetic engineering). It is added to milk to speed up the clotting of milk to form cheese.

3 *Brewing:* Enzymes from barley are used in beer production; they convert starches and proteins into sugars and amino acids. Enzymes in yeast ferment sugars to ethanol.

4 *Sweets:* Have you ever wondered how they get a liquid centre into a chocolate? To start with, the centre is solid. Enzymes incorporated in the solid centre convert the solid into a liquid.

CHECKPOINT

1 Explain the terms: catalyst, enzyme, substrate.

2 What is meant by the statement that an enzyme is 'specific' for a certain substrate?

3 Why is it important to industrial chemists to find catalysts which allow reactions to proceed (a) at lower temperatures (b) at lower pressures?

4 Give two examples of enzyme-catalysed reactions in the food industry.

5 What is an immobilised enzyme? Explain the advantage of using an immobilised enzyme, rather than a free enzyme, to catalyse an industrial process.

6 Answer the question in Figure 25.8A.

25.9 Chemical equilibrium

Reversible reactions

When calcium carbonate is heated strongly, it decomposes:

$$\text{Calcium carbonate} \rightarrow \text{Calcium oxide} + \text{Carbon dioxide}$$
$$CaCO_3(s) \rightarrow CaO(s) + CO_2(g)$$

However, the base calcium oxide reacts with the acidic gas carbon dioxide to form calcium carbonate.

$$CaO(s) + CO_2(g) \rightarrow CaCO_3(s)$$

We say that this reaction is **reversible**; it can go from right to left as well as from left to right. When a substance decomposes to form products which can recombine, we say that the substance **dissociates**. When the dissociation is brought about by heat, it is called **thermal dissociation**.

Sometimes the products of a reaction can combine to form the reactants. Such a reaction is a reversible reaction. An example is the thermal decomposition of calcium carbonate. In a reversible reaction a state of equilibrium is reached when
rate of forward reaction = rate of backward reaction

Think about what happens when calcium carbonate is heated inside a closed container. When heating starts, there is no carbon dioxide or calcium oxide, and the dissociation of calcium carbonate is the only reaction that takes place. Soon, the amounts of calcium oxide and carbon dioxide build up, and the combination of calcium oxide and carbon dioxide begins. As the concentration of carbon dioxide builds up inside the closed container, the rate of combination increases. Eventually a state is reached in which the rate of recombination is equal to the rate of dissociation of calcium carbonate. Then the amounts of calcium carbonate, calcium oxide and carbon dioxide remain constant. The system is described as being at **equilibrium**. We use reverse arrows to denote an equilibrium:

$$CaCO_3(s) \rightleftharpoons CaO(s) + CO_2(g)$$

The reaction from left to right (as written) is called the forward reaction. The reaction from right to left is called the backward reaction.

In a system at equilibrium, both the forward reaction and the backward reaction are taking place. The rate of the forward reaction is equal to the rate of the backward reaction. Therefore, once equilibrium is established, the amounts of the three substances remain constant.

A number of factors can disturb the equilibrium. If the container is opened, carbon dioxide escapes, and more calcium carbonate dissociates. When limestone is heated in a kiln (see Figure 21.1F) to make calcium oxide (quicklime) and carbon dioxide, the kiln is open, and a draft of air is blown through the kiln to remove carbon dioxide and prevent equilibrium from becoming established. Then more and more calcium carbonate dissociates in an attempt to reach equilibrium, and more quicklime is produced.

The Haber process

The Haber process for making ammonia uses a reversible reaction which comes to equilibrium. The right conditions move the position of equilibrium towards the product side of the equation. High pressure and low temperature favour the forward reaction.

The conditions employed are high pressure, moderate temperature and catalyst

Nitrogen and hydrogen combine to form ammonia (See Topic 22.2):

$$N_2(g) + 3H_2(g) \rightleftharpoons 2NH_3(g)$$

Equilibrium is reached as ammonia dissociates into nitrogen and hydrogen. Nitrogen is an unreactive gas, and the position of equilibrium is very far towards the left-hand side. Fritz Haber developed this reaction to manufacture ammonia. He had to try a number of ways to get the equilibrium to move towards the right-hand side and give a bigger yield of ammonia. He took note of **Le Chatelier's Principle**. This states that: **When conditions are changed, a system in a state of equilibrium adjusts itself in such a way as to minimise the effects of the change**. You can see that when the reaction goes from left to right there is a decrease in the number of molecules of gas. As a result (see Topic 26.8) if the reaction went to completion, the volume of ammonia would be only half that of the mixture of nitrogen and hydrogen used. So if the pressure on the mixture is increased, the system can absorb the increase in pressure by reducing its volume, that is, by reacting to form ammonia. The reaction is exothermic; that is, heat is given out in going from left to right. The reverse reaction, the dissociation of ammonia, is endothermic. If the temperature is raised, the system adjusts so as to absorb heat. It does this through the dissociation of ammonia because this reaction is endothermic. Running the process at high temperature therefore reduces the percentage conversion of the elements into ammonia. If the reactants are at a very low temperature, the system takes a long time to come to equilibrium. In practice, a high pressure (about 200 atm) and a moderate temperature (about 450 °C) are employed. A catalyst (iron or iron(III) oxide) is used to increase the speed of the reaction.

The contact process

In the contact process for making sulphuric acid, conditions are chosen to push equilibrium towards the product.

In the manufacture of sulphuric acid (See Topic 22.4), an important reaction is:

$$\text{Sulphur dioxide} + \text{Oxygen} \rightleftharpoons \text{Sulphur trioxide}$$
$$2SO_2(g) + O_2(g) \rightleftharpoons 2SO_3(g)$$

The formation of sulphur trioxide involves a decrease in volume as three moles of gas form two moles of gas. High pressure should favour the forward reaction. However, sulphur dioxide is easily liquefied, and industry uses a pressure of only 2 atm. The reaction from left to right is exothermic. If the temperature is raised, the reverse reaction, the dissociation of sulphur trioxide is favoured. Therefore the industry uses a moderate temperature (about 450 °C). The use of a catalyst, vanadium(V) oxide, helps to compensate for not using a very high temperature.

CHECKPOINT

1. Reactions take place more rapidly as the temperature rises. In view of this, why is the manufacture of ammonia carried out at 450 °C, rather than at a very high temperature?
2. A gas, A, reacts to form a mixture of two gases, B and C. The reaction is reversible, and the system reaches equilibrium.

   $$A(g) \rightleftharpoons B(g) + 2C(g)$$

   (a) Will an increase in pressure increase or decrease the yield of B and C?

   (b) What information would you need before you could predict the effect of raising the temperature on the rate of equilibrium?
3. An aqueous solution of bromine is called 'bromine water'. Some bromine molecules react with water molecules:

   $$\text{Bromine} + \text{Water} \rightleftharpoons \text{Hydrobromic acid} + \text{Bromic(I) acid}$$
   $$Br_2(aq) + H_2O(l) \rightleftharpoons HBr(aq) + HBrO(aq)$$

   The products are both strong acids. The reaction is reversible, and a solution of bromine in water reaches an equilibrium state in which the concentrations of all the species are constant.

   Predict what change will happen if you add a small amount of sodium hydroxide. In which direction will the equilibrium be displaced, from left to right or from right to left? Predict what colour change you will see. How could you reverse the colour change?

   Do an experiment to check your predictions.

Topic 26 Chemical calculations

26.1 Relative atomic mass

FIRST THOUGHTS

Every dot of ink on this page is big enough to have a million hydrogen atoms fitted across it from side to side.

www.keyscience.co.uk

SUMMARY

The relative atomic mass, A_r, of an element is the mass of one atom of the element compared with $^1/_{12}$ the mass of one atom of carbon-12.

The masses of atoms are very small. Some examples are:

- Mass of hydrogen atom, H = 1.4×10^{-24} g.
- Mass of mercury atom, Hg = 2.8×10^{-22} g.
- Mass of carbon atom, C = 1.7×10^{-23} g.

Chemists find it convenient to use **relative atomic masses**. The hydrogen atom is the lightest of atoms, and the masses of other atoms can be stated *relative to* that of the hydrogen atom. On the original version of the relative atomic mass scale:

- Relative atomic mass of hydrogen = 1.
- Relative atomic mass of mercury = 200 (a mercury atom is 200 times heavier than a hydrogen atom).
- Relative atomic mass of carbon = 12 (a carbon atom is 12 times heavier than a hydrogen atom).

Chemists now take the mass of one atom of carbon-12 as the reference point for the relative atomic mass scale. On the present scale

$$\text{Relative atomic mass of element} = \frac{\text{Mass of one atom of the element}}{(^1/_{12})\ \text{Mass of one atom of carbon-12}}$$

Since relative atomic mass (symbol A_r) is a ratio of two masses, the mass units cancel, and relative atomic mass is a number without a unit.

The relative atomic masses of some common elements are listed in Table 26.1.

Table 26.1 Some relative atomic masses

Element	A_r	Element	A_r
Aluminium	27	Magnesium	24
Barium	137	Mercury	200
Bromine	80	Nitrogen	14
Calcium	40	Oxygen	16
Chlorine	35.5	Phosphorus	31
Copper	63.5	Potassium	39
Hydrogen	1	Sodium	23
Iron	56	Sulphur	32
Lead	207	Zinc	65

CHECKPOINT

1 Refer to Table 26.1.

(a) How many times heavier is one atom of nitrogen than one atom of hydrogen?

(b) What is the ratio
Mass of one atom of mercury/Mass of one atom of bromine?

(c) How many atoms of oxygen are needed to equal the mass of one atom of bromine?

(d) How many atoms of sodium are needed to equal the mass of one atom of lead?

26.2 Relative molecular mass

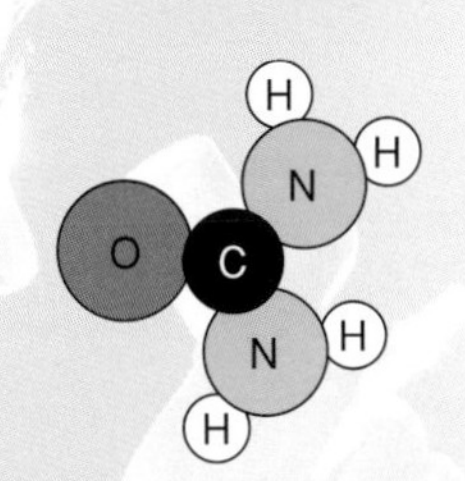

Figure 26.2A Atoms in a molecule of urea

SUMMARY

The relative molecular mass of a compound is equal to the sum of the relative atomic masses of all the atoms in one molecule of the compound or in one formula unit of the compound.

M_r = Sum of A_r values

The mass of a molecule is the sum of the masses of all the atoms in it. The relative molecular mass (symbol M_r) of a compound is the sum of the relative atomic masses of all the atoms in a molecule of the compound (see Figure 26.2A).

Worked example Find the relative molecular mass of urea.

Solution Formula of compound is CON_2H_4

1 atom of C (A_r 12) = 12
1 atom of O (A_r 16) = 16
2 atoms of N (A_r 14) = 28
4 atoms of H (A_r 1) = 4
Total = 60
Relative molecular mass, M_r, of urea = 60

Many compounds consist of ions, not molecules. The formula of an ionic compound represents a **formula unit** of the compound; for example, $CaSO_4$ represents a formula unit of calcium sulphate, not a molecule of calcium sulphate. The term relative molecular mass can be used for ionic compounds as well as molecular compounds.

CHECKPOINT

1 Work out the relative molecular masses of these compounds:
CO CO_2 SO_2 SO_3 $NaOH$ $NaCl$ CaO $Mg(OH)_2$ Na_2CO_3 $CuSO_4$
$CuSO_4.5H_2O$ $Ca(HCO_3)_2$

26.3 Percentage composition

How are you at percentages? 'Per cent' means 'per hundred'. What is the percentage of boys in your class? Say the class is 21 boys and 9 girls. This means that 21 out of 30 students are boys. If the class were 100 students, there would be
(100/30) × no. of boys = (100/30) × 21 = 70% or
% of boys = (no. of boys/total no of students) × 100%

From the formula of a compound you can find the percentage by mass of the elements in the compound.

Worked example 1 Find the percentage by mass of (a) calcium (b) chlorine in calcium chloride.

Solution First find the relative molecular mass of calcium chloride, formula $CaCl_2$.

$$M_r = A_r(\text{Ca}) + 2A_r(\text{Cl}) = 40 + (2 \times 35.5) = 111$$

$$\text{Percentage of calcium} = \frac{40}{111} \times 100 = 36\%$$

$$\text{Percentage of chlorine} = \frac{71}{111} \times 100 = 64\%$$

You can see that the two percentages add up to 100%.

Worked example 2 Find the percentage of water in crystals of magnesium sulphate-7-water.

Solution Find the relative molecular mass of $MgSO_4.7H_2O$.

1 atom of magnesium (A_r 24) = 24
1 atom of sulphur (A_r 32) = 32
4 atoms of oxygen (A_r 16) = 64
7 molecules of water = $7 \times [(2 \times 1) + 16]$ = 126
Total = M_r = 246

$$\text{Percentage of water} = \frac{\text{Mass of water in formula}}{\text{Relative molecular mass}} \times 100$$

$$= \frac{126}{246} \times 100 = 51.2\%$$

The percentage of water in magnesium sulphate crystals is 51%.

SUMMARY

You can calculate the percentage by mass composition of a compound from its formula.

CHECKPOINT

You do not need calculators for these problems.

1 Find the percentage by mass of;
   (a) calcium in calcium bromide, $CaBr_2$,
   (b) iron in iron(III) oxide, Fe_2O_3,
   (c) carbon and hydrogen in ethane, C_2H_6,
   (d) sulphur and oxygen in sulphur trioxide, SO_3,
   (e) hydrogen and fluorine in hydrogen fluoride, HF,
   (f) magnesium, sulphur and oxygen in magnesium sulphate, $MgSO_4$.

2 Calculate the percentage by mass of water in:
   (a) copper(II) sulphate-5-water, $CuSO_4.5H_2O$ (take A_r (Cu) = 64),
   (b) sodium sulphide-9-water, $Na_2S.9H_2O$.

26.4 The mole

FIRST THOUGHTS

Chemical equations tell us which products are formed when substances react. Equations can also be used to tell us what mass of each product is formed. The key to success is the mole concept.

A reaction of industrial importance is

Calcium carbonate → Calcium oxide + Carbon dioxide
$CaCO_3(s) \rightarrow CaO(s) + CO_2(g)$

Cement manufacturers use this reaction to make calcium oxide (quicklime) from calcium carbonate (limestone). The mole concept makes it possible to calculate what mass of calcium oxide will be formed when a certain mass of calcium carbonate dissociates.

The mole concept dates back to the nineteenth century Italian chemist called Avogadro. This is how he argued:

You are familiar with **mass** of substance and **volume** of substance. These are physical quantities. Mass is stated in the unit g or kg. Volume may be stated in the unit cm^3, dm^3 or l (litre) or m^3.

The relative atomic masses of magnesium and carbon are: A_r (Mg) = 24, A_r (C) = 12.

Therefore we can say:

Since one atom of magnesium is twice as heavy as one atom of carbon, then one hundred Mg atoms are twice as heavy as one hundred C atoms, and five million Mg atoms are twice as heavy as five million C atoms.

Imagine a piece of magnesium that has twice the mass of a piece of carbon. It follows that the two masses must contain equal numbers of atoms:

two grams of magnesium and one gram of carbon contain the same number of atoms; ten tonnes of magnesium and five tonnes of carbon contain the same number of atoms.

The same argument applies to the other elements. Take the relative atomic mass in grams of any element:

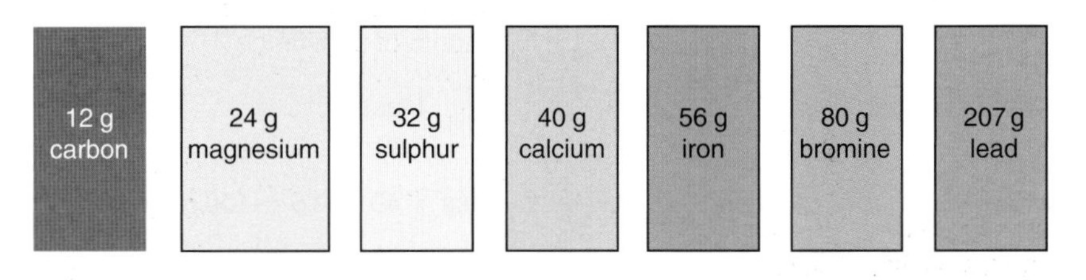

Another physical quantity is **amount** of substance. It depends on the number of particles present. The unit in which it is stated is the mole. Equal amounts of elements contain equal numbers of atoms.

All these masses contain the same number of atoms. The number is 6.022×10^{23}.

The amount of an element that contains 6.022×10^{23} atoms (the same number of atoms as 12 g of carbon-12) is called one **mole** of that element.

The symbol for mole is **mol**. The ratio 6.022×10^{23}/mol is called the Avogadro constant. When you weigh out 12 g of carbon, you are counting out 6×10^{23} atoms of carbon. This amount of carbon is one mole (1 mol) of carbon atoms. Similarly, 46 g of sodium is two moles (2 mol) of sodium atoms. You can say that the **amount** of sodium is two moles (2 mol). One mole of the compound ethanol, C_2H_6O, contains 6×10^{23} molecules of C_2H_6O, that is, 46 g of C_2H_6O (the molar mass in grams). To write 'one mole of oxygen' will not do. You must state whether you mean one mole of oxygen atoms, O (with a mass of 16 grams) or one mole of oxygen molecules, O_2 (with a mass of 32 grams).

Molar mass

The mass of one mole of a substance is called the **molar mass**, symbol *M*. The molar mass of carbon is 12 g/mol. The molar mass of sodium is 23 g/mol. The molar mass of a compound is the relative molecular mass expressed in grams per mole. Urea, CON_2H_4, has a relative molecular mass of 60; its molar mass is 60 g/mol. Notice the units: relative molecular mass has no unit; molar mass has the unit g/mol.

$$\text{Amount (in moles) of substance} = \frac{\text{Mass of substance}}{\text{Molar mass of substance}}$$

Molar mass of element = Relative atomic mass in grams per mole
Molar mass of compound = Relative molecular mass in grams per mole

Worked example 1 What is the amount (in moles) of sodium present in 4.6 g of sodium?

Solution A_r of sodium = 23. Molar mass of sodium = 23 g/mol

$$\text{Amount of sodium} = \frac{\text{Mass of sodium}}{\text{Molar mass of sodium}} = \frac{4.6\text{ g}}{23\text{ g/mol}} = 0.2\text{ mol}$$

The amount (moles) of sodium is 0.2 mol.

SUMMARY

The number of atoms in 12.000 g of carbon-12 is 6.022×10^{23}. The same number of atoms is present in a mass of any element equal to its relative atomic mass expressed in grams. This amount of any element is called **one mole** (1 mol) of the element. The ratio 6.022×10^{23}/mol is called the Avogadro constant. The number of moles of a substance is called the **amount** of that substance. The mass of one mole of an element or compound is the **molar mass**, *M*, of that substance.

M of an element = A_r expressed in g/mol

M of a compound = M_r expressed in g/mol

Worked example 2 If you need 2.5 mol of sodium hydroxide, what mass of sodium hydroxide do you have to weigh out?

Solution Relative molecular mass of NaOH = 23 + 16 + 1 = 40

Molar mass of NaOH = 40 g/mol

$$\text{Amount of substance} = \frac{\text{Mass of substance}}{\text{Molar mass of substance}}$$

$$2.5 = \frac{\text{Mass}}{40}$$

Mass = 40 × 2.5 = 100 g

You need to weigh out 100 g of sodium hydroxide.

CHECKPOINT

1 State the mass of:
   (a) 1.0 mol of aluminium atoms,
   (b) 3.0 mol of oxygen molecules, O_2,
   (c) 0.25 mol of mercury atoms,
   (d) 0.50 mol of nitrogen molecules, N_2,
   (e) 0.25 mol of sulphur atoms, S,
   (f) 0.25 mol of sulphur molecules, S_8.

2 Find the amount (moles) of each element present in:
   (a) 100 g of calcium,
   (b) 9.0 g of aluminium,
   (c) 32 g of oxygen, O_2,
   (d) 14 g of iron.

3 State the mass of:
   (a) 1.0 mol of sulphuric acid, H_2SO_4,
   (b) 0.5 mol of nitrogen dioxide molecules, NO_2,
   (c) 2.5 mol of magnesium oxide, MgO,
   (d) 0.10 mol of calcium carbonate, $CaCO_3$.

26.5 The masses of reactant and product

FIRST THOUGHTS

Have you grasped the mole concept? In this section you will find out how it is used to obtain information about chemical reactions.

As well as knowing what products are formed in a chemical reaction, chemists want to know **what mass** of product is formed from a given mass of starting material. For example, calcium oxide (quicklime) is made by heating calcium carbonate (limestone). Manufacturers need to know what mass of limestone to heat to yield the mass of quicklime they want.

Worked example 1 What mass of limestone (calcium carbonate) must be decomposed to yield 10 tonnes of calcium oxide (quicklime)?

Solution First write the equation for the reaction

Calcium carbonate → Calcium oxide + Carbon dioxide

$$CaCO_3(s) \rightarrow CaO(s) + CO_2(g)$$

The equation tells us that

1 mol of calcium carbonate forms 1 mol of calcium oxide.

You have met calculations of this type before.

For example,
To make 12 buns a cook needs 3 eggs. How many eggs does she need to make 20 buns?

The method you would use is:
To make 1 bun needs 3/12 eggs, therefore to make 20 buns needs 20 × 3/12 eggs = 5 eggs.

This chemical calculation uses the same method.

Using the molar masses $M(CaCO_3) = 100\,g/mol$, $M(CaO) = 56\,g/mol$,

100 g of calcium carbonate forms 56 g of calcium oxide.

The mass of calcium carbonate needed to make 10 tonnes of calcium oxide is therefore given by

$$\text{Mass of } CaCO_3 = \frac{100}{56} \times (\text{Mass of CaO}) = \frac{100}{56} \times 10 = 17.8\ \text{tonnes}$$

Worked example 2 What mass of aluminium can be obtained by the electrolysis of 60 tonnes of pure aluminium oxide, Al_2O_3?

Solution The equation comes first.

$$\text{Aluminium oxide} \rightarrow \text{Aluminium} + \text{Oxygen}$$
$$2Al_2O_3(s) \rightarrow 4Al(s) + 3O_2(g)$$

From the equation you can see that

1 mol of aluminium oxide forms 2 mol of aluminium.

Using the molar masses $M(Al) = 27\,g/mol$, $M(Al_2O_3) = 102\,g/mol$,

102 g of aluminium oxide form 54 g of aluminium.

The mass of aluminium obtained from 60 tonnes of aluminium oxide is therefore given by

$$\text{Mass of aluminium} = \frac{54}{102} \times \text{Mass of aluminium oxide}$$
$$= \frac{54}{102} \times 60 = 31.8\ \text{tonnes}$$

SUMMARY

The equation for a chemical reaction shows how many moles of product are formed from one mole of reactant. Using the equation and the molar masses of the chemicals, you can find out what mass of product is formed from a certain mass of reactant. In chemical calculations, the balanced equation for the reaction is the key to success.

CHECKPOINT

You do not need calculators for these problems.

1. What mass of magnesium oxide, MgO, is formed when 4.8 g of magnesium are completely oxidised?
2. Hydrogen will reduce hot copper(II) oxide, CuO, to copper:

   Hydrogen + Copper(II) oxide → Copper + Water

   (a) Write the balanced chemical equation for the reaction.

   (b) Calculate the mass of copper that can be obtained from 4.0 g of copper(II) oxide. Use $A_r(Cu) = 64$
3. What mass of sodium bromide, NaBr, must be electrolysed to give 8 g of bromine, Br_2?
4. Ammonium chloride can be made by neutralising hydrochloric acid with ammonia:

   $$HCl(aq) + NH_3(aq) \rightarrow NH_4Cl(aq)$$

   What mass of ammonium chloride is formed when 73 g of hydrochloric acid are completely neutralised by ammonia?

26.6 Percentage yield

If you managed the work on percentage composition you will have no difficulty here.

Calculations based on chemical equations give the **theoretical yield** of product to be expected from a reaction. Often the actual yield is less than the calculated yield of product. The reason may be that some product has remained in solution or on a filter paper and has not been weighed with the final yield. The percentage yield of a product is given by:

$$\text{Percentage yield} = \frac{\text{Actual mass of product}}{\text{Calculated mass of product}} \times 100$$

Worked example A student calculates that a certain reaction will yield 7.0 g of a salt. Her product weighs 6.3 g. What percentage yield has she obtained?

Solution

$$\text{Percentage yield} = \frac{\text{Actual mass of product}}{\text{Calculated mass of product}} \times 100$$

$$= \frac{6.3}{7.0} \times 100 = 90\%$$

CHECKPOINT

You will need a calculator to solve some of these problems.

1 When 6.4 g of copper were heated in air, 7.6 g of copper(II) oxide, CuO, were obtained.

$$2Cu(s) + O_2(g) \rightarrow 2CuO(s)$$

(a) Calculate the mass of copper(II) oxide that would be formed if the copper reacted completely. (Use A_r(Cu) = 64)

(b) Calculate the percentage yield that was actually obtained.

2 When 28 g of nitrogen and 6 g of hydrogen were mixed and allowed to react, 3.4 g of ammonia formed.

$$N_2(g) + 3H_2(g) \rightarrow 2NH_3(g)$$

(a) What is the maximum mass of ammonia that could be formed?

(b) What percentage of this yield was obtained?

3 A student passed chlorine over heated iron until all the iron had reacted. He collected 16.0 g of iron(III) chloride, $FeCl_3$. What percentage yield had he obtained?

4 A student neutralised 98 g of sulphuric acid, H_2SO_4, with ammonia, NH_3. On evaporating the solution until the salt crystallised, she obtained 120 g of ammonium sulphate, $(NH_4)_2SO_4$.

(a) Write a balanced chemical equation for the reaction.

(b) Calculate the theoretical yield of ammonium sulphate.

(c) Calculate the actual percentage yield.

(d) What do you think happened to the rest of the ammonium sulphate?

5 Copper(II) sulphate can be made by neutralising sulphuric acid with copper(II) oxide:

$$CuO(aq) + H_2SO_4(aq) \rightarrow CuSO_4(aq) + H_2O(l)$$

The salt crystallises as copper(II) sulphate-5-water, $CuSO_4.5H_2O$.

(a) Calculate the mass of crystals that can be made from 8.0 g of copper(II) oxide and an excess (more than enough) of sulphuric acid. Use A_r (Cu) = 64.

(b) A student obtained 22 g of crystals from this preparation. What percentage yield was this?

26.7 Finding formulas

FIRST THOUGHTS

All those fascinating formulas in chemistry books – where do they come from? In this section you can find out.

The formula of a compound is worked out from the percentage composition by mass of the compound.

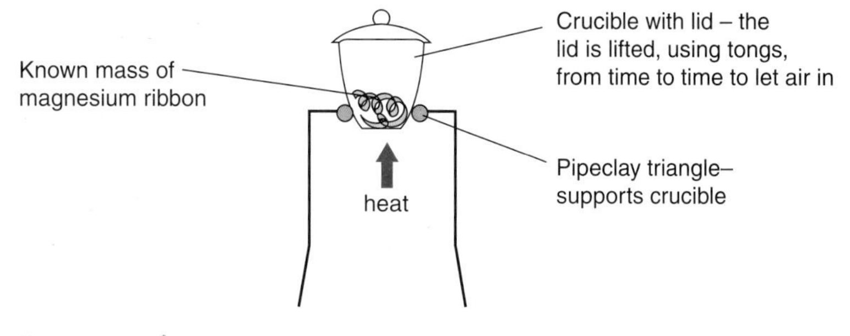

Figure 26.7A ▲ Heating magnesium

The steps in finding the formula are

- Find the mass of each element in a sample of the compound.
- Calculate the amount of each element present.
- Deduce the number of atoms of each element present, which gives the formula.

Worked example Finding the formula of magnesium oxide.
First, an experiment must be done to find the mass of oxygen that combines with a weighed amount of magnesium. Figure 26.7A shows a weighed quantity of magnesium being heated until it has been converted completely into magnesium oxide. Then the magnesium oxide must be weighed. The mass of oxygen that has combined with the magnesium is found by subtraction. A typical set of results is given below.

[1] Mass of crucible = 19.24 g

[2] Mass of crucible + magnesium = 20.68 g

[3] **Mass of magnesium** = [2] − [1] = 1.44 g

[4] Mass of crucible + magnesium oxide = 21.64 g

[5] **Mass of oxygen combined** = [4] − [2] = 0.96 g

Solution The results are used in this way:

Element	*Magnesium*	*Oxygen*
Mass	1.44 g	0.96 g
A_r	24	16
Amount in moles	1.44/24	0.96/16
	= 0.060	= 0.060
Divide through by 0.060	1 mole Mg	1 mole O
Formula is	MgO	

The formula MgO is the simplest formula which fits the results. Other formulas, such as Mg_2O_2, Mg_3O_3, etc. also fit the results. MgO is the **empirical formula** for magnesium oxide.

SUMMARY

The empirical formula of a compound shows the symbols of the elements present and the ratio of the number of atoms of each element present in the compound.

The empirical formula of a compound is the simplest formula which represents the composition by mass of the compound.

Empirical formula of hydrate

The formula of a hydrate is worked out in the same way.

Worked example Copper(II) sulphate crystallises as the well-known blue hydrate.

The percentage of water in crystals is 36%. Find the formula of the hydrate $CuSO_4.nH_2O$.

Method

	Molar mass	*Mass*	*Amount*	
$CuSO_4$ (taking Cu = 64)	160	64%	64/160	= 0.4 mol
H_2O	18	36%	36/18	= 2.0 mol
Ratio = 0.4 $CuSO_4$: 2.0 H_2O = 1 $CuSO_4$: $5H_2O$				

Answer Formula = $CuSO_4.5H_2O$

CHECKPOINT

You do not need calculators for these problems.

1 Find the empirical formulas of the following compounds:
(a) a compound of 3.5 g of nitrogen and 4.0 g of oxygen,
(b) a compound of 14.4 g of magnesium and 5.6 g of nitrogen,
(c) a compound of 5.4 g of aluminium and 9.6 g of sulphur.

2 Calculate the percentage by mass of water in
(a) sodium sulphide-9-water, $Na_2S.9H_2O$
(b) chromium(III) nitrate-9-water, $Cr(NO_3)_3.9H_2O$.

3 Calculate the empirical formulas of the compounds which have the following percentage compositions by mass:
(a) 40% sulphur, 60% oxygen,
(b) 50% sulphur, 50% oxygen,
(c) 20% calcium, 80% bromine,
(d) 39% potassium, 1% hydrogen, 12% carbon, 48% oxygen.

4 Find the empirical formulas of the compounds formed when
(a) 18 g of beryllium form 50 g of beryllium oxide,
(b) 11.2 g of iron form 16.0 g of an oxide of iron,
(c) 2.800 g of iron form 8.125 g of an iron chloride.

5 The element titanium (A_r = 48.0) combines with oxygen to form two different oxides. The mass of oxygen combining with 1.00 g of titanium in oxide A is 0.50 g and in oxide B is 0.67 g. Deduce the empirical formulas of the two oxides.

Finding the molecular formula from the empirical formula

The molecular formula is a multiple of the empirical formula:

Molecular formula = $n \times$ Empirical formula

The molecular formula gives the correct relative molecular mass.

The empirical formula is the simplest formula that represents the composition of a compound. It may not be the **molecular formula**, that is the formula that shows the number of atoms present in a molecule of the compound. The molecular formula of ethanoic acid is CH_3CO_2H, which can be written as $C_2H_4O_2$. The empirical formula is CH_2O. To work out the relative molecular mass for ethanoic acid, you have to use the molecular formula $C_2H_4O_2$, and then you obtain a value of 60.

Worked example 1 What is the molecular formula of a compound which has an empirical formula of CH_2 and a relative molecular mass of 70?

Method Relative molecular mass = 70
Empirical formula mass = 12 + 2 = 14
The relative molecular mass is 5 × the relative empirical formula mass
The molecular formula is 5 × the empirical formula

Answer The molecular formula is C_5H_{10}.

Worked example 2 What is the molecular formula of a compound which has an empirical formula of C_2H_6O and a relative molecular mass of 46?

Method Relative molecular mass = 46
Relative empirical formula mass = 24 + 6 + 16 = 46
The empirical formula gives the correct relative molecular mass. Therefore the molecular formula is the same as the empirical formula.

Answer The molecular formula is C_2H_6O.

CHECKPOINT

1 Find the molecular formula of each of the following compounds from the empirical formula and the relative molecular mass.

Compound	*Empirical formula*	M_r	*Compound*	*Empirical formula*	M_r
A	CH_4O	32	B	CH_2	42
C	C_2H_4O	88	D	CH_3O	62
E	CH_3	30	F	CH_2Cl	99
G	CH	78	H	C_2HNO_2	213

2 Calculate (a) the empirical formula and (b) the molecular formula of:

(i) a hydrocarbon which contains 80% by mass of carbon and has a relative molecular mass of 30

(ii) a hydrocarbon which contains 85.7% of carbon and has a relative molecular mass of 28.

3 What is (a) the empirical formula and (b) the molecular formula of a compound which contains 4.04% H, 24.24% C, 71.72% Cl and has a relative molecular mass of 99?

26.8 Reacting volumes of gases

FIRST THOUGHTS

Calculations on reacting volumes of gas are simple if you understand one fact.

Whatever the gas, the volume occupied by one mole of the gas is the same:
24.0 dm^3 at r.t.p. (the gas molar volume).

The volume of a certain mass of gas depends on its temperature and on its pressure. We therefore state the temperature and the pressure at which a volume was measured. It is usual to state gas volumes either at standard temperature and pressure (s.t.p., 0 °C and 1 atm) or at room temperature and pressure (r.t.p., 20 °C and 1 atm).

One mole of a gas is the amount of the gas that contains 6×10^{23} molecules of that gas. To measure one mole of gas, you take the molar mass expressed in grams, e.g. 2 g of hydrogen, H_2. Measurements show that one of mole of gas occupies 24.0 dm^3 at r.t.p. For all gases, the volume is the same.

- 2 g of hydrogen
- 28 g of nitrogen
- 64 g of sulphur dioxide

All these are one mole of gas (the molar mass expressed in grams) and all occupy 24 dm^3 at r.t.p.

The volume of one mole of gas, 24.0 dm^3 at r.t.p., is called the **gas molar volume**.

Calculations on the reacting volumes of gases start with the equation for the reaction. After that, it's as easy as one, two, three. Take the reaction:

$$A(g) + 3B(g) \rightarrow 2C(g)$$

The equation tells you that

1 mole of A reacts with 3 moles of B to form 2 moles of C

therefore, at r.t.p.,

24 dm^3 of A react with 3×24 dm^3 of B to form 2×24 dm^3 of C

and in general,

1 volume of A reacts with 3 volumes of B to form 2 volumes of C.

Write the equation.
Then the calculation becomes a simple ratio-type calculation.

Worked example 1 Nitrogen and hydrogen combine to form ammonia:

$$N_2(g) + 3H_2(g) \rightleftharpoons 2NH_3(g)$$

If hydrogen is fed into the plant at 12 m^3/second, at what rate should nitrogen be fed in?

Method From the equation, 1 mole of nitrogen combines with 3 moles of hydrogen, therefore 1 volume of nitrogen combines with 3 volumes of hydrogen, therefore, the rate of flow of nitrogen should be one third that of hydrogen, that is $\frac{1}{3} \times 12\ m^3/s = 4\ m^3/s$.

Answer Nitrogen should be fed in at a rate of 4 m^3/s.

Worked example 2 What volume of carbon dioxide (at r.t.p.) is formed by the complete combustion of 3 g of carbon?

Method From the equation,

$$C(s) + O_2(g) \rightarrow CO_2(g)$$

we can see that 1 mole of carbon forms 1 mole of carbon dioxide
that is 12 g carbon form 24 dm^3 of carbon dioxide
therefore 3 g carbon form $\frac{3}{12} \times 24\ dm^3 = 6\ dm^3$ of carbon dioxide.

Answer The combustion of 3 g of carbon produces 6 dm^3 of carbon dioxide at r.t.p.

SUMMARY

Calculations on reacting volumes of gases always start with the equation for the chemical reaction. One mole of any gas occupies 24.0 dm^3 at r.t.p. This volume is called the gas molar volume.

CHECKPOINT

1. Calculate the volume at r.t.p. of oxygen needed for the complete combustion of 8 g of sulphur.

2. Calculate the volume at r.t.p. of oxygen needed for the complete combustion of 250 cm^3 of methane, CH_4. What volume of carbon dioxide is formed?

3. A power station burns coal which contains sulphur. The sulphur burns to form sulphur dioxide, SO_2. If the power station burns 28 tonnes of sulphur in its coal every day, what volume of sulphur dioxide does it send into the air? (1 tonne = 1000 kg)

4. What volume of hydrogen at r.t.p. is formed when 7.0 g of iron react with an excess of sulphuric acid? The equation is

$$Fe(s) + H_2SO_4(aq) \rightarrow H_2(g) + FeSO_4(aq)$$

5. A cook puts 3 g of sodium hydrogencarbonate into a cake mixture. Calculate the volume at r.t.p. of carbon dioxide that will be produced in the reaction:

$$2NaHCO_3(s) \rightarrow Na_2CO_3(s) + CO_2(g) + H_2O(g)$$

6. The equation for the reaction between marble and hydrochloric acid is

$$CaCO_3(s) + 2HCl(aq) \rightarrow CaCl_2(aq) + CO_2(g) + H_2O(l)$$

What mass of marble is needed to give 6.0 dm^3 of carbon dioxide?

26.9 Calculations on electrolysis

If you need to refresh your grasp of electrolysis, refer to Topic 7.

When a current passes through a solution of a salt of a metal which is low in the electrochemical series, metal ions are discharged and metal atoms are deposited on the cathode (see Figure 26.9A).

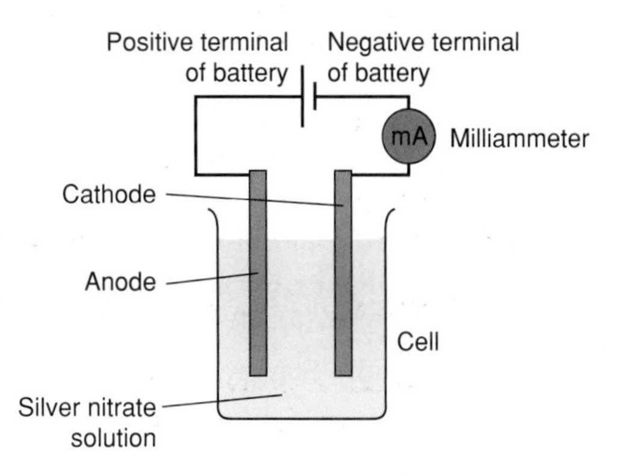

Figure 26.9A The electrolysis of silver nitrate solution

The cathode process is

$$Ag^+(aq) + e^- \rightarrow Ag(s)$$

This equation tells us that 1 silver ion accepts 1 electron to form 1 atom of silver. Therefore 1 mole of silver ions accept 1 mole of electrons to form 1 mole of silver atoms.

When 1 mole of silver has been deposited, the cathode has increased in mass by 108 g (the mass of 1 mole of silver). This process must have needed the passage of 1 mole of electrons through the cell. We can measure the quantity of electric charge that passes through the cell when 108 g of silver are deposited. This quantity is the charge on 1 mole of electrons. Electric charge is measured in coulombs (C). One coulomb is the electric charge that passes when 1 ampere flows for 1 second.

$$\text{Charge in coulombs} = \text{Current in amperes} \times \text{Time in seconds}$$
$$Q = I \times t$$

Using a cell such as that in Figure 26.9A, one can pass a current through a milliammeter in the circuit for a known time. By weighing the cathode before and after the passage of a current, the mass of silver deposited can be found. Experiments of this kind show that to deposit 108 g of silver requires the passage of 96 500 coulombs. This quantity of electricity must be the charge on one mole of electrons (6×10^{23} electrons). The value 96 500 C/mol is called the Faraday constant after Michael Faraday who did much of the early work on electrolysis.

In each type of calculation

- writing the equation for the reaction is the first step
- applying the mole concept is the second step.

For the mass of 1 mol of substance, write the molar mass in g/mol.
For the charge on 1 mol of electrons, write 96 500 coulombs/mol.
The mass of an element deposited can be measured by weighing.
The charge can be measured because
charge = current × time
(coulombs) (amperes) (seconds)

Now in the deposition of copper in electrolysis, the cathode process is

$$Cu^{2+}(aq) + 2e^- \rightarrow Cu(s)$$

1 copper ion needs 2 electrons to become 1 atom of copper;
1 mole of copper ions need 2 moles of electrons for discharge;
1 mole of copper ions need $2 \times 96\,500$ coulombs for discharge.

When gold is deposited during the electrolysis of gold(III) salts, the cathode process is:

$$Au^{3+}(aq) + 3e^- \rightarrow Au(s)$$

1 gold ion needs 3 electrons to form 1 mole of gold atoms;
1 mole of gold ions need 3 moles of electrons for discharge;
$3 \times 96\,500$ coulombs are needed to deposit 1 mole of gold.

It's as easy as one-two-three. First work out whether

1 mole of electrons discharge 1 mole of the element, e.g. silver, or

$\frac{1}{2}$ mole of the element, e.g. copper, or

$\frac{1}{3}$ mole of the element, e.g. gold.

$$\text{No. of moles of element discharged} = \frac{\text{No. of moles of electrons}}{\text{No. of charges on one ion of the element}}$$

$$= \frac{\text{No. of coulombs}/96\,500}{\text{No. of charges on one ion of the element}}$$

Worked example 1 What masses of the following elements are deposited by the passage of one mole of electrons through solutions of their salts? (a) silver (b) copper (c) gold

Method Start with the equations.

$$Ag^+(aq) + e^- \rightarrow Ag(s)$$
$$Cu^{2+}(aq) + 2e^- \rightarrow Cu(s)$$
$$Au^{3+}(aq) + 3e^- \rightarrow Au(s)$$

As already argued,

1 mol electrons deposit 1 mol silver = 108 g silver

1 mole electrons deposit $\frac{1}{2}$ mol copper = $\frac{1}{2} \times 63.5$ g = 31.8 g copper

1 mole electrons deposit $\frac{1}{3}$ mol gold = $\frac{1}{3} \times 197$ g = 65.7 g gold

Answer (a) 108 g silver, (b) 31.8 g copper, (c) 65.7 g gold

Worked example 2 A current of 10.0 milliamps (mA) passes for 4.00 hours through a solution of silver nitrate, a solution of copper(II) sulphate and a solution of gold(III) nitrate connected in series. What mass of metal is deposited in each?

Method Charge in coulombs = Current in amperes × Time in seconds

Charge = $0.010 \times 4.00 \times 60 \times 60 = 144\,C$

Charge = $\frac{144}{96\,500}$ moles of electrons

Equations are:

$$Ag^{+}(aq) + e^{-} \rightarrow Ag(s)$$
$$Cu^{2+}(aq) + 2e^{-} \rightarrow Cu(s)$$
$$Au^{3+}(aq) + 3e^{-} \rightarrow Au(s)$$

If you work patiently through the worked examples, you will overcome any difficulties. Guaranteed!

1 mol electrons deposit 1 mol silver; therefore $\frac{144}{96\,500}$ mol electrons deposit $\frac{144}{96\,500}$ mol silver = $144 \times \frac{108}{96\,500}$ g silver = 0.161 g silver

1 mol electrons deposit $\frac{1}{2}$ mol copper; therefore $\frac{144}{96\,500}$ mol electrons deposit $\frac{1}{2} \times \frac{144}{96\,500}$ mol copper = $\frac{1}{2} \times 144 \times \frac{63.5}{96\,500}$ g copper = 0.0474 g copper

1 mol electrons deposit $\frac{1}{3}$ mol gold; therefore $\frac{144}{96\,500}$ mol electrons deposit $\frac{1}{3} \times \frac{144}{96\,500}$ mol gold = $\frac{1}{3} \times 144 \times \frac{197}{96\,500}$ g gold = 0.0980 g gold

Answer 0.161 g silver, 0.0474 g copper, 0.0980 g gold

Worked example 3 Aluminium is extracted by the electrolysis of molten aluminium oxide. How many coulombs of electricity are needed to produce 1 tonne of aluminium (1 tonne = 1000 kg)?

Method The equation

$$Al^{3+}(l) + 3e^{-} \rightarrow Al(s)$$

shows that 3 mol electrons are needed to give 1 mol aluminium

$1.00 \times 10^{6} \times \frac{3}{27}$ mol electrons are needed to give 1 tonne of aluminium

$$1.00 \times 10^{6} \times \frac{3}{27} \text{ mol of electrons} = 1.00 \times 10^{6} \times \frac{3}{27} \times 96\,500\,C = 1.07 \times 10^{10}\,C$$

Answer 1.07×10^{10} C will deposit 1 tonne of aluminium.

Worked example 4 A metal of relative atomic mass 27 is deposited by electrolysis. If 0.201 g of the metal is deposited when 0.200 A flow for 3.00 hours, what is the charge on the ions of this element?

Method
Charge = $0.200 \times 3.00 \times 60 \times 60\,C = 2160\,C$

If 2160 C deposit 0.201 g of the metal,
then 96 500 C deposit $96\,500 \times \frac{0.201}{2160}\,g = 8.98\,g$

Since 8.98 g metal are deposited by 1 mol electrons, 26.0 g of metal are deposited by $\frac{27}{8.98}$ mol electrons = 3 mol electrons

Answer The charge on the metal ions is +3.

Calculate the volume of gas evolved during electrolysis

Hydrogen

When a current is passed through a solution of a salt of a metal which is high in the electrochemical series, hydrogen ions are discharged at the cathode.

$$H^+(aq) + e^- \rightarrow H(g)$$
$$\text{followed by } 2H(g) \rightarrow H_2(g)$$

Each hydrogen molecule needs 2 electrons for its evolution, and 1 mole of hydrogen needs 2 moles of electrons for its evolution. Thus 2 moles of electrons ($2 \times 96\,500\,C$) will result in the evolution of 24 dm^3 at r.t.p. (the gas molar volume) of hydrogen.

When the product of electrolysis is a gas, for 1 mol of gas write 24.0 dm^3 at r.t.p.

The only way you can go wrong is to forget that, e.g. 1 mol of hydrogen is 1 mol of H_2 (not H).

1 mol H needs 1 mol of electrons
1 mol of H_2 needs 2 mol of electrons
1 mol of OH^- needs 1 mol of electrons
1 mol of O_2 needs 4 mol of electrons.

Chlorine

When chlorine is evolved at the anode,

$$Cl^-(aq) \rightarrow Cl(g) + e^-$$
$$\text{followed by } 2Cl(g) \rightarrow Cl_2(g)$$

Each chlorine molecule gives 2 electrons when it is evolved, and 1 mole of chlorine gives 2 moles of electrons when it is evolved. Thus 2 moles of electrons ($2 \times 96\,500\,C$) accompany the evolution of 24 dm^3 at r.t.p. (the gas molar volume) of chlorine.

Oxygen

When oxygen is evolved at the anode,

$$OH^-(aq) \rightarrow OH(g) + e^-$$
$$\text{followed by } 4OH(g) \rightarrow O_2(g) + 2H_2O(l)$$

Thus 4 moles of electrons must pass with the evolution of 1 mole of oxygen (24 dm^3 at r.t.p.).

Worked example 5 Name the gases formed at each electrode when 15.0 mA of current passes for 6.00 hours through a solution of sulphuric acid and calculate their volumes (at r.t.p.).

Method
At the cathode hydrogen is evolved

$$H^+(aq) + e^- \rightarrow H(g)$$
$$\text{followed by } 2H(g) \rightarrow H_2(g)$$

so that 2 moles of electrons discharge 1 mole of hydrogen gas.

At the anode oxygen is evolved

$$OH^-(aq) \rightarrow OH(g) + e^-$$
$$\text{followed by } 4OH(g) \rightarrow O_2(g) + 2H_2O(l)$$

so that 4 moles of electrons discharge 1 mole of oxygen gas.

$$\text{Charge (C)} = \text{Current (A)} \times \text{Time (s)}$$
$$= 15.0 \times 10^{-3} \times 6.00 \times 60 \times 60 = 324\,C$$

$$\text{No. of mol electrons} = \frac{324\,C}{96\,500\,C/mol} = 3.36 \times 10^{-3}\,mol$$

Amount of hydrogen discharged $= \frac{1}{2} \times 3.36 \times 10^{-3}\,mol$
Volume of hydrogen $= 1.68 \times 10^{-3} \times 24.0\,dm^3 = 40.4\,cm^3$ at r.t.p.
Amount of oxygen $= \frac{1}{4} \times 3.36 \times 10^{-3}\,mol = 0.84 \times 10^{-3}\,mol$
Volume of oxygen $= 0.84 \times 10^{-3} \times 24.0\,dm^3 = 20.2\,cm^3$ at r.t.p.

Answer At the cathode $40.4\,cm^3$ (at r.t.p.) of hydrogen are evolved. At the anode $20.2\,cm^3$ (at r.t.p.) of oxygen are evolved.

CHECKPOINT

(See relative atomic masses given opposite.)

Some relative atomic masses	
Ag	108
Al	27
Br	80
Cd	112
Cl	35.5
Cu	63.5
Fe	56
H	1
Ni	59
O	16
Pb	207
Sn	119

1. Calculate the mass of each element discharged when 0.250 mol of electrons passes through each of the solutions listed.
   (a) copper from copper(II) sulphate solution
   (b) nickel from nickel chloride solution
   (c) lead from lead(II) nitrate solution
   (d) bromine from potassium bromide solution
   (e) tin from tin(II) nitrate solution
2. Calculate the volume (at r.t.p.) of each gas evolved when 48 250 C pass through a solution of (a) dilute hydrochloric acid (b) dilute nitric acid.
3. A current passes through two cells in series. The cells contain solutions of silver nitrate and lead(II) nitrate. In the first, 0.540 g of silver is deposited. What mass of lead is deposited in the second?
4. When a current passes through a solution of copper(II) sulphate, 0.635 g of copper is deposited on the cathode. What volume of oxygen is evolved at the anode?
5. A current passes through two cells in series. In the first, 0.2160 g of silver is deposited. In the second, 0.1125 g of cadmium is deposited. Use this information to calculate the charge on the cadmium ion.
6. A current of 2.00 A passes for 96.5 hours through molten aluminium oxide. What mass of aluminium is deposited?
7. A current of 2.01 A passed for 8.00 minutes through aqueous nickel sulphate. The mass of the cathode increased by 0.295 g.
   (a) How many coulombs of electricity passed during the experiment?
   (b) How many moles of nickel were deposited?
   (c) How many moles of electrons are needed to discharge 1 mole of nickel ions?
   (d) Write an equation for the cathode reaction.

8 Choose the correct answers.

A student electrolysed a solution of silver nitrate in series with a solution of sulphuric acid. A steady current passed for 30 minutes, and 24 cm^3 of hydrogen collected at the cathode of the sulphuric acid cell.

(a) The volume of oxygen evolved at the anode is
(i) 6 cm^3 (ii) 12 cm^3 (iii) 24 cm^3 (iv) 32 cm^3 (v) 48 cm^3

(b) If the student doubled the current and passed it for 15 minutes, the volume of hydrogen evolved would be
(i) 6 cm^3 (ii) 12 cm^3 (iii) 24 cm^3 (iv) 48 cm^3 (v) 96 cm^3

(c) The mass of silver deposited on the cathode of the silver nitrate cell is
(i) 0.108 g (ii) 0.216 g (iii) 0.0270 g (iv) 0.0540 g (v) 0.0135 g

9 Which one of the following requires the largest quantity of electricity for discharge at an electrode?
(a) 1 mol Ni^{2+} ions (b) 2 mol Fe^{3+} ions (c) 3 mol Ag^+ ions
(d) 4 mol Cl^- ions (e) 5 mol OH^- ions

10 A current of 0.010 A passed for 5.00 hours through a solution of a salt of the metal M, which has a relative atomic mass of 52. The mass of M deposited on the cathode was 0.0323 g. What is the charge on the ions of M?

26.10 Concentrations of solutions

FIRST THOUGHTS

Remember:

One litre (1 l) is the same as one cubic decimetre (1 dm^3).

1 l = 1000 cm^3 = 1 dm^3

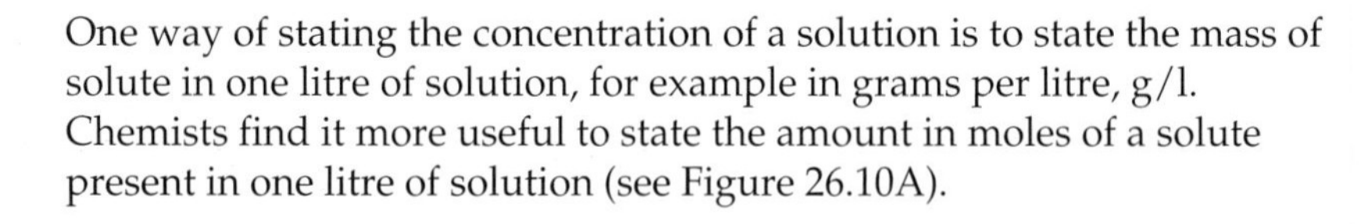

One way of stating the concentration of a solution is to state the mass of solute in one litre of solution, for example in grams per litre, g/l. Chemists find it more useful to state the amount in moles of a solute present in one litre of solution (see Figure 26.10A).

EXTENSION FILE ACTIVITY

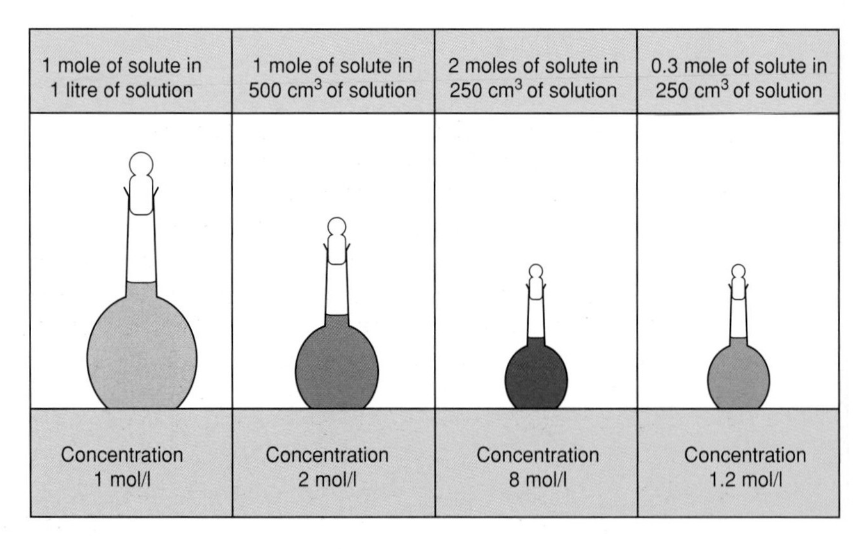

Figure 26.10A Concentrations of solutions in moles per litre (mol/l)

A solution of concentration 1 mol/l (1 mol l^{-1}) or 1 mol/dm^3 (1 mol dm^{-3}) is called a 1 M solution.

$$\text{Concentration in moles per litre} = \frac{\text{Amount of solute in moles}}{\text{Volume of solution in litres}}$$

Rearranging,

Amount of solute (mol) = Volume of solution (l) × Concentration (mol/l)

Be sure you know that for a solution

$$\text{concentration} = \frac{\text{amount of solute}}{\text{volume of solution}}$$

Usually, with amount in mol and volume in l or dm^3, concentration is stated in mol/l or mol/dm^3. The worked examples will make everything clear. Have patience; do work through them.

SUMMARY

The concentration of a solute in a solution can be stated as:
(a) grams of solute per litre of solution, g/l,
(b) moles of solute per litre of solution, mol/l.

$$\text{Concentration (mol/l)} = \frac{\text{Amount of solute (mol)}}{\text{Volume of solution (l)}}$$

Worked example 1 Calculate the concentration of a solution that was made by dissolving 100 g of sodium hydroxide and making the solution up to 2.0 litres.

Solution Molar mass of sodium hydroxide, NaOH = 23 + 16 + 1 = 40 g/mol

$$\text{Amount in moles} = \frac{\text{Mass}}{\text{Molar mass}} = \frac{100\,\text{g}}{40\,\text{g/mol}} = 2.5\,\text{mol}$$

Volume of solution = 2.0 litres

$$\text{Concentration} = \frac{\text{Amount in moles}}{\text{Volume in litres}} = \frac{2.5\,\text{mol}}{2.0\,\text{l}} = 1.25\,\text{mol/l}$$

Worked example 2 Calculate the amount in moles of solute present in 75 cm^3 of a solution of hydrochloric acid which has a concentration of 2.0 mol/l.

Solution Amount (mol) = Volume (l) × Concentration (mol/l)

Amount of solute, HCl = $(75 \times 10^{-3}) \times 2.0 = 0.15$ mol

Note that the volume in cm^3 has been changed into litres to make the units correct:

Amount (**moles**) = Volume (**litres**) × Concentration (**moles per litre**)

Worked example 3 What mass of sodium carbonate must be dissolved in 1 l of solution to give a solution of concentration 1.5 mol/l (a 1.5 M solution)?

Solution Amount (mol) = Volume (l) × Concentration (mol/l)
= 1.00 l × 1.5 mol/l
= 1.5 mol

Molar mass of sodium carbonate, Na_2CO_3 = (2 × 24) + 12 + (3 × 16)
= 106 g/mol

Mass of sodium carbonate = 1.5 mol × 106 g/mol = 159 g

CHECKPOINT

1. Calculate the concentrations of the following solutions.
   (a) 30 g of ethanoic acid, $C_2H_4O_2$, in 500 cm^3 of solution.
   (b) 8.0 g of sodium hydroxide in 2.0 l of solution.
   (c) 8.5 g of ammonia in 250 cm^3 of solution.
   (d) 12.0 g of magnesium sulphate in 250 cm^3 of solution.
2. Find the amount of solute in moles present in the following solutions.
   (a) 1.00 l of a hydrochloric acid solution of concentration 0.020 mol/l.
   (b) 500 cm^3 of a solution of potassium hydroxide of concentration 2.0 mol/l.
   (c) 500 cm^3 of sulphuric acid of concentration 0.12 mol/l.
   (d) 100 cm^3 of a 0.25 mol/l solution of sodium hydroxide.
3. Solutions which are injected into a vein must be isotonic with the blood. They contain 8.43 g/l of sodium chloride.
   (a) What is the concentration of sodium chloride in mol/l?
   (b) What does 'isotonic' mean? Why must the solutions be isotonic with blood?
4. A woman mixes a drink containing 9.2 g of ethanol, C_2H_6O, in 100 cm^3 of solution. What is the concentration of ethanol in the solution (in mol/l)?
5. Calculate the concentrations of the following solutions.
   (a) 4.0 g of sodium hydroxide in 500 cm^3 of solution
   (b) 7.4 g of calcium hydroxide in 5.0 l of solution
   (c) 49.0 g of sulphuric acid in 2.5 l of solution
   (d) 73 g of hydrogen chloride in 250 cm^3 of solution

6 Find the amount of solute in moles present in the following solutions.
(a) 1.00 l of a solution of sodium hydroxide of concentration 0.25 mol/l
(b) 500 cm^3 of hydrochloric acid of concentration 0.020 mol/l
(c) 250 cm^3 of 0.20 mol/l sulphuric acid
(d) 10 cm^3 of a 0.25 mol/l solution of potassium hydroxide

26.11 Volumetric analysis

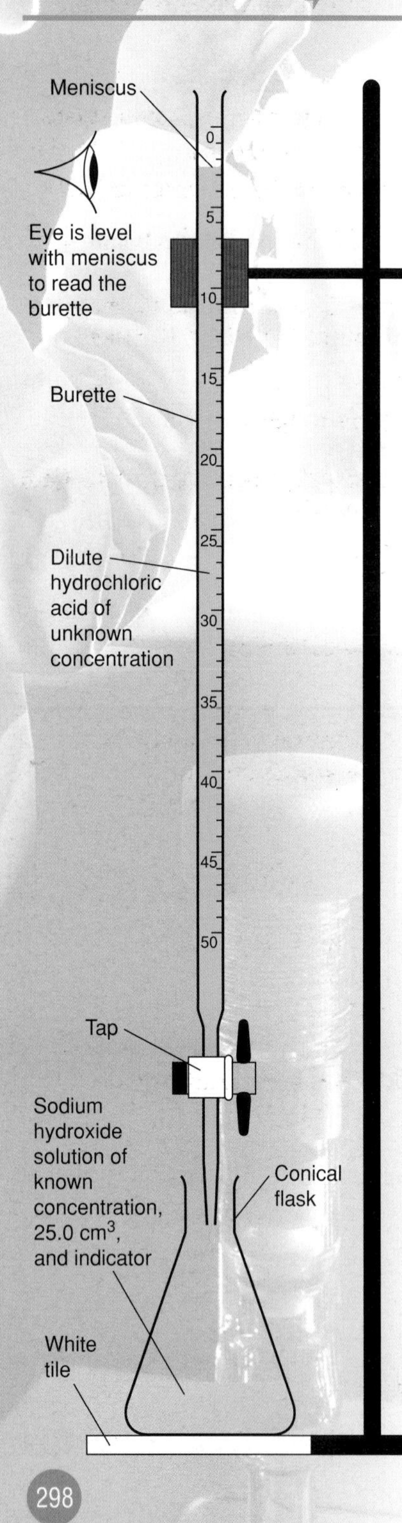

FIRST THOUGHTS

In your laboratory periods, you will learn how to do careful practical work in order to make precise measurements of concentration. This topic summarises the technique you will learn.

Volumetric analysis is a means of finding out the concentration of a solution. The method is to add a solution of, say an acid, to a solution of, say a base in a careful way until there is just enough of the acid to react with the base. This method is called **titration**. The concentration of one of the two solutions must be known, and the volumes of both must be measured.

You can use a standard solution of a base to find out the concentration of a solution of an acid. You have to find out what volume of the acid solution of unknown concentration is needed to neutralise a known volume, usually 25 cm^3, of the standard solution of a base. The steps in volumetric analysis are as follows.

1 Pipette 25.0 cm^3 of the alkali solution into a clean conical flask. Add a few drops of indicator.
2 Read the burette, V_1 cm^3. Run the acid solution from the burette drop by drop into the alkali until the indicator just changes colour. This is the 'end-point' of the titration.
3 Read the burette again, V_2 cm^3. The volume of acid used, $(V_2 - V_1)$ cm^3, is the volume of acid needed to neutralise 25.0 cm^3 of the alkali.

Worked example 1 A solution of hydrochloric acid is titrated against a standard sodium hydroxide solution. 15.0 cm^3 of hydrochloric acid neutralise 25.0 cm^3 of a 0.100 mol/l solution of sodium hydroxide. What is the concentration of hydrochloric acid?

Solution

(1) First write the equation for the reaction:

Hydrochloric acid + Sodium hydroxide → Sodium chloride + Water
$HCl(aq)$ + $NaOH(aq)$ → $NaCl(aq)$ + $H_2O(l)$

The equation tells you that 1 mole of HCl neutralises 1 mole of NaOH.

(2) Now work out the amount in moles of base. You must start with the base because you know the concentration of base, and you do not know the concentration of acid.

Amount (mol) = Volume (l) × Concentration (mol/l)
Amount (mol) of NaOH = Volume (25.0 cm^3) × Concentration (0.100 mol/l)
$= 25.0 \times 10^{-3}$ l × 0.100 mol/l = 2.50×10^{-3} mol

Figure 26.11A Titration. Hydrochloric acid is added to the conical flask until the indicator shows that the contents of the flask are neutral. The volume of acid added can be read off the burette

Here we go again with the worked examples. They will make everything clear if you will work through them.

EXTENSION FILE ACTIVITY

SUMMARY

A standard solution is a solution of known concentration. The concentration of a solution can be found by volumetric analysis. The concentration of a solution of an acid can be found by titrating it against a standard solution of a base. Similarly, the concentration of a solution of a base can be found by titrating it against a standard solution of an acid.

(3) Now work out the concentration of acid

Amount (mol) of HCl = Amount (mol) of NaOH = 2.50×10^{-3} mol

also Amount (mol) of HCl = Volume of HCl(aq) × Concentration of HCl(aq)

Therefore, if c is the concentration of HCl,

$$2.50 \times 10^{-3}\,\text{mol} = 15.0 \times 10^{-3}\,\text{l} \times c$$

$$c = \frac{2.50 \times 10^{-3}\,\text{mol}}{15.0 \times 10^{-3}}$$

$$= 0.167\,\text{mol/l}$$

The concentration of hydrochloric acid is 0.167 mol/l.

Worked example 2 In a titration, 25.0 cm^3 of sulphuric acid of concentration 0.150 mol/l neutralised 30.0 cm^3 of potassium hydroxide solution. Find the concentration of the potassium hydroxide solution.

Solution

(1) First write the equation:

Sulphuric acid + Potassium hydroxide → Potassium sulphate + Water

$$H_2SO_4(aq) + 2KOH(aq) \rightarrow K_2SO_4(aq) + 2H_2O(l)$$

The equation tells you that 1 mole of H_2SO_4 neutralises 2 moles of KOH.

(2) Now work out the amount (moles) of acid. You start with the acid because you know the concentration of the acid.

Amount (mol) of acid = Volume (25.0 cm^3) × Concentration (0.150 mol/l)

$= 25.0 \times 10^{-3}\,\text{l} \times 0.150\,\text{mol/l} = 3.75 \times 10^{-3}\,\text{mol}$

(3) Now work out the concentration of base.

Amount (mol) of KOH = 2 × amount (mol) of H_2SO_4

$= 7.50 \times 10^{-3}$ mol

also Amount (mol) of KOH = Volume of KOH(aq) × Concentration of KOH(aq)

Therefore, if c mol/l is the concentration of KOH

$$c\,\text{mol/l} = \frac{7.50 \times 10^{-3}\,\text{mol}}{30.0 \times 10^{-3}\,\text{l}}$$

$$= 0.250\,\text{mol/l}$$

The concentration of potassium hydroxide is 0.250 mol/l.

CHECKPOINT

1. 25.0 cm^3 of sodium hydroxide solution are neutralised by 15.0 cm^3 of a solution of hydrochloric acid of concentration 0.25 mol/l. Find the concentration of the sodium hydroxide solution.
2. A solution of sodium hydroxide contains 10 g/l.
   (a) What is the concentration of the solution in mol/l?
   (b) What volume of this solution would be needed to neutralise 25.0 cm^3 of 0.10 mol/l hydrochloric acid?
3. 25.0 cm^3 of hydrochloric acid are neutralised by 20.0 cm^3 of a solution of 0.15 mol/l sodium carbonate solution.
   (a) How many moles of sodium carbonate are neutralised by 1 mol HCl?
   (b) What is the concentration of the hydrochloric acid?
4. The class decide to test some antacid indigestion tablets. They dissolve tablets and titrate the alkali in them against a standard acid. The table shows their results.

Brand	*Price of 100 tablets* (£)	*Volume of 0.01 mol/l acid required to neutralise 1 tablet* (cm^3)
Paingo	0.60	2.8
Relievo	0.70	3.0
Alko	0.80	3.3
Clearo	0.90	3.6

   (a) Which antacid tablets offer the best value for money?
   (b) What other factors would you consider before choosing a brand?

Topic 27 Chemical energy changes

27.1 Exothermic reactions

FIRST THOUGHTS

In many places in this book, we study the importance of energy. One means by which energy is converted from one form into another is by chemical reactions.

Check that you know the meanings of these terms:

- exothermic reaction
- oxidation
- combustion
- burning
- fuel
- respiration
- neutralisation
- hydration.

www.keyscience.co.uk

Our bodies obtain energy from the combustion of foods, and our vehicles obtain energy from the combustion of hydrocarbons. Reactions of this kind, which give out energy, are **exothermic reactions** (ex = out; therm = heat). You will meet many such reactions. Some are described below.

Combustion

Topic 24.3 described the combustion of the fuels:

- coal (carbon + hydrocarbons),
- natural gas (largely methane, CH_4),
- petrol (a mixture of hydrocarbons, e.g. octane, C_8H_{18})

We use the heat from these exothermic reactions to warm our homes and buildings and to drive our vehicles.

Figure 27.1A ▲ They all use exothermic reactions

- An oxidation reaction in which heat is given out is **combustion**. The oxidation of sugars in our bodies is combustion.
- Combustion accompanied by a flame is **burning**. When coal is burned in a fireplace, you see the flames.
- A substance which is oxidised with the release of energy is a fuel. Both sugars and coal are fuels.

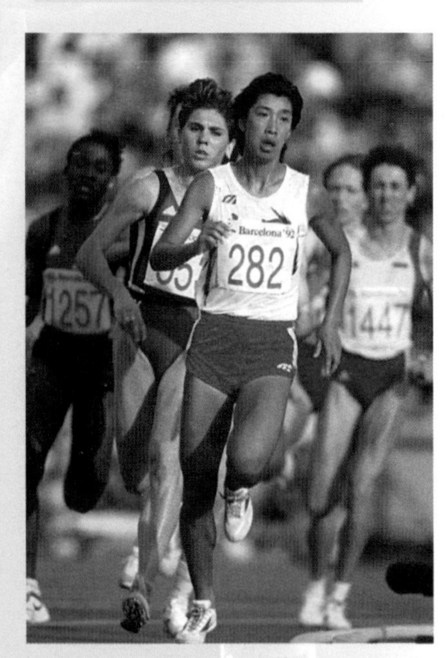

Figure 27.1B ▲ Energy from respiration

Respiration

We use the combustion of foods to supply our bodies with energy. Important among 'energy foods' are **carbohydrates**. Glucose, a sugar, is oxidised to carbon dioxide and water with the release of energy. This exothermic reaction, which takes place inside plant cells and animal cells, is called **aerobic respiration**.

Neutralisation

When an acid neutralises an alkali, heat is given out.

Hydrogen ion + Hydroxide ion	→	Water	Heat is given out
$H^+(aq)$ + $OH^-(aq)$	→	$H_2O(l)$	

SUMMARY

In exothermic reactions, energy is released. Examples are:

- the combustion of hydrocarbons,
- respiration,
- neutralisation,
- hydration of anhydrous salts.

Hydration

Many anhydrous salts react with water to form hydrates. During hydration, heat is given out.

Anhydrous copper (II) sulphate	+	Water	→	Copper(II) sulphate-5-water	Heat is given out
$CuSO_4(s)$	+	$5H_2O(l)$	→	$CuSO_4.5H_2O(s)$	

To reverse the reaction, heat must be taken in.

Then

copper(II) sulphate-5-water → anhydrous copper (II) sulphate + water.

27.2 Endothermic reactions

Check that you know the meanings of these terms:

- endothermic reaction
- photosynthesis
- thermal decomposition.

Photosynthesis

Plants manufacture sugars in the process of **photosynthesis**. They convert the energy of sunlight into the energy of the chemical bonds in sugar molecules. Photosynthesis is an **endothermic reaction**: it takes in energy.

Carbon dioxide + Water —(catalysed by chlorophyll, in the leaves of green plants)→ Glucose + Oxygen   Energy is taken in

$$6CO_2(g) + 6H_2O(l) \rightarrow C_6H_{12}O_6(aq) + 6O_2(g)$$

Thermal decomposition

Many substances decompose when they are heated. An example is calcium carbonate (limestone).

Calcium carbonate	→	Calcium oxide	+	Carbon dioxide	Heat is taken in
$CaCO_3(s)$	→	$CaO(s)$	+	$CO_2(g)$	

This is an important reaction because it yields **calcium oxide** (quicklime).

SUMMARY

In endothermic reactions, energy is taken in from the surroundings. Examples are:

- photosynthesis,
- thermal decomposition,
- the reaction between steam and coke.

The reaction between steam and coke

The endothermic reaction between steam and hot coke produces carbon monoxide and hydrogen. This mixture of flammable gases is used as fuel.

Carbon (coke)	+	Steam	→	Carbon monoxide	+	Hydrogen	Heat is taken in
$C(s)$	+	$H_2O(g)$	→	$CO(g)$	+	$H_2(g)$	

CHECKPOINT

1. Give an example of a reaction which is of vital importance in everyday life and which is (a) exothermic and (b) endothermic. Explain why the reactions you mention are so important.
2. The reaction shown below is exothermic.

   Anhydrous copper(II) sulphate + Water → Copper(II) sulphate-5-water

   Describe or illustrate an experiment by which you could find out whether this statement is true.

27.3 Energy change of reaction

FIRST THOUGHTS

Why is energy (heat and other forms of energy) given out or taken in during a chemical reaction? That is the question for this section.

Chemical bonds are forces of attraction between the atoms or ions or molecules in a substance. To break these bonds, energy must be supplied. When bonds are created, energy is given out. In a chemical reaction, bonds are broken, and new bonds are made.

Figure 27.3A shows the breaking and making of bonds when methane burns.

Methane	+	Oxygen	→	Carbon dioxide	+	Water
$CH_4(g)$	+	$2O_2(g)$	→	$CO_2(g)$	+	$2H_2O(l)$

The energy given out when the new bonds are made is greater than the energy taken in to break the old bonds. This reaction is therefore **exothermic**.

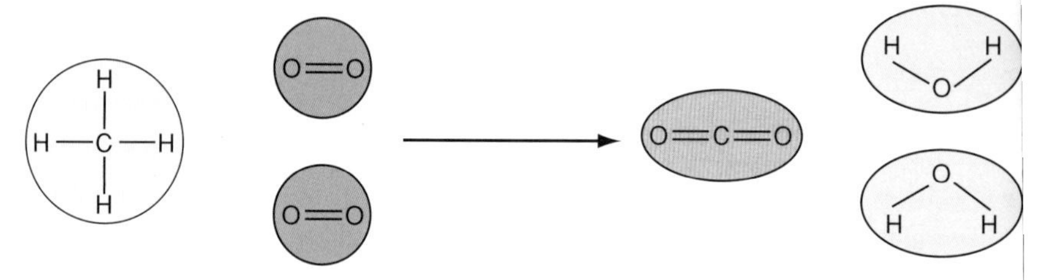

These bonds are broken. Energy must be taken in

These new bonds are made. Energy is given out

In this reaction, the energy given out is greater than the energy taken in – this reaction is exothermic

Figure 27.3A Bonds broken and made when methane burns

To break chemical bonds in a compound, energy must be supplied. Heat or light or electricity may be supplied. The other side of the coin is that when bonds are made energy is given out.

The difference:
energy supplied – energy given out is the heat of reaction.

Figure 27.3B shows the breaking and making of bonds in the reaction

Carbon (coke)	+	Water (steam)	→	Carbon monoxide	+	Hydrogen
C(s)	+	$H_2O(g)$	→	CO(g)	+	$H_2(g)$

In this reaction, the energy taken in to break the old bonds is greater than the energy given out when the new bonds form. The reaction is therefore **endothermic**.

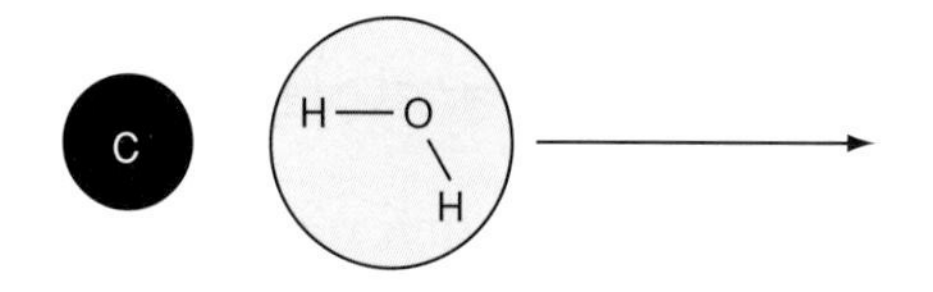

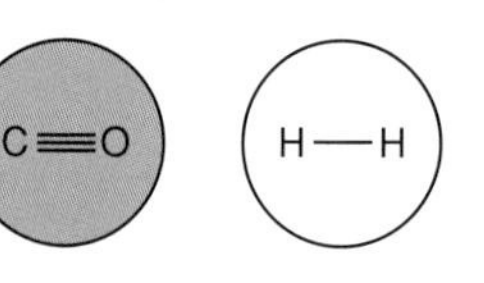

The bonds in H_2O must be broken. Energy must be taken in

As the bonds in CO and H_2 are made, energy is given out

In this reaction, the energy taken in is greater than the energy given out. This reaction is endothermic

Figure 27.3B Bonds broken and made when steam reacts with hot coke

Energy diagrams

In a chemical reaction, the reactants and the products possess different chemical bonds. They therefore possess different amounts of energy. A diagram which shows the energy content of the reactants and the products is called an **energy diagram**.

Figure 27.3C is an energy diagram for an exothermic reaction. It shows the energy content of the reactants and the products.

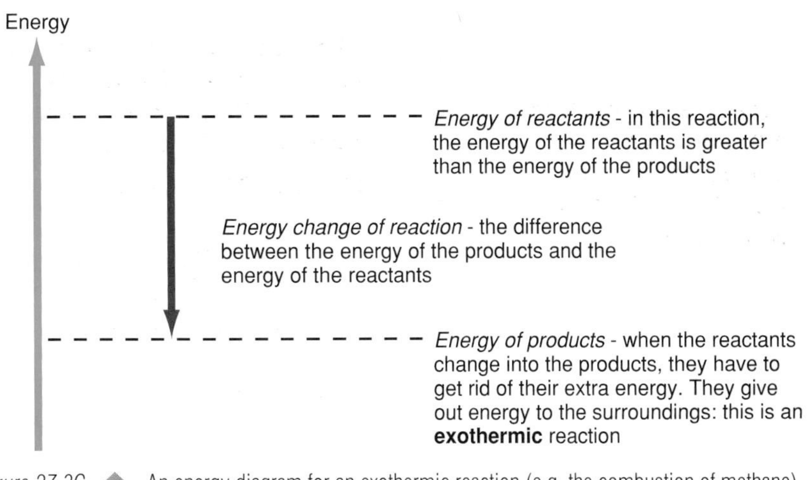

Figure 27.3C ▲ An energy diagram for an exothermic reaction (e.g. the combustion of methane)

An energy diagram aids our understanding by showing what is happening in pictorial form.

Figure 27.3D shows an energy diagram for an endothermic reaction. Marked on both energy diagrams is the energy change of reaction

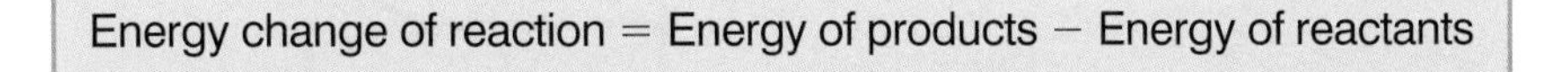

Energy change of reaction = Energy of products − Energy of reactants

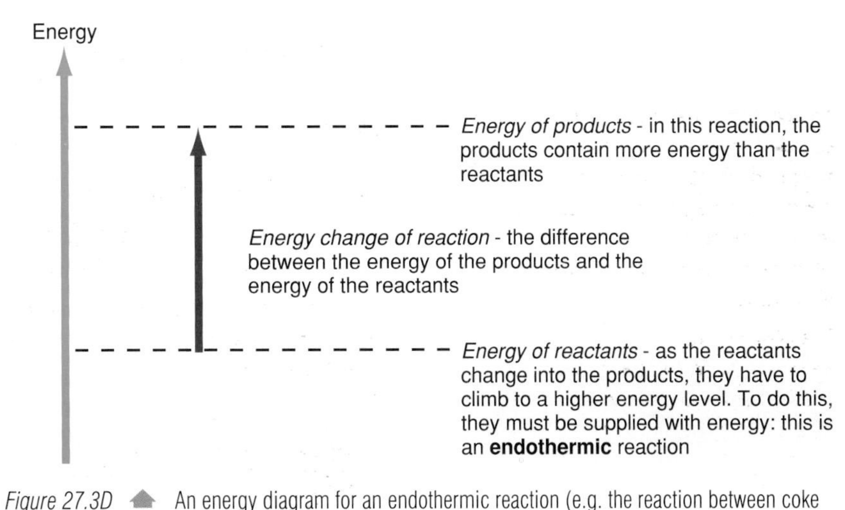

Figure 27.3D ▲ An energy diagram for an endothermic reaction (e.g. the reaction between coke and steam)

SUMMARY

In a chemical reaction, energy must be supplied to break chemical bonds in the reactants. Energy is given out when new chemical bonds are made during the formation of the products. The difference between the energy content of the products and the energy content of the reactants is the energy change of reaction.

CHECKPOINT

1. Draw an energy diagram for the neutralisation of hydrochloric acid by sodium hydroxide solution. If you have forgotten whether it is exothermic or endothermic, see Topic 27.1. Mark the energy change of reaction on your diagram.
2. Draw an energy diagram for the combustion of petrol to form carbon dioxide and water. Mark the energy change of reaction on your diagram.
3. Is photosynthesis exothermic or endothermic? What are the reactants and what are the products? Illustrate your answer by an energy diagram.
4. Describe the reaction that takes place when magnesium ribbon is heated in air. What do you see that shows the reaction is exothermic? What forms of energy are released? Draw an energy diagram for the reaction.

27.4 Activation energy

The figure shows how the energy of the reactants changes during the course of a chemical reaction. This particular reaction is exothermic; the products are at a lower energy level than the reactants. The reactants do not simply slide down an energy hill, however. Before the reactants can be converted into the products, they must overcome an energy barrier, and gain an amount of energy equal to that shown at P on the diagram. Once the reactants have acquired energy equal to that at the peak P they are automatically converted into the products. The energy which the reactants must gain to overcome the energy barrier is called the **activation energy**.

Perhaps you are wondering why exothermic reactions do not happen spontaneously – without any help from heat or electricity.

The reason is that there is a barrier to be overcome: a barrier called the activation energy.

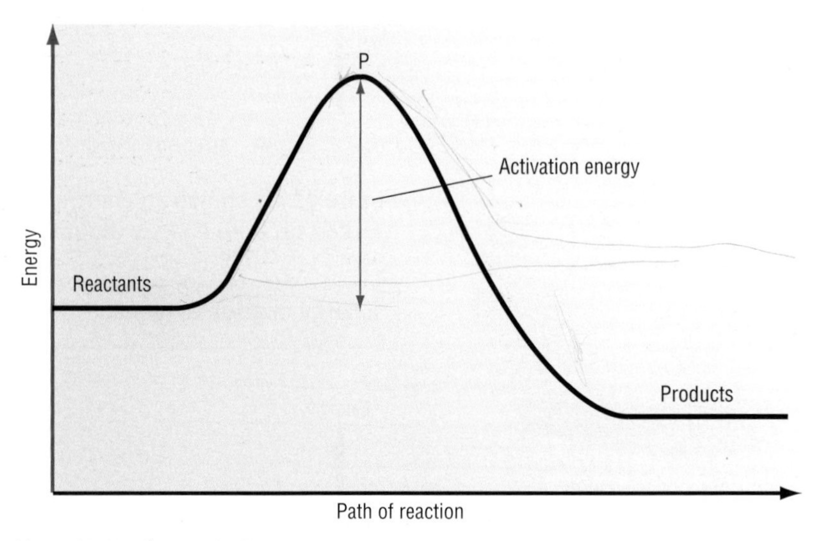

Figure 27.4A Activation energy

Molecules of the reactants must collide before they can react. If they are moving fast enough, part of their kinetic energy is converted into the activation energy.

The way in which a catalyst increases the speed of a reaction is by lowering the activation energy. Colliding molecules then have a lower energy barrier to climb, and more of the collisions between molecules result in reaction.

If there were no energy barrier, all exothermic reactions would happen spontaneously. A piece of iron would turn into iron oxide before your very eyes. A can of petrol would burst into flames as soon as you unscrewed the cap. The world would be a very unstable place.

27.5 Using bond energy values

Chemists have drawn up tables which tell exactly how much energy it takes to break different chemical bonds. To break 1 mole of C—C bonds requires 348 kJ. We can say that the bond energy of the C—C bond is 348 kJ/mol. Table 27.1 lists the bond energies of some bonds in kJ/mol. To break a chemical bond, energy must be supplied. There are no bonds that fly apart by magic and release energy.

We can use bond energy values to calculate the energy change of reaction.

Table 27.1 ▼ Bond energy values

Bond	Bond energy (kJ/mol)
H—H	436
O—O	496
C—C	348
C=C	612
C—H	412
C—O	360
C=O	743
H—O	463
Br—Br	193
C—Br	276

Worked example Use bond energies to calculate the energy change for the reaction,

$$CH_4(g) + 2O_2(g) \rightarrow CO_2(g) + 2H_2O(l)$$

Solution First show the bonds that are broken and the bonds that are made.

```
      H
      |
  H—C—H + 2O=O  →  O=C=O + 2H—O—H
      |
      H
```

Next list the bonds that are broken and the bonds that are made and their bond energies.

Bonds broken are 4(C—H) bonds; energy = 4 × 412 = 1648 kJ/mol
2(O=O) bonds; energy = 2 × 496 = 992 kJ/mol
Total energy required = 2640 kJ/mol

Bonds made are 2(C=O) bonds; energy = 2 × 743 = 1486 kJ/mol
4(H—O) bonds; energy = 4 × 463 = 1852 kJ/mol
Total energy given out = 3338 kJ/mol

Energy change of reaction = Energy given out when new bonds are made − Energy required to break old bonds
= 3338 − 2640 = 698 kJ/mol

The energy given out when the new bonds are made is greater than the energy taken in to break the old bonds so this reaction is exothermic.

CHECKPOINT

1 Find the energy change of reaction for the reaction

```
  H  H                   H  H
  |  |                   |  |
H—C=C—H + H₂  →  H—C—C—H
                         |  |
                         H  H
```

(Hint: add the energies of the bonds that are broken, and the energies of the bonds created, and then find the difference.

2 Find the energy change of reaction for the reaction

$$2H_2(g) + O_2(g) \rightarrow 2H_2O(l)$$

3 Calculate the energy change of reaction for the reaction

```
       O                           O—H
       ||                          |
CH₃—C—CH₃(g) + H₂(g)  →  CH₃—C—CH₃(g)
                                   |
                                   H
```

4 Calculate the energy change of reaction for the reaction

$$CH_2{=}CHCH_3(g) + Br_2(g) \rightarrow CH_2BrCHBrCH_3(g)$$

(Hint: draw structural formulas to show all the bonds.)

Theme Questions

1 The energy sources used in the UK are as follows:
Coal 31%
Oil 30%
Natural gas 21%
Nuclear power 17%
Hydroelectric power 1%
(a) Show these figures in the form of a pie-chart or a bar graph.
(b) Say which of these fuels are fossil fuels.
(c) Explain the meaning of the term 'fossil fuel', and say why such fuels are non-renewable.
(d) Name a product which is formed when all fossil fuels burn.
(e) Why is the contribution of hydroelectric power small in the UK?

2 You are provided with a solution of dilute hydrochloric acid of concentration 0.100 mol/l and a dilute solution of sodium hydroxide of unknown concentration.
(a) Briefly describe how you would find out what volume of the acid solution is needed to neutralise 25.0 cm^3 of the sodium hydroxide solution.
(b) A student following your method finds that 35.0 cm^3 of 0.100 mol/l hydrochloric acid neutralise 25.0 cm^3 of sodium hydroxide solution. Calculate the concentration of the sodium hydroxide solution.

3 (i) What amount (in moles) of hydroxide ion is present in 25.0 cm^3 of 0.050 mol/l sodium hydroxide solution?
(ii) Which of the following will neutralise exactly 25.0 cm^3 of 0.050 mol/l sodium hydroxide solution? (More than one answer is correct.)
(a) 25.0 cm^3 of 0.050 mol/l hydrochloric acid
(b) 25.0 cm^3 of 0.050 mol/l nitric acid
(c) 25.0 cm^3 of 0.050 mol/l sulphuric acid
(d) 50.0 cm^3 of 0.025 mol/l hydrochloric acid
(e) 25.0 cm^3 of 0.025 mol/l sulphuric acid

4 You are given a sample of copper(II) oxide mixed with carbon. You are asked to find the percentage purity of the copper oxide. You decide to add an excess of dilute sulphuric acid to 2.00 g of the mixture to react with all the copper(II) oxide. You filter, wash and dry the carbon that remains. The mass is 0.090 g. What is the percentage purity of the sample of copper(II) oxide?

5 The manufacture of ammonia uses the reaction:

$$N_2(g) + 3H_2(g) \rightleftharpoons 2NH_3(g)$$

The reaction is reversible, and the forward reaction is exothermic.
(a) What can be done to increase the rate of the reaction?
(b) What steps can be taken to increase the yield of ammonia?
(c) Is there any conflict between the requirements of (a) and (b)?
(d) What conditions does the industrial plant employ?

6 In a set of experiments, zinc was allowed to react with sulphuric acid. Each time, 0.65 g of zinc was used. The volume of acid was different each time. The volume of hydrogen formed each time is shown in the table.

Volume of sulphuric acid (cm^3)	*Volume of hydrogen* (cm^3)
5	45
15	135
20	180
25	215
30	235
35	240
40	240

Plot, on graph paper, the volume of hydrogen produced against the volume of acid used. From the graph, find out:
(a) where the reaction is most rapid (and explain why),
(b) what volume of sulphuric acid will produce 100 cm^3 of gas,
(c) what volume of gas is produced if 10 cm^3 of sulphuric acid are used,
(d) what volume of sulphuric acid is just sufficient to react with 0.65 g of zinc.

7 The graph (*A*) below shows the volume of carbon dioxide formed during a reaction between *excess* marble chips (calcium carbonate) and dilute hydrochloric acid.

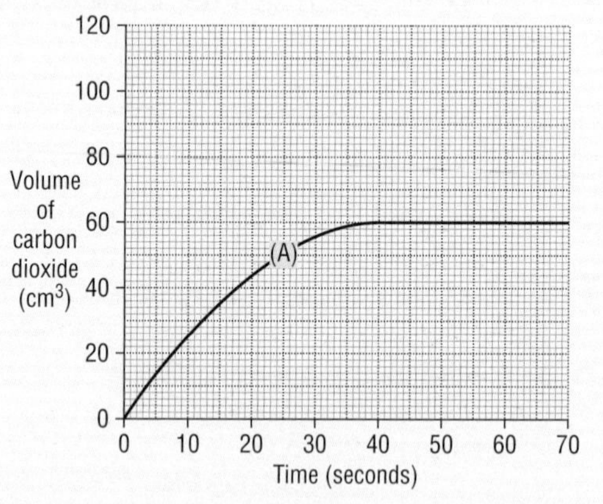

(i) On a copy of the grid sketch carefully the graph that would be obtained if the acid had been replaced by
I. an equal volume and concentration of hydrochloric acid at a **lower** temperature, with the marble chips still in excess. Label this graph *B*. [2]
II. an equal volume of hydrochloric acid of **double** the concentration with the marble chips still in excess. Label this graph *C*. [2]
(ii) I. On your grid sketch carefully the graph that would be obtained if the marble chips had been ground to a powder with the volume and concentration, of the acid the same as for graph *A*. Label this graph *D*. [2]
II. Explain your answer to part **(ii) I**. [2] (WJEC)

The solubility of ammonia gas in water at various temperatures is shown in the table below.

Temperature (°C)	0	10	20	30	40	50
Solubility of ammonia (g per 100 g water)	90.0	69.0	53.0	41.0	31.0	

(a) Plot a graph of solubility against temperature on a copy of the grid below and use it to predict the solubility of ammonia at 50 °C. [4]

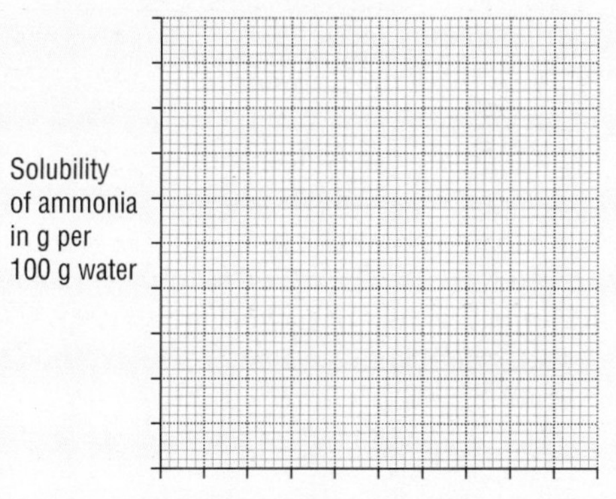

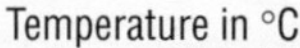

(b) (i) Suggest how you would obtain ammonia gas from ammonia solution. [1]

(ii) Draw a diagram to show how you would collect a sample of this gas. [1]

(c) At 20 °C, a saturated solution of ammonia contains 53.0 g of ammonia in 100 g of water.
Calculate the concentration of ammonia (NH_3) in the solution in mol dm^{-3}.
You should assume that 100 g water has a volume of 100 cm^3 and that the volume does not change on dissolving the ammonia.
(Relative atomic masses: N = 14; H = 1.) [3]

(d) At 0 °C, a saturated solution of ammonia contains 90.0 g of ammonia dissolved in 100 g of water. Calculate the volume, measured at 0 °C and atmospheric pressure, of 90.0 g of ammonia.
(1 mol of gas occupies 22.4 dm^3 at 0 °C and atmospheric pressure.) [2] (Edexcel)

9 Write a **brief** account of **three** of the following. Relevant chemical equations and/or diagrams **may** be included in your answer where appropriate.

(a) Explain how addition polymers can be made from alkenes.
Discuss whether the advantages of plastics in everyday life outweigh the disadvantages. [5]

(b) Describe and explain the purification of copper.
How do the uses of copper in everyday life relate to its properties? [5]

(c) Describe how the appearance of rocks is used as evidence to explain their formation and hence to classify rock type. [5]

(d) Outline the manufacture of sulphuric acid. [5] (WJEC)

10 (a) Write the formula for:
(i) a sodium ion; (ii) a chloride ion;
(iii) a copper(II) ion. [3]

(b) Copy and complete the following tables which show the tests for some ions.

(i) Flame tests

Name of ion	Colour of flame
Potassium	Lilac
______	Yellow
Calcium	______

[2]

(ii) Tests for ions in solution

Name of ion in solution	Reagent added to the solution	Positive result
Copper(II)	______	Light blue precipitate
______	Dilute nitric acid + silver nitrate solution	White precipitate
Sulphate	____ + ____	______

[5]

(c) Describe a test to show the presence of ammonium ions in ammonium chloride. [4] (Edexcel)

11 (a) Iron is extracted from iron(III) oxide in a blast furnace.
One of the main reactions in the furnace is

$$Fe_2O_3 + 3CO \rightarrow 2Fe + 3CO_2.$$

A_r (O) = 16; A_r (Fe) = 56.

(i) Calculate the relative molecular mass (M_r) of iron(III) oxide, Fe_2O_3. [2]

(ii) Use the given equation for the extraction of iron to calculate how many tonnes of metal could be obtained from 80 tonnes of iron(III) oxide. [3]

(b) 14.4 g of another oxide of iron was found to contain 11.2 g of iron. Calculate the simplest formula for this oxide of iron. *Show your working*.
A_r (O) = 16; A_r (Fe) = 56. [3]

(c) Iron(II) and iron(III) salts in solution form different coloured precipitates with sodium hydroxide solution.
State the *colour* of the precipitate formed with
(i) iron(II) salts, (ii) iron(III) salts. [2] (WJEC)

12 (i) The relative molecular mass (M_r) of sodium hydroxide, NaOH, is 40. If 8.0 g of sodium hydroxide are present in 1000 cm^3 (1 dm^3) of aqueous solution

I. what is the concentration of this solution in mol/dm^3? [1]

II. how many moles would be present in 25 cm^3 of the sodium hydroxide solution? [2]

(ii) Ethanoic acid, CH_3COOH, and sodium hydroxide react in a 1 : 1 ratio. When ethanoic acid was neutralised by sodium hydroxide solution, it was found that 25 cm^3 of the sodium hydroxide solution required 20 cm^3 of ethanoic acid solution.

I. How many moles of ethanoic acid are present in 20 cm^3 of ethanoic acid solution? [1]

II. How many moles would be present in 1 dm^3 of ethanoic acid solution? [2]

(iii) If the relative molecular mass (M_r) of ethanoic acid is 60, calculate the number of *grams* of ethanoic acid present in 1 dm^3 of the solution. [2] (WJEC)

13 Jo and Andy are finding out about rates of reaction. They react hydrochloric acid with marble chips (calcium carbonate).

Hydrochloric acid + calcium carbonate → calcium chloride + carbon dioxide + water

They use this apparatus.

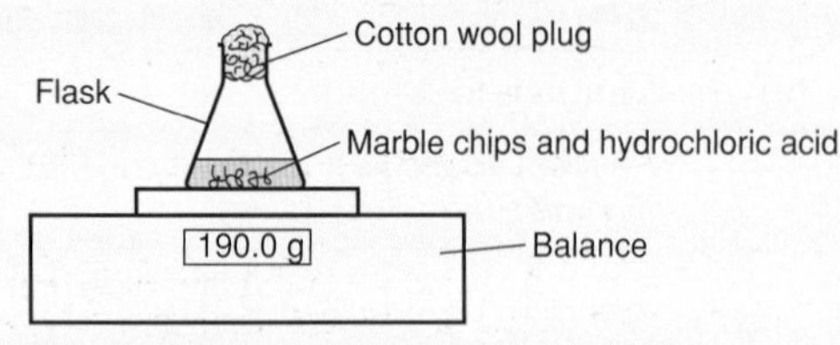

(a) The mass of the flask and its contents decreases during the experiment. Suggest why this happens. [1]

(b) Jo and Andy measure the total mass of the flask and its contents as the reaction takes place. The table shows their results.

Time in minutes	*Mass of flask and contents in grams*
0	190.0
2	188.0
4	187.0
6	186.3
8	186.1
10	186.0
12	186.0

(i) Plot their results on a copy of the grid. [2]

(ii) Finish the graph by drawing the best curve through the points. [1]

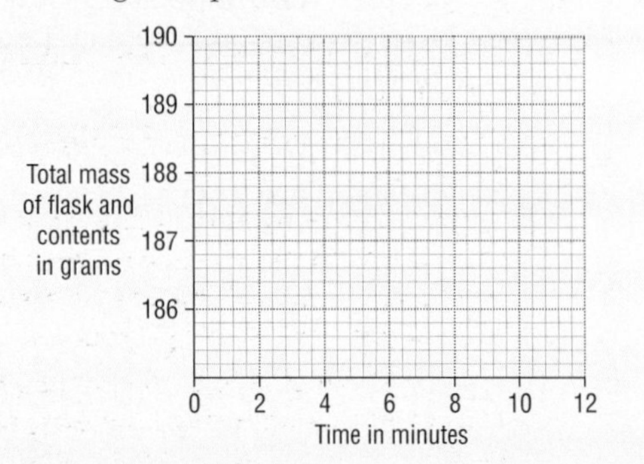

(iii) Jo says that the reaction was faster between 0 and 2 minutes than between 2 and 4 minutes. How do the results show this? [2]

(c) In the first experiment the hydrochloric acid was at room temperature.
Jo and Andy repeat the experiment. The only difference is that the hydrochloric acid is at a higher temperature. Sketch a curve on the grid to show the results they get. [2]

(d) Jo and Andy do the original experiment at room temperature again.
This time they add an equal volume of water to the hydrochloric acid before adding the marble chips.
How would the rate of reaction be different from the first experiment?
Use your knowledge of particles to explain your answer. [3] (OCR)

14 A series of experiments was carried out on a given solution of sulphuric acid.

(a) In the first of these experiments a titration was carried out to determine the concentration of the acid. 20.0 cm^3 of the acid required 32.0 cm^3 of 0.15 moles per litre (mol/dm^3) sodium hydroxide solution.

(i) Calculate the number of moles of sodium hydroxide used in the titration.
The equation of the titration reaction is
$2NaOH + H_2SO_4 \rightarrow Na_2SO_4 + 2H_2O$ [2

(ii) Calculate the number of moles of H_2SO_4 titrated. [3

(iii) Calculate the concentration of the sulphuric acid used in moles per litre (mol/dm^3). [2

(b) In a second experiment, a group of pupils proceeded to make nickel sulphate crystals by adding an excess of nickel carbonate to 50.0 cm^3 of the *same* sulphuric acid. The equation for the reaction is:

$H_2SO_4 + NiCO_3 \rightarrow NiSO_4 + H_2O + CO_2$

(i) Calculate the number of moles of sulphuric acid used. [2]

(ii) What is the *minimum* mass of $NiCO_3$ which must be used to react with all the sulphuric acid? (Ni = 59, C = 12, O = 16.) [4]

(iii) Nickel sulphate crystallises as $NiSO_4.7H_2O$. What is the *maximum* mass of crystals which could be obtained from this preparation? [4] (Ni = 59, S = 32, O = 16, H = 1.)

(iv) At the end of the experiment, one student collected 1.20 g of crystals. Suggest *one* reason why the mass collected was less than that calculated in part **(b) (iii)** above. [1]

(c) A possible way to make ammonium sulphate is to react ammonia gas with sulphuric acid.
If 25.0 cm^3 of the sulphuric acid from part **(a)** were used, what volume, in cm^3, of ammonia gas would be needed to react exactly with the acid. (1 mole of any gas occupies 24 dm^3 at room temperature.)
The equation for the reaction is

$H_2SO_4 + 2NH_3 \rightarrow (NH_4)_2SO_4$ [6] (CCEA)

15 The industrial **electrolysis** of an aqueous solution of sodium chloride is shown.

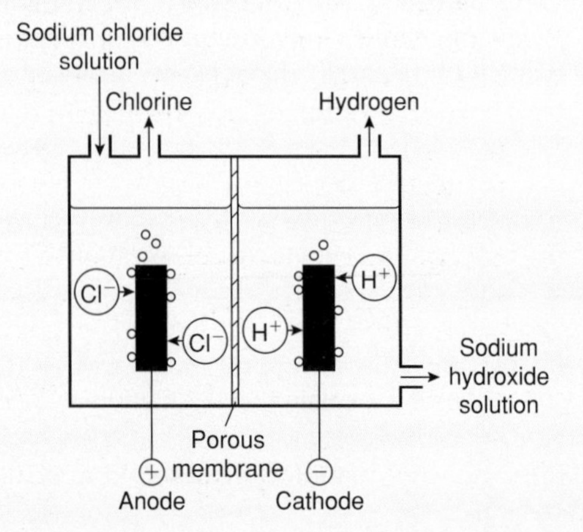

(a) What is meant by **electrolysis**? [2]

(b) Explain how hydrogen and sodium hydroxide solution are produced. [5]

(c) The equation for the anode reaction is:

$$2Cl^- \rightarrow Cl_2 + 2e^-$$

A current of 0.2 amps was passed through an aqueous solution of sodium chloride for 2 hours.

(i) Calculate the number of coulombs of electricity passed. [2]

(ii) How many coulombs would be needed to form 71 g of chlorine?
Relative atomic mass: Cl 35.5.
96 500 coulombs is the amount of electricity carried by a mole of electrons. [2]

(iii) Using your answers in **(i)** and **(ii)** calculate the mass, in grams, of chlorine formed.
[2] AQA(SEG) 2000

16 (a) Ammonia is manufactured by the Haber process.

$$N_2 + 3H_2 \rightleftharpoons 2NH_3$$

The graph below shows the yield of ammonia at different temperature and pressure conditions.

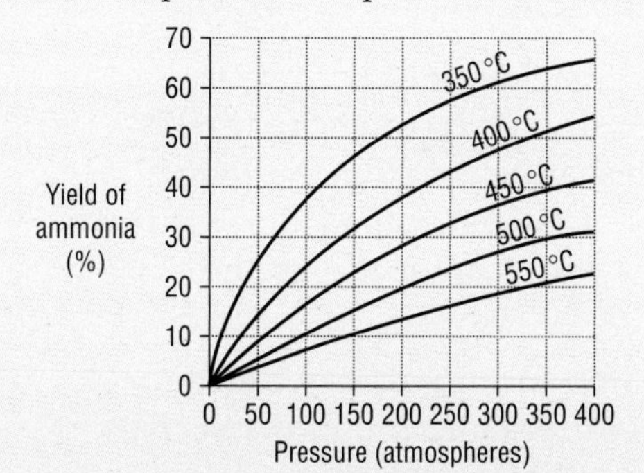

Use the graph to answers parts **(i)** and **(ii)**.

(i) State what happens to the yield of ammonia as the

I. temperature increases, [1]

II. pressure increases. [1]

(ii) Find the

I. pressure needed to obtain 40% yield of ammonia at 450 °C, [1]

II. temperature needed to obtain 20% yield of ammonia at a pressure of 200 atmospheres. [1]

(b) State how the *rate* of ammonia production is increased, apart from changing the temperature and pressure conditions. [1]

(c) Ammonia is used to make the nitrogenous fertiliser, ammonium nitrate, NH_4NO_3. Calculate the relative molecular mass (M_r) of ammonium nitrate.
(A_r(H) = 1; A_r(N) = 14; A_r(O) = 16.) [2]

(d) The formation of ammonia from nitrogen and hydrogen can also be represented by the equation:

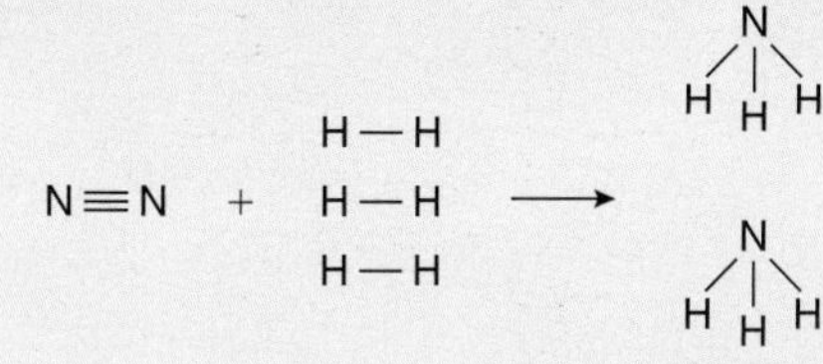

The relative amounts of energy needed to break the bonds in the above reaction are shown in the table.

Bond	*Amount of energy needed to break the bond* (kJ)
N ≡ N	945
H – H	436
N – H	391

NOTE: The amount of energy needed to make a bond is equal and opposite to that needed to break the bond.

(i) Using the energy values in the table calculate the relative energy

I. needed to break *all* the bonds in the reactants, [2]

II. evolved when bonds in the product are formed [2]

(ii) Using your answers to parts **I** and **II** *explain* why the relative overall energy change is exothermic. [1] (WJEC)

17 The balanced equation for the fermentation of glucose is:

$$C_6H_{12}O_6 \rightarrow 2C_2H_5OH + 2CO_2$$

(a) 9.0 g of glucose are fermented completely.

(i) Calculate the mass of ethanol formed.
(Relative atomic masses: H = 16; C = 2; O = 16.) [3]

(ii) Calculate the volume of carbon dioxide, measured at room temperature and pressure, evolved.
(1 mol of any gas occupies 24 000 cm^3 at room temperature and pressure.) [2]

(b) Name the process used to obtain a concentrated solution of ethanol from the fermentation mixture. [2]

(c) Glucose burns in excess oxygen to form carbon dioxide and water.

(i) Write the balanced equation for this reaction. [2]

(ii) How does the volume of oxygen used compare with the volume of carbon dioxide produced? Both volumes are measured at room temperature and pressure. [1]

(iii) The total energy of the reactants, glucose and oxygen, is shown on the diagram.
Draw a line on a copy of the diagram to show the total energy of the products.

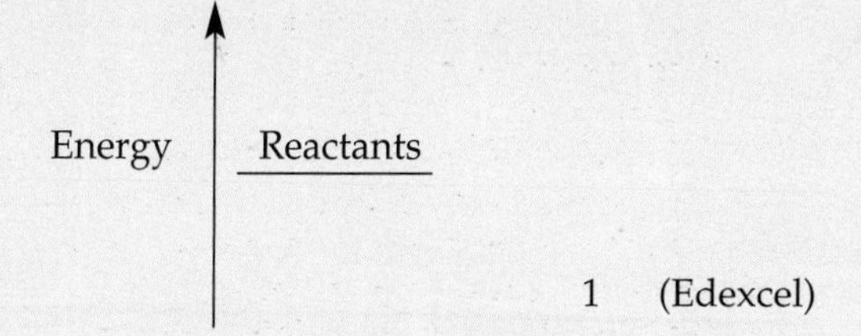

1 (Edexcel)

Organic Chemistry

The chemistry of carbon compounds is known as organic chemistry. Many of the substances which it includes are obtained from living organisms. This explains how this branch of the subject got its name.

Organic chemistry includes plastics and fibres. Many of our leisure activities need equipment such as footballs, tennis racquets, golf balls, climbing ropes, canoes and sailing dinghies, all made from synthetic organic chemicals. Much of the clothing we wear is composed of synthetic fibres such as nylon, polyester and acrylics which are made by the chemical industry.

Organic chemistry includes all the foodstuffs: carbohydrates, fats and proteins. When we are ill we need the help of organic chemicals such as painkillers, antiseptics, antibiotics and anaesthetics. The chemical industry provides all these substances by using the Earth's resources and human ingenuity.

FIRST THOUGHTS

The toys you see in Figure 28A are made of plastics. The clothes you see in Figure 28B are made of synthetic fibres. Both types of material are polymers. In this topic, you will find out what a polymer is.

Figure 28A ▲ Plastics

Figure 28B ▲ Fibres

28.1 Alkenes

You studied alkanes in Topic 23.2. They are saturated hydrocarbons with single carbon-carbon bonds in the molecules. They take part in substitution reactions. Some hydrocarbons have carbon–carbon double bonds in the molecule. They are unsaturated hydrocarbons. They take part in addition reactions.

www.keyscience.co.uk

Ethene is a hydrocarbon of formula C_2H_4 (see Figure 28.1A). There is a double bond between the carbon atoms.

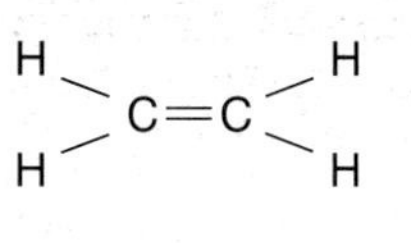

Figure 28.1A ▲ (a) A model of ethene (b) The formula of ethene

Ethene and other hydrocarbons which contain double bonds between carbon atoms are described as **unsaturated** hydrocarbons. The double bond will open to allow another molecule to add on. Unsaturated hydrocarbons will add hydrogen to form **saturated** hydrocarbons. For example, ethene adds hydrogen to form ethane, which is an **alkane**. Reactions of this kind are called **addition reactions**; see Figure 28.1B Saturated hydrocarbons, such as alkanes, contain only single bonds between carbon atoms.

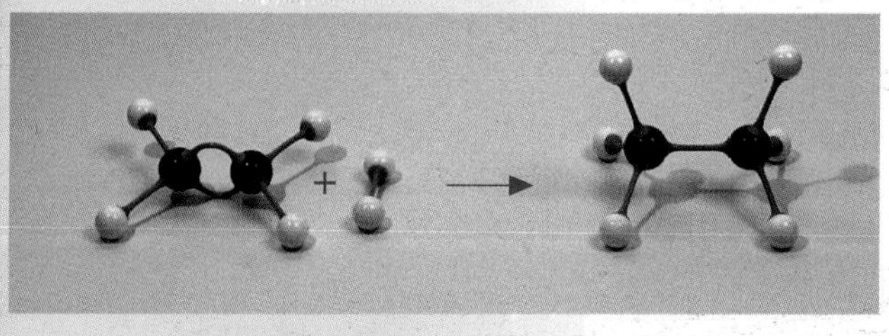

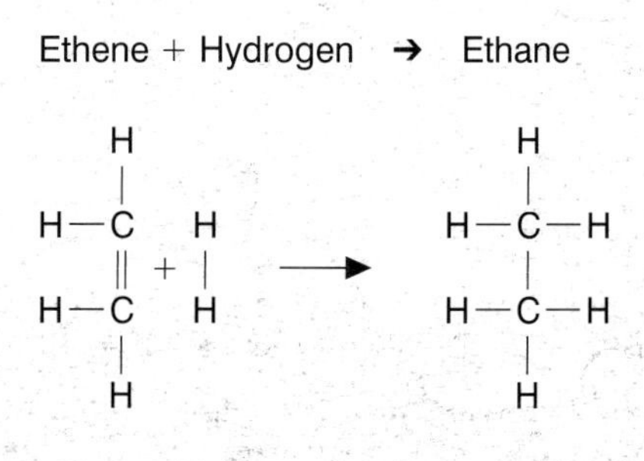

Figure 28.1B ▲ The addition reaction between ethene and hydrogen

SUMMARY

Alkenes are unsaturated hydrocarbons. They possess a double bond between carbon atoms. They are a homologous series of general formula C_nH_{2n}.

Propene is an unsaturated hydrocarbon which resembles ethene (see Figure 28.1C). Ethene and propene are members of a **homologous series**, that is, a set of similar compounds whose formulas differ by CH_2. They are called **alkenes** (see Table 28.1). The general formula is C_nH_{2n}.

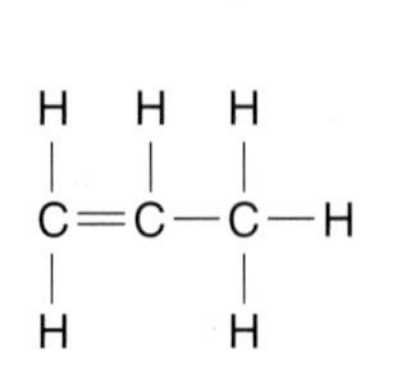

Figure 28.1C The formula of propene

Table 28.1 The first members of the alkene series

Alkene	Formula, C_nH_{2n}
Ethene	C_2H_4
Propene	C_3H_6
Butene	C_4H_8
Pentene	C_5H_{10}
Hexene	C_6H_{12}
Heptene	C_7H_{14}
Octene	C_8H_{16}

Reactions of alkenes

Hydrogenation

Animal fats, such as butter, are solid. Vegetable oils, such as sunflower seed oil, are liquid. More vegetable oil is produced than we need for cooking, and insufficient butter is produced to satisfy the demand for solid fat. It is therefore profitable to convert vegetable oils into solid fats. Manufacturers make use of the fact that fats are saturated, while oils are unsaturated. Hydrogenation (the addition of hydrogen) is used to convert an unsaturated oil into a saturated fat. The vapour of an oil is passed with hydrogen over a nickel catalyst.

$$\text{Vegetable oil (unsaturated)} + \text{Hydrogen} \xrightarrow[\text{nickel catalyst}]{\text{pass over heated}} \text{Solid fat (saturated)}$$

The product is margarine. The process can be modified to leave some of the double bonds intact and yield soft margarine.

Hydration

Water will add to alkenes. A molecule of water can add across the double bond. Combination with water is called **hydration**. The product formed by the hydration of ethene is ethanol, C_2H_5OH.

Ethene + Water → Ethanol

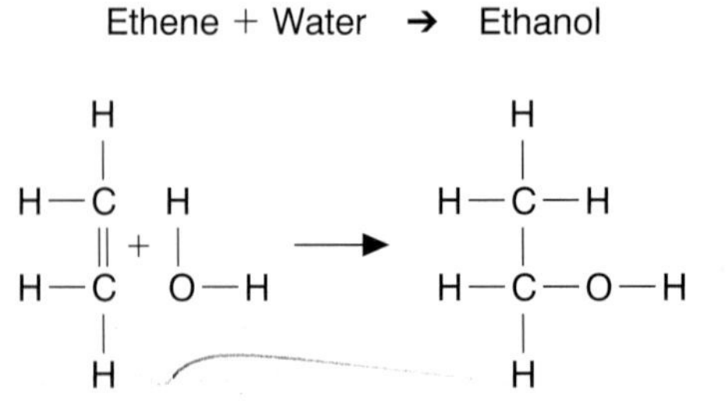

Ethanol is the compound we commonly call **alcohol**. It is an important industrial solvent. The industrial manufacture is

$$\text{Ethene} + \text{Steam} \xrightarrow[\text{(phosphoric acid), under pressure}]{\text{pass over a heated catalyst}} \text{Ethanol}$$

These unsaturated hydrocarbons are called alkenes. Addition reactions include
- hydrogenation of vegetable oils to form solid fats, e.g. margarine
- hydration to form alcohols, e.g. ethene → ethanol and ...

Only about 10% of ethene is converted to ethanol. The unreacted gases are recycled over the catalyst.

...
- *addition polymerisation in which molecules of monomer, e.g. ethene, join to form a molecule of polymer, a poly(alkene), e.g. poly(ethene) and ...*

Addition polymerisation

The double bond in ethene enables many molecules of ethene to join together to form a large molecule.

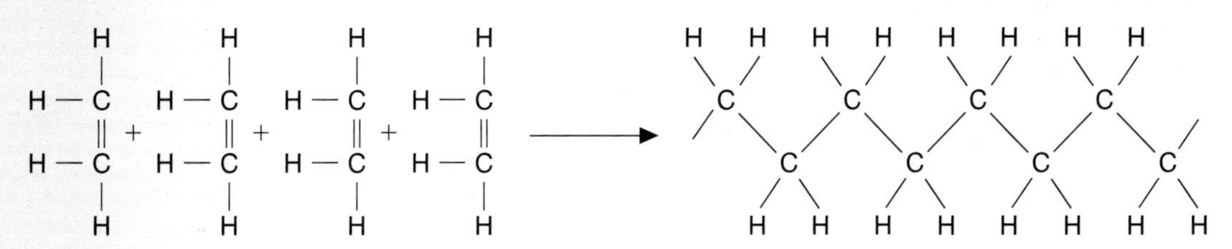

This reaction is called **addition polymerisation**. Many molecules (30 000–40 000) of the **monomer**, ethene, **polymerise** (i.e. join together) to form one molecule of the **polymer**, poly(ethene). The conditions needed for polymerisation are high pressure, a temperature of room temperature or higher and a catalyst.

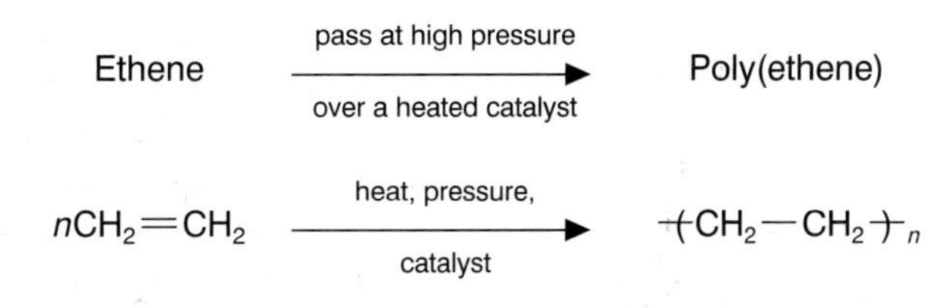

Poly(ethene) is better known by its trade name of **polythene**. It is used for making plastic bags, kitchenware (buckets, bowls, etc.), laboratory tubing and toys. It is flexible and difficult to break.

SUMMARY

The double bonds makes alkenes reactive. They take part in addition reactions, e.g. with hydrogen to form alkanes and with water to form alcohols. Hydrogenation is used to turn vegetable oils into saturated fats.

Addition of bromine

A solution of bromine in an organic solvent or in water is brown. If an alkene is bubbled through a solution of bromine, the solution loses its colour. Bromine has added to the alkene to form a colourless compound. The reaction can be shown as

Ethene + Bromine → 1,2-Dibromoethane

...
- *addition of bromine. The decolourisation of a bromine solution is used as a test for a carbon–carbon double bond.*
- *addition of chlorine*

The product has single bonds: it is a saturated compound. With two carbon atoms in the molecule, it is named after ethane. With two bromine atoms in the molecule, it is a dibromo-compound, 1,2-dibromoethane. The numbers 1,2- tell you that one bromine atom is bonded to one carbon atom and the second bromine atom is bonded to the second carbon atom. The decolourisation of a bromine solution is used to distinguish between an alkene and an alkane. Chlorine adds to alkenes in a similar way.

28.2 Plastics

Science at work

When someone breaks a bone, the broken ends must be held in place by a plaster cast until new bone grows. Sometimes, metal pins are used to join the broken ends together. Now there is an easier way. Poly(methylmethacrylate) can be used to 'glue' the ends of a broken bone together without the need for pins or plaster.

Thermosoftening plastics and thermosetting plastics. Again the uses to which they are put depend on their properties. And the properties depend on the structures.

Polymers such as poly(ethene) and other poly(alkenes) are **plastics**. Plastics are materials which soften on heating and harden on cooling. They are therefore useful materials from which to mould objects. There are two kinds of plastics: **thermosoftening plastics** and **thermosetting plastics**.

Thermosoftening plastics can be softened by heating, cooled and resoftened many times. Thermosetting plastics are plastic during manufacture, but once moulded they set and cannot be resoftened.

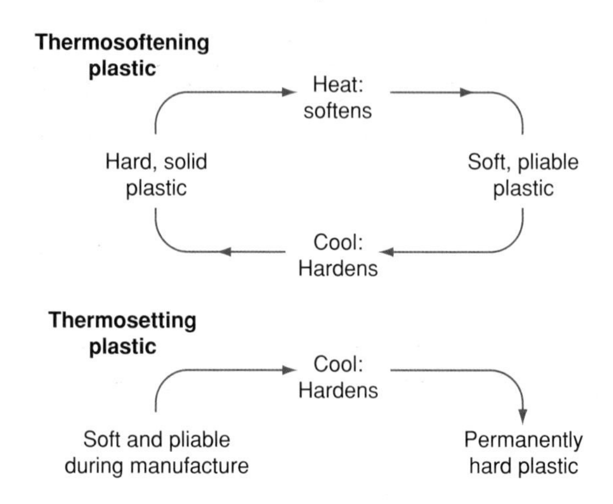

The reason for the difference in behaviour is a difference in structure (see Figure 28.2A).

Thermosoftening plastics consist of long polymer chains. The forces of attraction between chains are weak

(a) Part of the structure of a thermosoftening plastic

When a thermosetting plastic is softened and moulded, the chains react with one another. Cross-links are formed and a huge three-dimensional structure is built up. This is why thermosetting plastics can be formed only once

(b) Part of the structure of a thermosetting plastic

Figure 28.2A

SUMMARY

Addition polymerisation is the addition of many molecules of monomer to form one molecule of polymer. Thermosoftening and thermosetting plastics have advantages for different uses.

Condensation polymerisation

The chief thermosetting polymers are not poly(alkenes). Many of them are made by another type of polymerisation called **condensation polymerisation**. For this to occur, the monomer must possess two groups of atoms which can take part in chemical reactions. When the reactive groups in one molecule of monomer react with the groups in other molecules of monomer, large polymer molecules are formed (see Figure 28.2B). In the reaction, small molecules are eliminated, e.g. H_2O, HCl, NH_3. The elimination of water gave the name **condensation** to this type of polymerisation.

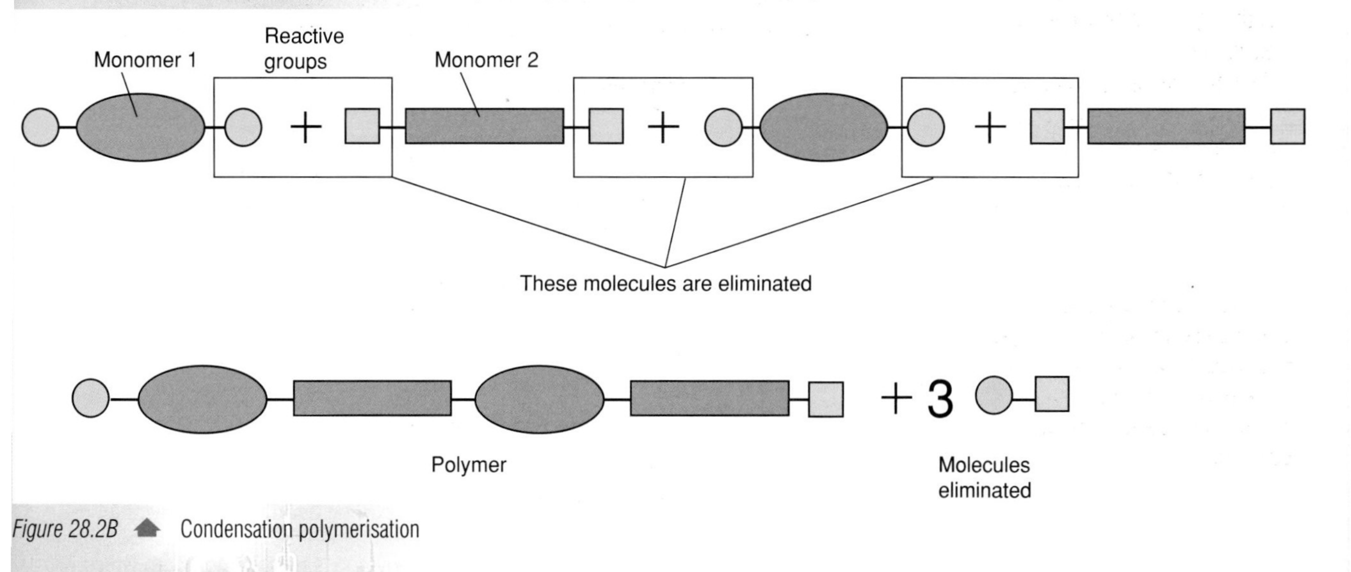

Figure 28.2B ▲ Condensation polymerisation

Examples of condensation polymers are:

- epoxy resins, which are used in glues,
- polyester resins, which are used in glass-reinforced plastics,
- polyurethanes, which are used in varnishes
- nylon, terylene and rayon (see Table 28.3)
- polysaccharides, Topic 30.2,
- polypeptides, Topic 30.4.

SUMMARY

Condensation polymerisation is the reaction between many molecules of monomer to form one molecule of polymer, with the elimination of many small molecules such as water molecules.

Methods of moulding plastics

Different methods are used for moulding thermosoftening and thermosetting plastics.

Manufacturers find thermosoftening plastics very convenient to use. They can buy thermosoftening plastic in bulk in the form of granules, then melt the material and press it into the shape of the object they want to make. Thermosoftening plastics are easy to colour. When a pigment is added to the molten plastic and thoroughly mixed, the moulded objects are coloured all through. This is much better than a coat of paint which can become chipped. Thermosetting plastics have important uses too. Materials used for bench tops must be able to withstand high temperatures without softening. 'Thermosets' are ideal for this use.

Some manufacturers mix gases with plastics to make low density plastic foams. These foams are used in packaging, for thermal insulation of buildings, for insulation against sound and in the interior of car seats. Sometimes it is necessary to strengthen plastics by the addition of other materials. Boat hulls, plastic panels in cars, instrument panels and wall-mounted hand-driers are just a few of the articles which are made of GRP (glass-fibre-reinforced plastic).

SUMMARY

Different methods are used for moulding thermosoftening plastics (which can be softened by heat many times) and thermosetting plastics (which are softened by heat during manufacture and then set permanently).

Polyalkenes

You have met all these plastics many times. You will see even from the small selection listed here how many of the objects we use in daily life are made from plastics.

Table 28.2 Some poly(alkenes) and their uses

Poly(ethene); trade name Polythene

Monomer Polymer

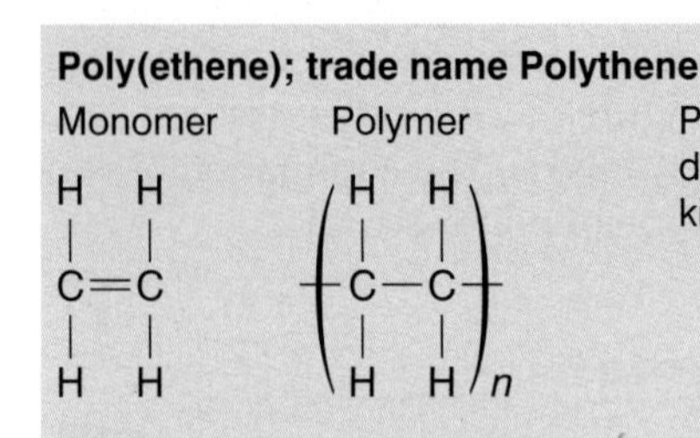

Polythene is used to make plastic bags. High density polythene is used to make kitchenware, laboratory tubing and toys.

Poly(chloroethene); trade name PVC

Monomer Polymer

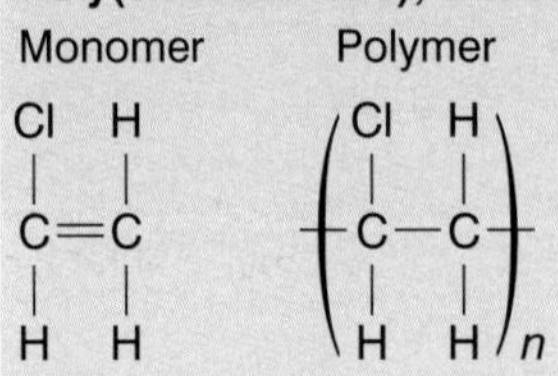

PVC is used to make wellingtons, raincoats, floor tiles, insulation for electrical wiring, gutters and drainpipes.

Poly(propene); trade name Polypropylene

Monomer Polymer

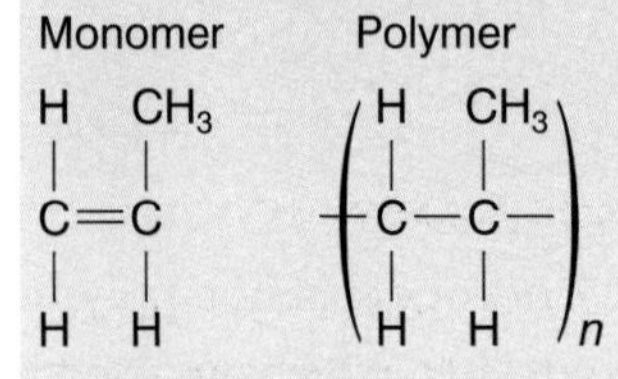

Polypropylene is resistant to attack by chemicals. Since it does not soften in boiling water, it can be used to make hospital equipment which must be sterilised. Polypropylene is drawn into fibres which are used to make ropes and fishing nets.

Poly(tetrafluoroethene); trade names PTFE and Teflon

Monomer Polymer

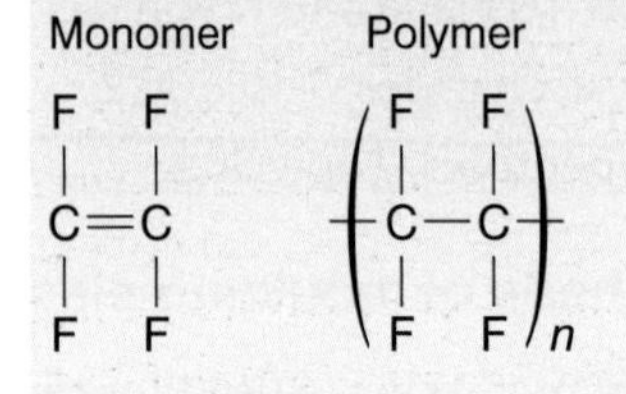

PTFE is a hard plastic which is not attacked by most chemicals. Few substances can stick to its surface. It is used to coat non-stick pans and skis.

Perspex

Monomer

H CO_2CH_3
C=C
H CH_3

Perspex finds many applications because it is transparent and can be used instead of glass. It is more easily moulded than glass and less easily shattered.

Polystyrene

Monomer

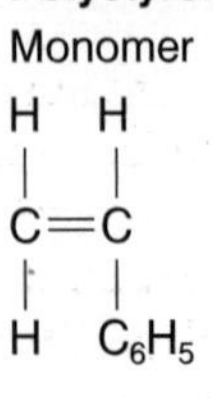

Polystyrene is a hard, brittle plastic used for making construction kits. Polystyrene foam is made by blowing air into the softened plastic. It is used for making ceiling tiles, insulating containers and packaging materials.

SUMMARY

Some important poly(alkenes) are:

- poly(ethene),
- poly(chloroethene),
- poly(propene),
- poly(tetrafluoroethene),
- perspex,
- polystyrene.

Table 28.3 Some condensation polymers and their uses

Polymer	Properties	Uses
Nylon	A polyamide Thermosoftening m.p. 200 °C High tensile strength	Textile fibre used in clothes; also in ropes, fishing lines and nets
Terylene	A polyester Thermosoftening	Important in clothing manufacture
Bakelite	Thermosetting An electrical insulator Insoluble in water and organic solvents Not attacked by chemicals	Electrical appliances, e.g. switches, sockets, plugs. Casings for radios and telephones
Melamine	Similar to bakelite	Kitchen surfaces, bench tops
Rayon	Made from cellulose by reshaping, that is breaking the long chains of sugar molecules in cellulose into shorter chains and then rejoining them	Important in clothing manufacture

SUMMARY

Some important condensation polymers are nylon, terylene, bakelite, melamine and rayon.

Some drawbacks

Nylon and terylene do not absorb water well. Clothes made of these fibres do not absorb perspiration and allow it to evaporate. Nylon, terylene and other polyesters are usually mixed with natural fibres such as wool and cotton. The mixtures absorb water and are therefore more comfortable to wear.

Many refreshment stands serve coffee and soft drinks in disposable cups. These are often made of polystyrene. All the plastic cups, plates and food containers which people use and throw away have to be disposed of. This is difficult to do because plastics are **non-biodegradable**. They are not decomposed by natural biological processes. Some plastic waste is burned in incinerators and the heat generated is used. Other plastics cannot be disposed of in this way because they burn to form poisonous gases, for example, hydrogen chloride, carbon monoxide and hydrogen cyanide. Much plastic waste is buried in landfill sites. One third of the plastics manufactured is used for packaging: it is made to be used once and thrown away. As the mass of non-biodegradable plastic waste increases, more land is used up to bury the waste.

Chemists are working on the problem. They have invented some **biodegradable** plastics, that is plastics which can be broken down by micro-organisms. One type has starch incorporated into the plastic. When bacteria feed on the starch, the plastic partially breaks down. Other new types of plastic are completely biodegradable.

There are some dangers in the use of plastics. Plastic foams are used as insulation in many buildings. Furniture is often stuffed with plastic foam. If there is a fire, burning plastics spread the fire rapidly. This is because plastics have lower ignition temperatures than materials like wood, metal, brick and glass. There is another danger too: some burning plastics give off poisonous gases.

SUMMARY

Some disadvantages in the use of plastics are:

- Clothes made of synthetic fibres do not absorb perspiration.
- Most plastics are non-biodegradable.
- They ignite easily, and some burn to form toxic products.

SUMMARY

Plastics are petrochemicals obtained from oil. Earth's resources of oil are limited. Should we use oil as fuel or save it for the petrochemicals industry?

Oil: a fuel and a source of petrochemicals

Industry is constantly finding new uses for plastics. The raw materials used in their manufacture come from oil. At first the price of oil was low and plastics were cheap materials. As the price of oil has risen, this is no longer true. The Earth's resources of oil will not last for ever. We should be thinking about whether we ought to be burning oil as fuel when we need it to make plastics and other petrochemicals.

CHECKPOINT

1 Say what materials the following articles were made out of before plastics came into use. Say what advantage plastic has over the previous material, and state any disadvantage.

(a) gutters and drainpipes
(b) toy soldiers
(c) dolls
(d) motorbike windscreens
(e) lemonade bottles
(f) buckets
(g) electrical plugs and sockets
(h) wellingtons
(i) furniture stuffing
(j) electrical cable insulation
(k) dustbins

2 (a) What does the word 'plastic' mean?
(b) Plastics can be divided into two types, which behave differently when heated. Name the two types. Describe how each behaves when heated. Say how the difference in behaviour is related to (i) the use made of the plastics and (ii) the molecular nature of the plastics.

3 Study the following list of substances.

nylon sucrose ethane styrene olive oil starch
margarine glass silk rubber melamine

(a) List the naturally occurring polymers.
(b) List the synthetic polymers.
(c) Name the substance which is not a polymer but can easily be converted into one.
(d) List the substances which are not polymers and which cannot easily polymerise.

4 PVC is used in the manufacture of drainpipes and plastic bags.

(a) Calculate the relative molecular mass of the monomer, $CH_2{=}CHCl$.
(b) The M_r of the polymer is 40 000. Calculate the number of molecules of monomer which combine to form one molecule of polymer.
(c) What method could be used to mould PVC pipes?
(d) If you needed to mould a straight length of PVC pipe to make it fit round a curve in a drainage system, how would you do this?
(e) Used PVC bags have to be disposed of. Burning PVC bags in an incinerator would cause pollution. Name two pollutants that would be released into the atmosphere.

28.3 Review of materials

FIRST THOUGHTS

You have now studied materials of many types. In this section, we compare the properties of different solid materials. We look at how these properties make different materials useful for different purposes. We look at the types of structure which give rise to the different properties.

Metals

Metals are crystalline materials. Metals:

- are generally hard, tough and shiny
- change shape without breaking (i.e. are ductile and malleable)
- are generally strong in tension and compression
- are good thermal and electrical conductors
- are in many cases corroded by water and acids.

These properties result from the nature of the metallic bond (see Topic 19.2). Metals are giant structures in which some electrons are free to move. These electrons hold the atoms together and also allow atoms to slide over one another when the metal is under stress and to conduct heat and electricity. The bonds are non-directed in space.

Ceramics

Ceramics are crystalline compounds of metallic and non-metallic elements. Ceramics:

- are very hard and brittle
- are strong in compression but weak in tension
- are electrical insulators
- have very high melting points
- are chemically unreactive.

The bonds are mainly covalent, but some are ionic. The bonds are directed in space, and the structure of a ceramic is therefore more rigid and less flexible than a metallic structure. As a result of this structure, ceramics are both harder than metals and also more brittle. Ceramics have lower densities and higher melting points than metals.

It's a fact!

The ceramic of formula $YBa_2Cu_3O_7$ is a superconductor. (A superconductor has practically zero resistance to the flow of an electric current.) The ceramic is metallic in appearance and becomes a superconductor at 90 K. It has a potential for use in zero-loss power transmission lines. Schemes for cooling underground superconducting cables in liquid air have been put forward.

Glasses

Glasses are similar to ceramics. Glasses:

- are generally transparent
- have lower melting points than ceramics.

Plate glass is transparent because it is non-crystalline so there are no reflecting surfaces to make the material opaque. It is brittle because of the rigid bonds.

Plastics

Plastics consist of a tangled mass of a very long polymer molecules. Plastics:

- are usually strong in relation to their mass
- are usually soft, flexible and not very elastic
- soften easily when heated and melt or burn
- are thermal and electrical insulators.

The structure is described as *amorphous* (shapeless) because the polymer chains take up a random arrangement. The bonds between chains are weak in thermosoftening plastics and strong in thermosetting plastics. Most polymers are poor conductors due to the lack of free electrons.

You will remember that what makes metals strong and ductile is the metallic bond; Topic 19.2.

Ceramics find important uses because of their very high melting points and very low chemical reactivity.

Glasses are used when a transparent material is needed.

Plastics are used when the demand is for a material that can be easily moulded. There is a variety of plastics with a range of properties.

It's a fact!

Polymers can be made to behave like metals. Polymers which conduct electricity are made by including salts in certain polymers. They have up to 10^{11} times the conductivity of typical polymers.

In some polymers there are regions where the chains are packed together in a regular way (see Figure 28.3A). Many polymers are a mixture of ordered regions and amorphous regions (see Figure 28.3B). In high-density poly(ethene), the chains pack closely together to give a material which is stronger than low-density poly(ethene) and is not as easily deformed by heat. High-density poly(ethene) is used to make water tanks and kitchen equipment, e.g. buckets and food containers. Articles made from high-density poly(ethene) can be heated to sterilise them so they are used for hospital equipment.

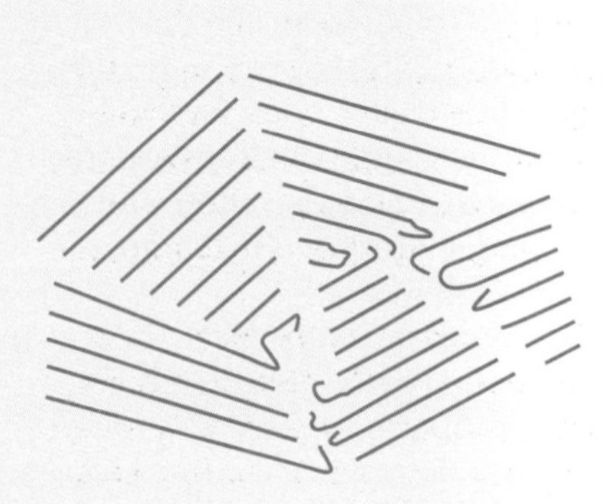

Figure 28.3A ▲ The arrangement of chains in high-density poly(ethene)

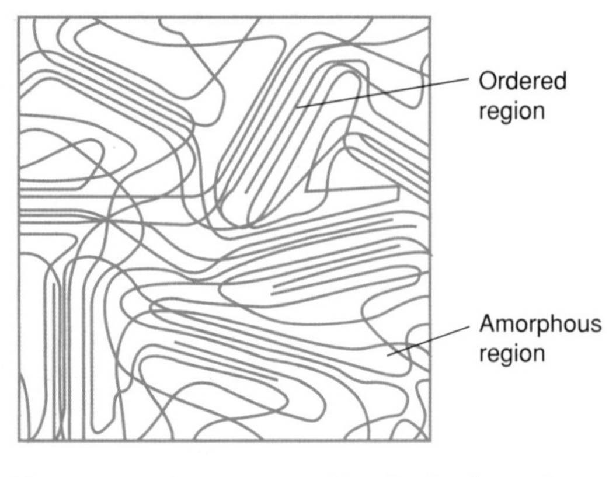

Figure 28.3B ▲ A polymer with ordered and amorphous regions

Synthetic fibres

Fibres are polymers which have been stretched to increase their tensile (stretching) strength.

Fibres are made by drawing (stretching) plastics. Molten plastic is forced through a fine hole. As the fibre cools and solidifies, it is stretched to align its molecules along the length of the fibre. Fibres therefore have greater tensile strength along the length of the fibre than plastics. Some polymers can be deformed by straightening out the polymer chains. Examples are rubber, polyesters, polyamides and polycarbonates.

Concrete and most types of rock

Concrete and most types of rock are:

- strong in compression (loading)
- weak in tension (stretching).

Reinforced concrete is described below.

Composite materials

Composite materials combine the properties of their components.

Concrete reinforced with steel girders is strengthened.

Fibre reinforced plastics are stronger than plastics but can still be moulded,
e.g. glass-fibre-reinforced-polyester
e.g. carbon-fibre-reinforced epoxy resin.

Some materials are not homogeneous. They include composite materials, e.g. reinforced concrete, and cellular materials, e.g. coral. Composite materials combine the properties of more than one material. The resulting material is more useful for particular purposes than the individual components.

Reinforced concrete

Concrete is weak in tension, strong in compression and brittle. It can be reinforced with steel wires or bars. Steel is strong in tension, and the combination is a relatively cheap, tough material. It is essential for the construction of large structures, e.g. high-rise buildings, bridges and oil platforms.

The reason why unreinforced concrete has low tensile strength is that it contains microscopic pores. To reduce porosity and increase tensile strength, one solution is to drive air out of the powder by vibrating it before mixing with water. Another method is to add to the cement-water mixture materials, e.g. sulphur or resin, which will fill the pores. Alternatively, a water-soluble polymer can be added to the cement-water mixture to fill the spaces left between cement particles. Other filler materials, e.g. glass fibre, silicon carbide, aluminium oxide particles or fibres have been used in the polymer-cement mixture.

The strengths of the different materials depend on the types of chemical bond in them: the metallic bond, covalent bond, ionic bond, intermolecular forces of attraction.

Fibre-reinforced plastics

Fibre composites such as epoxy resin and polyester resin contain thermosetting plastics. These are chosen because they are stronger than thermoplastics. However, thermoplastics have the advantage of being easy to mould, and this encouraged chemists to develop fibre-reinforced thermoplastics. These can be used for some applications which were traditionally filled by metals. Examples of fibre-reinforced plastics are:

1. *Glass-fibre-reinforced polyester*: The glass fibres are extremely strong, though relatively brittle. The polymer matrix is weaker but relatively flexible. The combination of materials and properties results in a tough, strong material. The composite has many applications, including small boats, skis and motor vehicle bodies.

Science at work

Kevlar fibres are five times as strong as steel. They are flexible, fire-resistant and of low density. Long, rigid, linear molecules of the polymer

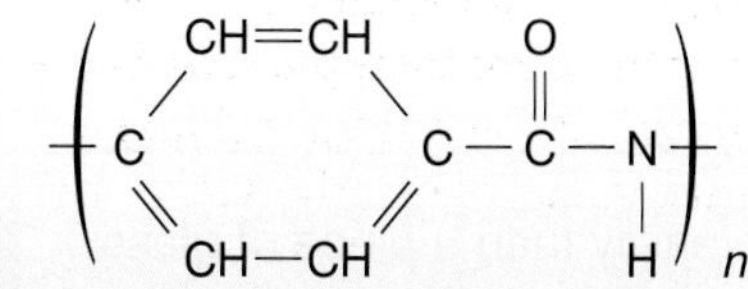

are packed tightly together by forces of attraction between >C=O groups in one chain and >N—H groups in a parallel chain. Kevlar fibres are used to strengthen car tyres.

Figure 28.3C ▲ Glass-fibre-reinforced polyester (GRP)

2. *Carbon-fibre-reinforced epoxy resin*: This is used in tennis racquets and in aircraft. It is stronger and lower in density than conventional materials, e.g. aluminium alloys.

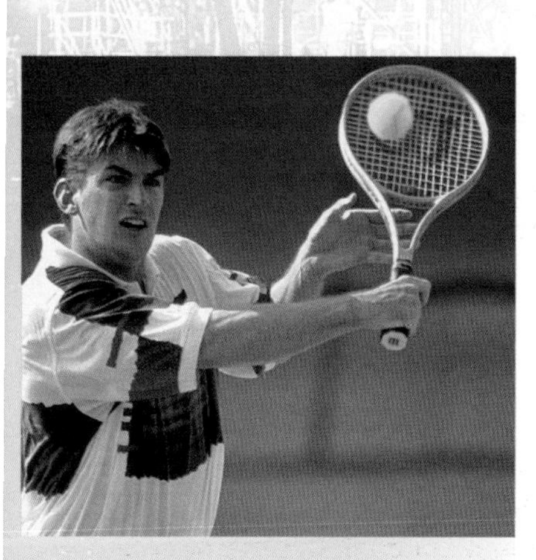

Figure 28.3D ▲ Carbon-fibre-reinforced resin in use

Fibres

The fibres may be glass, carbon, silicon carbide, poly(ethene), kevlar and other substances. The tensile strength of carbon fibres is not high but they are stiff and therefore included in materials used for the frames of aircraft.

Economics

Composites are expensive. However, a reduction in fuel consumption by lightweight transport vehicles could increase the use of composites.

Why do metals bend, polymers bend, stretch or break and ceramics and glasses break?

Metals are strong in tension and in compression because the bonds between atoms are strong. Metals can change shape without breaking because the metallic bond allows layers of atoms to slide over one another when the metal is under stress. Metal crystals contain dislocations which will move under stress and allow the metal to change shape. Metals can withstand smaller loads without changing shape than can glass.

Glass shatters easily. Glass should be a strong material because the bonds between atoms are strong, and it does not contain dislocations. It is able to withstand mechanical loading without breaking. The brittleness of glass is due to surface defects or scratches. Glass breaks easily if a scratch is first made on its surface. Cracks find it harder to grow in metals because of the movement of dislocations.

Glass fibres resist cracking when they are bent because they are free of surface scratches.

Polymers. If you break a plastic ruler, made of poly(methyl methacrylate), only 10% of the fracturing is due to the breaking of covalent bonds; the rest is due to individual chains being pulled out of the material. In other polymers chain pull-out occurs less because the molecular chains are more densely packed and more entangled. Polymers such as rubber are elastic because the polymer chains can adjust in position and shape relative to one another.

SUMMARY

The properties of metals, ceramics, glasses, plastics, fibres and concrete are revised. Composite materials, e.g. reinforced concrete, combine the properties of their components. The physical properties of the different materials are related to their structures.

CHECKPOINT

1 Explain why it is possible to bend a piece of poly(ethene) tubing more easily than a piece of glass tubing.

2 Explain why a piece of metal is dented by a hammer blow but a piece of pottery shatters.

3 Explain why a rubber band stretches more than a length of metal wire.

4 For each of the materials below list the uses to which it can be put. Each material may have more than one use.

Material	*Use*
1. Metal	**A** Windows
2. Ceramic	**B** Ovenware
3. Glass	**C** Electrical plugs
4. Thermosoftening plastic	**D** Carrier bags
5. Thermosetting plastic	**E** Ropes
6. Synthetic fibre	**F** Bridge
7. Concrete	**G** Rifle

29.1 Alcohols

FIRST THOUGHTS

Timothy's idea of a good time is to go down to the pub and have a few beers with his friends. Sometimes, however, he wakes up the next day with a throbbing headache and a feeling of nausea. He has a 'hangover'. The substance which has produced these effects is **ethanol**, a liquid which we usually call **alcohol**.

Ethanol

Ethanol is the best-known member of a series of compounds called alcohols. Ethanol is a drug. It is classified as a depressant. This means that it depresses (suppresses) feelings of fear and tension and therefore makes people feel relaxed. The body can absorb ethanol quickly because it is completely soluble in water.

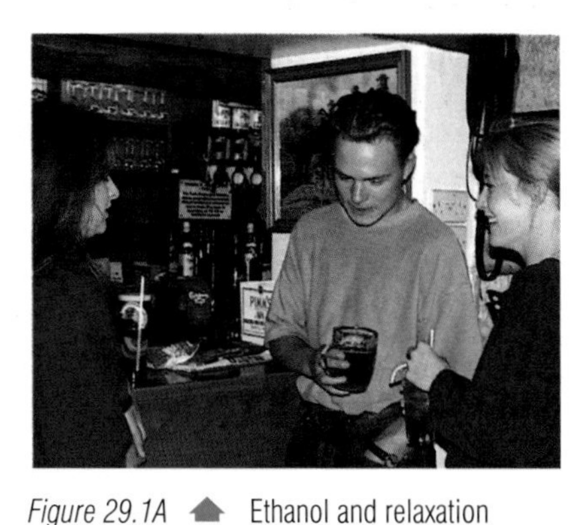

Figure 29.1A ▲ Ethanol and relaxation

Figure 29.1B ▲ Ethanol

The alcohols

The compounds of carbon are called organic compounds, with the exception of some simple compounds such as carbonates. Organic compounds were originally obtained only from natural sources.

Alcohols are a homologous series with the formula $C_nH_{2n+1}OH$.

Ethanol is a member of a homologous series of compounds called **alcohols**. Alcohols possess the group

$$
\begin{array}{c} | \\ -C-O-H \\ | \end{array}
$$

and have the general formula $C_nH_{2n+1}OH$ (see Table 29.1). The formulas are written as CH_3OH, etc. rather than CH_4O to show the —OH group and show that they are alcohols. The members of the series have similar physical properties and chemical reactions. Ethanol is the only alcohol that is not poisonous. Methanol is very toxic: drinking only small amounts of methanol can lead to blindness and death.

Table 29.1 ▼ The alcohols

Alcohol	Formula, $C_nH_{2n+1}OH$
Methanol	CH_3OH
Ethanol	C_2H_5OH
Propanol	C_3H_7OH
Butanol	C_4H_9OH
Pentanol	$C_5H_{11}OH$

www.keyscience.co.uk

SUMMARY

Alcohols are a homologous series of formula $C_nH_{2n+1}OH$. Ethanol is the only alcohol which is safe to drink in moderate quantities. Regular abuse of alcohol ruins your health. The alcohols have important industrial uses.

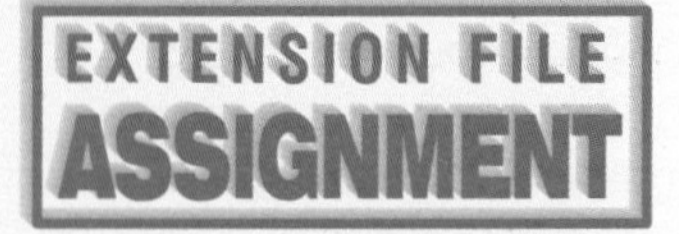

SUMMARY

Alcohols have the general formula $C_nH_{2n+1}OH$. The group C_nH_{2n+1} is called an alkyl group, e.g. $—CH_3$ the methyl group and $—C_2H_5$, the ethyl group. The group —OH is called the hydroxyl group. This is the group which gives alcohols their reactions: it is the **functional group** of the alcohols.

Drinking ethanol

Ethanol is the only alcohol that can be safely drunk (in moderation). Methanol, CH_3OH, is very toxic. and drinking even small amounts can cause permanent harm. Ethanol dissolves completely in water and is therefore rapidly absorbed through the stomach and intestines. It can take up to 6 hours for the ethanol in a single drink to be absorbed when the stomach is full, but only about 1 hour when the stomach is empty. This is why people feel the effects of a drink faster on an empty stomach than on a full one. After ethanol enters the bloodstream, it diffuses rapidly into the tissues until the concentration of ethanol in the tissues equals that in the blood. This is why measuring the concentration of ethanol in the breath or in urine will indicate the level of ethanol in the blood. As the concentration of ethanol in the blood increases, speech becomes slurred, vision becomes blurred and reaction times increase. This is why it is so dangerous to drive 'under the influence' of alcohol. A driver needs short reaction times. Drinking large amounts of ethanol regularly causes damage to the liver, kidneys, arteries and brain. Many people abuse alcohol; they do not use it properly, that is, in moderation. Such people become addicted to alcohol, and their health suffers.

Uses of alcohols

The use of ethanol is not restricted to drinking. Ethanol is an important solvent. It is used in cosmetics and toiletries, in thinners for lacquers and printing inks. Being volatile (with b.p. 78 °C), the solvent evaporates and leaves the solute behind. Other alcohols also are used as solvents for paints, lacquers, shellacs and industrial detergents. The big advantages of alcohols as solvents is that they are miscible with water and many organic liquids.

CHECKPOINT

1. (a) Write the structural formulas of (i) methane and methanol (ii) ethane and ethanol (iii) propane and propanol.
   (b) What is the difference in structure between the members of each pair of compounds in (a)?
   (c) What general formula can be written for the members of (i) the alkane series and (ii) the alcohol series?

Manufacture of ethanol

Fermentation

People have known for centuries how to obtain ethanol from sugars and starches. These substances are carbohydrates. They are compounds of carbon, hydrogen and oxygen, e.g., the sugar glucose, $C_6H_{12}O_6$, and starch, $(C_6H_{10}O_5)_n$. Glucose can be converted into ethanol by an enzyme called zymase, which is found in yeast. The conversion of glucose into ethanol is called **fermentation**.

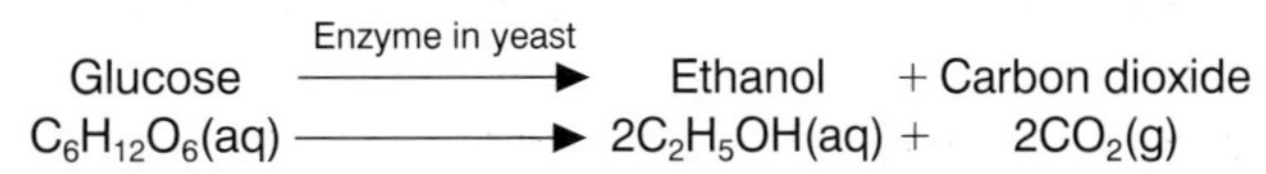

Wine is made by adding yeast to fruit juices, which contain sugars. When the content of ethanol produced by fermentation reaches 14%, it kills the yeast. More concentrated solutions of ethanol (up to 96% ethanol) can be obtained by fractional distillation.

Beer is made from a number of starchy foods, such as potatoes, rice, malt, barley, hops and others. Germinated barley (malt) contains an enzyme which hydrolyses starch to a mixture of sugars. The addition of malt is followed by yeast which ferments the sugars.

$$\text{Starch} + \text{Water} \xrightarrow{\text{Enzyme in malt}} \text{Sugar}$$
$$(C_6H_{10}O_5)_n(aq) + nH_2O(l) \longrightarrow nC_6H_{12}O_6(aq)$$

Ethanol is sold in four main forms.

- Absolute alcohol: 96% ethanol, 4% water.
- Industrial alcohol or methylated spirit: 85% ethanol, 10% water, 5% methanol. The methanol is added to make the liquid unfit to drink.
- Spirits: gin, rum, whisky, brandy, etc. which contain about 35% ethanol.
- Fermented liquors: wines (12–14% ethanol), beers and ciders (3–7% ethanol). These contain flavourings and colouring matter.

Baking, like wine-making and brewing, involves fermentation. In baking, the important product of fermentation is carbon dioxide. In bread-making, flour, water, yeast, salt, sugar and fat are mixed to form a dough. The yeast starts to act on the sugar in the dough. The dough is divided into portions and allowed to stand in a warm container while the carbon dioxide produced makes it 'rise'. Then it is passed on a conveyor belt through a hot oven, from which emerge baked loaves.

It's a fact!

One of the effects of drinking alcohol is to increase reaction times. This is why it is dangerous to drive 'under the influence' of alcohol. A survey of 17- and 18-year-old drivers involved in road accidents showed that half of them had drunk too much.

Figure 29.1C ▲ Ethanol in beer, wine, brandy, after-shave lotion

SUMMARY

Alcoholic drinks are made from sugars by fermentation. The reaction is catalysed by an enzyme in yeast. Starches can be hydrolysed to sugars and then fermented to give ethanol. Fermentation also takes place in baking when carbon dioxide makes dough rise. Industrial ethanol is made by the hydration of ethene.

Hydration

Ethanol which is to be used as an industrial solvent is made from ethene by **catalytic hydration**.

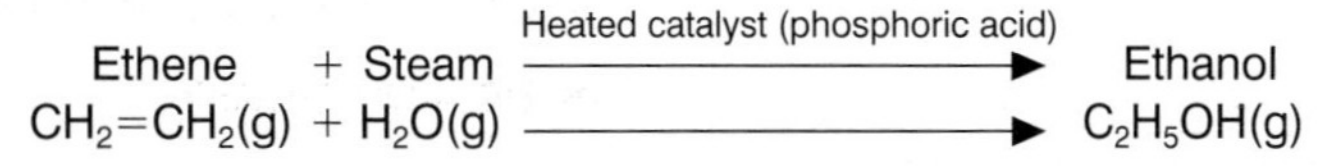

About 10% of the ethene is converted; the unreacted ethene and steam are recycled.

The two chief methods of manufacture of ethanol are the fermentation of sugars and the hydration of ethene.

You should weigh up whether it is better to use a renewable resource (sugars) or a finite resource (oil).

Another question is whether it is more convenient to use a batch process or a continuous process.

Table 29.2 shows a comparison of the two methods.

Table 29.2 A comparison of two methods of manufacturing ethanol

Method	Raw materials	Rate	Quality of product	Type of process
Fermentation	Sugars (a renewable resource)	Slow	Impure ethanol	Batch process
Catalytic hydration	Ethene (from oil, a finite resource)	Faster, but needs heat and pressure	Pure ethanol	Continuous process

Reactions of alcohols

Oxidation

Ethanol is oxidised by air if the right micro-organisms are present. The reason why wine goes sour if it is open to the air is that the ethanol in it is oxidised to ethanoic acid. Vinegar is 3% ethanoic acid.

$$\text{Ethanol} + \text{Oxygen} \xrightarrow{\text{Certain micro-organisms}} \text{Ethanoic acid} + \text{Water}$$
$$C_2H_5OH(aq) + O_2(g) \longrightarrow CH_3CO_2H(aq) + H_2O(l)$$

This reaction is used as an industrial method of making ethanoic acid.

A faster method of oxidising ethanol is to use the powerful oxidising agent, acidified potassium dichromate(VI), $K_2Cr_2O_7$. Dichromate(VI) ions, $Cr_2O_7^{2-}$, which are orange, are reduced by ethanol and other reducing agents to chromium(III) ions, which are blue. As the reaction proceeds, the colour changes from orange through green to blue. This colour change was the basis of the first 'breathalyser' test. A motorist suspected of having too much ethanol in his or her blood had to breathe out through a tube containing some orange potassium dichromate(VI) crystals. If they turned green, or even blue, he or she was 'over the limit' (Figure 29.1D).

Now you can explain why wine goes sour. Ethanol is oxidised by air to ethanoic acid.

Powerful oxidising agents, e.g. acidified potassium dichromate(VI), also oxidise alcohols.

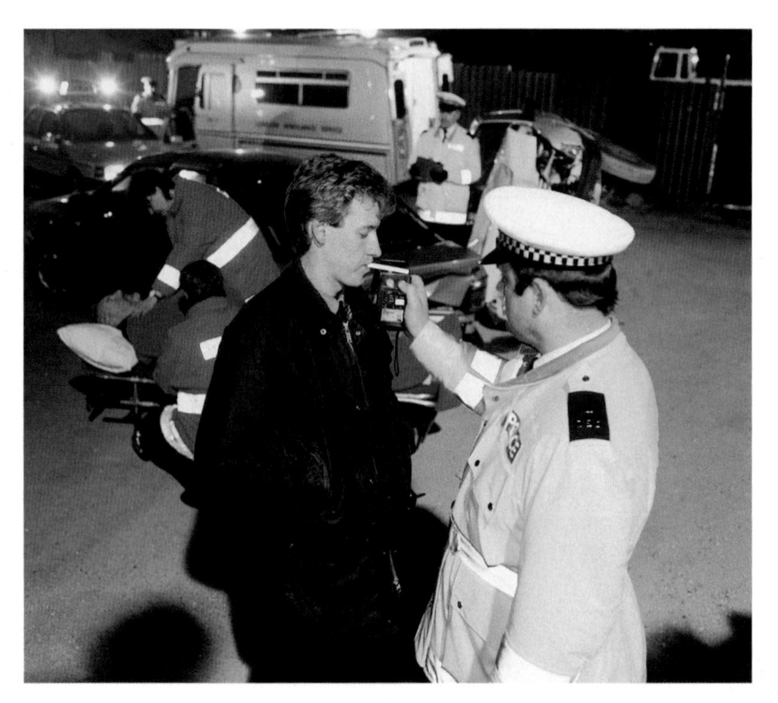

Figure 29.1D A modern 'breathalyser'

Ethanol can be dehydrated to ethene.

Alcohols and organic acids react to form esters.

Figure 29.1F gives a summary of the reactions of ethanol.

Dehydration

When ethanol is dehydrated, it gives ethene (see Figure 29.1E). Catalysts for the reaction include aluminium oxide, unglazed porcelain, porous pot and pumice stone.

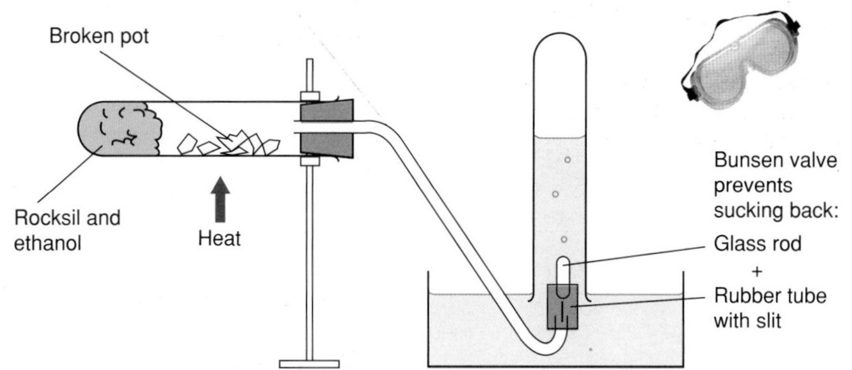

Figure 29.1E ▲ Making ethene by the dehydration of ethanol

Esterification

Alcohols react with organic acids to form sweet-smelling liquids called **esters** (see Topic 30.3). For example,

$$\text{Ethanol} + \text{Ethanoic acid} \xrightarrow{\text{Conc. sulphuric acid}} \text{Ethyl ethanoate} + \text{Water}$$

$$C_2H_5OH(l) + CH_3CO_2H(l) \longrightarrow CH_3CO_2C_2H_5(l) + H_2O(l)$$

Concentrated sulphuric acid is added to catalyse the **esterification** reaction and to combine with the water that is formed. The structural formula for ethyl ethanoate is:

```
     H     O
     |    //
 H — C — C       H    H
     |    \      |    |
     H     O  —  C  — C — H
                 |    |
                 H    H
```

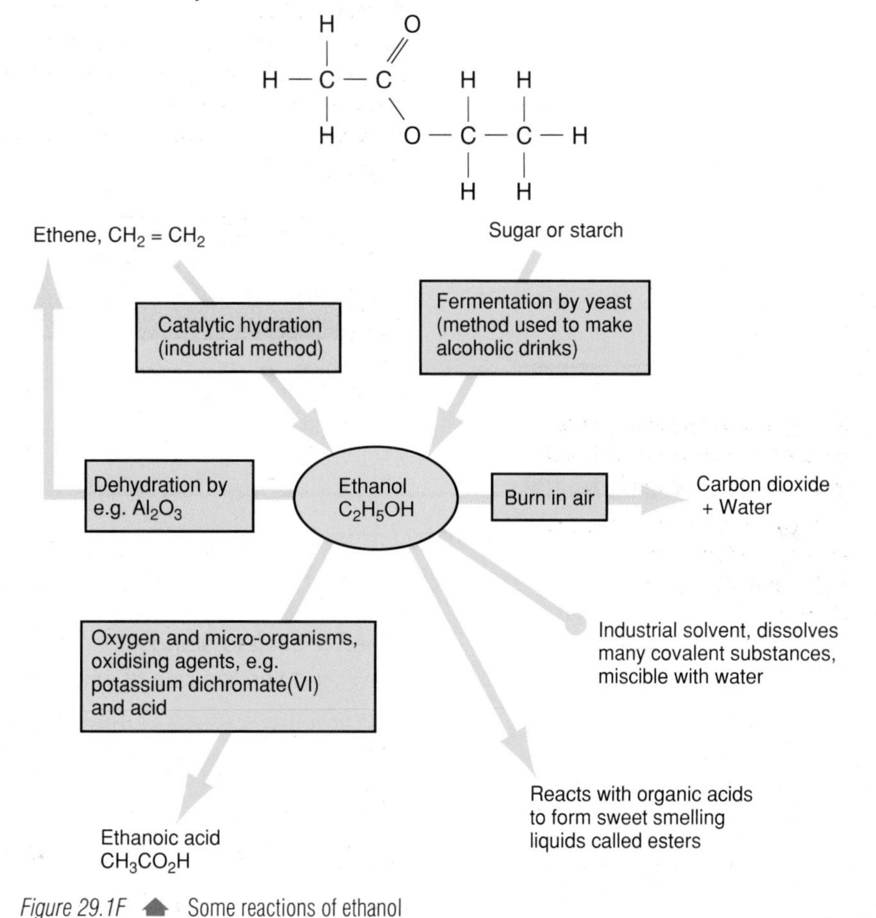

Figure 29.1F ▲ Some reactions of ethanol

SUMMARY

Ethanol is an important industrial solvent. For this use, ethanol is made from ethene by catalytic hydration. Ethanol is oxidised to ethanoic acid. Alcohols react with organic acids to form esters, e.g. ethyl ethanoate, $CH_3CO_2C_2H_5$.

Gasohol

The cost of petroleum oil rose sharply after the oil crisis of 1973. Countries which have to import oil want to find alternatives. The major use of petroleum oil fractions is in vehicle engines. Petrol engines are designed to operate over a temperature range at which petrol will vaporise. Any fuel added to petrol must vaporise at the engine temperature and it must dissolve in petrol. Ethanol has the same boiling point as heptane, and it dissolves in petrol. Ethanol burns well in vehicle engines, producing 70% as much heat per litre as petrol does. A petrol engine will take 10% ethanol in the petrol without any adjustments to the carburettor (which controls the ratio of air to fuel in the cylinders). A mixture of petrol and ethanol is described as **gasohol**. The combustion of ethanol produces carbon dioxide and water: there is little atmospheric pollution.

To invest in the production of ethanol by fermentation, a country needs land available for growing suitable crops. It needs plenty of sunshine to ripen the crops quickly and supply sugar or starch for fermentation. Brazil has already started using ethanol as a vehicle fuel. Brazil has very little oil, but plenty of land and sunshine. Most of the petrol sold there contains 10% ethanol. This reduces Brazil's expenditure on oil imports. By the year 2000, Brazil hopes to provide for 75% of its motor fuel needs by using 2% of its land for growing crops for fermentation (Figure 29.1G).

The photograph shows 'alcool', a mixture of ethanol and petrol, for sale in Brazil. Sugar cane is grown as a source of sugar for fermentation to produce ethanol. To obtain more agricultural land for growing sugar cane, Brazil is clearing rain forest. Topic 13.8 discusses how cutting down forests will increase the greenhouse effect.

Figure 29.1G ▲ Petrol pumps in Brazil selling 'Alcool' gasohol

SUMMARY

Ethanol can be added to the petrol in vehicle engines. Brazil is a country with no oil but with plenty of arable land and a sunny climate. Brazil is growing crops which can be fermented to give ethanol. Ethanol is a clean fuel: it burns to form carbon dioxide and water.

CHECKPOINT

1 (a) Explain what is meant by fermentation.
(b) Name a commercially important substance which is made by this method. Say what it is used for.

2 (a) Why does wine turn sour when it is left to stand?
(b) Suggest two methods of slowing down the rate at which wine turns sour.

3 What are the dangers of drinking (a) ethanol and (b) methanol?

4 Petrol is produced by distillation and by 'cracking'.
(a) Explain what cracking is.
(b) Say where the energy used in (i) distillation and (ii) cracking comes from.
(c) Say where the energy used in fermentation comes from.

▶ **5** Europe has a surplus of grain. Someone proposes building plants to obtain ethanol from the surplus cereals. Say what advantages this would bring to
(a) the environment (b) the farmer (c) the motorist (d) industry (e) the tax payer.
Can you see anything wrong with the idea? Explain your answers.

▶ **6** Ethanol is made by fermentation and by catalytic hydration.
(a) Why is it an advantage to make ethanol for solvent use by a continuous process, rather than a batch process?
(b) Why are alcoholic drinks not made by catalytic hydration?
(c) Which method is more economical with energy?
(d) Which method is more economical with the Earth's resources?

29.2 ▶ Carboxylic acids

The reactions of carboxylic acids are due to the carboxyl group, $-CO_2H$. In most organic acids this group is only slightly ionised and they are therefore weak acids. Their reactions are summarised in Figure 29.2A.

Ethanoic acid has the structural formula:

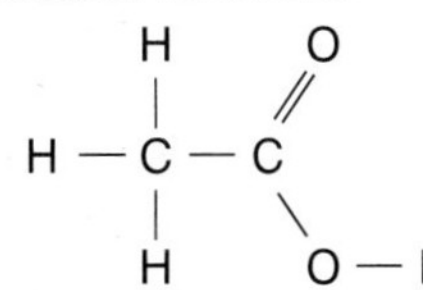

The formula is usually written as CH_3CO_2H. Ethanoic acid is a member of a homologous series called **carboxylic acids** (or alkanoic acids). Other members of the series are

methanoic acid, HCO_2H
propanoic acid, $C_2H_5CO_2H$
general formula, $C_nH_{2n+1}CO_2H$.

The carboxylic acids possess the group $—CO_2H$, which is called the carboxyl group. This is the functional group of the carboxylic acids. The carboxyl group ionises in solution to give hydrogen ions:

$$\text{Carboxylic acid} + \text{Water} \rightleftharpoons \text{Hydrogen ions} + \text{Carboxylate ions}$$
$$RCO_2H(aq) \rightleftharpoons H^+(aq) + RCO_2^-(aq)$$

R is an alkyl group, e.g. CH_3, C_2H_5. Since they are only partially ionised, carboxylic acids are weak acids. Some of the reactions of ethanoic acid are shown in Figure 29.2A.

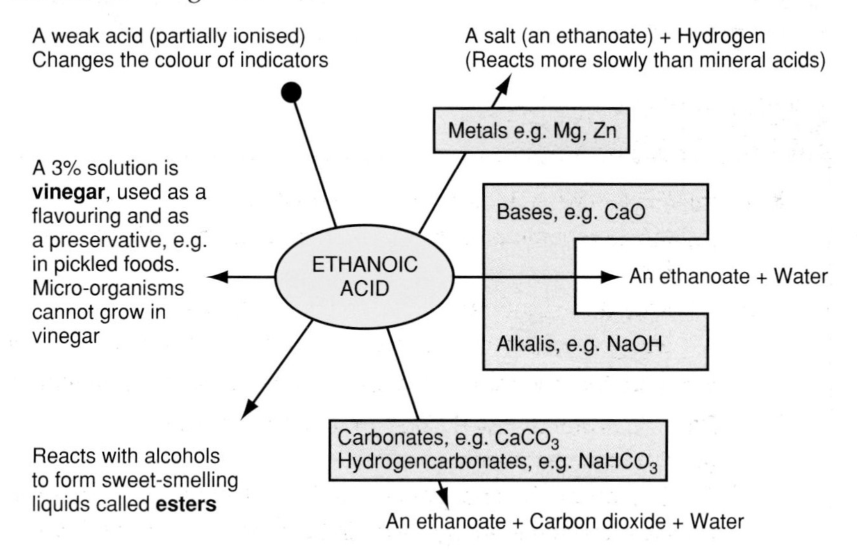

Figure 29.2A ▲ Reactions of ethanoic acid

SUMMARY

Ethanoic acid is a carboxylic acid. It is a weak acid. It reacts in the same way as mineral acids but more slowly. Carboxylic acids react with alcohols to form esters.

CHECKPOINT

1 (a) Name three substances which will react with ethanoic acid. Name the products of the reactions.
(b) Explain why a solution of ethanoic acid is less reactive than a solution of hydrochloric acid of the same concentration.
(c) Describe an experiment which you could do to demonstrate that ethanoic acid is weaker than hydrochloric acid.

29.3 Esters

Many esters are liquids with fruity smells. They occur naturally in fruits. They are used as food additives to improve the flavour and smell of processed foods. Other esters are used as solvents, for example in glues. People can get 'high' by inhaling esters. Some people enjoy the sensation so much that they become 'glue sniffers'. It is really the solvent, which may be a hydrocarbon or an ester, that they want to inhale. This dangerous habit is called **solvent abuse**. It produces the same symptoms as ethanol abuse. In addition, sniffers who are 'high' on solvents become disoriented. They may believe that they can jump out of windows or off bridges or walk through traffic. Many deaths occur from solvent abuse both through disoriented behaviour and from sniffers passing out and suffocating on their own vomit.

SUMMARY

Esterification is the reaction between carboxylic acids and alcohols to form esters. These compounds are used as food additives and as solvents. Solvent abuse is a dangerous habit.

Figure 29.3A ▲ Esters in thinner, nail varnish remover, UHU glue

Animal fats and vegetable oils are esters. They are liquids or solids, depending on the size and shape of their molecules. They are esters of glycerol, an alcohol which has three hydroxyl groups:

$$\begin{array}{c} CH_2OH \\ | \\ CHOH \\ | \\ CH_2OH \end{array}$$

Each —OH group can esterify with a molecule of a carboxylic acid. The acids hexadecanoic acid, $C_{15}H_{31}CO_2H$, octadecanoic acid, $C_{17}H_{35}CO_2H$ and others combine with glycerol to form fats. These acids are known as 'fatty acids'. (Hexadecane means 16, and octadecane means 18. Count the number of carbon atoms, and you will see how they get their systematic names.)

SUMMARY

Fats and oils are esters of glycerol and fatty acids. Saponification converts fats and oils into soaps, the sodium salts of fatty acids.

In soap-making, fats and oils are boiled with a concentrated solution of sodium hydroxide. The reaction can be represented by:

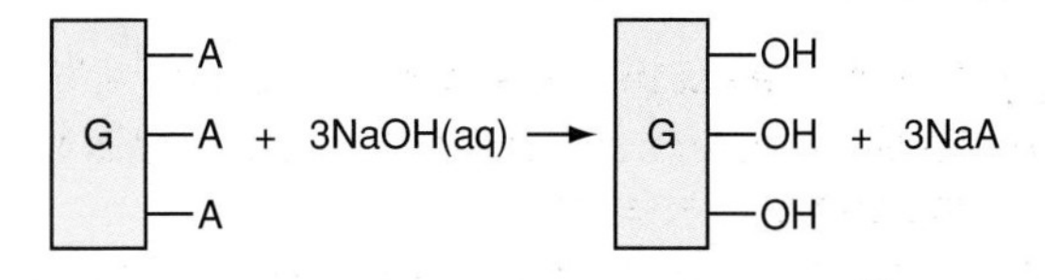

where $G(OH)_3$ represents glycerol and HA represents the fatty acid. The sodium salt of the fatty acid, e.g. sodium hexadecanoate or sodium octadecanoate, is a soap. This important reaction is called **saponification**.

CHECKPOINT

1 Explain the terms 'esterification' and 'solvent abuse'.

2 (a) How are soaps manufactured?
(b) How do soaps emulsify oil and water?

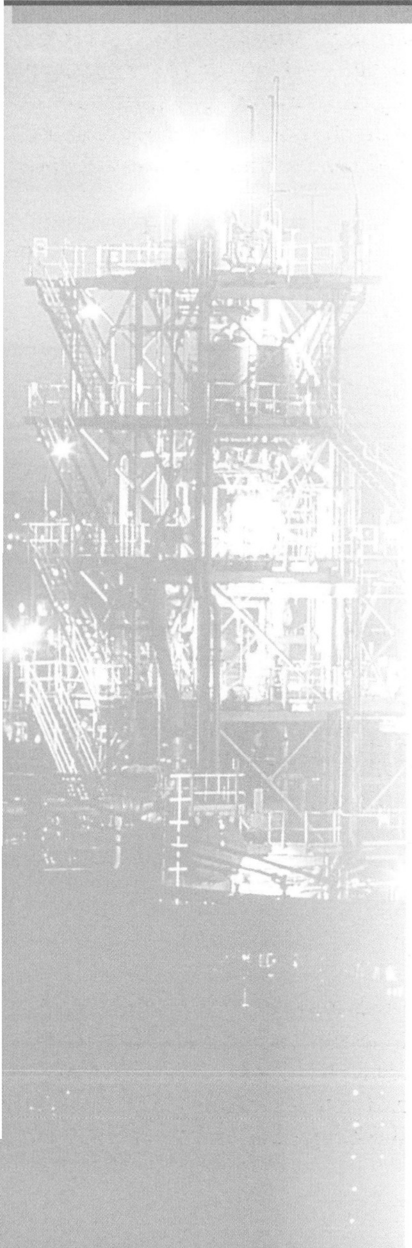

30.1 Food

FIRST THOUGHTS

You hear some people talking about 'chemicals' as if all chemicals were noxious substances. Little do they know that they eat platefuls of chemicals every day!

Why do they need food (see Figure 30.1A)? They need food

- to supply the energy they use in all their activities
- to build new tissues
- to repair damaged tissues, such as cuts and broken bones

All foods are chemical compounds. There are three main classes of foods: carbohydrates, proteins and fats. In addition we need vitamins, mineral salts, water and fibre.

Figure 30.1A Children in a school canteen

30.2 Carbohydrates

FIRST THOUGHTS

What do sugar and starch, the cell walls of plants and the exoskeletons of insects have in common? This topic will tell you!

Sugars

The sugar in the sugar bowl is a compound called **sucrose**. There are many other sugars. One of the simplest is **glucose**, $C_6H_{12}O_6$. It is a **carbohydrate**, that is, a compound of carbon, hydrogen and oxygen only, in which the ratio (number of hydrogen atoms/number of oxygen atoms) = 2. A molecule of glucose contains a ring of six atoms, five carbon atoms and one oxygen atom; see Figure 30.4A(a). A shorthand way of writing the formula is shown in Figure 30.4A(b).

You will understand the formula better if you construct a model.

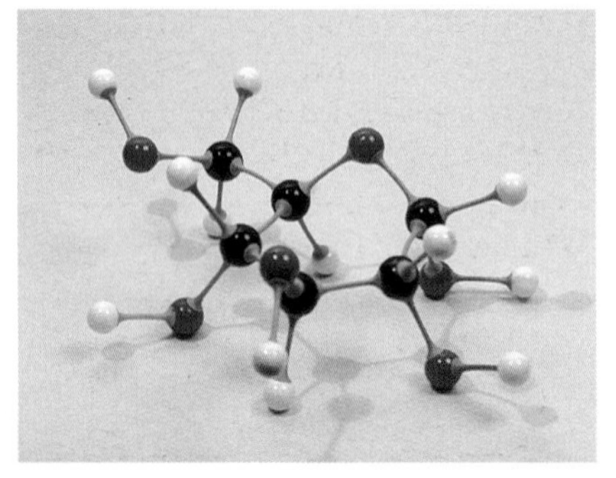

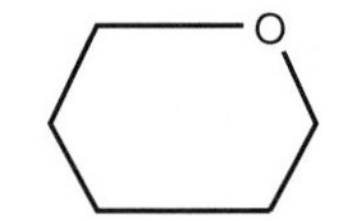

Figure 30.2A (a) Model of a glucose molecule (b) The formula in shorthand form

Fructose, $C_6H_{12}O_6$, is another sugar. It is an isomer of glucose and, as with glucose, its molecules contain a ring of six atoms. Sugars whose molecules contain one ring structure are called **monosaccharides** (mono = one).

Sucrose, $C_{12}H_{22}O_{11}$, is a **disaccharide**: its molecules contain two ring structures. It is formed from glucose and fructose:

$$\text{Glucose} + \text{Fructose} \rightarrow \text{Sucrose} + \text{Water}$$
$$C_6H_{12}O_6(aq) + C_6H_{12}O_6(aq) \rightarrow C_{12}H_{22}O_{11}(aq) + H_2O(l)$$

A monosaccharide has one sugar ring per molecule, e.g. glucose and sucrose. A disaccharide has two sugar rings per molecule, e.g. sucrose and maltose.

This is an example of condensation polymerisation (Topic 28.2).

Maltose, $C_{12}H_{22}O_{11}$, is a disaccharide formed from glucose:

$$\text{Glucose} \rightarrow \text{Maltose} + \text{Water}$$
$$2C_6H_{12}O_6(aq) \rightarrow C_{12}H_{22}O_{11}(aq) + H_2O(l)$$

The shorthand form of the formula of maltose is shown in Figure 30.2B.

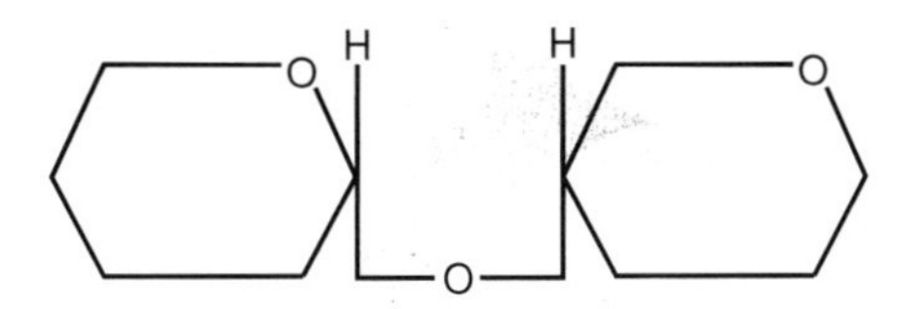

Figure 30.2B ▲ The formula for maltose in shorthand form

The cellular respiration of monosaccharides and disaccharides provides energy in all plants and animals.

Other biologically important carbohydrates are starches, glycogen and cellulose. All these are polysaccharides with many sugar rings per molecule.

Carbohydrates have a vital function in all living organisms: they provide energy. When carbohydrates are oxidised, carbon dioxide and water are formed, and energy is released.

$$\text{Glucose} + \text{Oxygen} \rightarrow \text{Carbon dioxide} + \text{Water}$$
$$C_6H_{12}O_6(aq) + 6O_2(aq) \rightarrow 6CO_2(g) + 6H_2O(l); \text{ Energy is released}$$

If you burn glucose in the laboratory, you will observe a very rapid release of energy. Inside the cells of a plant or animal, however, the oxidation takes place in a controlled manner so that energy is made available to the organism as needed. The process is called cellular respiration. Living organisms also respire fats, oils and proteins.

Starch, glycogen and cellulose

Starch is a food substance which is stored in plant cells. Glycogen is a food substance which is stored in animal cells. Cellulose is the substance of which plant cell walls are composed. All these substances are carbohydrates. They are **polysaccharides**, that is, their molecules contain a large number (several hundred) of sugar rings. These are glucose rings (see Figure 32.4C). Polysaccharides differ in the length and structure of their chains (see Figure 30.2D).

Figure 30.2C ▲ Part of a starch molecule

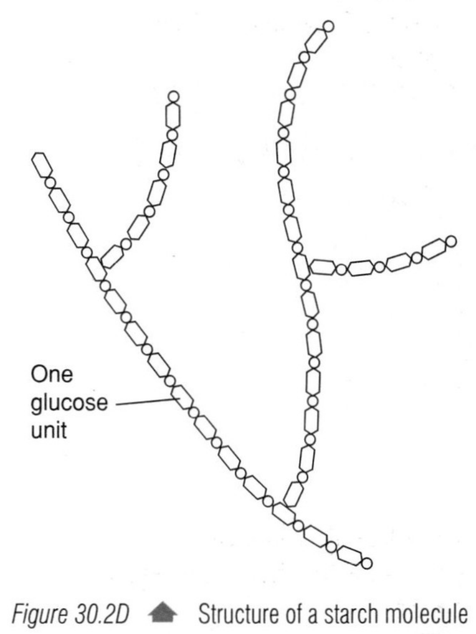

Figure 30.2D ▲ Structure of a starch molecule

SUMMARY

Carbohydrates contain carbon, hydrogen and oxygen only. Monosaccharides are sugars, including hexoses, with six-atom rings in the molecule, e.g. glucose and fructose, and pentoses, with five-atom rings, e.g. ribose. Disaccharides have two sugar rings in the molecule, e.g. sucrose and maltose. Polysaccharides have molecules with a large number of sugar rings. They include starch and glycogen (food stores), cellulose (in plant cell walls) and chitin (in insect exoskeletons).

Cellulose is the chief component of wood. For thousands of years, wood was our chief structural material, used to build houses, ships and carriages. It is still an important building material. The cellulose in wood is the raw material from which paper is made.

Starch and glycogen are only slightly soluble in water. They can remain in the cells of the organism without dissolving and therefore make good food stores. When energy is needed, cells hydrolyse starch and glycogen into glucose, which is soluble, and then oxidise the glucose. Since the hydrolysis is enzyme-catalysed, it takes place rapidly at the temperature of the plant or animal.

Cellulose is insoluble. The walls of plant cells consist of a tough framework of cellulose fibres (see Figure 30.2E). Mammals have no enzymes for digesting cellulose. We need cellulose in our diet, however, to provide 'dietary fibre' or 'roughage'. This adds bulk to the food and helps it to pass through the gut.

Another polysaccharide, chitin, forms the exoskeleton that encloses the body of some invertebrates (see Figure 30.2F and Key Science Biology Topic 10.1.

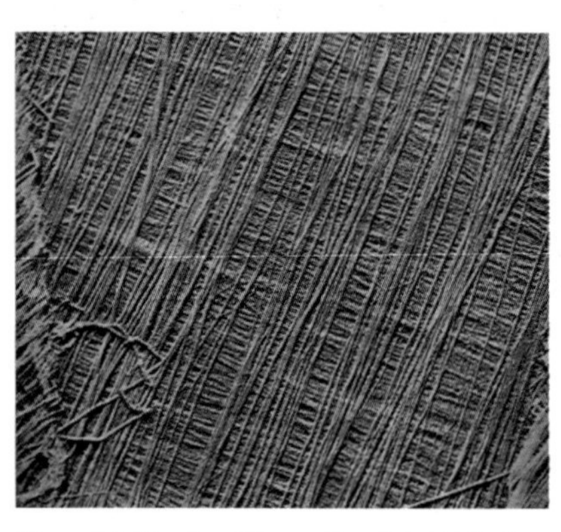

Figure 30.2E ▲ Plant cell wall (×10 000) showing the framework of cellulose fibres

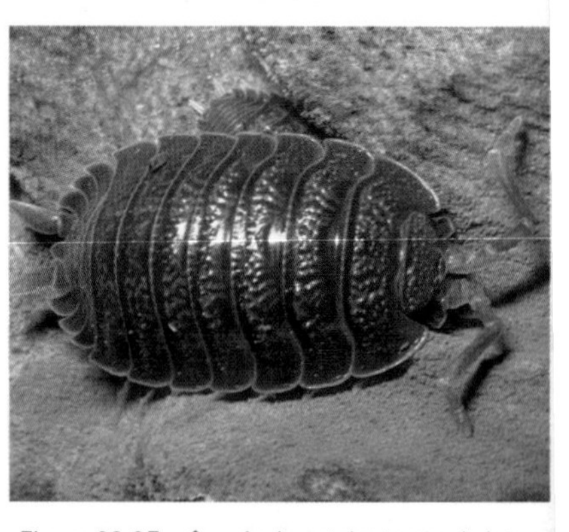

Figure 30.2F ▲ An invertebrate exoskeleton of chitin

Bread making

Much of the carbohydrate in our diet comes from bread. It is made principally from wheat. Grains of wheat are ground to make flour. Then flour is mixed with water, yeast, salt, sugar and fat to make dough. Enzymes in the yeast start to act on the sugars in the dough. The sugars ferment to produce ethanol and carbon dioxide.

$$\text{Glucose} \rightarrow \text{Ethanol} + \text{Carbon dioxide}$$
$$C_6H_{12}O_6(aq) \rightarrow 2C_2H_5OH(aq) + 2CO_2(g)$$

The carbon dioxide and water vapour produced both leaven the dough (make it rise). Chemical raising agents are added to help bread dough and cake mixtures to rise. The most important is baking powder, which contains sodium hydrogencarbonate, $NaHCO_3$, (called baking soda) and weak acids. Both on heating and on reaction with acids sodium hydrogencarbonate releases carbon dioxide and water vapour.

When the risen dough is put into an oven, it continues to rise as the gases inside it expand. Eventually the oven temperature kills the yeast, and fermentation stops. Proteins and starches solidify to form a rigid structure. This traps the gases so that the bread remains risen when cool.

CHECKPOINT

1. Runners in the London marathon ate pasta the evening before. Why did they think this would give them energy?
2. Give examples of carbohydrates which are used as structural materials in plants and animals.
3. Glucose, maltose and starch are carbohydrates. Answer the following questions about them.
   (a) Which one tastes sweet?
   (b) Which two are very soluble in water?
   (c) List the elements that make up all three.
   (d) The formula of glucose is $C_6H_{12}O_6$. What is the ratio number of atoms of hydrogen : number of atoms of oxygen?
   (e) The formula of maltose is $C_{12}H_{22}O_{11}$. What is the ratio number of atoms of hydrogen : number of atoms of oxygen?
   (f) Using your answers to (d) and (e), try to explain where the name 'carbohydrate' comes from.
   (g) The 'shorthand' formula for glucose is ⬡O
   Draw the shorthand formulas for maltose and starch.

30.3 Fats and oils

FIRST THOUGHTS

Have you seen brands of margarine and cooking oil described as 'high in polyunsaturates' and wondered what it means? This section will tell you.

SUMMARY

Fats and oils are together called lipids. As foods, they provide energy and dissolved vitamins. They also provide thermal insulation and protection for delicate organs.

Fats and oils are together called **lipids**. They contain carbon, hydrogen and oxygen only. At room temperature, most fats are solid and most oils are liquid. Fats and oils are insoluble in water. Their functions are as follows:

Source of energy: Lipids are important stores of energy in living organisms. The oxidation of fats and oils gives carbon dioxide and water with the release of energy. Lipids provide about twice as much energy per gram as do carbohydrates.

Thermal insulation: Mammals have a layer of fat under the skin. This acts as a thermal insulator.

Protection: Delicate organs, e.g. kidneys, are protected by a layer of fat.

Food: Some vitamins, e.g. vitamins A, D and E, are insoluble in water and soluble in lipids. Foods containing lipids provide these vitamins.

Cell membranes: Lipids are incorporated in cell membranes.

Saturated and unsaturated fats and oils

Do you watch your diet? Many people are concerned about having too much fat in their diet, particularly **saturated fats**. Let us look at what the term 'saturated' means here.

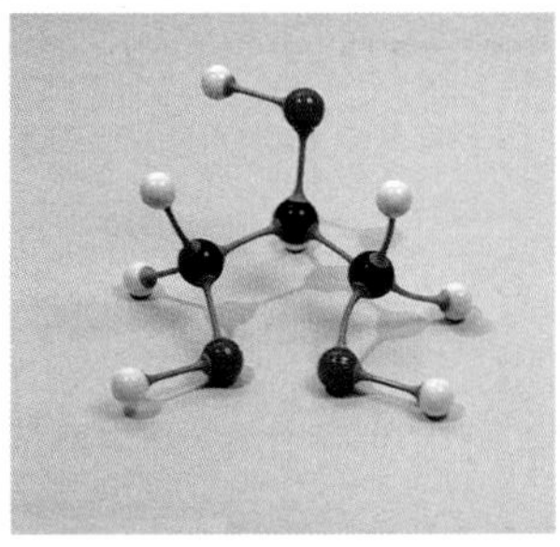

(a) Model of a molecule of glycerol

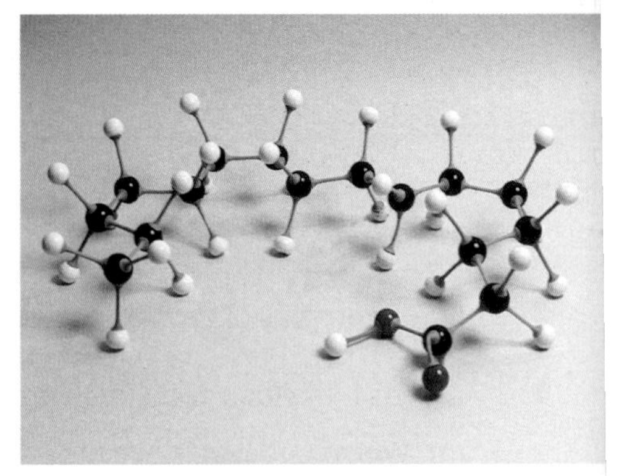

(b) Model of a molecule of a fatty acid (hexadecanoic acid

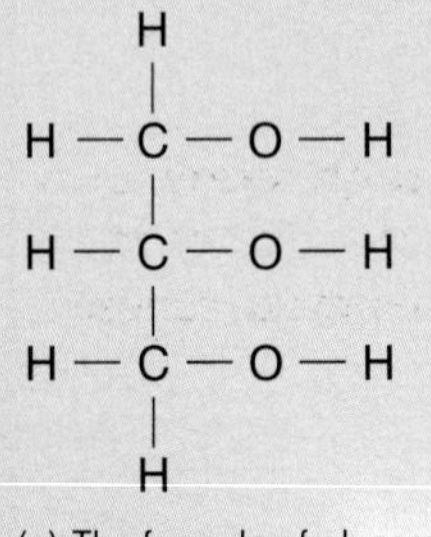

(c) The formula of glycerol

```
    H   H   H   H   H   H   H   H   H   H   H   H   H   H   H   O
    |   |   |   |   |   |   |   |   |   |   |   |   |   |   |  //
H — C — C — C — C — C — C — C — C — C — C — C — C — C — C — C — C
    |   |   |   |   |   |   |   |   |   |   |   |   |   |   |   \
    H   H   H   H   H   H   H   H   H   H   H   H   H   H   H    O — H
```

(d) The formula of hexadecanoic acid

Figure 30.3A

Fats and oils contain esters of glycerol and fatty acids (carboxylic acids with long alkyl groups) (see Topic 29.3). The fatty acid may be saturated or unsaturated. If there is one carbon-carbon double bond in the molecule, the compound is described as **unsaturated**. If there is more than one carbon-carbon double bond, the compound is described as **polyunsaturated**. Below is part of the carbon chain in (a) a saturated fatty acid, (b) an unsaturated fatty acid and (c) a polyunsaturated fatty acid.

You read a lot of discussion about the value of polyunsaturated fats in the diet. Now you know what they are!

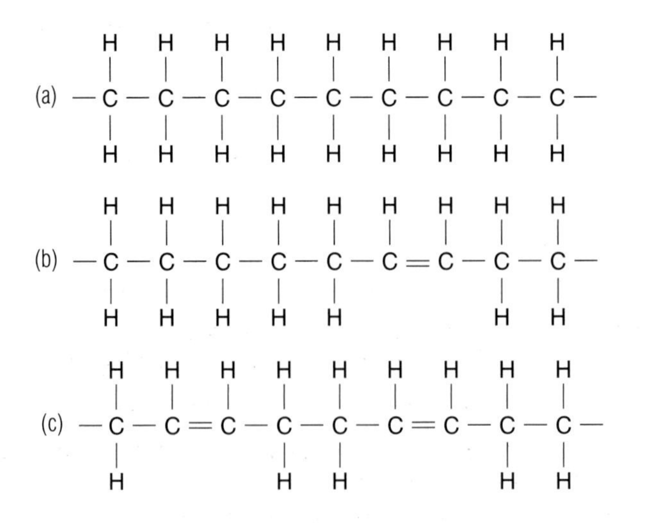

Fats and oils are mixtures of esters. Fats which contain esters of glycerol and saturated fatty acids are called saturated fats. Fats which contain esters of glycerol and unsaturated fatty acids are called unsaturated fats or polyunsaturated fats, depending on the number of double bonds in a

SUMMARY

Lipids (fats and oils) are mixtures of esters of glycerol with carboxylic acids (called fatty acids). The carboxylic acid may be saturated or unsaturated or polyunsaturated. Unsaturated fats and oils can be converted into saturated fats and oils by catalytic hydrogenation.

molecule of the acid. Animal fats contain a large proportion of saturated esters and are solid. Plant oils contain a large proportion of unsaturated esters and have lower melting points. Many scientists believe that eating a lot of saturated fats increases the risk of heart disease.

Unsaturated oils can be converted into saturated fats by catalytic hydrogenation. The vapour of the oil is passed with hydrogen over a nickel catalyst (see Topic 28.1).

Having three hydroxyl groups, glycerol can combine with three carboxyl groups. The esters formed are called **triglycerides.**

glycerol	a triglyceride
CH_2OH	CH_2OCOR^1
|	|
$CHOH$	$CHOCOR^2$
|	|
CH_2OH	CH_2OCOR^3

glycerol — a triglyceride

The groups R^1, R^2 and R^3 may be the same (in a simple triglyceride) or different (in a mixed triglyceride). An example of a simple triglyceride is the ester of glycerol with palmitic acid (systematic name hexadecanoic acid; see Figure 30.3A), $C_{15}H_{31}CO_2H$:

$CH_2OCOC_{15}H_{31}$
|
$CHOCOC_{15}H_{31}$
|
$CH_2OCOC_{15}H_{31}$

glyceryl tripalmitate

Fats, like other esters, are hydrolysed by alkalis (see Topic 29.3). Glyceryl tripalmitate is hydrolysed by sodium hydroxide solution to form glycerol and sodium palmitate, which is a soap.

$$\begin{array}{lcl} CH_2OCOC_{15}H_{31} & & CH_2OH \\ | & & | \\ CHOCOC_{15}H_{31}\,(s) + 3NaOH\,(aq) & \longrightarrow & CHOH\,(l) + 3C_{15}CO_2Na\,(aq) \\ | & & | \\ CH_2OCOC_{15}H_{31} & & CH_2OH \end{array}$$

glyceryl tripalmitate — glycerol + sodium palmitate (soap)

CHECKPOINT

▶ **1** (a) Say which are the fats and which are the oils.
soft margarine, butter, lard, hard margarine, cooking oil

(b) What is the main physical difference between fats and oils?

▶ **2** The manufacturers of *Flower* margarine claim that their product encourages healthy eating because it is 'low in saturated fats'.

(a) What do they mean by 'low in saturated fats'?

(b) Name a product which contains much saturated fat.

(c) What particular benefit to health is claimed for products such as *Flower*?

3 The following equation represents the formation of a fat.

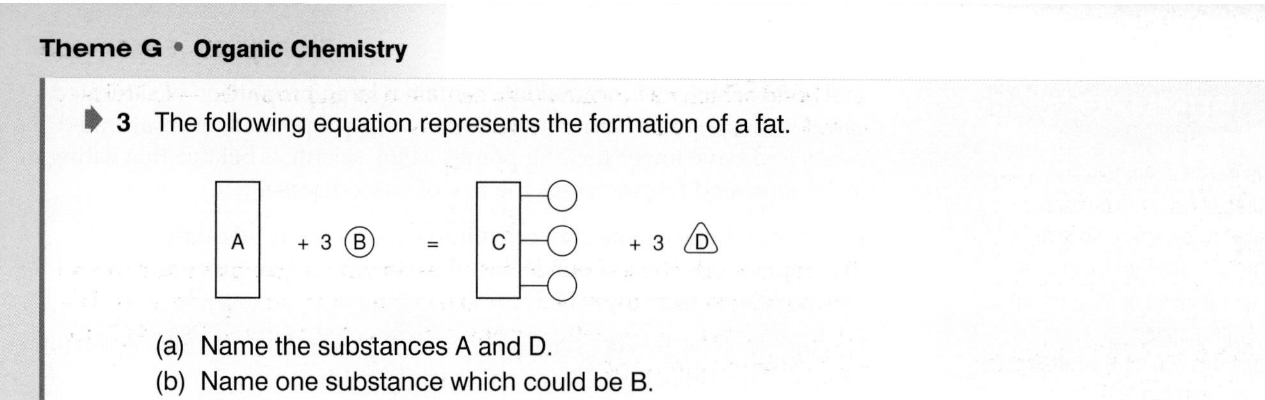

(a) Name the substances A and D.
(b) Name one substance which could be B.
(c) Name one fat and one oil which could be C.

30.4 Proteins

FIRST THOUGHTS

Muscle, skin, hair and haemoglobin; what do they have in common? You can find out in this section.

EXTENSION FILE ACTIVITY

Amino acids have two functional groups: the amino group $-NH_2$ and the carboxyl group $-CO_2H$. These groups enable them to polymerise to form peptides, polypeptides and proteins.

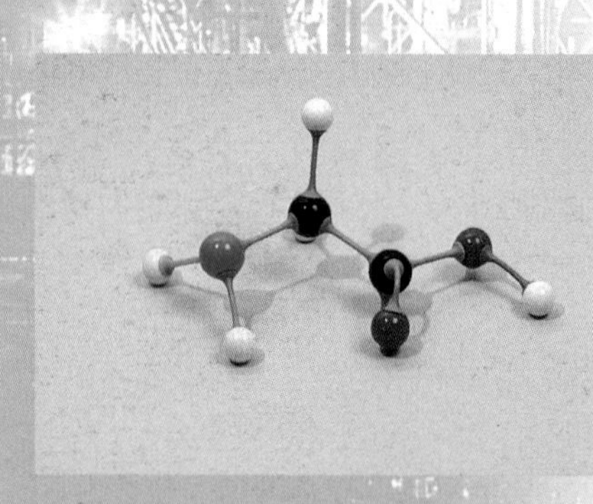

Figure 30.4A Model of a glycine molecule

Proteins are compounds of carbon, hydrogen, oxygen, nitrogen and sometimes sulphur. They are vitally important in living things for the following reasons:

1 Proteins are the compounds from which new tissues are made. When organisms need to grow or to repair damaged tissues, they need proteins.

2 Enzymes are proteins which catalyse reactions in living things.

3 Hormones are proteins which control the activities of plants and animals.

4 Proteins can be used as a source of energy in respiration.

Proteins have large molecules. A protein molecule consists of a large number of amino acid groups. There are about 20 **amino acids**. The simplest is glycine, of formula

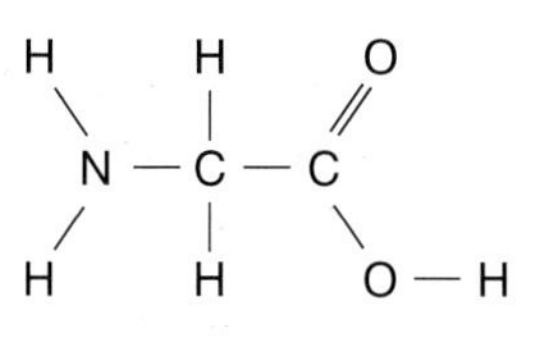

The general formula of an amino acid is

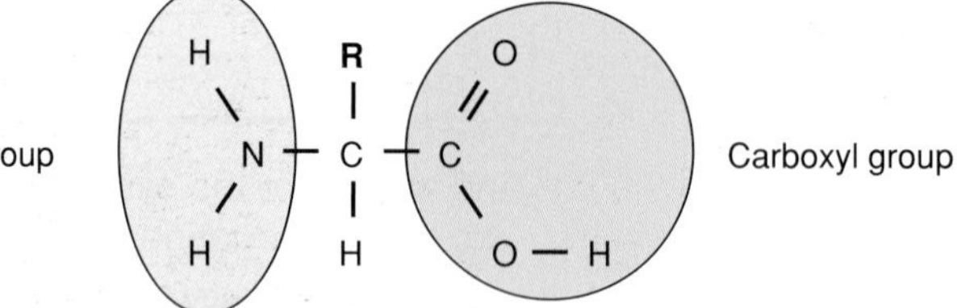

R is a different group in each amino acid; it can be $-H$, $-CH_3$, $-CH_2OH$, $-SH$ and many other groups. The carboxyl group ($-CO_2H$) is acidic, and the amino group ($-NH_2$) is basic (like ammonia, NH_3). The amino group of one amino acid can react with the carboxyl group of another amino acid. When this happens, the two amino acids combine to form a **peptide** and water.

$$H_2NCH_2CO_2H + H_2NCH_2CO_2H \rightarrow H_2NCH_2CONHCH_2CO_2H + H_2O$$

The bond which has formed between the two amino acids, —CONH—, is called the **peptide link**. Its structure is

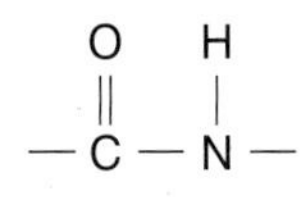

The peptide which has been formed has an amino group at one end and a carboxyl group at the other end so it can form more peptide links. We can show this by an equation, using a different shape, e.g. ○ and □, for each different amino acid.

□ + ○ → □—○ + H_2O
□—○ + ▲ + ○ → ▲—□—○—○ + $2H_2O$
▲—□—○—○ + ▲ + □ → ▲—▲—□—○—○—□ + $2H_2O$

In this way, many amino acids combine by condensation polymerisation to form a long chain. Each link in the chain is a peptide bond.

Peptides have molecules with up to 15 amino acid groups.
Polypeptides have molecules with 15–100 amino acid groups.
Proteins have still larger molecules.

Proteins, polypeptides and peptides are hydrolysed to amino acids. This reaction is the reverse of condensation polymerisation. The hydrolysis can be carried out in the lab. In the body it is enzyme-catalysed so it takes place rapidly at body temperature.

SUMMARY

Many tissues in living organisms are made of proteins. The molecules of proteins are long chains of amino acid groups. Peptides and polypeptides have smaller molecules.

It's a fact!

Fifteen million children a year die of starvation and disease. You will have seen pictures of children with swollen abdomens. They are suffering from a disease called kwashiorkor. They are starved of protein. In parts of the world where rice is the staple diet, kwashiorkor is common.

Diet

An animal can break down the proteins in its food into amino acids and then use the amino acids to build the proteins which it needs. Animals are able to make some amino acids in their bodies. The other amino acids which they need must be supplied in the diet. These are called **essential amino acids**. A protein which contains all the essential amino acids is called a **first class protein**. Meat, fish, cheese and soya beans supply first class protein. A protein which lacks some essential amino acids is called a **second class protein**. Such proteins are in flour, rice and oatmeal.

KEY SCIENTIST

Dr Max Perutz began work on the structure of haemoglobin in 1936. It took over 20 years to find out that the molecule contains 574 amino acid groups, arranged in four chains. With a total of 10 000 atoms, the molecule has a mass 65 000 times that of a hydrogen atom.

Enzymes are proteins; see Topic 25.8.

Structure

A protein molecule is twisted into a three-dimensional shape (see Figures 30.4B and C). It is kept in this shape by bonding between groups in different parts of the chain.

Figure 30.4B ▲ Insulin. In 1953, Dr Frederick Sanger of the UK reported the complete sequence of amino acids in the protein insulin. This was the first time that a protein structure had been worked out. Insulin controls blood sugar levels. Its molecule is one of the smallest protein molecules.

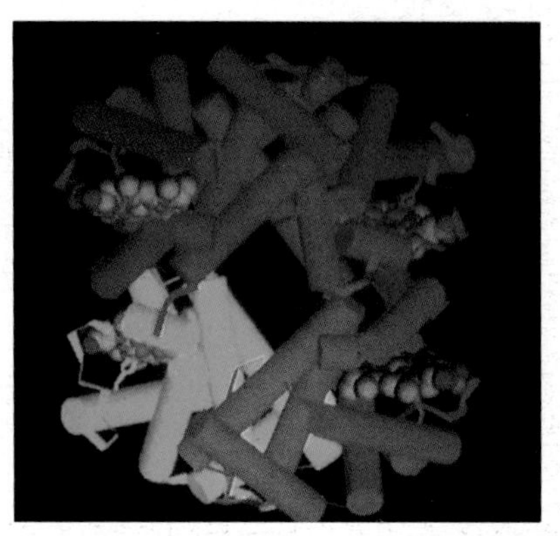

Figure 30.4C ▲ Haemoglobin. In 1959, Dr Max Perutz worked out the shape of the haemoglobin molecule. Haemoglobin is the red pigment in blood which transports oxygen round the body. The haemoglobin molecule is larger than the insulin molecule.

CHECKPOINT

1. What is the difference between (a) a peptide and a polypeptide (b) a polypeptide and a protein?
2. What compounds are formed when molecules of the following compounds combine?
   (a) monosaccharides (b) amino acids (c) glycerol and fatty acids?
3. (a) What type of compound is this?

$$
\begin{array}{c}
\quad\;\; H \\
\quad\;\; | \\
H_2N-C-CO_2H \\
\quad\;\; | \\
\quad\;\; CH_2-SH
\end{array}
$$

   (b) Draw the structure of the compound (other than water) that is formed when two molecules of the above substance react.
   (c) What type of bond is formed in the reaction?
   (d) What type of compound is the substance in (b)?
   (e) What further reaction or reactions can this substance undergo?

30.5 Vitamins

Vitamins are substances which the body needs in small quantities. They are not digested and are not used as a source of energy. They work with enzymes to enable chemical reactions to take place in the cells. Vitamins are either

- fat-soluble, e.g. vitamins A, D, E and K or
- water-soluble, e.g. vitamins B and C

Vitamin B12 is present in meat, fish and dairy products. A shortage causes anaemia.

Vitamin C is present in most fresh fruits and vegetables. A shortage lowers a person's resistance to infection. A severe shortage causes scurvy (which results in bleeding gums). Vitamin C works by destroying harmful free radicals (for free radicals see Topic 16.11).

Some of the vitamins in food are partially destroyed by the high temperature when food is cooked. Also, when food is boiled, vitamins B and C dissolve in the cooking water. When vegetables are boiled, enzymes in their cells attack the vitamin C content. If vegetables are plunged straight into boiling water, these enzymes are destroyed and less vitamin C is lost. Vitamins A and D are fat-soluble and are not dissolved out by water or destroyed by heat.

Vitamin D is present in foods containing fats and oils. A severe shortage causes rickets (soft bones in children and fragile bones in adults; see Key Science Biology Topic 14.1).

30.6 Mineral salts

We need small quantities of mineral salts in our diet. Extremely important are the following:

Iron salts are essential for the manufacture of haemoglobin in red blood cells. A shortage causes anaemia. They are obtained from red meat, eggs, spinach and other vegetables.

Calcium salts are essential for the formation of bones and teeth. They are obtained from cheese and milk.

30.7 Food additives

FIRST THOUGHTS

We eat some foods exactly as they are harvested, e.g. apples and cucumbers. Most foods, however, go to the food industry to be processed, that is changed in some way. Three quarters of the food eaten in industrialised countries is processed.

Why use additives?

In food processing, many substances are added. They are called **food additives**. A food additive is defined as a substance which is not normally eaten or drunk as a food either by itself or as a typical ingredient of food. Salt and sugar are added to foods but are not called additives. No substance may be used as an additive unless the food manufacturer can give a good reason for its use. The reasons for using additives are:

- to flavour food
- to colour food
- to alter the texture of food
- to preserve food.

Additives are chemicals, but so are proteins, vitamins and all the other substances which occur naturally in foodstuffs. In 1950, 50 additives were in general use; now 3500 additives are used in our food.

Types of food additives

Additives which alter the taste of food

Flavourings Flavourings are the largest group of food additives, with about 3000 in use. The large number is not so surprising when it takes a mixture of up to 50 substances to produce a natural flavour such as apple or peach.

Sweeteners The commonest sweetener is sucrose (sugar); this is a food, not an additive. Some people want to cut down on sucrose either because it causes tooth decay or because they are overweight. Diabetics cannot cope with sucrose. Substitutes are saccharin, sorbitol and mannitol.

Flavour enhancers Flavour enhancers are not flavourings; they are substances which make existing flavours seem stronger. The best known is monosodium glutamate, MSG. It stimulates the taste buds.

Figure 30.7A ▶ They all contain MSG

Additives which alter the colour of food

When food is processed, it may lose some of its colour; then the manufacturer will want to restore the original colour of the food. Different countries vary widely in the number of colourings allowed. Colourings are not added to baby foods.

Additives which alter the texture of food

Emulsifiers and stabilisers When oil and water are added, they form two layers. Some substances, called **emulsifiers**, can make oil and water mix. The mixture is called an emulsion. Any substance which helps to prevent the emulsion from separating out again is called a **stabiliser**. Margarine, ice cream and salad dressings all use emulsifiers and stabilisers.

SUMMARY

Food additives are used to alter
- the taste of food
- the colour of food
- the texture of food.

Figure 30.7B ▲ All these contain emulsifiers and stabilisers

Thickeners You will see that 'modified starch' appears on many labels. It is used to thicken foods. It can be a main ingredient of instant soups and puddings.

Figure 30.7C ▲ These contain modified starch

Anti-caking agents and humectants Anti-caking agents are substances which can absorb water without becoming wet. They are added to powdery or crystalline foods, such as cake mixes and table salt, to prevent lumps from forming. Humectants keep products moist. They are added to products such as bread and cakes.

Figure 30.7D ▲ Which products contain an anti-caking agent? Which contain a humectant?

Gelling agents To make jams, desserts etc. set, a gelling agent is added. Pectin is the commonest.

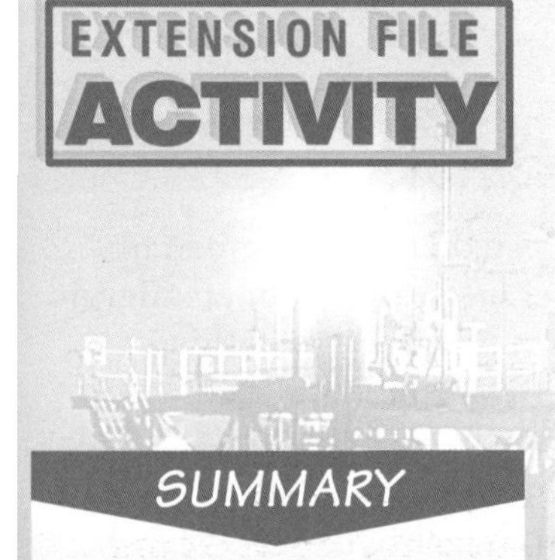

Anti-oxidants

Foods which contain fats and oils can turn rancid on exposure to the air. The fats and oils are oxidised to unpleasant-smelling acids. Anti-oxidants are added to prevent oxidation. Two common ones are BHA and BHT (butylated hydroxyanisole and butylated hydroxytoluene). Sulphur dioxide and sulphites are widely used as anti-oxidants. They have two effects: as well as preventing oxidation of fats and oils, they also deprive micro-organisms of the oxygen they need and delay their growth.

SUMMARY

Food additives used to preserve food include antioxidants and preservatives, e.g. BHA, BHT, sulphur dioxide and sulphites.

Figure 30.7E ▲ Preserved by sulphur dioxide and sulphites

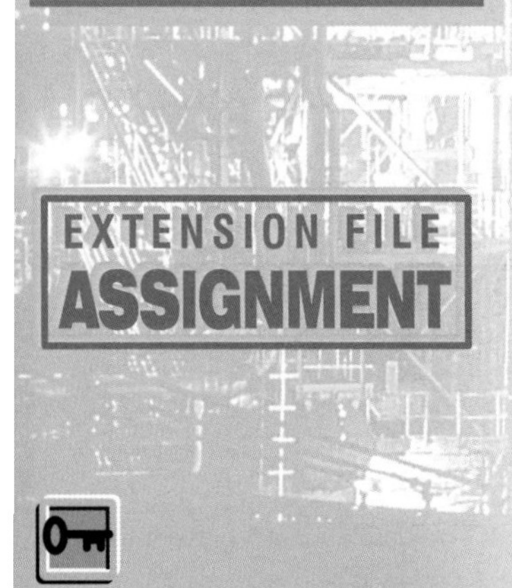

Preservatives

Food is stored in the warehouse, in the shop and in the home before it is eaten. Chemical preservatives are added to stop the growth of micro-organisms. Preservatives increase the food's shelf-life, i.e. the length of time the food will keep before it deteriorates. Longer shelf-lives mean less wastage on the shelves. This allows the shopkeeper to charge lower prices and to stock a wider range of foods. Some people are doubtful about whether the customer receives all the benefit of the lower prices. Some people believe that the savings from the widespread use of additives go to the food manufacturers rather than the customers.

Controls on additives

It is illegal to put anything harmful into food. Before an additive may be used, it must be approved by the Government. By law, all additives must be safe, that is, safe for almost everyone. Some people are made ill by some additives, but then some people are made ill by some natural foods.

E Numbers

Food additives are classified by E numbers.

On the labels of some foods you will see E numbers. E numbers are a device of the European Community (EC). The EC has drawn up a list of 314 safe additives. This is the numbering system:
Colourings: E number begins with 1, e.g. E150, caramel
Preservatives: E number begins with 2, e.g. E221 sodium sulphite
Anti-oxidants: E numbers 310–321, e.g. E320, butylated hydroxyanisole, BHA
Texture controllers: E numbers 322–494, e.g. E461, methylcellulose.
Additives which have been passed in the country of origin but not yet in the EC have a number only, e.g. 107 yellow 2G; 524 sodium hydroxide; 925 chlorine.

It's a fact!

We eat between 3 and 7 kg of additives a year. Some people think that over a long period some additives could be a threat to health. Medical doctors believe that additives make little, if any, contribution to serious illness.

Are additives good for you?

Some foods make some people ill. When a person reacts to a food by becoming ill, the reaction is called an **intolerance reaction** or an **allergic reaction**. Allergic reactions can take the form of asthma (breathing difficulty), eczema (a skin complaint), digestive troubles, rhinitis (like hay fever), headaches, migraines and hyperactivity. Putting hyperactive children on a diet free from additives often produces a dramatic improvement. Tartrazine (E102), a yellow dye, is the one that is most under suspicion. It is used in sweets, fizzy drinks and packet desserts. If you know that you are allergic to tartrazine, you can read the labels on the foods you fancy and reject any which list E102. Many people prefer to buy additive-free foods. Many of the large supermarket chains are reducing the number of additives in their products, and some firms are offering additive-free items.

CHECKPOINT

1 Which types of additives are present in this Apricot Pie?

APRICOT PIE

Ingredients: Flour, Sugar, Apricot, Animal and Vegetable fat, Dextrose, Modified starch, Sorbitol syrup, Glucose syrup, Salt, Citric acid, Preservative E202, Flavouring, Emulsifiers E465, E471, E475, Whey powder, Colours E102, E110

2 A ham manufacturer wants to increase the weight of his hams by injecting water into them. He needs an additive to keep the water and other ingredients well mixed. What type of additive does he use? Who benefits from the use of this additive, the manufacturer or the consumer?

▶ **3** The labels on two soft drink bottles are shown below.

Kooler-cola	**diet cola**
Soft drink with vegetable extracts	*Low calorie soft drink with vegetable extracts*
Ingredients: Carbonated water, Sugar, Colour (caramel), Phosphoric acid, Flavourings, Caffeine	Ingredients: Carbonated water, Colour (caramel), Artificial sweetener (Aspartame), Phosphoric acid, Flavourings, Citric acid, Preservative (E211) Contains phenylamine
BEST SERVED ICE COLD	***TASTES GREAT ICE COLD!***

(a) Which sweetener is present (i) in A (ii) in B?
(b) Which preservative is present in both A and B?
(c) Why is a preservative needed?
(d) Why do the labels recommend serving the drinks ice-cold?
(e) What is the difference between A and B that enables one to describe 'diet cola' as 'low calorie'?
(f) What pH would you expect the two drinks to have?

30.8 Drugs

FIRST THOUGHTS

Chemotherapy is the treatment of illness by chemical means. Chemicals can cure diseases and relieve pain. Advances in chemotherapy have made life safer, longer and more free from pain than it was a century ago. Some of the chemicals which are used in medicine are mentioned in this section.

Painkillers

Aspirin is the most popular pain-killer. It was first sold in 1899. Each person in the UK swallows an average of 200 aspirins a year. Many other medicines, e.g. APC, contain aspirin. When you swallow aspirin, there is slight bleeding of the stomach wall. To reduce irritation of the stomach wall, you should always swallow plenty of water with aspirin. **Codeine** is a stronger pain-killer, used in headache tablets and in cough medicines.

Aspirin is a a **drug** and a **medicine**. A **drug** is a substance from *outside* the body which affects chemical reactions *inside* the body. A **medicine** is a substance used for the treatment or prevention or diagnosis of disease so medicines are a subset of drugs. They include analgesics, antibiotics, anaesthetics and others. There are also drugs with adverse effects on the body. They include drugs which affect the mind and alter behaviour, e.g. alcohol, cocaine and heroin. Many of these drugs are habit-forming, e.g. cocaine, and their use is illegal.

Aspirin is an extraordinary drug: it acts

- as an analgesic – that relieves pain
- as an antipyretic – that reduces fever
- as an anti-inflammatory agent – taken by people with painful joints, e.g. rheumatism
- by reducing the formation of blood clots – taken by people who have had a heart attack.

Aspirin is an ester of salicylic acid.

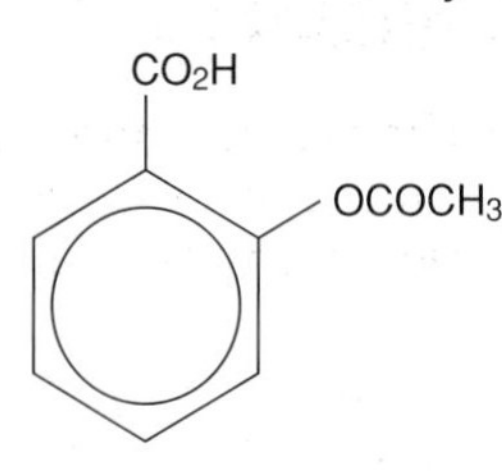

Aspirin (ethanoyl salicylic acid)

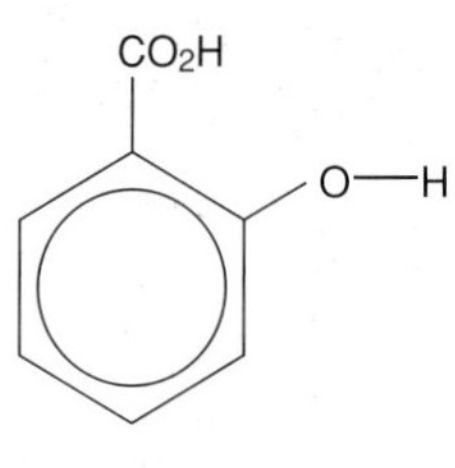

Salicylic acid

Neither aspirin nor salicylic acid occurs in nature, but several closely related compounds do. The bark of the willow tree has been known for centuries to have the power to relieve pain and reduce fever. In 1827 chemists extracted *salicin* from willow bark. It is a compound of salicyl alcohol and glucose. Aspirin can be made from it and also from other sources. In 1893 Felix Hofmann, a German chemist working for the firm Baeyer was concerned about his father's rheumatism. He demonstrated that aspirin is more powerful than *salicin* and less irritating to the stomach and intestines than salicylic acid. Baeyer started to market aspirin, and aspirin has never looked back!

Tablets of aspirin contain about 300 mg of aspirin powder and an ingredient which binds the powder into a tablet. Aspirin does not begin to do its work until it reaches the bloodstream. In the alkaline contents of the intestine, aspirin is hydrolysed to salicylic acid. This is the active analgesic which is absorbed into the bloodstream. To speed up absorption, some companies sell **buffered aspirin** tablets. These contain a base, e.g. magnesium carbonate, magnesium hydroxide.

Soluble aspirin is the sodium salt or the calcium salt of aspirin. When these salts meet the acidic contents of the stomach, they immediately form aspirin in the form of very fine crystals. These irritate the stomach less than swallowing aspirin.

Paracetamol® is comparable to aspirin as a pain reliever but is gentler to the stomach. Ibuprofen® has similar effects to aspirin but is less irritating to the stomach. Codeine®, which is a stronger analgesic than aspirin, is used in headache tablets and in cough medicines.

Morphine is a substance with an amazing ability to relieve intense pain. The problem with morphine is that it is **addictive**. It is only used in cases of dire necessity, for example when soldiers are wounded in battle. Morphine can only be obtained by doctors. **Heroin** is a pain-reliever which is even more potent and more addictive than morphine. Heroin is not used in medicine because it is so addictive. Drug-dealers like to get their customers to try heroin because they will soon become 'hooked' and come back to buy again and again.

Tranquillisers and sedatives

One person in twenty takes **tranquillisers** every day. Tranquillisers are substances which relieve tension and anxiety. With so many people taking tranquillisers over long periods of time, some doctors are worried about long-term effects. Research workers are now investigating whether there is any danger in taking tranquillisers for a long time.

Many people take **sedatives** (sleeping tablets). Drugs called **barbiturates** are used for this purpose. They are habit-forming. People who rely on barbiturates sometimes kill themselves accidentally by taking an overdose. These drugs are also used in the treatment of high blood pressure and mental illness.

It's a fact!

Many of the victims of AIDS are drug addicts who used infected needles to inject themselves with heroin. Never be tempted to give drugs a try. The dangers of addiction and AIDS are very real. It's not worth the risk to satisfy your curiosity.

KEY SCIENTIST

Laughing gas was discovered by Sir Humphry Davy. At first it was used only as a curiosity at parties. One of these parties was attended by a dentist, Horace Morton. He decided to try laughing gas as an anaesthetic on his patients. Morton obtained better results by using ether. The use of chloroform spread rapidly after Queen Victoria took it during the birth of one of her children in 1853.

Stimulants

Adrenalin is a substance which the body produces when it needs to prepare for action: for 'fight or flight'. It is a **stimulant**: it makes the heart beat faster and makes a person ready for strenuous action.

Amphetamines (pep pills) have similar effects on the body. Some people take amphetamines to keep themselves awake; other people want to pep themselves up and make themselves more entertaining. People can become addicted to amphetamines. A person taking amphetamines becomes excitable and talkative, with trembling hands and enlarged pupils.

Anaesthetics

Anaesthetics made modern surgery possible. Before the days of anaesthetics, a surgeon asked his patient to drink some brandy or smoke some opium to deaden the pain. Then the surgeon did the operation as quickly as he could. Sometimes patients died from pain and shock. After anaesthetics came into use, surgeons were able to take time to do the best operation they could, rather than the fastest. They were able to explore new techniques.

The first anaesthetics to be used were chloroform (1846), ether (1847) and dinitrogen oxide (laughing gas). There were some drawbacks to using these gases. In large doses, chloroform is harmful. Ether is very flammable, and it has been the cause of fires in hospitals in the past. Dinitrogen oxide (laughing gas) does not produce a very deep anaesthesia; however it is suitable for use in dentistry. Research workers have found a better anaesthetic, **fluothane**, which was first used in 1956. It has been so successful and free from side-effects that most operations are now done under fluothane.

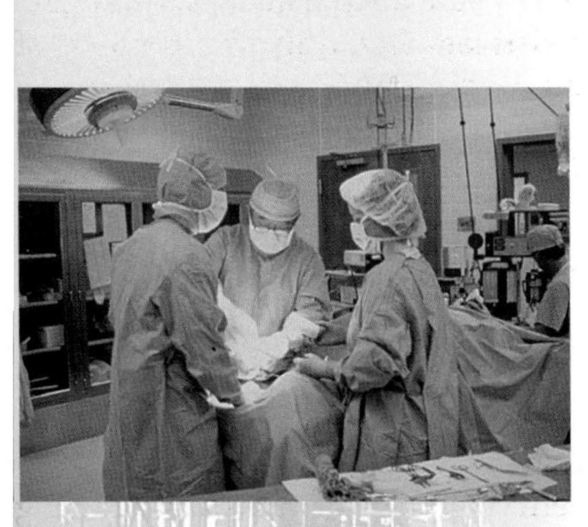

Figure 30.8A ▲ A modern operating theatre

Antiseptics

Another advance in surgery came with antiseptics. A century ago, many patients survived operations but died later in the wards. A surgeon called James Lister realised that the patients' wounds were becoming infected. He sprayed the operating theatre and the wards with a mist of phenol and water. His experiment produced immediate results: the death rate fell. Phenol had killed micro-organisms that would otherwise have infected surgical wounds. Phenol is an **antiseptic**. It is unpleasant to use. Solid phenol will burn the skin, and its vapour is toxic. Research chemists made other compounds which work as well as phenol and are safer to use. TCP® and Dettol® are antiseptics which contain trichlorophenol.

KEY SCIENTIST

The power of sulphonamide drugs was discovered by Gerhardt Domagh. The first human being on whom he tested the drug was his own daughter. She was dying of 'child-bed fever', a bacterial infection which can attack women who have just had a baby. Happily, the drug worked. Domagh received the Nobel prize in 1939.

Antibiotics

Infectious diseases, such as tuberculosis and pneumonia, used to kill thousands of people every year. The grim picture changed in 1935 with the discovery of the **sulphonamides**. They are antibiotics: substances which fight disease carried by bacteria. Infectious diseases are no longer a serious threat. In surgery too, powerful antibiotics have made having an operation safer than it was at the beginning of this century.

KEY SCIENTIST

The old treatment for syphilis was to apply a paste of heavy metal salts (salts of arsenic, lead, mercury and antimony) to the skin. The metal salts killed the micro-organism that caused the disease, but many patients experienced metal poisoning. Paul Ehrlich reasoned that, if he could attach one of these metals to a dye which was able to stain the micro-organism, the metal–dye compound might target the micro-organism 'like a magic bullet' without attacking body tissues. Ehrlich and his team of research workers set to work. At the 606th attempt they were successful. They called their new compound salvarsan. Ehrlich received the Nobel prize in 1908.

Alexander Fleming was a research bacteriologist in a London hospital. When the First World War began, he joined the Medical Corps. He was distressed by what he saw in his field hospital. Soldiers who did not seem to be mortally wounded when they arrived in the hospital died later from infections. Bacteria in mud and dirty clothing had infected their wounds, and gangrene set in. Watching men die a slow, painful death, Fleming wished that his work as a bacteriologist would enable him to discover a substance that would kill bacteria: a **bactericide**. After the war, Fleming went back to his research. One day, he found a mould called *Penicillium* on a dish of bacteria which he was culturing. To his amazement, he saw that the mould had killed bacteria. From the mould, he prepared an extract which he called **penicillin**. Would this be the powerful bactericide he had been hoping for? Sadly, although penicillin worked in the laboratory, it did not work in patients. Substances in the blood made the bactericide inactive.

In 1940, work on penicillin recommenced. The Second World War had started, and a powerful bactericide was needed urgently. Two chemists, Howard Florey and Ernst Chain, succeeded in making a stable extract of penicillin. They tested it, first on mice and then on human patients. The tests were successful, and the USA built a plant for the mass-production of penicillin. In 1942, penicillin was used in hospitals on the battlefield. The results were spectacular. No longer did soldiers die from minor wounds. In 1944, Fleming was knighted. Later, Fleming, Florey and Chain shared the Nobel Prize for medicine.

Penicillin has been widely used to treat a variety of infections. One disadvantage is that penicillin cannot be taken by mouth. It is broken down by acids, in this case the hydrochloric acid in the stomach. A more recent antibiotic, **tetracycline**, does not share this drawback. Tetracycline is a 'broad spectrum' antibiotic which can be used against many kinds of bacteria.

CHECKPOINT

1. Your Uncle Bert is always swallowing aspirins. Explain to him (a) why he should not take too many aspirins and (b) why he should take water with aspirins.
2. (a) A singer you know feels so tired that she is thinking of taking pep pills before giving a performance. Explain to her why this is not a good idea in the long run.
   (b) What is the body's natural stimulant? How does it work?
3. Briefly explain how surgery has changed as a result of (a) the discovery of anaesthetics and (b) the discovery of antiseptics.
4. Morphine is a very powerful pain reliever. Why do doctors prescribe it for so few patients? For what types of patient is morphine prescribed?
5. Why do doctors never prescribe heroin as a pain-killer?
6. Someone you know takes barbiturates to help her to get to sleep. Suggest to her what else she could do, instead of taking pills, to get to sleep.
7. (a) Before the time of James Lister, surgeons did not change their operating gowns between patients. What was wrong with this practice?
   (b) What did James Lister do to improve surgery?
8. (a) What did Sir Alexander Fleming do to merit a Nobel prize for medicine?
   (b) He had a piece of luck in his research, but his success was not due to luck. What else went into his discovery?

30.9 The chemical industry

The chemical industry can be divided roughly into ten sections.

1 **Heavy chemical industry:** oils, fuels, etc. (see Topic 23).
2 **Agriculture:** fertilisers, pesticides, etc. (see Topic 22).
3 **Plastics:** poly(ethene), poly(styrene), PVC, etc. (see Topic 28.2).
4 **Dyes**.
5 **Fibres:** nylon, rayon, courtelle, etc. (see Topic 28.2).
6 **Paints, varnishes,** etc.
7 **Pharmaceuticals:** medicines, drugs, cosmetics, etc. (see Topic 30.8).
8 **Metals:** iron, aluminium, alloys, etc. (see Topic 19).
9 **Explosives:** dynamite, TNT, etc. (see Topic 22.3).
10 **Chemicals from salt:** sodium hydroxide, chlorine, hydrogen, hydrochloric acid, etc. (see Topic 7.6 and Topic 10.1).

Figure 30.9A shows some of the petrochemicals which are obtained by the route

Petroleum oil → Naphtha → Ethene → Petrochemical

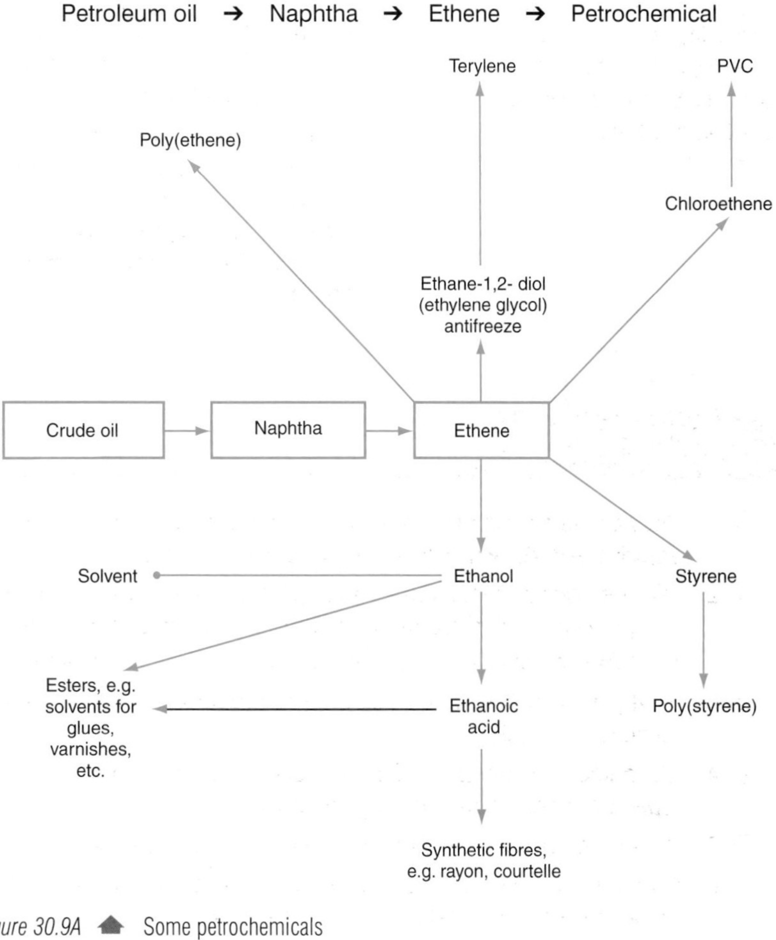

Figure 30.9A ▲ Some petrochemicals

Theme Questions

1 (a) Describe the manufacture of aqueous ethanol from a carbohydrate. Name the starting material, state the conditions required and give a chemical equation for the reaction.
(b) Name a process which can be used to concentrate the ethanol formed. Explain the principle on which the process depends.
(c) Suggest one disadvantage of drinking ethanol.
(d) What is 'methylated spirit'? Why must it never be drunk?

2 (a) Ethene is an unsaturated hydrocarbon. Describe how you could test it for unsaturation.
(b) Ethene polymerises to form poly(ethene). Write an equation for this reaction.
(c) Give two examples of articles commonly made from poly(ethene). In each case name an alternative material and suggest why poly(ethene) has been chosen in preference to the other material.

3 (a) State the conditions under which ethene is formed from ethanol.
(b) State the conditions under which ethanol is manufactured from ethene.
(c) Give a brief description of the fermentation process for the manufacture of ethanol.
(d) Compare the ethene process of (b) with the fermentation process of (c) with regard to
(i) the purity of the product
(ii) whether the process can be run continuously
(iii) the availability of the starting materials
(iv) the rate of formation of ethanol.

4 Tennis racquets were traditionally made of wood. Starting in the 1970s, aluminium alloys have been used for tennis racquets. Many popular tennis racquets are now made from composite materials, consisting of plastic reinforced with graphite fibres and mounted on a metal core. Suggest how such a material compares with (a) wood and (b) aluminium alloy with respect to (i) weight of racquet, (ii) strength of racquet and (iii) retaining its shape in all weathers.

5 The structures of plastics A and B are shown below.

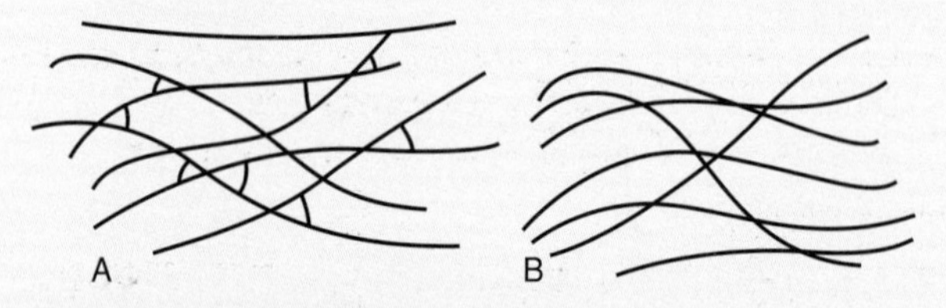

(a) Say what will happen when each of the plastics is heated strongly. Explain the reason for the difference in behaviour.
(b) Suggest a use for which plastic A would be more suitable than plastic B.
(c) Suggest a use for which plastic B would be more suitable than plastic A.

6 (a) Explain what is meant by
(i) a food additive
(ii) a food preservative
(iii) an emulsifier
(iv) an anti-oxidant
(v) a permitted colouring
(vi) an E number
(b) Suggest reasons why the number of food additives in use has grown rapidly during the last 20 years.

7 (a) (i) Ethane, C_2H_6, is a *saturated* hydrocarbon. Explain the meaning of the term *saturated*. [2]
(ii) When ethane burns it produces carbon dioxide and water vapour. Balance the equation below by supplying the correct numbers for the boxes.

$2C_2H_6 + \square\, O_2 \longrightarrow + \square\, CO_2 + \square\, H_2O$ [1]

(b) (i) Ethene reacts with bromine water, ethane does not. Give the reason for this difference. [1]
(ii) Complete and balance the *symbol* equation for the reaction between ethene and bromine.

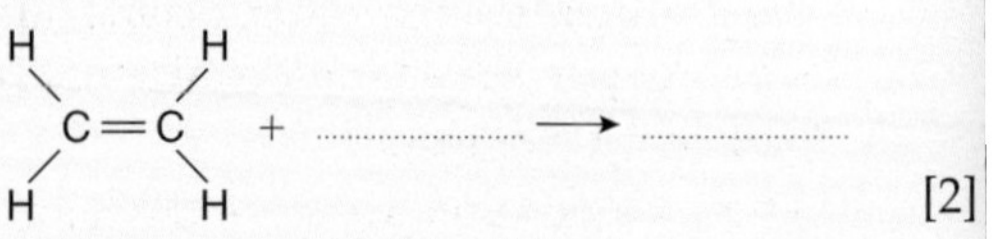

[2]

(c) C_4H_{10} has two isomers. Draw the structural formula of each isomer. [2] (WJEC)

8 Propene (C_3H_6) can be obtained by cracking alkanes.
(a) Draw the structure of a molecule of propene showing *all* the bonds. [1]
(b) One molecule of the alkane decane ($C_{10}H_{22}$) was cracked to give two molecules of propene and one molecule of another alkane.
Write the balanced equation for this reaction. [2]
(c) Propene is used to make poly(propene).
(i) What features of a propene molecule enables it to form poly(propene)? [1]
(ii) Draw the structure of the repeating unit in poly(propene). [2]
(iii) Poly(ethene) is used to make many types of bottle.
Suggest why the more expensive poly(propene) is used to make bottles for fizzy drinks. [1] (Edexcel)

9 Write a *brief* account of *three* of the following. Relevant chemical equations and/or diagrams *may* be included in your answer where appropriate.

(a) Outline the industrial extraction of aluminium.
State and explain the major factors that determine the siting of a new aluminium extraction plant. [5]

(b) Outline the similarities and differences between alkanes and alkenes.
State and explain the importance of alkanes and alkenes in everyday life. [5]

(c) Substances can be classified as simple molecular or giant ionic. Give an example for each type of structure and show how the physical properties are related to each structure. [5]

(d) State and explain how ethanol can be obtained from sugar.
Discuss the dangers of alcohol abuse.
[5] (WJEC)

10 Octane is a saturated hydrocarbon. Its graphical (displayed) formula is

```
    H   H   H   H   H   H   H   H
    |   |   |   |   |   |   |   |
H — C — C — C — C — C — C — C — C — H
    |   |   |   |   |   |   |   |
    H   H   H   H   H   H   H   H
```

(a) Name the family to which octane belongs: alkanes, alkenes, carbohydrates or carbonates. [1]

(b) The products formed when octane burns in air depend upon the amount of air present. [4]
Explain this statement.

(c) When octane vapour is passed over a heated catalyst a reaction takes place.
Ethene and hydrogen are the only products.
(i) What type of reaction is taking place? [1]
(ii) Finish and balance the symbol equation for this reaction.

Octane → Ethene + Hydrogen
C_8H_{18} → ______ C_2H_4 + ______ [2]

(iii) Excess ethene is bubbled through bromine water.
What colour change would you see? [2]

(d) Ethene is used as the raw material for making poly(ethene).
(i) Draw the graphical (displayed) formula of ethene and of poly(ethene). [3]
(ii) Poly(ethene) has replaced paper and cardboard for many packaging uses. Suggest one advantage and one disadvantage of poly(ethene) compared to paper and cardboard. Do not consider the relative costs of the materials.
[2] (OCR)

11 (a) Glucose solution was fermented under **anaerobic** conditions. The glucose solution was boiled and then cooled before the yeast was added.

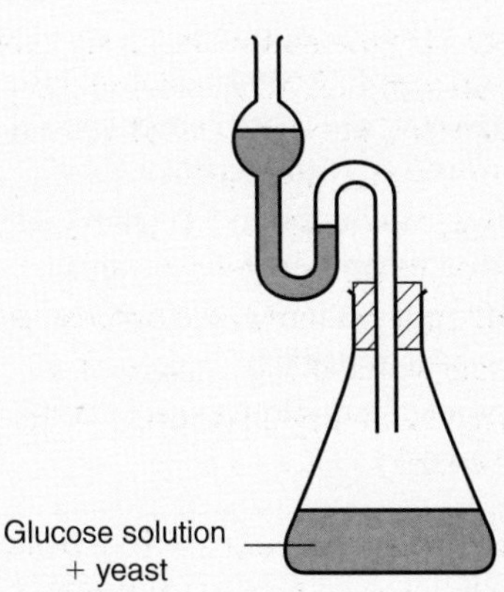

(i) Why was the glucose solution boiled and then cooled before the yeast was added? [1]
(ii) What does **anaerobic** mean? [1]
(iii) Give *one* expected observation during the fermentation. [1]
(iv) Give *one* use of ethanol, apart from alcoholic drinks. [1]

(b) The flow chart shows a reaction of ethanol.

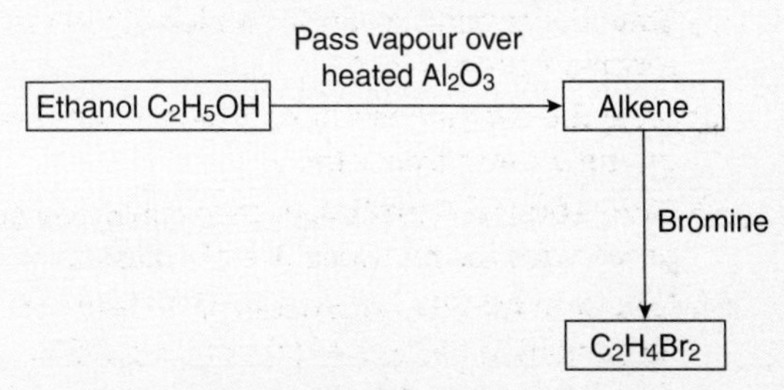

(i) Draw the displayed formula for ethanol. Draw a circle round the functional group of this alcohol. [2]
(ii) Name the type of reaction that occurs when alkenes are formed from alcohols. [1]
(iii) Name the alkene formed from ethanol. [1]
(iv) Name the type of reaction that occurs when bromine reacts with the alkene. [1]
(v) Name the compound formed, $C_2H_4Br_2$. [1]

(c) Ethanol can be oxidised to ethanoic acid, CH_3COOH.
(i) An aqueous solution of ethanoic acid is a **weak acid**. Explain the term weak acid. [2]
(ii) Describe *two* different chemical reactions that would show the acidic properties of ethanoic acid solution. [4] AQA(SEG) 1998

12 Cars in Brazil use ethanol as a fuel instead of petrol (octane). The ethanol is produced by the fermentation of sugar solution from sugar cane.

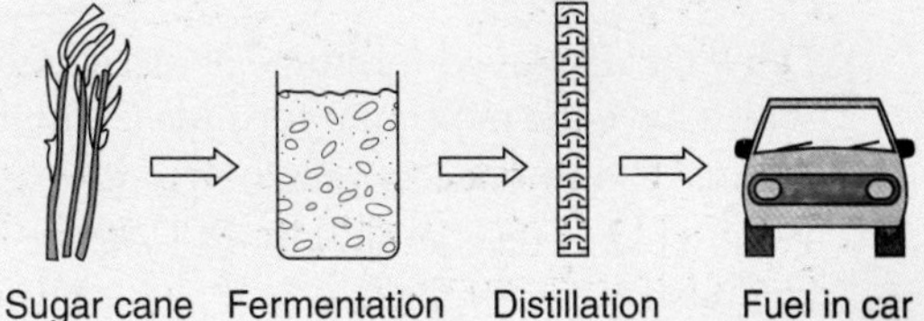

(a) What must be added to sugar solution to make it ferment? [1]

(b) State the most suitable temperature for a fermentation. 0 °C; 10 °C; 30 °C; 70 °C; 100 °C. [1]

(c) (i) What compounds are formed by the complete combustion of ethanol? [2]

(ii) Why are these compounds *not* harmful to the environment? [1]

(d) Suggest why pollution from cars is less when using ethanol instead of petrol. [1]

(e) Give *one* reason why *ethanol* is *not* used as a fuel for cars in Britain. [1]

(f) Some information about octane and ethanol is shown.

Property	*Octane*	*Ethanol*
Melting point in °C	−57	−113
Boiling point in °C	125	78.5
Density in g/cm³	0.70	0.79
Heat produced in kJ/mol	5512	1367

Explain a similarity between octane and ethanol that allows ethanol to be used as a fuel in cars.

[2] AQA(SEG) 2000

13 The table contains some data about a series of alcohols.

Name	*Molecular formula*	*Boiling point in °C*	*Difference in boiling point between adjacent alcohols in °C*
Ethanol	C_2H_5OH	78	
Propan-1-ol	C_3H_7OH	98	20
Butan-1-ol	C_4H_9OH	117	(a)
Pentan-1-ol	$C_5H_{11}OH$	138	(b)

(a) (i) Calculate the differences (a) and (b) in boiling points between adjacent alcohols. The difference in boiling point between the first two alcohols has been done for you. [1]

(ii) The boiling point of another alcohol in the same series is 177 °C. Suggest its molecular formula. Give your reason. [2]

(b) The displayed (graphical) formulae of ethanol and propan-1-ol are shown below.

```
    H  H                 H  H  H
    |  |                 |  |  |
H — C — C — O — H    H — C — C — C — O — H
    |  |                 |  |  |
    H  H                 H  H  H
```

Draw the displayed formulae of *two* isomers which have the molecular formula C_4H_9OH. [2]

(c) Ethanol can be used as a fuel. Write a balanced equation for its *complete* combustion in oxygen. [2] (OCR)

14 The molecular formulae of two hydrocarbons *M* and *N* are given.

$$M = C_4H_{10}$$
$$N = C_4H_8$$

(a) *M* reacts with chlorine to form C_4H_9Cl.

(i) Write a balanced chemical equation for the reaction of chlorine with *M*. [2]

(ii) Name this type of reaction. [1]

(b) A displayed structural formula for *N* is:

```
    H   H   H   H
    |   |   |   |
H — C — C = C — C — H
    |           |
    H           H
```

Draw a displayed structural formula of a compound which is an isomer of *N*. [1]

(c) Show the displayed structural formula for each of the products formed.

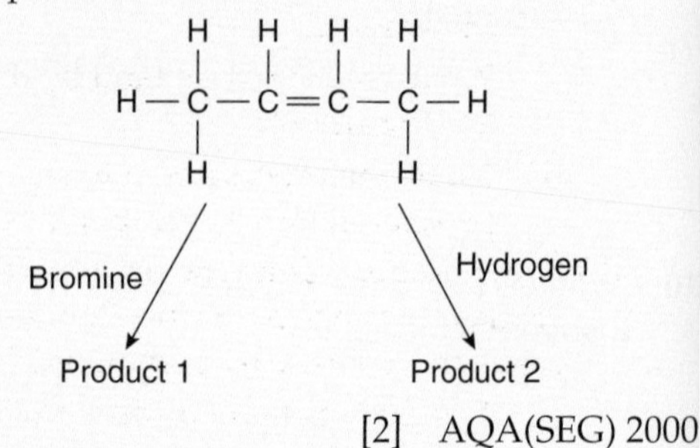

[2] AQA(SEG) 2000

15 (a) The following table shows how some plastics are used in many houses. Complete a copy of the table by giving *two* properties of each plastic which explain why it is used in the way stated.

Plastic	*Use*	*Property*
Polystyrene	Roof insulation	1 ______ 2 ______
PVC	Plastic guttering	1 ______ 2 ______
Polythene	Plastic bucket	1 ______ 2 ______
Melamine	Kitchen work top	1 ______ 2 ______

[8]

(b) (i) Plastics can be classified as thermosoftening or thermosetting. Explain precisely what is meant by these two terms. [4]

(ii) Using only those plastics in the table in part **(a)** give *one* example of a thermosoftening plastic and *one* example of a thermosetting plastic. [2]

(c) (i) Many plastic articles, e.g. polythene bags, create litter problems because they are non-biogredable. What do you think the term 'non-biodegradable' means? [2]

(ii) Suggest *two* methods which would help minimise the litter problems plastic articles create. [2] (CCEA)

16 Ethanol can be made from sugar solution by the process of fermentation.

Fermentation is the method used to make alcoholic drinks.

(i) State why distillation can be used to separate the ethanol from the fermented mixture. [1]

(ii) I. Give *one* health problem associated with alcohol abuse. [2]

II. Give *one* social problem associated with alcohol abuse. [2]

(iii) Ethanol in wine is easily changed to ethanoic acid (vinegar).

I. What causes this change? [1]

II. Give the name for the chemical process taking place. [1] (WJEC)

17 (a) Petroleum (crude oil) is a very important source of hydrocarbons.

(i) What do you understand by the term 'hydrocarbon'? [2]

(ii) Name the process used to separate petroleum into its components. [2]

(b) Some products from the process in **(a) (ii)** are cracked to produce ethene. Polythene may be made from ethene by polymerisation.

(i) Explain what you understand by the terms: Cracking; Polymerisation. [4]

(ii) Give *one* advantage of the cracking of hydrocarbons. [1]

(c) (i) Draw the full structural formula of the ethene molecule. [2]

(ii) Describe what would be observed when bromine water is added to ethene. [2]

(d) (i) Polyvinyl chloride (PVC) is a very important and useful plastic. By using an equation show how vinyl chloride molecules link together to form part of a PVC molecule. [3]

(ii) Give *two* uses of polyvinyl chloride and state the property of polyvinyl chloride which makes it suitable for the use given. [4]

(e) It has been proposed to construct a factory for the production of plastics on a derelict site close to Belfast.

Give *one* advantage and *one* disadvantage of siting the factory in this area. [2] (CCEA)

18 (a) Petrol is a fossil fuel obtained from oil by fractional distillation.

(i) Name the element which all fossil fuels contain. [1]

(ii) Why can petrol be described as a non-renewable energy source? [1]

(iii) Give *two* other examples of non-renewable energy sources. [2]

(iv) Explain how petrol is separated from the other fractions in crude oil by distillation. [2]

(b) The table below gives information about some alkanes and alkenes. Write the missing formulae and name. [3]

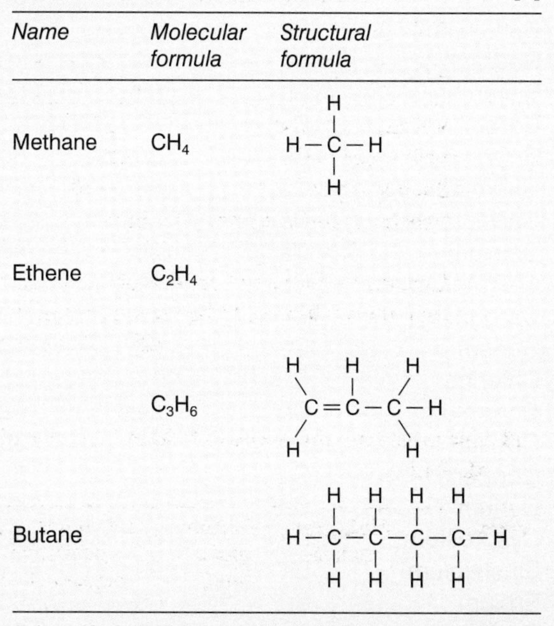

Name	*Molecular formula*	*Structural formula*
Methane	CH_4	H H–C–H H
Ethene	C_2H_4	
	C_3H_6	H H H C=C–C–H H H
Butane		H H H H H–C–C–C–C–H H H H H

(c) Ethanol, C_2H_5OH, is an alternative fuel to petrol.

(i) Give *one other* use of ethanol. [1]

(ii) Draw the structural formula of ethanol. [1]

(iii) Write a symbol equation to show how ethanol can be manufactured using ethene and steam. [1]

(iv) The combustion of ethanol is an exothermic reaction. The equation of this reaction is shown below.

$$C_2H_5OH + 3O_2 \rightarrow 3H_2O + 2CO_2$$

Explain with reference to bond making and breaking why the combustion of ethanol is exothermic. [3] (CCEA)

Table of symbols, atomic numbers and relative atomic masses

Element	Symbol	Atomic number	Relative atomic mass
Actinium	Ac	89	227
Aluminium	Al	13	27
Americium	Am	95	243
Antimony	Sb	51	122
Argon	Ar	18	40
Arsenic	As	38	75
Astatine	At	85	210
Barium	Ba	56	137
Berkelium	Bk	97	247
Beryllium	Be	4	9
Bismuth	Bi	83	209
Boron	B	5	11
Bromine	Br	35	80
Cadmium	Cd	48	112
Caesium	Cs	55	133
Calcium	Ca	20	40
Californium	Cf	98	251
Carbon	C	6	12
Cerium	Ce	58	140
Chlorine	Cl	17	35.5
Chromium	Cr	24	52
Cobalt	Co	27	59
Copper	Cu	29	63.5
Curium	Cm	96	247
Dysprosium	Dy	66	162.5
Einsteinium	Es	99	254
Erbium	Er	68	167
Europium	Eu	63	152
Fermium	Fm	100	253
Fluorine	F	9	19
Francium	Fr	87	223
Gadolinium	Gd	64	157
Gallium	Ga	31	70
Germanium	Ge	32	72.5
Gold	Au	79	197
Hafnium	Hf	72	178
Helium	He	2	4
Holmium	Ho	67	164
Hydrogen	H	1	1
Indium	In	49	115
Iodine	I	53	127
Iridium	Ir	77	192
Iron	Fe	26	56
Krypton	Kr	36	84
Lanthanum	La	57	139
Lawrencium	Lw	103	257
Lead	Pb	82	207
Lithium	Li	3	7
Lutecium	Lu	71	175
Magnesium	Mg	12	24
Manganese	Mn	25	55
Mendelevium	Md	101	256
Mercury	Hg	80	201
Molybdenum	Mo	42	96
Neodymium	Nd	60	144
Neon	Ne	10	20
Neptunium	Np	93	237
Nickel	Ni	28	59
Niobium	Nb	41	93
Nitrogen	N	7	14
Nobelium	No	102	254
Osmium	Os	76	190
Oxygen	O	8	16
Palladium	Pd	46	106
Phosphorus	P	15	31
Platinum	Pt	78	195
Plutonium	Pu	94	242
Polonium	Po	84	210
Potassium	K	19	39
Praesodymium	Pr	59	141
Promethium	Pm	61	147
Protactinium	Pa	91	231
Radium	Ra	88	226
Radon	Rn	86	222
Rhenium	Re	75	186
Rhodium	Rh	45	103
Rubidium	Rb	37	85
Ruthenium	Ru	44	101
Samarium	Sm	62	150
Scandium	Sc	21	45
Selenium	Se	34	79
Silicon	Si	14	28
Silver	Ag	47	108
Sodium	Na	11	23
Strontium	Sr	38	87
Sulphur	S	16	32
Tantalum	Ta	73	181
Technetium	Tc	43	99
Tellurium	Te	52	127
Terbium	Tb	65	159
Thallium	Tl	81	204
Thorium	Th	90	232
Thulium	Tm	69	169
Tin	Sn	50	119
Titanium	Ti	22	48
Tungsten	W	74	184
Uranium	U	92	238
Vanadium	V	23	51
Xenon	Xe	54	131
Ytterbium	Yb	70	173
Yttrium	Y	39	89
Zinc	Zn	30	65
Zirconium	Zr	40	91

The Periodic Table of the Elements

Group																	
1	2											3	4	5	6	7	0
					1 **H** Hydrogen 1												4 **He** Helium 2
7 **Li** Lithium 3	9 **Be** Beryllium 4											11 **B** Boron 5	12 **C** Carbon 6	14 **N** Nitrogen 7	16 **O** Oxygen 8	19 **F** Fluorine 9	20 **Ne** Neon 10
23 **Na** Sodium 11	24 **Mg** Magnesium 12											27 **Al** Aluminium 13	28 **Si** Silicon 14	31 **P** Phosphorus 15	32 **S** Sulphur 16	35·5 **Cl** Chlorine 17	40 **Ar** Argon 18
39 **K** Potassium 19	40 **Ca** Calcium 20	45 **Sc** Scandium 21	48 **Ti** Titanium 22	51 **V** Vanadium 23	52 **Cr** Chromium 24	55 **Mn** Manganese 25	56 **Fe** Iron 26	59 **Co** Cobalt 27	59 **Ni** Nickel 28	64 **Cu** Copper 29	65 **Zn** Zinc 30	70 **Ga** Callium 31	73 **Ge** Germanium 32	75 **As** Arsenic 33	79 **Se** Selenium 34	80 **Br** Bromine 35	84 **Kr** Krypton 36
85 **Rb** Rubidium 37	88 **Sr** Strontium 38	89 **Y** Yttrium 39	91 **Zr** Zirconium 40	93 **Nb** Niobium 41	96 **Mo** Molybdenum 42	99 **Tc** Technetium 43	101 **Ru** Ruthenium 44	103 **Rh** Rhodium 45	106 **Pd** Palladium 46	108 **Ag** Silver 47	112 **Cd** Cadmium 48	115 **In** Indium 49	119 **Sn** Tin 50	122 **Sb** Antimony 51	128 **Te** Tellurium 52	127 **I** Iodine 53	131 **Xe** Xenon 54
133 **Cs** Caesium 55	137 **Ba** Barium 56	139 **La** Lanthanum 57 *	178 **Hf** Hafnium 72	181 **Ta** Tantalum 73	184 **W** Tungsten 74	186 **Re** Rhenium 75	190 **Os** Osmium 76	192 **Ir** Iridium 77	195 **Pt** Platinum 78	197 **Au** Gold 79	201 **Hg** Mercuy 80	204 **Tl** Thallium 81	207 **Pb** Lead 82	209 **Bi** Bismuth 83	**Po** Polonium 84	**At** Astatine 85	**Rn** Radon 86
Fr Francium 87	226 **Ra** Radium 88	227 **Ac** Actinium 89 †															

*58–71 Lanthanum series
†90–103 Actinium series

140 **Ce** Cerium 58	141 **Pr** Praseodymium 59	144 **Nd** Neodymium 60	147 **Pm** Promethium 61	150 **Sm** Samarium 62	152 **Eu** Europium 63	157 **Gd** Gadolinium 64	159 **Tb** Terbium 65	162 **Dy** Dysprosium 66	165 **Ho** Holmium 67	167 **Er** Erbium 68	169 **Tm** Thulium 69	173 **Yb** Ytterbium 70	175 **Lu** Lutetium 71
232 **Th** Thorium 90	**Pa** Protactinium 91	238 **U** Uranium 92	**Np** Neptunium 93	**Pu** Plutonium 94	**Am** Americium 95	**Cm** Curium 96	**Bk** Berkelium 97	**Cf** Californium 98	**Es** Einsteinium 99	**Fm** Fermium 100	**Md** Mendelevium 101	**No** Nobelium 102	**Lr** Lawrencium 103

Key

a **X** b

a = relative atomic mass
X = atomic symbol
b = atomic number

CHECKPOINT 1.3
1 2.7 g/cm^3
2 7500 g (7.5 kg)
3 2720 g (2.72 kg)
4 A 0.86 g/cm^3: floats B 4.3 g/cm^3: sinks

CHECKPOINT 1.6
8 (a) (i) 55 g (ii) 100 g (b) 45 g crystallise (c) 200 g (d) 300 g (e) (i) 3 g (ii) 25 g

CHECKPOINT 2.4
3 (a) E (b) C (c) D (d) A, B

THEME A QUESTIONS
6 (b) 48 g/100 g (c) 50 °C (d) 55 g KNO_3 crystallise

CHECKPOINT 6.2
2 9p, 9e, 10n
3 (a) 17, 35 (b) 27, 59 (c) 50, 119

CHECKPOINT 6.4
3 N 7, 7, 7; Na 11, 11, 12; K 19, 20, 19; U 92, 143, 92

THEME B QUESTIONS
12 11p, 12n
13 82p, 124n, No

CHECKPOINT 10.3
3 (a) 4 (b) 9 (c) 3 (d) 11 (e) 18
4 (a) C (b) A (c) B

THEME C QUESTIONS
2 (b) 45%
8 (a) blue (b) orange (c) orange (d) 4–6 (e) 7.5 (f) purple (g) orangè
18 (a) 1.0 (c) (i) 7.0 (ii) 25.0 cm^3

CHECKPOINT 15.7
1 20 g KCl crystallises
2 20 g NaCl crystallises
3 20 g KCl crystallises
4 10 g K_2SO_4 + 10 g KBr crystallise

CHECKPOINT 16.8
6 (a) (i) 80–100 km/h (ii) 30 km/h (iii) 100 km/h (b) (i) 80–100 km/h (ii) 50–60 mph

CHECKPOINT 18.7
1 (a) D (b) C (c) F

THEME D QUESTIONS
17 (b) (iii) 79 cm^3

CHECKPOINT 22.6
7 (a) 65 tonnes (b) 18 tonnes
8 2 months

THEME E QUESTIONS
6 (a) (i) Alkaline manganese dry cell gives 400 flashes/£; zinc–carbon cell gives approximately 100 flashes/£. (ii) Zinc–carbon cell gives 35 hours/£; alkaline manganese cell gives 31 hours/£

CHECKPOINT 26.2
1 28, 44, 64, 80, 40, 58.5, 56, 58, 106, 159.5, 249.5, 162

CHECKPOINT 26.3
1 (a) 20% (b) 70% (c) 80% C, 20% H (d) 40% S, 60% O (e) 5.0% H, 95.0% F (f) 20% Mg, 27% S, 53% O
2 (a) 36% (b) 67.5%

CHECKPOINT 26.4
1 (a) 27 g (b) 96 g (c) 50 g (d) 14 g (e) 8 g (f) 64 g
2 (a) 2.5 mol (b) 0.33 mol (c) 1.0 mol (d) 0.25 mol
3 (a) 98 g (b) 23 g (c) 100 g (d) 10 g

CHECKPOINT 26.5
1 8.0 g
2 (b) 3.2 g
3 10.3 g
4 107 g

CHECKPOINT 26.6
1 (a) 8.0 g (b) 95%
2 (a) 34 g (b) 10%
3 98%
4 (b) 132 g (c) 91%
5 (a) 25 g (b) 88%

CHECKPOINT 26.7A
1 (a) NO (b) Mg_3N_2 (c) Al_2O_3
2 (a) 67.5% (b) 40.5%
3 (a) SO_3 (b) SO_2 (c) $CaBr_2$ (d) $KHCO_3$
4 (a) BeO (b) Fe_2O_3 (c) $FeCl_3$
5 Ti_2O_3 and TiO_2

CHECKPOINT 26.7B
1 A CH_4O, B C_3H_6, C $C_4H_8O_2$, D $C_2H_6O_2$, E C_2H_6, F $C_2H_4Cl_2$, G C_6H_6, H $C_6H_3N_3O_6$
2 (i) (a) CH_3 (b) C_2H_6 (ii) (a) CH_2 (b) C_2H_4
3 (a) CH_2Cl (b) $C_2H_4Cl_2$

CHECKPOINT 26.8
1 6 dm^3
2 500 cm^3 O_2, 250 cm^3 CO_2
3 21 million dm^3
4 3 dm^3
5 0.43 dm^3
6 25 g

CHECKPOINT 26.9
1 (a) 15.9 g (b) 14.8 g (c) 52.0 g (d) 20.0 g (e) 29.8 g
2 (a) 6.00 dm^3 H_2, 6.00 dm^3 Cl_2 (b) and (c) 6.00 dm^3 H_2, 3.00 dm^3 O_2
3 0.518 g
4 120 cm^3
5 +2
6 64.8 g
7 (a) 965 C (b) 5×10^{-3} mol (c) 2 mol (d) $Ni^{2+}(aq) + 2e \rightarrow Ni(s)$
8 (a) 12 cm^3 (b) 24 cm^3 (c) 0.216 g
9 (b)
10 +3

CHECKPOINT 26.10
1 (a) 1.00 mol/l (b) 0.10 mol/l (c) 2.0 mol/l (d) 0.40 mol/l
2 (a) 0.02 mol (b) 1.00 mol (c) 0.06 mol (d) 0.025 mol
3 (a) 0.144 mol/l
4 2 mol/l
5 (a) 0.2 M (b) 0.02 M (c) 0.2 M (d) 8 M
6 (a) 0.25 mol (b) 0.010 mol (c) 0.05 mol (d) 0.0025 mol

CHECKPOINT 26.11
1 0.15 M
2 (a) 0.25 M (b) 10.0 cm^3
3 (a) 0.5 mol (b) 0.24 M
4 (a) Paingo (b) speed of action, taste

CHECKPOINT 27.6
1 –124 kJ/mol
2 –484 kJ/mol
3 –56 kJ/mol
4 –95 kJ/mol

THEME F QUESTIONS
2 (b) 0.14 mol/l
3 (i) 1.25×10^{-3} mol (ii) a,b,d,e
4 (a) $Al^{3+} + 3e \rightarrow Al(l)$ (b) 162 kg
5 95.5%
8 (a) 24 g (c) 31.2 mol dm^{-3} (d) 119 dm^3
11 (a) (i) 160 (ii) 56 tonnes (b) FeO
12 (i) I 0.20 mol dm^{-3} II 5.00×10^{-3} mol (ii) I 5.00×10^{-3} mol (II) 0.25 mol (iii) 15 g
14 (a) (i) 4.8×10^{-3} mol (ii) 2.4×10^{-3} mol (iii) 1.2 mol/dm^3 (b) (i) 6.0×10^{-3} mol (ii) 0.714 g (iii) 1.69 g (c) 144 cm^3
15 (c) (i) 1440 C (ii) 193×10^3 C (iii) 0.530 g
16 (a) (ii) I 350 atm II 500 °C (c) 80 (d) (i) I 2253 kJ/mol II 2346 kJ/mol
17 (a) (i) 4.6 g (ii) 2400 cm^3

CHECKPOINT 28.2
4 (a) 62.5 (b) 640

Page numbers in bold indicate the main entry for that topic.

A

D

E

F

G

H

O

P

Q

R

S

T

U

V

W

X

Y

Z